GeoComplexity and the Physics of Earthquakes

Geophysical Monograph Series

Including

IUGG Volumes
Maurice Ewing Volumes
Mineral Physics Volumes

84 **Solar System Plasmas in Space and Time** *J. Burch, J. H. Waite, Jr. (Eds.)*

85 **The Polar Oceans and Their Role in Shaping the Global Environment** *O. M. Johannessen, R. D. Muench, and J. E. Overland (Eds.)*

86 **Space Plasmas: Coupling Between Small and Medium Scale Processes** *Maha Ashour-Abdalla, Tom Chang, and Paul Dusenbery (Eds.)*

87 **The Upper Mesosphere and Lower Thermosphere: A Review of Experiment and Theory** *R. M. Johnson and T. L. Killeen (Eds.)*

88 **Active Margins and Marginal Basins of the Western Pacific** *Brian Taylor and James Natland (Eds.)*

89 **Natural and Anthropogenic Influences in Fluvial Geomorphology** *John E. Costa, Andrew J. Miller, Kenneth W. Potter, and Peter R. Wilcock (Eds.)*

90 **Physics of the Magnetopause** *Paul Song, B.U.Ö. Sonnerup, and M.F. Thomsen (Eds.)*

91 **Seafloor Hydrothermal Systems: Physical, Chemical, Biological, and Geological Interactions** *Susan E. Humphris, Robert A. Zierenberg, Lauren S. Mullineaux, and Richard E. Thomson (Eds.)*

92 **Mauna Loa Revealed: Structure, Composition, History, and Hazards** *J. M. Rhodes and John P. Lockwood (Eds.)*

93 **Cross-Scale Coupling in Space Plasmas** *James L. Horwitz, Nagendra Singh, and James L. Burch (Eds.)*

94 **Double-Diffusive Convection** *Alan Brandt and H. J. S. Fernando (Eds.)*

95 **Earth Processes: Reading the Isotopic Code** *Asish Basu and Stan Hart (Eds.)*

96 **Subduction Top to Bottom** *Gray E. Bebout, David Scholl, Stephen Kirby, and John Platt (Eds.)*

97 **Radiation Belts: Models and Standards** *J. F. Lemaire, D. Heynderickx, and D. N. Baker (Eds.)*

98 **Magnetic Storms Bruce** *T. Tsurutani, Walter D. Gonzalez, Yohsuke Kamide, and John K. Arballo (Eds.)*

99 **Coronal Mass Ejections** *Nancy Crooker, Jo Ann Joselyn, and Joan Feynman (Eds.)*

100 **Large Igneous Provinces** *John J. Mahoney and Millard F. Coffin (Eds.)*

101 **Properties of Earth and Planetary Materials at High Pressure and Temperature** *Murli Manghnani and Takehiki Yagi (Eds.)*

102 **Measurement Techniques in Space Plasmas: Particles** *Robert F. Pfaff, Joseph E. Borovsky, and David T. Young (Eds.)*

103 **Measurement Techniques in Space Plasmas: Fields** *Robert F. Pfaff, Joseph E. Borovsky, and David T. Young (Eds.)*

104 **Geospace Mass and Energy Flow: Results From the International Solar-Terrestrial Physics Program** *James L. Horwitz, Dennis L. Gallagher, and William K. Peterson (Eds.)*

105 **New Perspectives on the Earth's Magnetotail** *A. Nishida, D. N. Baker, and S. W. H. Cowley (Eds.)*

106 **Faulting and Magmatism at Mid-Ocean Ridges** *W. Roger Buck, Paul T. Delaney, Jeffrey A. Karson, and Yves Lagabrielle (Eds.)*

107 **Rivers Over Rock: Fluvial Processes in Bedrock Channels** *Keith J. Tinkler and Ellen E. Wohl (Eds.)*

108 **Assessment of Non-Point Source Pollution in the Vadose Zone** *Dennis L. Corwin, Keith Loague, and Timothy R. Ellsworth (Eds.)*

109 **Sun-Earth Plasma Interactions** *J. L. Burch, R. L. Carovillano, and S. K. Antiochos (Eds.)*

110 **The Controlled Flood in Grand Canyon** *Robert H. Webb, John C. Schmidt, G. Richard Marzolf, and Richard A. Valdez (Eds.)*

111 **Magnetic Helicity in Space and Laboratory Plasmas** *Michael R. Brown, Richard C. Canfield, and Alexei A. Pevtsov (Eds.)*

112 **Mechanisms of Global Climate Change at Millennial Time Scales** *Peter U. Clark, Robert S. Webb, and Lloyd D. Keigwin (Eds.)*

113 **Faults and Subsurface Fluid Flow in the Shallow Crust** *William C. Haneberg, Peter S. Mozley, J. Casey Moore, and Laurel B. Goodwin (Eds.)*

114 **Inverse Methods in Global Biogeochemical Cycles** *Prasad Kasibhatla, Martin Heimann, Peter Rayner, Natalie Mahowald, Ronald G. Prinn, and Dana E. Hartley (Eds.)*

115 **Atlantic Rifts and Continental Margins** *Webster Mohriak and Manik Talwani (Eds.)*

116 **Remote Sensing of Active Volcanism** *Peter J. Mouginis-Mark, Joy A. Crisp, and Jonathan H. Fink (Eds.)*

117 **Earth's Deep Interior: Mineral Physics and Tomography From the Atomic to the Global Scale** *Shun-ichiro Karato, Alessandro Forte, Robert Liebermann, Guy Masters, Lars Stixrude (Eds.)*

118 **Magnetospheric Current Systems** *Shin-ichi Ohtani, Ryoichi Fujii, Michael Hesse, and Robert L. Lysak (Eds.)*

119 **Radio Astronomy at Long Wavelengths** *Robert G. Stone, Kurt W. Weiler, Melvyn L. Goldstein, and Jean-Louis Bougeret (Eds.)*

Geophysical Monograph 120

GeoComplexity and the Physics of Earthquakes

John B. Rundle
Donald L. Turcotte
William Klein
Editors

American Geophysical Union
Washington, DC

Library of Congress Cataloging-in-Publication Data
GeoComplexity and the physics of earthquakes / John B. Rundle, Donald L. Turcotte, William Klein, editors.
p. cm -- (Geophysical monograph series ; no. 120)
Includes bibliographical references.
ISBN 0-87590-978-7
1. Earthquakes--Computer simulation. I. Rundle, John. II Turcotte, Donald Lawson. III. Klein, William, 1943- IV. Series.

QE539.2.D36 G46 2000
551.22--dc21 00-030618

ISBN 0-87590-978-7
ISSN 0065-8448

Back Cover: An example of a complex pattern of earthquakes in southern California. The figure shows the relative intensity of seismic activity during the years 1932-December 31, 1991, ending just prior to the M ~ 7.4, June 28, 1992 Landers, California earthquake. Relative intensity is plotted on a scale from 0+ (lightest red) to 1 (darkest red). Analysis using conventional techniques failed to indicate that any special attention should be given to the Mojave Desert region where the earthquake occurred.

Printed in the United States of America.

CONTENTS

CONTENTS

PREFACE

Earthquakes in urban centers are capable of causing enormous damage. The January 16, 1995 Kobe, Japan earthquake was only a magnitude 6.9 event and yet produced an estimated $200 billion loss. Despite an active earthquake prediction program in Japan, this event was a complete surprise. Similar scenarios are possible in Los Angeles, San Francisco, Seattle, and other urban centers around the Pacific plate boundary. The development of forecast or prediction methodologies for these great damaging earthquakes has been complicated by the fact that the largest events repeat at irregular intervals of hundreds to thousands of years, resulting in a limited historical record that has frustrated phenomenological studies. The papers in this book describe an emerging alternative approach, which is based on a new understanding of earthquake physics arising from the construction and analysis of numerical simulations. With these numerical simulations, earthquake physics now can be investigated in numerical laboratories. Simulation data from numerical experiments can be used to develop theoretical understanding that can be subsequently applied to observed data. These methods have been enabled by the information technology revolution, in which fundamental advances in computing and communications are placing vast computational resources at our disposal.

Many of our colleagues have expressed to us the feeling that an exciting new kind of science is being born, a science of *GeoComplexity* that focuses on the temporal and spatial evolution of strongly correlated earth systems. Within this new science of GeoComplexity, the study of earthquakes will play a central role, encompassing as it does ideas of self-organization, nucleation, critical phenomena, self-organized criticality, and emergent structures and phenomena. Many of the results and methods that are discussed in this book are beginning to have a major impact on other areas of physics well beyond earth science. In particular, ideas about scaling regimes and friction are being used as paradigms in explaining the behavior of other kinds of threshold systems that display sudden transitions and "avalanche" phenomena, such as neural networks in biology, driven foams, charge density wave semiconductor devices, superconductors, and even the expansion of the early universe.

In this monograph, we have collected a number of articles that describe recent advances in the understanding of earthquake physics, together with suggestions for new research directions and new ideas that might be exploited by subsequent investigations. The great majority of these articles use the concept that earthquake fault systems are characterized by strongly correlated dynamics, giving rise to the observed scaling laws, correlations, and space-time patterns. We hope that the papers in this volume, which grew out of a series of AGU special sessions on the Physics of Earthquakes, will help to communicate some of the excitement we feel about the emergence and promise of this new field of research. We believe that the papers that we have collected here will be accessible to a broad range of earth scientists. We further hope that among the readers of this book will be scientists who, although not familiar with GeoComplexity, will nonetheless be stimulated to find out more. We encourage readers of this monograph, particularly students and young scientists, to explore these ideas on their own, and to contribute to this rapidly growing body of literature.

Most of our work and that of our colleagues has been funded by a variety of federal agencies, including the National Science Foundation, the U.S. Department of Energy, the National Aeronautics and Space Administration, the National Oceanic and Atmospheric Administration, and the Office of Naval Research. Without the generous support provided by these agencies, the work described in this monograph would not have been possible.

J. B. Rundle
Colorado Center for Chaos and Complexity, and CIRES

D. L. Turcotte
Cornell University

W. Klein
Boston University

Introduction

John B. Rundle
Colorado Center for Chaos & Complexity, & CIRES

Donald L. Turcotte
Cornell University

William Klein
Boston University

Earthquake physics is a very broad topic, and only a subset of subjects is covered in this monograph. The major focus is the contributions that studies in the new field of GeoComplexity, which deals with the physics of chaotic and complex earth systems, have made to the understanding of the physics of earthquakes. The relevant aspects of GeoComplexity are fractals and chaos, together with nucleation and critical phenomena, self-organized criticality, and more generally, self-organizing complex systems. The dynamics of the strongly correlated systems studied in GeoComplexity are characterized by a large multiplicity of spatial and temporal scales, and earthquakes are no exception. Strong correlations lead directly to space-time clustering of earthquakes, self-organization and scaling. Examples of the latter include the Gutenberg-Richter magnitude-frequency relation, and the Omori law for aftershocks (and foreshocks). More generally, nonlinear fault instability processes qualitatively similar to friction are known to be physically significant from the molecular scale, on time intervals of less than 10^{-8} seconds and lengths of Angstroms (Å), to plate motion scales, on time intervals of 10^6 years and lengths in excess of 1000 km. The physical processes directly associated with faulting and seismology, including nucleation and quasi-static crustal deformation, occur over time intervals of fractions of seconds to thousands of years, and lengths of meters to hundreds of kilometers.

By contrast, laboratory experiments on frictional sliding of rock samples are usually carried out over bench-top scales extending over time intervals of 10^{-3} seconds to days, and over length scales of centimeters to meters. We therefore have a critical need to understand how these physical processes that operate on the laboratory scales can be connected to the processes that we observe on much larger and longer scales in field situations. Some of the spatial scales for physical fault geometries include:

The microscopic scale (~ 10^{-6} m to 10^1 m) associated with static and dynamic friction (the primary nonlinearities associated with the earthquake process).

The fault-zone scale (~ 10^{-1} m to 10^2 m) that features complex structures containing multiple fractures and crushed rock.

The fault-system scale (10^2 m to 10^4 m), in which faults are seen to be neither straight nor simply connected, but in which bends, offsetting jogs and sub-parallel strands are common and known to have important mechanical consequences during fault slip.

The regional fault-network scale (10^4 m to 10^5 m), where seismicity on an individual fault cannot be understood in isolation from the seismicity on the entire regional network of surrounding faults, and where concepts such as *correlation length* and *critical state* borrowed from statistical physics have led to new approaches to understanding regional seismicity.

The tectonic plate-boundary scale (10^5 m to 10^7 m), at which Planetary Scale boundaries between plates can be approximated as thin shear zones and the motion is uniform at long time scales.

In GeoComplexity, the dynamical systems under study, such as earthquakes and earthquake faults, cannot be understood deterministically. Instead the scientific questions

GeoComplexity and the Physics of Earthquakes
Geophysical Monograph 120

are focussed on using methods of nonlinear and statistical mechanics to understand the stochastic processes involved, and upon characterizing and understanding the space-time patterns of activity that occur. The usual scientific methods are still applied, in which hypotheses are formulated based on data, predictions are made and tested with new data, and the results fed back into formulating new hypotheses. In the new methods associated with GeoComplexity, there are now often intermediate steps that involve the construction and analysis of numerical simulations, together with consideration of universal behavior and processes that arise in other diverse nonlinear dynamical systems. As applied to earthquakes, these questions would relate to:

Cataloging and understanding the nature and configurations of space-time patterns of earthquakes, and examining whether these are scale-dependent or scale-invariant in space and time. Certain characteristic patterns may indicate that a given event is a candidate foreshock of a future, larger event.

Developing and testing potential earthquake forecast algorithms, based primarily upon the use of space-time patterns in the fault system of interest.

Understanding the physical conditions that allow space-time coarse-graining of sub-grid scale processes, and whether these processes can be represented as correlated or uncorrelated noise.

Developing the theoretical framework to integrate diverse data and extrapolate existing data in space and time so that model predictions can be tested with new observations ("Model-Based Inference").

Developing an understanding of hardware, software and algorithmic issues in computational support for multi-scale science and engineering simulations which are of pervasive importance in many fields.

Past earthquake research has often focused on problems that display only a limited range of spatial and temporal scales. The examples discussed previously include laboratory studies of friction at the bench-top scale and crustal deformation at the plate boundary scale. However, the Gutenberg-Richter and Omori scaling laws, the scale-invariant physical structure of fault zones, and observations that seismic events cluster at all space and time scales investigated, all imply that multi-scale processes are at work. In most other areas of science in which multi-scale processes dominate, large-scale computing approaches have had a significant and lasting impact. It is highly likely that a computational approach to the earthquake problem in the spirit of GeoComplexity will enable major advances in earthquake science. A classic example of fractal statistics, or scaling laws, is the Gutenberg-Richter frequency-magnitude relation for earthquakes. First observed in the United States in 1942 (a similar observational relation had been observed earlier in Japan by Ishimoto and Iida), the frequency-magnitude relation can actually be interpreted as a power-law scaling relation for cumulative frequency of occurrence of earthquakes having a seismic moment larger than some value M_o. Similar scaling laws are the Omori relation for aftershocks, and the more recently discovered Bufe-Varnes relation for pre-seismic activation. During much of the 20th century, the focus in earthquake investigations was of necessity on observationally oriented research. However, as the 21st century dawns, and with it the advent of the Information Age, the focus is shifting in part towards the implementation and analysis of numerical simulations of earthquakes and other tectonic phenomena. Since all of these systems are fundamentally nonlinear in character, and most of them are high-dimensional complex systems, a variety of new techniques must be developed to make sense of the simulation data, then used to understand observed data. Some of these methodologies are applications of existing analytical techniques from similar areas of statistical physics and chaotic systems, but others are new.

One of the most rapidly developing areas is the analysis of slider-block models of earthquake faults. Since it was first introduced by Burridge and Knopoff in 1967, the nearest-neighbor slider-block model has grown in popularity as a simple model for the behavior of faults. Depending on the parameters used, even the simplest of these models have been shown to exhibit chaotic behavior. For example, a pair of interacting slider blocks demonstrates classical deterministic chaos with a period doubling route to chaos. Slider-block models having many interacting blocks, both with and without inertia, have now been analyzed extensively. Slider-block models and related approaches form a major part of this volume. For example, Becker examines in detail the chaotic behavior of a slider-block model. Pelletier argues that realistic simulations of earthquakes using slider-block models require both structural heterogeneity and a viscous interaction. Klein et al. show that the scaling statistics and nonlinear behavior of large arrays of slider blocks with heterogeneous friction can be understood and predicted in the limit when each block interacts with many other blocks. These models having long-range interactions are generically called *mean field models* and lead to the appearance of the same kind of spinodal that is seen in first-order thermal phase transitions. Gross uses a slider-block model, together with a slip-dependent frictional model (traveling density wave model) to understand the earthquake rupture process. Turcotte et al. explain the behavior of slider-block models in terms of an inverse cascade and a mapping of a fractal network to percolation clusters. Roberts and Turcotte argue that a plastic instability mechanisms proposed for deep earthquakes is also applicable to crustal earthquakes. Finally, an alternative simulation model is the lattice solid model, which is a particle-based computation that models the physics of fault gouge. Mora et al. discuss the ability of this model to reproduce observed seismic phenomena. One of the fundamental observations of earthquake physics is the tendency of seismicity to form space-time clusters of activity. Rundle et al. discuss methods of analysis that rely on the decomposition of seismicity into sums over space-time eigenvectors (eigenpatterns) of a correlation operator. They further show that the temporal behavior of the seismicity defines an apparent "pattern dynamics" that can be used to forecast future activity in relatively realistic simulations of earthquake fault systems.

Explaining and predicting the inter-relationships of physical variables such as stress, strain, displacements, temperatures, changes in gravity or density observed in field and laboratory data is the primary goal of models and simulations. The problem with which we are frequently faced in earthquake science is that the data that we can observe are not these physical variables. For example, it is easy to observe epicenters, occurrence times, and magnitudes of earthquakes, but not the failure thresholds, space-time variations in stress field, loading rates and other factors that lead to the earthquakes. Selected observations and laboratory data that serve to constrain the physical processes in earthquakes are critical to testing the success and validity of the models, and eventually for assimilation into the simulations. Kanamori and Heaton use observations and inferences for the energy budget of deep earthquakes to construct a model for friction that involves frictional heating and melting on the fault surface. They propose a two-stage process in which slip occurs at one value of kinetic friction until an effective sliding distance is reached, after which the kinetic friction coefficient falls to a new, much lower value as a result of melting on the sliding surface. Bilek and Lay find a systematic depth dependence of the source rupture duration of deep subduction earthquakes that any model for deep seismicity must explain. Karner and Marone report laboratory studies of stick-slip behavior on bare granite surfaces emphasizing the role of surface healing. One of the most promising approaches to earthquake forecasting is the observed accelerating precursory seismic moment/energy release. Jaumé discusses the relevant observations and uses models to interpret the data to define underlying causes. Finally, Tiampo et al. apply the methods developed by Rundle et al. and find that systematic variations in seismicity can be found to precede major earthquakes in southern California.

Friction laws of the kinds proposed by Kanamori and Heaton and by Karner and Marone are now finding their way into modeling codes, that are in turn being used to carry out realistic simulations of the fault systems in southern California. Papers in the final section point out that new advances in computational methods enable many new approaches to the modeling and simulation of earthquakes and other crustal processes. Fox et al. give a detailed summary of the state-of-the-art of developments in modern Web-based computing, focusing specifically on Object-Broker systems and computational "middleware." The emergence of these methods are a direct result of the rapidity with which the Web has come to dominate all aspects of computing and information technology. Bosl presents a detailed finite element discretization for the thermoporoelastic continuum equations applicable to the behavior of the crust. Bosl and Nur incorporate the pore pressure in considering the poroelastic response of the crust following the 1992 Landers, California earthquake.

Looking at progress in the recent past, it is clear that research in GeoComplexity and earthquake physics represents an extremely promising direction for the future. The new tools of analysis and interpretation that GeoComplexity has provided to us have revealed a wealth of new information on problems that had previously been resistant to understanding and solution. The papers in this volume describe only a small subset of recent results, but nevertheless provide a clear indication of the interest and vitality in the field. GeoComplexity, this new marriage of geoscience, nonlinear and statistical physics together with modern information technology, promises to revolutionize the way in which we view and understand the nonlinear dynamics of the complex earth system, as well as the roles of its many component parts. From a personal perspective, we feel extremely fortunate to have experienced the excitement of partcipating in the birth and development of this entirely new field of science.

Deterministic Chaos in two State-variable Friction Sliders and the Effect of Elastic Interactions

Thorsten W. Becker

Department of Earth and Planetary Sciences, Harvard University, Cambridge, Massachusetts, USA

It is demonstrated that a quasistatic slider model with two state-variable rate and state dependent friction shows chaotic dynamics in the deterministic sense. The irregular system behavior can be reduced to a one-dimensional, unimodal mapping which explains the existence of universal period doubling sequences in the stick-slip cycles *en route* to chaos. This property furthermore allows the approximate prediction of time intervals between sliding events. Lyapunov exponents and periodogram branches depend on the control parameter, the load point coupling, in a self-similar way. Thus, the single slider is a good example of low-dimensional chaos in a homogeneous system, possibly implying a microscopic source of irregularity for earthquakes in nature. However, the sliding patterns of interacting slider models are found to be dominated by perturbation wave phenomena. The wavelength of the asperities that are formed along a slider chain ("fault") increases with the strength of the spring coupling between sliders. This could imply a regularizing effect of interactions, but instabilities prohibit the exploration of the full parameter range for coupled sliders without damping.

1. INTRODUCTION

Laboratory rock sliding experiments initiated by *Dieterich* [1979] and *Ruina* [1980] have led to the establishment of semi-empirical laws that describe the dependence of the friction coefficient on the rate and history of sliding. These "rate and state dependent" laws can explain most of the laboratory observations for a range of materials and parameters [for a recent review, see *Marone*, 1998], and there has been progress in the application to seismicity in nature [e.g., *Dieterich*, 1994; *Scholz*, 1998]. Yet, it is still not entirely clear if and how the microscopic results from sample sizes of $\mathcal{O}$(cm) and sliding velocities of $\mathcal{O}(\mu\text{m/s})$ can be scaled to real faults with lengths of $\mathcal{O}$(km) and speeds of $\mathcal{O}$(m/s) [e.g., *Schmittbuhl et al.*, 1996]. However, I will proceed to examine rate and state friction under the assumption that these laws are also relevant for seismicity in nature. Only simplified quasistatic slider models and elastic interactions will be considered. My models, therefore, do not capture all important processes for faulting in nature, but this simplicity will allow us to gain some physical insight into what might be a part of the complex earthquake system.

In the laboratory experiments, complicated stick-slip sequences have been studied in which slow shear stress increases and vanishing sliding velocities alternate with rapid stress drop and slip. The underlying sliding and history dependence of the friction coefficient, which was formerly reduced to a "static" and a "dynamic" value, can be studied as a microscopic source leading to the observed seismic cycles in fault zones in nature [e.g., *Rice and Tse*, 1986]. Usually, one "state-variable" laws suffice to explain the observed deviations from the traditional concept of static and dynamic friction. The

GeoComplexity and the Physics of Earthquakes
Geophysical Monograph 120

simplest mathematical model which mimics stick-slip with one state-variable friction is a one degree of freedom slider that is spring-coupled to a steadily progressing load point. When inertia is neglected, the resulting quasistatic system of non-linear, ordinary differential equations is two-dimensional (2-D), i.e., it is fully determined by two initial conditions, for example, stress and sliding velocity.

The linear and non-linear stability of this slider system with respect to perturbations has been studied previously [*Rice and Ruina*, 1983; *Gu et al.*, 1984; *Gomberg et al.*, 1998; *Ranjith and Rice*, 1999], and I will consider these equations as a starting point for slightly increased complexity. More complicated system behavior can arise when inertia [e.g., *Rice and Tse*, 1986], continuum interactions [e.g., *Horowitz and Ruina*, 1989], or a second state-variable [*Gu et al.*, 1984; *Gu and Wong*, 1994; *Zhiren and Chen*, 1994] are considered. The latter approach results in the introduction of one additional degree of freedom, and it seems that this is required by some laboratory observations [e.g., *Ruina*, 1980; *Dieterich*, 1981; *Gu et al.*, 1984; *Blanpied and Tullis*, 1986; *Gu and Wong*, 1994] which show that a single state-variable friction law might be insufficient to interpret the data.

Here I will focus on a quasistatic slider model with two state-variables so that the effect of this constitutive relation can be studied in isolation from other processes such as dissipation or the effect of inertia. Also, I will not be considering pre-existing heterogeneities, be it in the initial conditions (pre-loading) or the material properties. Mathematically, the resulting three dimensional (3-D) phase space dynamics are no longer limited to fixed points (corresponding to stable sliding in the rock friction experiments) or limit cycle oscillations (stick-slip "earthquakes"). It was noted earlier that friction measurements from the laboratory and the corresponding numerical models show period doubling cascades toward irregularity for certain parameter values [*Ruina*, 1983; *Gu et al.*, 1984; *Gu and Wong*, 1994]. More recently, *Zhiren and Chen* [1994] studied a two state-variable slider numerically and suggested that the irregular state which is reached by successive period doubling bifurcations might be chaotic. However, they failed to demonstrate quantitative period doubling conclusively.

This paper will show how the irregular system behavior of a one slider, quasistatic, two state-variable system can be quantified and how certain parameter settings can lead to chaos in the strict deterministic sense. Further, it will be demonstrated that the 3-D dynamics of the chaotic state can be reduced to a one dimensional, unimodal mapping. This allows the approximate prediction of the time intervals between sliding events. As shown by *Feigenbaum* [1978], the existence of this low-dimensional structure is the reason that a universal period doubling route is realized. The convergence of the control parameter intervals for bifurcations is extracted using "periodograms" that are derived from Poincare sections of the phase space. The resulting plots resemble those for the logistic map [e.g., *May*, 1976] and demonstrate peculiar features of non-linear dynamics such as universal period doubling cascades and intermittent period-three windows. Both the period bifurcations and the Lyapunov exponents depend on the controlling load point stiffness in a self-similar way. The slider system thus turns out to be a good example of low-dimensional deterministic chaos. However, when next-neighbor interactions for coupled sliders are taken into account, the resulting perturbation waves dominate the sliding heterogeneities, and a regularizing effect is observed with increasing coupling stiffness.

2. THEORETICAL BACKGROUND

This section will discuss the mathematical properties and governing equations of the quasistatic two state-variable slider model and briefly review some measures of irregularity in non-linear dynamical systems.

2.1. Model Definition

We can write a constitutive law for rock interface friction at constant normal stress in terms of the shear stress, τ, as a function of sliding velocity, V, and two "state-variables", θ_1 and θ_2:

$$\tau(V, \theta_1, \theta_2) = \tau_* + A \ln\left(\frac{V}{V_*}\right) + \sum_{i=1}^{2} B_i \ln\left(\frac{V_* \theta_i}{L_i}\right). \quad (1)$$

The material and environment (e.g., temperature and gouge width) dependent parameters A and B_i control the character of the velocity dependence of friction (weakening or strengthening), and the critical slip lengths, L_i, are relevant for stability. For simplicity, we assume A, B_i, and L_i are constant. Further, V_* and τ_* denote a reference velocity and stress respectively, and the θ_i as well as V depend on time, t. When the friction law is cast in the form of eq. (1), the state-variables can be interpreted as average contact lifetimes for sliding surface roughness. The θ_i were introduced by the experimenters to fit the observed exponential decay processes together with an evolution law like:

$$\frac{d\theta_i}{dt} = -\frac{V\theta_i}{L_i}\ln\left(\frac{V\theta_i}{L_i}\right), \tag{2}$$

where we have used the Ruina ("slip") version [see, e.g., *Marone*, 1998, for discussion]. Inertia-free load point coupling with fixed load point velocity V_0 and coupling stiffness $K > 0$ can be written as:

$$\frac{d\tau}{dt} = K(V_0 - V), \tag{3}$$

and completes our model equation system for stick-slip (see Figure 1).

We can non-dimensionalize eqs. (1), (2), and (3) by introducing the following variables [slightly modified from *Gu et al.*, 1984]:

$$x = \ln\left(\frac{V}{V_*}\right) \qquad y = \frac{\tau - \tau_*}{A}$$
$$\beta_i = \frac{B_i}{A} \qquad \kappa = \frac{KL_1}{A}$$
$$z = \beta_2 \ln\left(\frac{V_*\theta_2}{L_2}\right) \qquad \rho = \frac{L_1}{L_2}$$
$$T = \frac{V_*}{L_1}t.$$

The resulting equations that constitute our model are:

$$\dot{x} = e^x\left((\beta_1 - 1)x + y - z\right) + \dot{y} - \dot{z} \tag{4}$$
$$\dot{y} = (1 - e^x)\,\kappa \tag{5}$$
$$\dot{z} = -e^x\rho\,(\beta_2 x + z)\,, \tag{6}$$

where dotted quantities are derived with respect to rescaled time T, and we have set the load point velocity V_0 to V_* without loss of generality. Equations (4) to (6) describe the friction system in terms of the non-dimensionalized quantities velocity, x, stress, y, and second state-variable, z. In shorthand we can write

$$\dot{\mathbf{x}} = \mathbf{F}(\mathbf{x}) \quad \text{with} \quad \mathbf{x} = (x, y, z). \tag{7}$$

Gu et al. [1984] state that the choice

$$\beta_1 = 1 \quad \beta_2 = 0.84 \quad \text{and} \quad \rho = 0.048 \tag{8}$$

for three of the four remaining free parameters is appropriate to reproduce experiments of *Ruina* [1980]. Keeping these numbers fixed, the system behavior for a certain initial value is fully characterized by the control parameter, the non-dimensionalized stiffness κ.

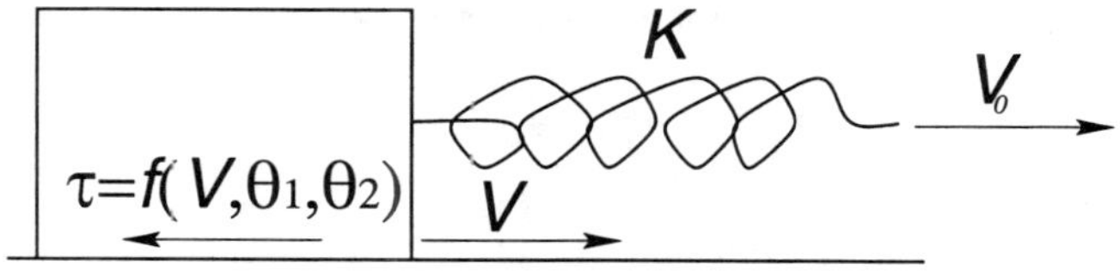

Figure 1. Cartoon of the one degree of freedom slider system. The block is pulled at constant rate V_0 and the friction at the contact interface follows a two state-variable rate and state dependent law.

2.1.1. Stability. The steady state for the model system corresponds to a fixed point in phase space for $\mathbf{x}$, and the only solution to

$$\mathbf{F}(\mathbf{x}) = \mathbf{0} \tag{9}$$

is $\mathbf{x} = \mathbf{0}$ with the physical interpretation of steady sliding along at the speed of the load point. *Rice and Ruina* [1983] give a general criterion for the linear stability of this fixed point for a class of rate and state type friction systems. Let us consider constant load point velocity, $\mathbf{x} = \mathbf{0}$ (corresponding to slow aseismic creep of a fault), and a sudden perturbation in sliding speed, x, say due to a passing seismic wave. For very stiff coupling with large values of κ, $\mathbf{x} = \mathbf{0}$ is a linearly stable state and small perturbations will die out. Yet, when κ is decreased below a critical value, κ_{cr}, the system undergoes a Hopf bifurcation. This means that the spiraling attraction of the fixed point is transformed into a limit cycle, and further to a repelling spiral where small perturbations from $\mathbf{x} = \mathbf{0}$ move away from the origin without bounds. The system has then become unstable.

Gu et al. [1984] have applied *Rice and Ruina* [1983]'s criterion to eqs. (4) to (6) and discuss several general properties of the two state-variable friction law. Here, I will only state that their formula for the critical stiffness, $\kappa_{cr}^{(2)}$, translates to my scaling as

$$\begin{aligned}\kappa_{cr}^{(2)} = \Big[&\kappa_{cr}^{(1)} + \rho\,(2\beta_1 + (\beta_2 - 1)\,(2+\rho)) \times \\ &\Big\{4\rho^2\left(\kappa_{cr}^{(1)} + \beta_2\right) + \\ &\left(\kappa_{cr}^{(1)} + \rho^2(\beta_2 - 1)\right)^2\Big\}^{\frac{1}{2}}\Big] \Big/ \Big[2 + 2\rho\Big].\end{aligned} \tag{10}$$

$\kappa_{cr}^{(1)}$ denotes the critical stiffness of a one state-variable law

$$\kappa_{cr}^{(1)} = \beta_1 - 1 \tag{11}$$

and eq. (10) transforms into eq. (11) for $\rho = 1$ and $\beta_2 = 0$. All control parameter values in this paper will be given in dimensionless form as a fraction of the critical stiffness

$$\kappa' = \frac{\kappa}{\kappa_{cr}^{(2)}}, \tag{12}$$

where $\kappa_{cr}^{(2)} \sim 0.08028$ for the choice of parameters (8). Only the normalized quantity κ' is independent of any non-dimensionalization scheme and

$$\kappa' < 1 \tag{13}$$

is the condition for a linearly unstable system.

There has not been a comprehensive parameter study for the two state-variable quasistatic slider equations yet, although *Blanpied and Tullis* [1986] explored stability surfaces and *Gu and Wong* [1994] conducted a range of laboratory and numerical experiments. It is thus not clear if the system properties under consideration here are general, intrinsic features or if they depend strongly on the fine tuning. In addition, other issues such as the quest for the right evolution law (eq. (2)) remain unresolved [e.g., *Marone*, 1998]. At this stage, it seems reasonable to be foremost consistent with the literature [*Gu et al.*, 1984], and I will assume that generic aspects of the friction law (eq. (1)) are captured by the particular parameter choice (8).

2.1.2. Numerical implementation. A step size controlled Cash-Karp Runge-Kutta scheme [*Press et al.*, 1993] was used to solve the system of equations numerically. The integration routine has been benchmarked, was compiled at double precision machine number representation, and set to a precision better than 10^{-8} and an accuracy better than 10^{-7} for single and coupled sliders respectively.

2.2. Measures of the Irregular System State

Various tools have been developed to quantify the irregularity of non-linear dynamical systems such as the one realized by our set of model equations [e.g., *Ott*, 1993]. Changes in the system variables with time can be described as Lagrangian flow of state points forming a trajectory in phase space from $\mathbf{x}$ to $\mathbf{x}'$. Calculating Lyapunov exponents for that flow then gives: a) a description of the dynamic stretching of a small sphere of radius r_0 around any initial condition $\mathbf{x}_0$, and, by inference, b) a way to determine the dimensionality of an attracting limit object if such a thing exists for bounded trajectories.

Considering a) first, the vigor of mixing and stretching of the initial conditions-sphere tells us about the irregularity of the system evolution from different starting points. The time evolution of the major axes (eigendirections) r_i of the ellipsoid that results from stretching in a 3-D flow can be written in the Floquet form as:

$$r_i(T) \propto r_0 \exp(h_i T) \quad \text{with} \quad i = 1, 2, 3. \tag{14}$$

This defines the Lyapunov exponents h_i, so that

$$h_i = \lim_{T \to \infty} \frac{1}{T} \ln \left(\frac{r_i(T)}{r_0} \right). \tag{15}$$

We will sort according to $h_1 > h_2 > h_3$. Under the assumption that the system is ergodic, the time-averaging of eq. (15) should be equivalent to an ensemble average over different initial conditions $\mathbf{x}_0$, and the values for h_i are taken as representative of the flow in general. $h_1 > 0$ corresponds to exponentially fast divergence of initial conditions in one direction, the so called "butterfly effect".

When the system equations are known the h_i can be approximated by averaging the singular values of the Jacobian matrix which gives a linearized version of the flow $\mathbf{F}(\mathbf{x})$. Since the properties of the system lead to rapid growth and shrinkage of matrix elements, a numerical realization of this method has to include frequent re-normalization to obtain accurate results. An alternative approach to quantify stretching has become standard and was proposed by *Benettin et al.* [1980]: for 3-D, the scheme is based on tracking the evolution of three orthogonal vectors, $\mathbf{y}_i$, which can be approximated by

$$\dot{\mathbf{y}}_i(T) = \mathbf{J}|_{\mathbf{x}(T)} \mathbf{y}_i(T) \quad \text{with} \quad i = 1, 2, 3, \tag{16}$$

where $\mathbf{J}|_{\mathbf{x}(T)}$ is the Jacobian of $\mathbf{F}$ at the position to which initial condition $\mathbf{x}_0$ has moved at time T. When a Gram-Schmidt ortho-normalization is applied to the $\mathbf{y}_i$ at time intervals δT to avoid overflow, approximations for h_i can be obtained by

$$h_i \approx \frac{1}{l\, \delta T} \sum_{k=1}^{l} \ln \left(\frac{\alpha_k^{(i)}}{\alpha_k^{(i-1)}} \right). \tag{17}$$

Here, l denotes the number of times the test vectors have been normalized. $\alpha_k^{(i)}$ is the i-dimensional "volume" of the parallelepiped spanned by the $\mathbf{y}_1 \ldots \mathbf{y}_i$ vectors (i.e., $|\mathbf{y}_1|$, $|\mathbf{y}_1 \times \mathbf{y}_2|$, and $(\mathbf{y}_1 \times \mathbf{y}_2) \cdot \mathbf{y}_3$ for $i = 1$, $i = 2$, and $i = 3$ respectively) before the kth normalization took place [see, e.g., *Ott*, 1993, p. 138]. The *Benettin et al.* [1980] method described above was implemented by analytically calculating the Jacobian, evaluating it at the $\mathbf{x}(T)$ location obtained by step size controlled Runge-Kutta and propagating the $\mathbf{y}_i$ by the Euler method. l and δT were usually on the order of 1000 and 50 respectively, and the integration along trajectories was stopped when changes for the h_i converged below 10^{-6}.

The connection between Lyapunov exponents and fractal dimension of objects attracting trajectories is the conjecture of *Kaplan and Yorke* [1979]. These authors demonstrated that it is likely that the information dimension and the quantity D_{KY} (the Lyapunov or Kaplan-Yorke dimension) are identical. D_{KY} is given by

$$D_{KY} = D + \frac{1}{|h_{D+1}|} \sum_{i=1}^{D} h_i, \tag{18}$$

where D is the largest integer for which $\sum_{i=1}^{D} h_i > 0$. D_{KY} is therefore a convenient geometrical measure of objects in phase space if Lyapunov exponents can be calculated readily.

3. RESULTS AND DISCUSSION

The next two sections are concerned with the discussion of model results, first from a single slider system and second from interacting spring-coupled slider chains.

3.1. *Single Slider Experiments*

3.1.1. Period doubling cascades. Previous studies have shown that the two state-variable quasistatic slider system can evolve into a stick-slip limit cycle below the stability bound $\kappa' = 1$ [e.g., *Gu et al.*, 1984]. When the stiffness κ' is continuously decreased, a period-doubling sequence is observed, eventually leading to irregular behavior. Figure 2 shows results from my numerical experiments to illustrate this behavior.

For a stable limit cycle of period two, a slow build-up of stress is followed by a rapid stress-drop in a sliding event, and this is repeated in a strictly periodic fashion. The characteristic zig-zag pattern of stick-slip in the stress versus time plots translates to a deformed limit-cycle in phase space, as shown in Figure 2a for $\kappa' = 0.9$. For the parameters given in (8), the system evolves into this state from the unstable fixed point $\mathbf{x} = \mathbf{0}$ when it is slightly perturbed. Large perturbations, on the other hand, lead to growing oscillations and unstable sliding since $\kappa' < 1$. When κ' is decreased further to $\kappa' = 0.86$, the system changes to period four oscillations as in Figure 2b. By looking at the phase space trajectories in Figure 2b, it becomes evident that the folded-loop structure that characterizes period four oscillations could not have been realized in a system with a single state-variable. In that model, the phase space is restricted to 2-D, where uniqueness requires that trajectories do not cross.

Another folding of trajectories forms the period eight cycle when κ' is down to ~ 0.856 (Figure 2c), period sixteen for $\kappa' \sim 0.8552$ (Figure 2d), and so on until an apparently chaotic state is reached at $\kappa' \sim 0.853$ (see, e.g., Figure 5). This period doubling behavior can be quantified using the frequency spectrum of the corresponding time series [*Zhiren and Chen*, 1994]. In the case of laboratory experiments with incomplete knowledge of the system equations, this is sometimes the only way to proceed in analyzing the system properties [e.g., *Libchaber et al.*, 1982]. Detecting the bifurcation values of κ' where new 2^n orbits are formed from the power spectrum can be complicated. For our model I propose the use of simple Poincare sections as a more accurate and straightforward way of quantifying the period doubling cascade.

If we plot the y position of trajectories intersecting the x-z–plane versus κ' (see Figure 2a), the "periodograms" of Figures 3a and 3b arise. Branching lines in these plots correspond to newly created cycles *en-route* to irregularity on the right hand side. Figure 3 shows all the features which have been discussed for the one-dimensional logistic equation [e.g., *May*, 1976] such as the period three windows (e.g., at $\kappa' \sim 0.854$) and geometrical self-similarity (compare Figure 3a with 3b). Period doubling bifurcations are now easily detected when the integration of the system equations is run for long enough times (on the order of 25000) to get rid of transients which introduce spurious higher order cycles.

If the two state-variable quasistatic friction system follows the universal period doubling route [*Feigenbaum*, 1978], the distance factor between the critical values of κ' for the bifurcation sequence $n-1$, n and $n+1$,

$$\delta_n = \frac{\kappa'_n - \kappa'_{n-1}}{\kappa'_{n+1} - \kappa'_n}, \tag{19}$$

should converge to one of the Feigenbaum numbers:

$$\hat{\delta} = 4.669201\ldots.$$

To test this hypothesis experimentally when κ' is continuously decreased, a bifurcation can be defined as the point where the number of Poincare intersections, 2^n, leaves the plateau of the current cycle of order n. It will then rise to the next level of 2^{n+1} after some numerical transient. Based on this definition, Table 1 was obtained by varying the stepsize in κ' and iteratively narrowing the intervals around the critical values. The δ_n can be observed to converge monotonically to a value $\sim$4.48. I interpret this as satisfactory agreement with the *Feigenbaum* [1978] theory and attribute the small

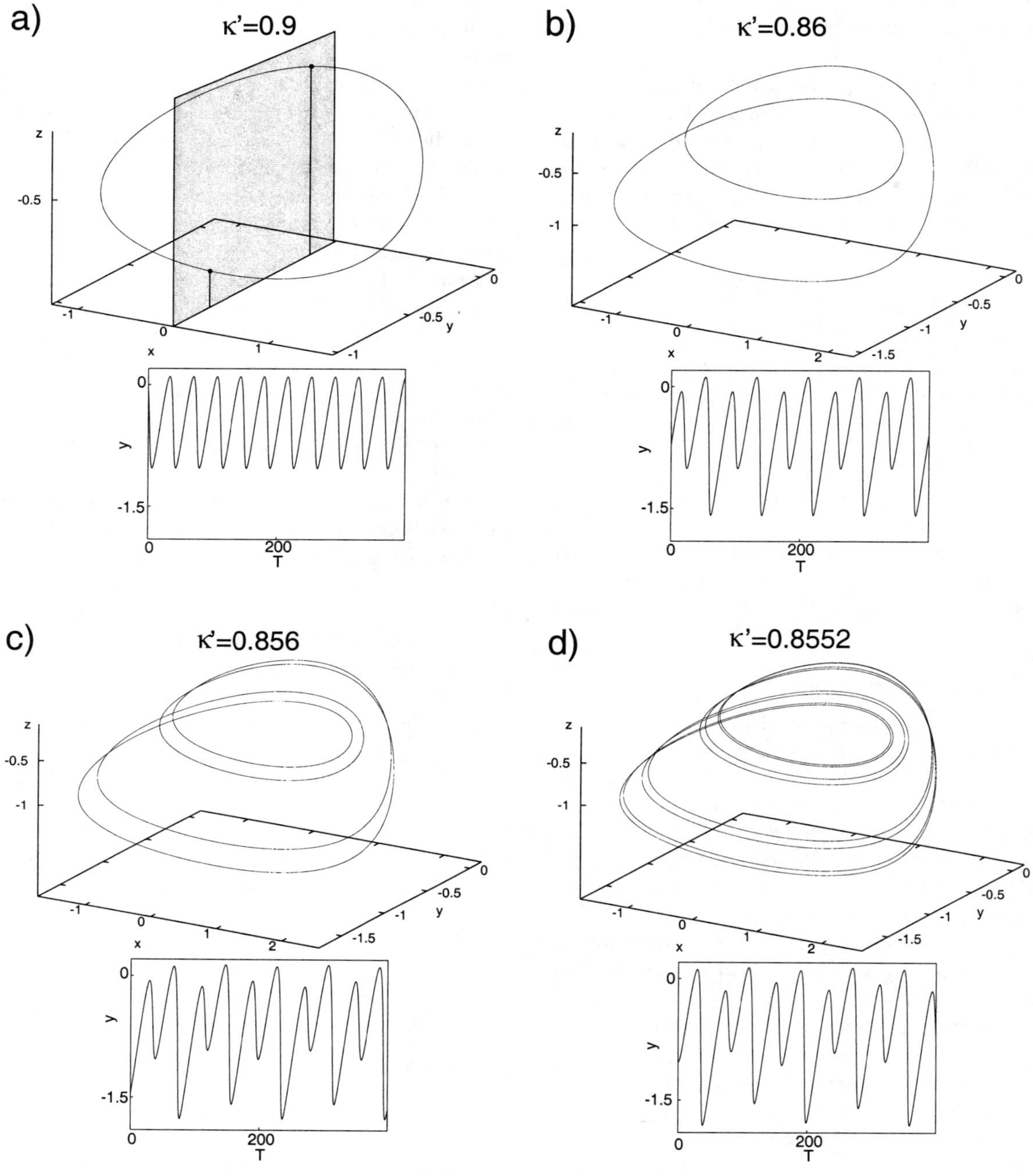

Figure 2. Phase space trajectories (top) and parts of the corresponding stress, y, versus time, T, series (bottom) for different values of the controlling stiffness κ'. (T scale is offset so that initial transients have decayed.) (a) shows period two, (b) period four, (c) period eight, and (d) period sixteen stick-slip limit cycles. The shaded x-z–plane for part a) illustrates how the Poincare sections for Figures 3a and 3b are obtained; the trajectories were reduced to a projection on $z = 0$ so that the period two orbit would result in two points at different y values. Trajectories were generated by numerically integrating eqs. (4), (5) and (6) from $\mathbf{x}_0 = (0.05, 0, 0)$ until $T = 2000$ and plotting the system evolution for the next 1000 timesteps.

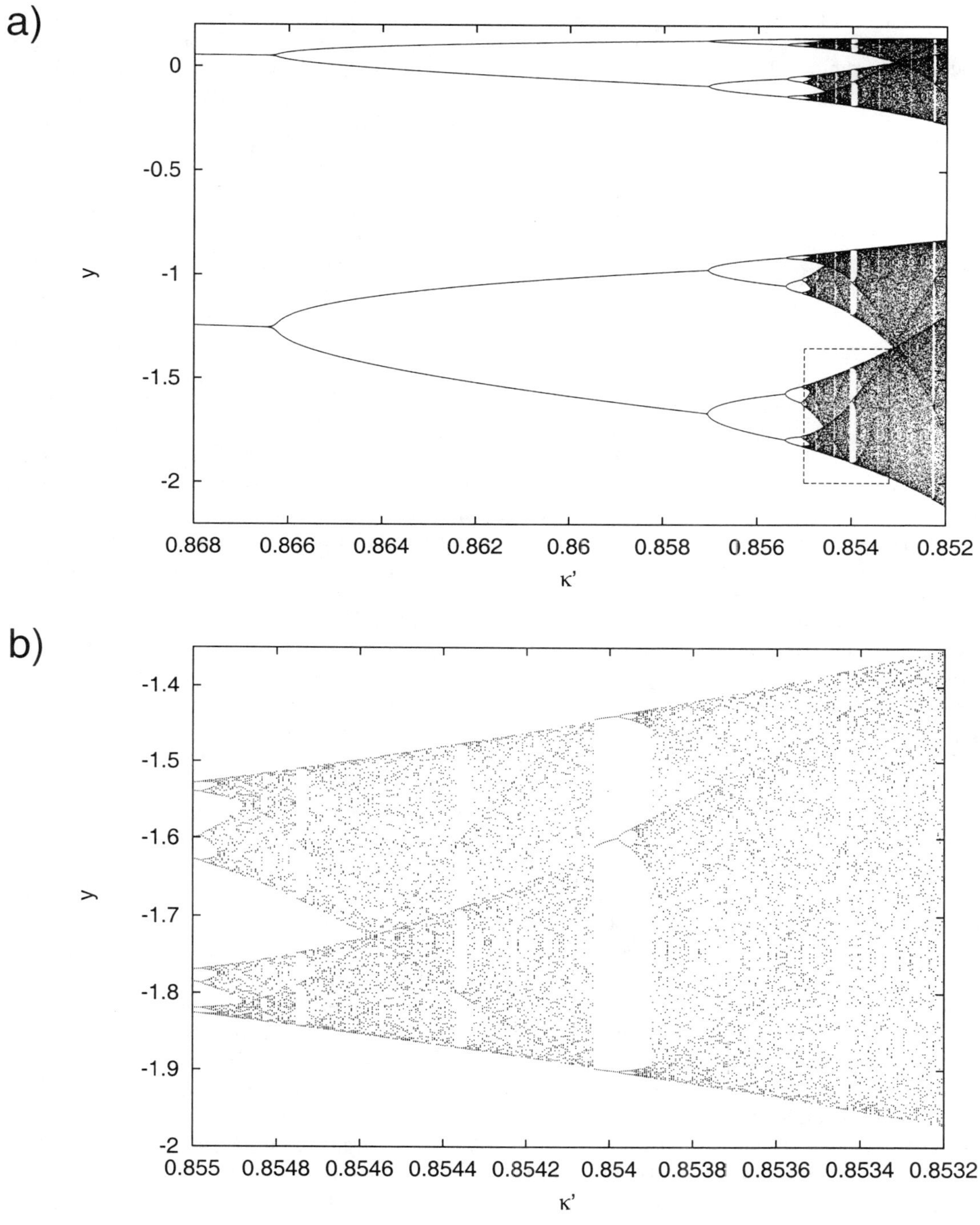

Figure 3. (a) Periodogram showing a montage of Poincare sections of the asymptotic system behavior. Obtained by integrating the model equations from $\mathbf{x}_0 = (0.05, 0, 0)$ until $T = 20000$ and then tracking all Poincare intersections until $T = 23000$. The apparent distortion of the y scale for the upper tree structure is due to the choice for the Poincare section and could be improved by adjusting the projection further toward the attractor. System behavior for $\kappa' \lesssim 0.852$ is unstable. (b) Magnification of the box marked by dashed lines in part a).

Table 1. Critical numbers for the controlling stiffness κ' at the creation of a 2^n-period limit cycle. The progression factor δ_n is calculated from eq. (19).

n	2^n	κ'	δ_n
1	2	1	
2	4	0.866592	14.09
3	8	0.857127	5.56
4	16	0.855424	4.69
5	32	0.8550609	4.50
6	64	0.8549803	4.48
7	128	0.8549623	

deviation to systematic errors, probably due to numerical noise or the way the bifurcations were picked.

My results are at odds with the study of *Zhiren and Chen* [1994] who did not find a clear cut convergence for a similar system of equations. These authors used a different parameter choice, though, and I could not directly reproduce their results, probably because of a misprint in their paper. The discrepancies might therefore be due to actually different system behavior or due to differences in the approach of quantifying the bifurcation locations. Spurious transients might have led to inaccuracies in their Fourier spectrum approach.

By using periodograms, we could therefore show that the quasistatic two state-variable slider follows a universal period doubling road to irregularity, as suggested by *Zhiren and Chen* [1994]. Previously, similar behavior has been found for asymmetrically coupled slider pairs with a simpler friction law [*Huang and Turcotte*, 1990; *Turcotte*, 1997, chap. 11], but this paper makes the first stringent case for a homogeneous friction system.

3.1.2. Unimodal Lorenz mappings. *Feigenbaum* [1978] demonstrated that quantitative universality in period doubling cascades arises because the dynamics of all qualifying systems can be reduced to a unimodal mapping. It is thus an obvious step to look for the existence of such a mapping in the search for an explanation of the period doubling we found.

As suggested by *Lorenz* [1963], an irregular time series, in our case the stress y, can be analyzed by plotting the amplitudes of sequential extrema n and $n+1$ against each other. If the resulting dots of, say, y^n_{min} versus y^{n+1}_{min} trace out a unique graph this indicates that the irregular system has hidden low dimensionality and can be reduced to a 1-D mapping. As a more practical aspect, we could then determine the next minimum value based on the knowledge of the current one, even if a strange looking time-series might suggest otherwise.

Proceeding to construct such a Lorenz-mapping, I plot the y^n_{min} versus y^{n+1}_{min} for the irregular system state at $\kappa' = 0.8525$ in Figure 4a as small dots ($-y_{min}$ is used for convenience). We find that the graph which is traced out is indeed almost unique, i.e. the dots lie basically on a line with small width that does not curve over. Second, the mapping is unimodal; there is only one maximum. The existence of this maximum where the derivative of the mapping goes to zero is in fact the reason for the darker streaks of accumulating points one can find in the irregular region of Figure 3a [e.g., *Strogatz*, 1994, p. 463]. Third, a fixed point for the 1-D mapping is found at the intersection of the dotted graph and the $y^{n+1}_{min} = y^n_{min}$ line at $y_{min} \sim -1.36$. Since the dot mapping has a slope with absolute value larger than unity at this point, small offsets from it will have grown by the next iteration. This means that the fixed point is unstable and we can expect aperiodic system behavior for all times.

A similar mapping exists for the maximum stress values y_{max}. Since the loading rate is constant, we might then infer that it is possible to predict the "quiet" time intervals ΔT between sliding events (see Figures 2 and 5) on the basis of a simple mapping as well. Figure 4b demonstrates that this is only approximately the case since the finite width of the stress minima and maxima mappings add up to substantial non-uniqueness for the "seismic period" ΔT, especially for $40 < \Delta T < 45$. However, even a rough chance of predicting the timing of the next sliding events might be considered a remarkable property that arises simply from the determinism behind the chaotic time-series of Figure 5.

Summing up, quantitative period doubling with monotonic convergence to $\hat{\delta}$ can be observed as a route to irregularity for the two state-variable friction slider. It was demonstrated that the reason for this universal behavior is the existence of a unimodal mapping between stress extrema. This property allows the approximate reduction of the 3-D dynamics to a 1-D mapping and confirms our suspicion that the system behavior will be aperiodic for all times in the irregular state.

3.1.3. Lyapunov exponents. Figure 5 shows the object that is traced out by trajectories in phase space for the irregular system state at $\kappa' = 0.8525$. It is characteristic for the whole irregular parameter range as indicated in Figure 3 and also found for different values of κ', say, 0.853. The suggestive interpretation is that it is a strange attractor. With the tools described in section 2.2 we can address the question of classification by calculating the Lyapunov exponents h_i. The numbers obtained for the typical irregular system state

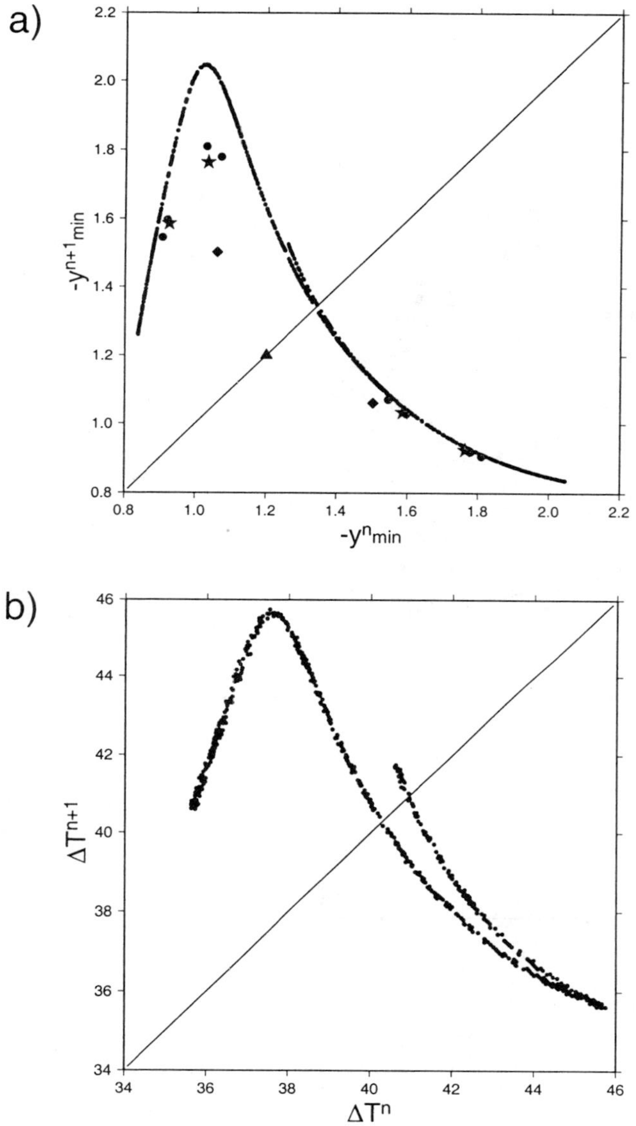

Figure 4. (a) Lorenz mappings of subsequent minima in stress amplitudes, y^n_{min} versus y^{n+1}_{min}, for different values of κ' ($-y_{min}$ is shown for visualization purposes). The single triangle for $\kappa' = 0.9$ corresponds to period two stick-slip (compare Figure 2a), all stress minima plot on the fixed point line for $y^{n+1}_{min} = y^n_{min}$. $\kappa' = 0.86$ (diamonds) is a period four limit cycle where y_{min} alternates in amplitude (compare Figure 2b). Further, $\kappa' = 0.856$ (stars) and $\kappa' = 0.8552$ (circles) symbols correspond to period eight and sixteen cycles, respectively (compare Figure 2c and 2d), and the numerous small dots for $\kappa' = 0.8525$ trace out the mapping for the irregular system state (compare Figure 5). Note that the resulting graph is almost unique with some indication of higher dimensionality at $y \sim -1.3$. (b) Mapping of subsequent "quiet" seismic periods, ΔT^n versus ΔT^{n+1}, between steep drops of y for $\kappa' = 0.8525$.

at $\kappa' = 0.8525$ are given in Table 2, where the error range has been estimated on the basis of convergence. The results for the Lyapunov exponents illustrate four points:

First, h_1 is small but positive in the irregular state, indicating exponential stretching of small deviations from a trajectory on the attractor. (We will assume that linear stretching –which could also lead to positive h_1 for finite times– can be ruled out and we ran the simulation long enough to insure this.) This sensitive dependence on initial conditions is one hallmark of deterministic chaos, the others being aperiodicity and determinism. The latter is fulfilled since we specifically set out not to consider any inhomogeneities or random fluctuations. Aperiodicity is indicated by the Lorenz mappings of Figure 4a, which demonstrate that there is no fixed point where the system might get hung up. We can therefore finally classify the irregular system state of the two state-variable slider as chaotic in the strict sense.

Second, $h_1 > 0$, $h_2 \sim 0$ and $h_3 < 0$ holds for chaotic control parameter values. This sign triple appears when trajectories form a strange attractor [e.g., *Ott*, 1993, p. 136]; then $h_1 > 0$ corresponds to stretching of flow perpendicular to the trajectory along the attractor shape, $h_3 < 0$ arises from the contracting properties of the flow toward the attractor, and the remaining vanishing Lyapunov exponent ($h_2 \sim 0$) corresponds to flow tangential to the trajectory (see also Figure 6).

Third, the sum of all h_i is negative. This means that the flow is on average contractive, corresponding to

$$\sum_{i=1}^{3} h_i \approx \lim_{T\to\infty} \frac{1}{T} \int_0^T \nabla \cdot \mathbf{F}(\mathbf{x}(T'))dT' < 0 \tag{20}$$

When we consider a volume V that encompasses all trajectories, we can also write

$$\int_V \nabla \cdot \mathbf{F}(\mathbf{x})dV < 0. \tag{21}$$

Eq. (21) could have been derived by integrating the divergence of $\mathbf{F}$ directly, but with the drawback that the resulting expression depends on the integration bounds. Contracting flow all over the phase space implies the existence of a limit state. Since we observe long excursions in phase space nevertheless (see Figure 5) we conclude that this limit state has to be an object with "holes" and can not be a simple geometric surface. We have thus

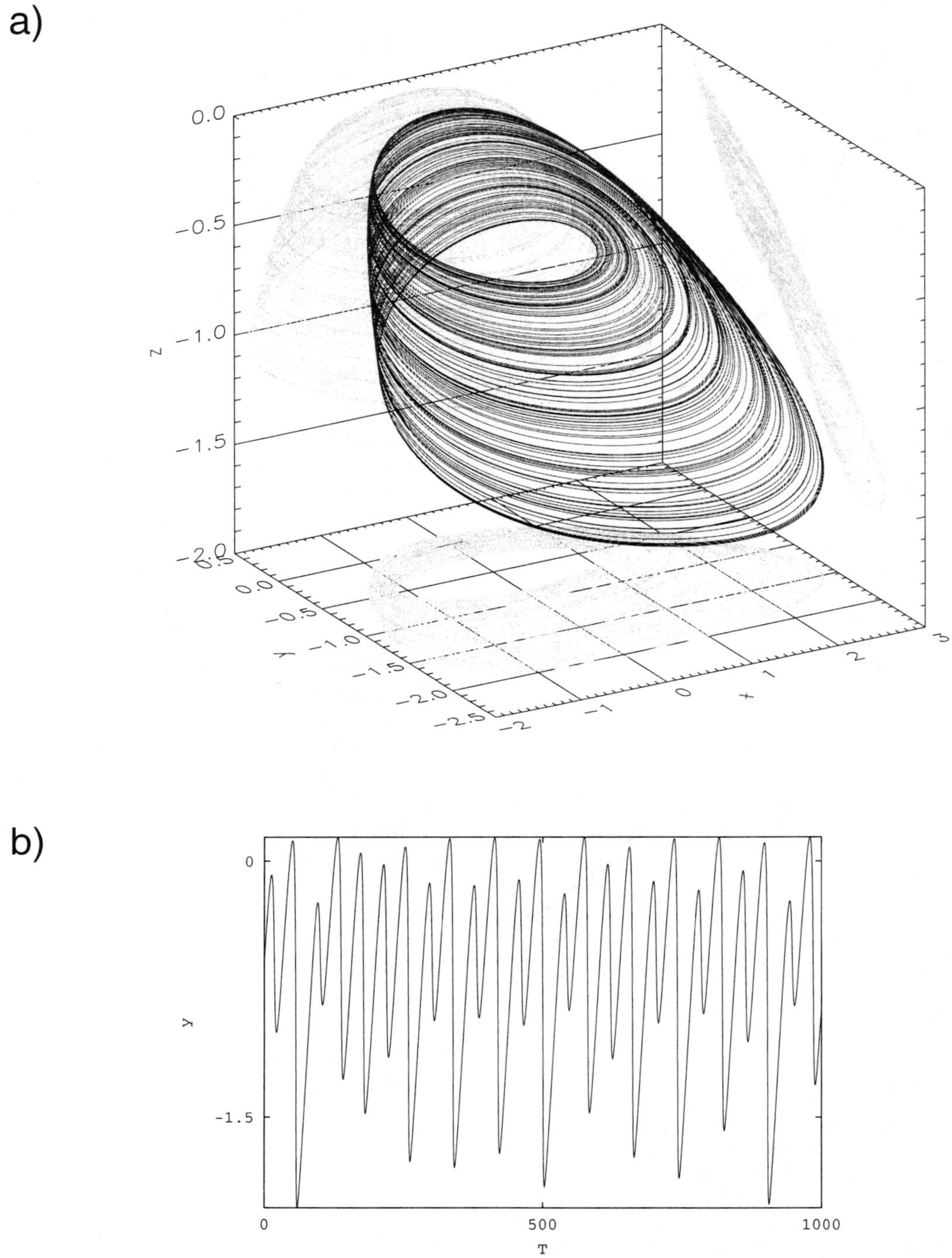

Figure 5. (a) Strange attractor for the chaotic system state at $\kappa' = 0.8525$ in phase space (side walls show projections). (b) Corresponding stress versus time series. Note how the trajectories almost form a surface resembling a Moebius ribbon in phase space. While the stick-slip sequence is strictly aperiodic and irregular overall, intermittent quasiperiodic sequences can be observed.

Table 2. Lyapunov exponents h_i, sum of all h_i, and Kaplan-Yorke dimension calculated from eq. (18) with $D = 2$.

Quantity	Value
h_1	0.01 ± 10^{-4}
h_2	0.0002 ± 10^{-4}
h_3	-0.086 ± 10^{-4}
$\sum_{i=1}^{3} h_i$	-0.0758 ± 10^{-4}
D_{KY}	2.119 ± 10^{-3}

shown that the object that is traced by trajectories in Figure 5 is an attractor.

Fourth, the Kaplan-Yorke dimension indicates that the attractor has a fractal strangeness, meaning a non-integer dimension of ~2.11. Preliminary calculations of the *correlation* dimension of the attractor, D_c, as defined by *Grassberger and Procaccia* [1983] indicate that D_c is substantially lower than 2.11, close to ~1.9. A puzzling and as yet unexplained result, since $D_{KY} \approx D_c \gtrsim 2$ is what is usually found for 3-D systems such as the Lorenz attractor [*Grassberger and Procaccia*, 1983].

After completing this study I became aware of the paper of *Niu and Chen* [1995], in which the authors calculated Lyapunov exponents in a similar fashion for a two state-variable slider system with different parameter values and a single value of κ'. *Niu and Chen*'s exponents can be rewritten as $h_1 = 0.0124$, $h_2 = 0$ and $h_3 = -0.1094$ in my notation. Hence, the h_i are roughly in agreement and a Lyapunov dimension of 2.11 follows as well. I take this as an indication that the chaotic dynamics might be a stable characteristic of the single slider system for a range of parameters.

3.1.4. The h_i as a function of κ'. This section will be completed by a discussion of the dependence of the Lyapunov exponents on κ' (see Figure 6). By comparing Figure 3 and Figure 6 the values of the h_i can be interpreted in terms of the asymptotic system behavior. For low values of $\kappa' < 0.855$, h_1 (solid line) is positive, indicating a sensitive dependence on initial conditions, hence chaos. The spikes in that parameter range where $h_1 > 0$ correlate with the period three windows of Figure 3 because h_2 (dashed line) is zero in chaotic regions but negative in the periodic windows. For higher values of κ', h_2 is in general negative but increases to zero repeatedly at the period doubling bifurcations. This κ' dependence is analogous to the first Lyapunov exponent in chaotic 1-D mappings such as the logistic map [e.g., *May*, 1976]. h_3 (dotted line) mirrors h_1 and h_2 since the sum $\sum h_i$ (dash-dotted line in Figure 6a) is constrained by the contracting property of the flow, and h_3 stays negative for all values of κ'. The overall patterns repeat themselves at different magnification scales (compare Figure 6a and 6b), suggesting that the h_i versus κ' plot shows self-similarity as well.

In summary, the results for the quasistatic single slider model with a two state-variable rate and state dependent friction law have demonstrated that the system is inherently chaotic in the deterministic sense. The model system is universal in that the road to chaos goes through period doubling cycles. If experiments and observations further substantiate the use of two state-variable friction laws to explain fault processes in nature, complex friction laws like the one examined here should be considered a microscopic source for irregular seismicity in the Earth.

3.2. Spring Coupled Sliders

With the results of the last section in mind, one can ask what effect elastic coupling between sliders has on the model seismicity. The inclusion of such interactions is a first step toward accounting for continuum effects in our model, attempting eventually to study the effect of microscopically chaotic friction laws in a homogeneous medium. In the following I will present observations on the resulting stress cycle oscillations.

Figure 7 shows the modified model set-up, a chain of coupled sliders with connecting springs. For mathematical simplicity, only next neighbor interactions are taken into account. This type of interacting slider block-model is similar to other studies in geophysics [e.g., *Burridge and Knopoff*, 1967; *Horowitz and Ruina*, 1989; *Carlson and Langer*, 1989; *Espanol*, 1994] or in tribology [e.g., *Weiss and Elmer*, 1996], but it is unique in the choice of friction law. Equation (3) for the i–th slider is modified to become

$$\frac{d\tau_i}{dt} = K(V_0 - V_i) + K_{cpl}(V_{i+1} + V_{i-1} - 2V_i). \quad (22)$$

V_i is the velocity of the slider where the force balance is taken, V_{i+1} and V_{i-1} are the neighboring slider speeds, and K_{cpl} is the spring constant between sliders $n-1$, n and $n+1$. For simplicity and symmetry K_{cpl} is assumed constant. K_{cpl} will further be non-dimensionalized in the same way as K and the resulting κ_{cpl} will be expressed as a fraction of κ' so that

$$\kappa'_{cpl} = \frac{\kappa_{cpl}}{\kappa'}. \quad (23)$$

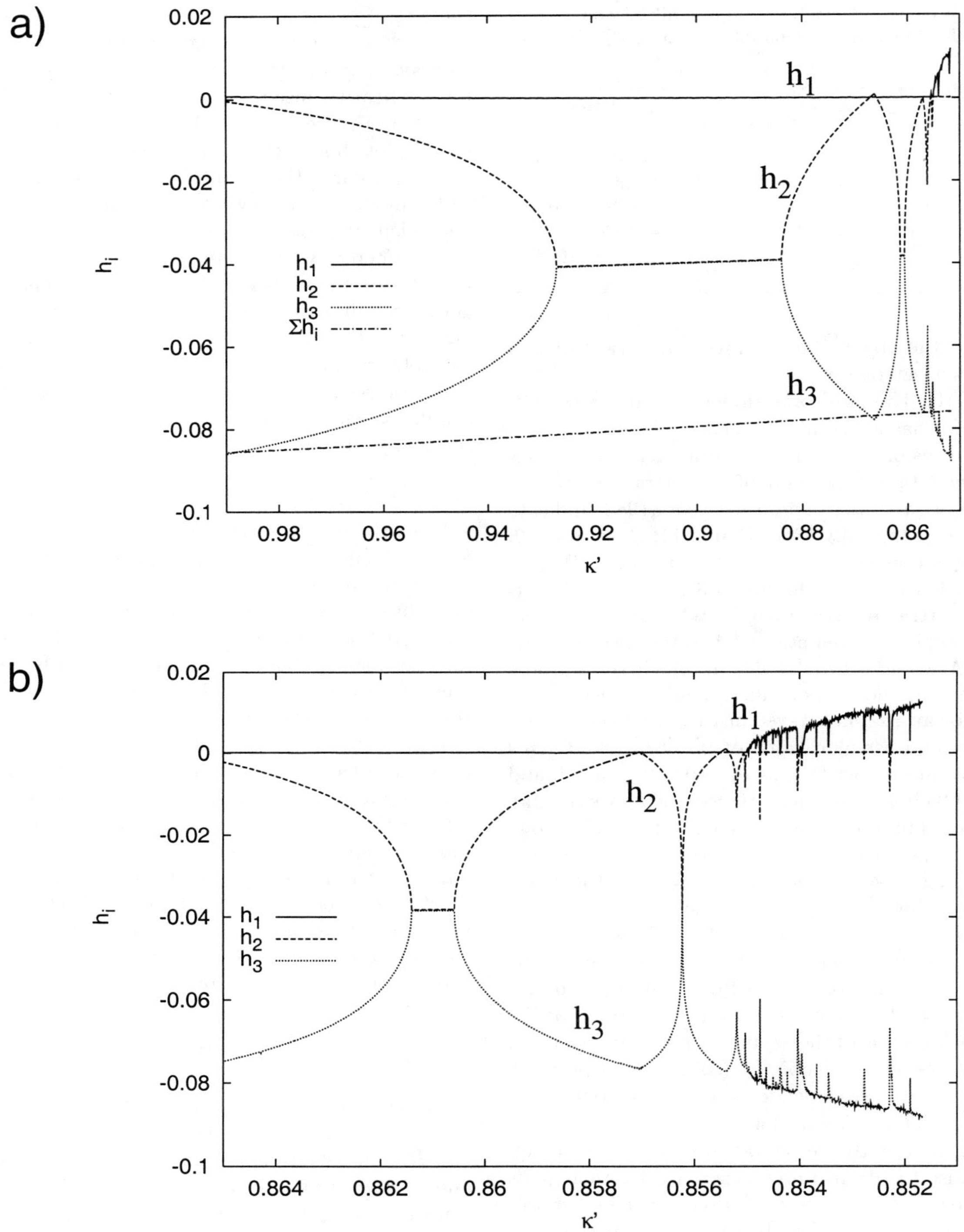

Figure 6. (a) Lyapunov exponents h_1, h_2, and h_3 (solid, dashed, and dotted line respectively) as well as $\sum h_i$ (dash-dotted line) versus control parameter κ'. (b) Magnification of part a), only the h_i are shown for clarity.

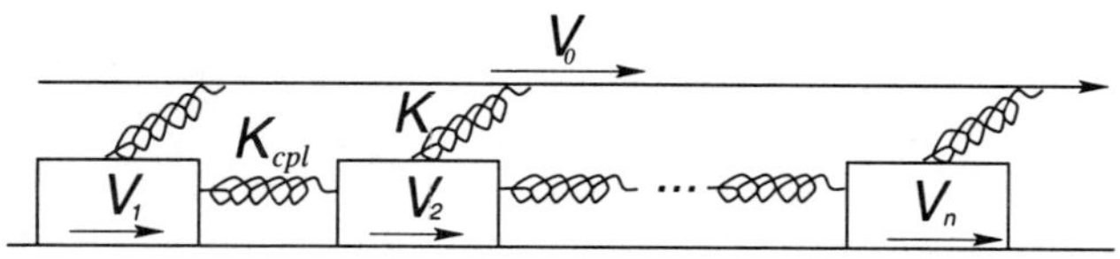

Figure 7. Cartoon of a coupled spring slider system mimicking a fault.

Therefore, if κ'_{cpl} is of order unity or larger the system can be expected to behave homogeneously like a single slider, possibly with a modified effective stiffness. The other case, $\kappa'_{cpl} \ll 1$, corresponds to isolated sliders with very weak interactions. In the continuum analogy, we can think of κ'_{cpl} as the elasticity modul of the material that ruptures along the slider plane, and κ' is a –possibly different– stiffness that connects the material to a fixed driving mechanism, corresponding to asthenospheric loading of a fault in nature. Two examples of coupled models with varying κ'_{cpl} will be described next.

3.2.1. Modulated stick-slip cycles. Figure 8a shows the stress, y, versus time, T, for 10 coupled slider blocks and three different values of the coupling stiffness κ'_{cpl}. The load point stiffness is $\kappa' = 0.965$, a value at which the single slider would show stable period two stick-slip. The initial condition of the coupled models is a y–perturbation of one slider at $T = 0$. As can be seen by comparing different traces in Figure 8a, stick-slip oscillations of varying regularity build up over time for every model, regardless of κ'_{cpl}. It is found, however, that a new type of irregularity was introduced and interaction has led to modulation of individual stick-slip oscillators. For weak coupling, $\kappa'_{cpl} = 0.05$, the changes in amplitude and phase ("beating") are strong and an irregular stress drop pattern results. Some traces show that varying amplitude stick-slip alternates with "quiet" periods during which individual sliders creep along with the loading.

Figure 8b shows the average frequency domain representation for the three different coupling experiments of part a), and the power spectrum of a synthetic saw-tooth timeseries for comparison. The plots were obtained by Bartlett-tapering all stress timeseries, taking the Fourier transform of the signal [FFT, e.g., *Press et al.*, 1993, p. 504], and averaging over all participating sliders. As expected, the power spectrum of the $\kappa'_{cpl} = 1$ slider chain has the characteristics of the saw-tooth (stick-slip) timeseries although the sliders have slightly less power in the side bands since the transition to sliding is not as abrupt in the friction models (compare the solid and the dashed lines in Figure 8b). The main frequency, ν^s, of the high coupling "fault" for $\kappa'_{cpl} = 1$ is $\nu^s \sim 0.0317$. This corresponds to a seismic period of $T \sim 31.5$ for sliding events that span the whole fault. From linear perturbation analysis we know that the circular frequency of the periodic orbit at the Hopf bifurcation for the single slider, ω^0, is equal to the imaginary part of the complex conjugate pair that is found in the three eigenvalues of the Jacobian of the flow $\mathbf{F}$ at the fixed point $\mathbf{x} = \mathbf{0}$. For $\kappa' = 1$ and the parameters given in (8), the eigenvalues $\Lambda_{1,2,3}$ of $\mathbf{J}$ are $\Lambda_{1,2} \sim \pm 0.2093i$, and $\Lambda_3 \sim -0.088$. Since we are still close to the bifurcation, the system frequency ν^s is therefore only slightly detuned from the corresponding eigenvalue frequency at $\kappa' = 1$, $\nu^0 \sim 0.033$, even with interactions.

Turning to the $\kappa'_{cpl} = 0.5$ slider signal (dashed line) in Figure 8b, we observe that its spectral power is richer in frequencies other than the base frequency of the seismic cycle which dominates the $\kappa'_{cpl} = 1$ spectrum. The beating that is observed in the stress drop amplitudes in Figure 8a for $\kappa'_{cpl} = 0.5$ corresponds to two discernible side band entries at AM-modulation periods of $T \sim 670$ and $T \sim 1430$ (small bumps, offset $\Delta\nu \sim \pm 0.0015$ and $\Delta\nu \sim \pm 0.0007$ from ν^s, respectively). Finally, the stronger irregularities observed in the timeseries of $\kappa'_{cpl} = 0.05$ show up as a broad, irregular band of modulation frequencies (dotted line in Figure 8b), as the absence of any dominating cycle modulation pattern in Figure 8a would have led us to suspect. The range of modulation frequencies that is observed for weak coupling corresponds to a broad distribution of the extent of lateral coherence for sliding events, analogous to a broad distribution of seismic event sizes.

In summary, Figure 8 demonstrates that the increase of κ'_{cpl} leads to greater uniformity in the time series until all sliders are almost perfectly synchronized for $\kappa'_{cpl} = 1.0$. This behavior can be interpreted intuitively along the lines of a typical physics textbook example: two spring-coupled pendula. If the coupling between the individual oscillators is weak, an amplitude modulation arises and kinetic energy is transferred back and forth between the pendula. For high coupling, both pendula will swing at the same amplitude and period, acting almost as one pendulum with a modified eigenfrequency.

3.2.2. Slip deficit asperities. Figure 9 presents an alternative view of the sliding history ("seismicity"), now for a larger system with 100 sliders. The plots are gray-shaded representations of the slider slip surplus, Σ, in a

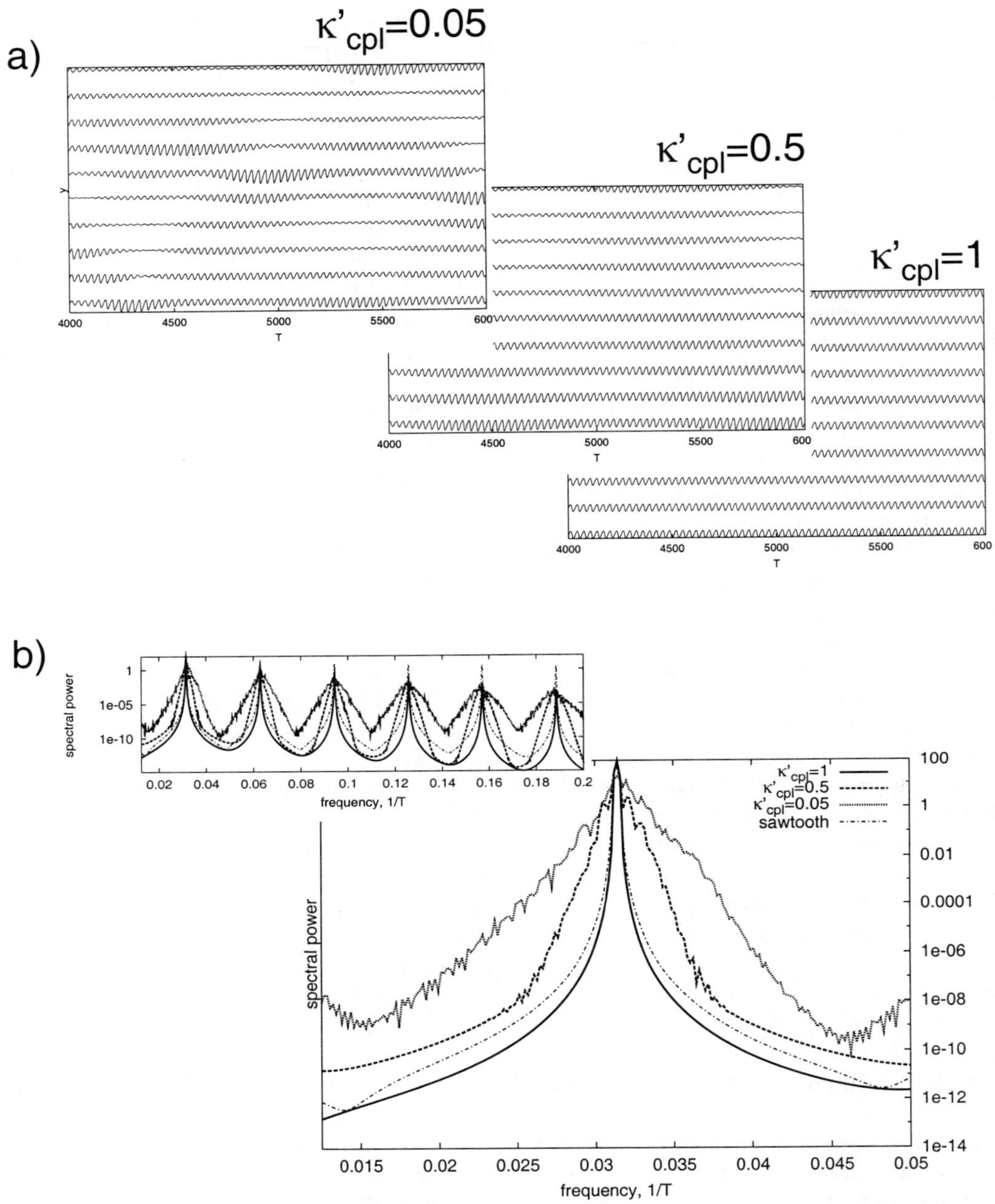

Figure 8. (a) Non-dimensionalized stress y versus time T for 10 coupled sliders and three different coupling stiffnesses, κ'_{cpl}. The individual slider traces were offset in y-direction according to their position in the chain, absolute values of y-oscillations are $\sim \pm 2$. κ' is 0.965 and open ended boundary conditions are applied, that is sliders 1 and 10 had only single sided interactions. (b) Fourier domain representation of the time series of part a). All y-traces were Bartlett-window tapered, Fourier transformed, and then averaged over all sliders. Both parts of the figure show the absolute value of the Fourier coefficients versus frequency, while the large plot is a blow-up of the small upper figure around the dominant stick-slip frequency of $\nu^s \sim 0.0317$. Solid, dashed, dotted, and dash-dotted lines are for slider time series with $\kappa'_{cpl} = 1$, $\kappa'_{cpl} = 0.5$, $\kappa'_{cpl} = 0.05$, and a synthetic sawtooth signal with $\nu = 0.0317$ respectively. The synthetic series was sampled at the same time intervals as the slider chain and added for comparison. (Note the log-scale which broadens the steep side bands in the FFT of a sawtooth.)

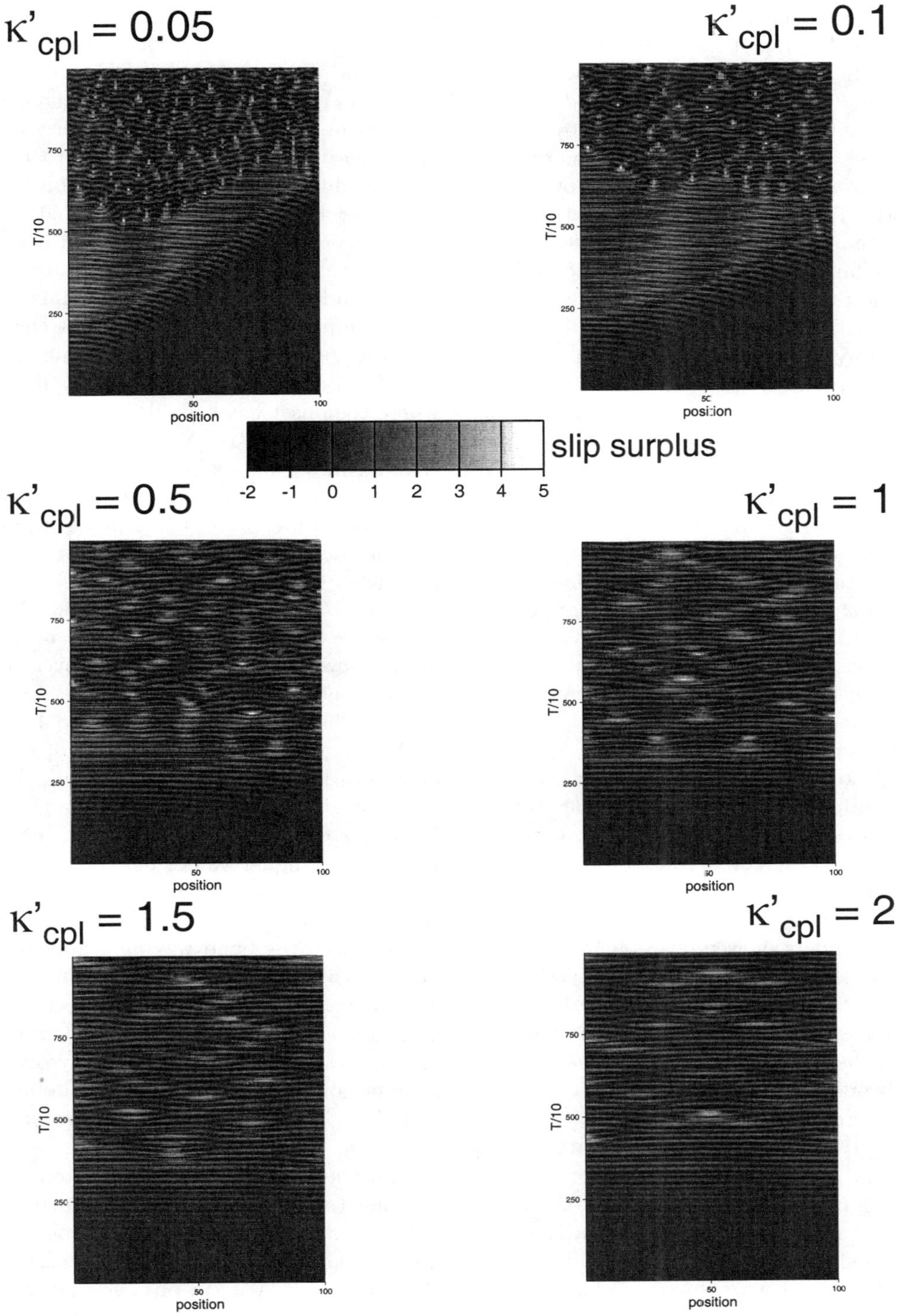

Figure 9. Slip surplus Σ for 100 coupled sliders and six different coupling stiffnesses, κ'_{cpl}, while κ' is 0.965. The abscissa indicates the slider location in the chain while the ordinate shows time with a scale of $T/10$. Center plot shows the gray-scale used, while individual slip surplus peaks are between -2 and ~ 20. The initial condition is a perturbation in sliding speed at the leftmost slider that can be traced as it spreads laterally as an oscillatory instability.

position versus time plane. The neutral state is shown in medium gray shades, denoting sliders whose slip is equal to the amount that would have accumulated by simple sliding along with the load point, V_0t. While darker shading denotes a slip deficit ($\Sigma < 0$) arising from long "healing" times, bright shading marks slip surplus ($\Sigma > 0$) in which individual sliders overshoot the average offset. The resulting irregular pattern can be considered as an analog to the stress field formed by asperities along a fault subjected to a seismic cycle in nature. (Here, the term asperities is used in a general sense for patches with varying pre-stress along a fault, not necessarily implying changes in the surface properties.) As for the experiment presented in Figure 8, it is evident that higher κ'_{cpl} values result in a more regular seismicity whereas weakly coupled sliders show more small scale irregularity.

Hence, introducing interactions, not surprisingly, has lead to a modulation of the slip characteristics along the coupled oscillator system. On the other hand, increasing the interaction to the same order as the load point coupling has been shown to have a regularizing effect. Similar results were presented by *Horowitz and Ruina* [1989] and *Espanol* [1994] for different types of fault models. That these modulations show up here as well hints at their common origin as an elasticity-coupled oscillator phenomenon.

It is important to note that it was not possible to reproduce the chaotic behavior of the one slider system discussed in the previous section with the undamped inertia-free multi-slider model of this study. Rather than revealing the same period doubling cascades for decreasing κ' as the single slider, numerical simulations of coupled chains of sliders showed unstable behavior for $\kappa < 0.9$. Fault zones in nature are clearly dissipative because of wear on the sliding interface and the radiation of seismic waves. However, the introduction of a regularizing term that might damp the aforementioned instabilities is beyond the scope of this study.

3.2.3. Spatial heterogeneity as a wave phenomenon. As the varying slope of the propagating perturbation wave front for different κ'_{cpl} in Figure 9 indicates, the group velocity with which perturbations in the stick-slip oscillations travel, u, increases with κ'_{cpl}. Since κ'_{cpl} serves as an analog to an elastic modulus for the slider chain, it can be expected that a wave speed analog depends on κ'_{cpl}. For constant material parameters and unit volume, we expect that the phase velocity should scale as $\sqrt{\kappa/m}$ for an elastic medium with inertia, where κ and m denote a stiffness (modul per unit length) and mass respectively. In our quasistatic analysis, the rate and state dependent friction alone plays a role similar to inertia. For example, the "direct effect" in eq. (1), $A\ln(V/V_*)$, gives instantaneous changes in stress for changes in velocity V, thereby representing resistance to acceleration. Assuming then that this virtual inertia effect is independent of κ'_{cpl} to first order and that there is no dispersion we would expect that $u \propto {\kappa'_{cpl}}^{0.5}$.

u can be estimated in numerical experiments based on the time it takes the initial perturbation to reach a certain slider starting from the initial condition applied at position one. However, it is not clear how this onset of a perturbation wave front should be determined exactly. Figure 10a shows data for five measurements where the onset has been defined as the time when the absolute slip surplus, $|\Sigma|$, of the middle slider is larger than c times the maximum slip surplus, $|\Sigma|_{max}$, that is reached by the slider in the remaining experiment. The data points are plotted in a log-log plot for different c values between 0.0001% and 1% together with linear regression lines.

We see that the u dependence on κ'_{cpl} can be fit by a power law regardless of the value of c. The exponent of the power law does vary, however, between ~ 0.8 for low values of c and ~ 0.2 for $c = 1\%$. We also observe that the slope of the fitted lines converges to ~ 0.8 toward the smallest values of c. I take this as an indication that u scales as $\sim {\kappa'_{cpl}}^{0.8}$ for the highest interaction frequencies, which might be expected to travel fastest. The velocity of the main perturbation wave front (higher values of c mean larger amplitudes of Σ) seems to scale with a smaller exponent and the slope of ~ 0.54 for $c = 0.1\%$ is close to the prediction of 0.5 based on the hand-waving argument above. Values of c larger than 1% are probably not meaningful if we are interested in determining the perturbation velocity u. It was also found that the exponent of the u–κ'_{cpl} relation does not depend significantly on the system size (number of sliders) or the type of boundary condition. This is in accordance with the wave speed interpretation of u. However, the results for scaling exponents demonstrate that quantitative statements about u are complicated by nonlinearities and dispersion.

We now turn to the lateral heterogeneities in the slip deficit that form after the initial transients in the models of Figure 9. We can observe that small κ'_{cpl} models show short wavelength asperities along the fault. Strong coupling, on the other hand, goes along with longer wavelength, larger scale heterogeneity. (For periodic boundary conditions (not shown), variations in sliding

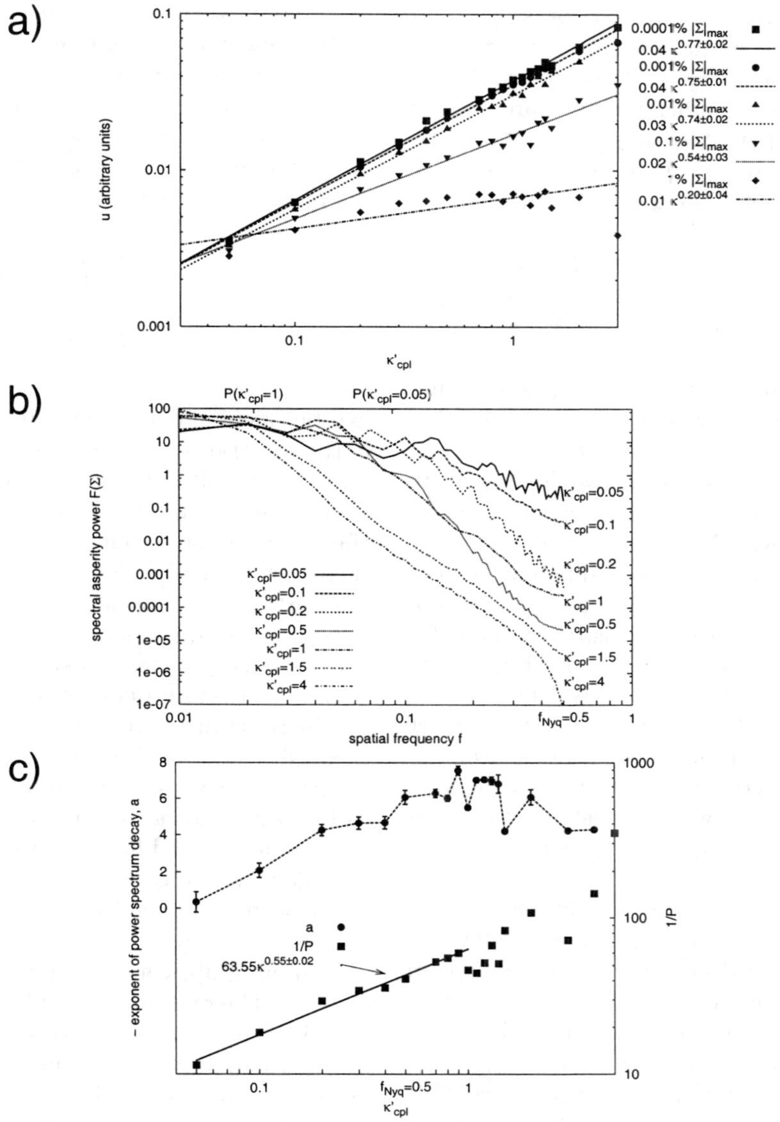

Figure 10. (a) Perturbation velocity u versus coupling stiffness κ'_{cpl}. u was obtained from the inverse of the time at which the middle slider in Figure 9 shows an absolute slip deficit, $|\Sigma|$, of magnitude $\geq c|\Sigma|_{max}$. $|\Sigma|_{max}$ denotes the maximum of $|\Sigma|$ during the remainder of the experiment and five measurements between $c = 0.0001\%$ and $c = 1\%$ are shown. The data was fit with a power law for each experiment and the exponents are given in the legend together with formal *a posterior* estimates of 1σ-uncertainty using χ^2. (b) Average spectral power for slip asperities in the spatial domain, $\bar{F}(\Sigma)$, versus spatial frequency, f. $\bar{F}(\Sigma)$ was computed from the models shown in Figure 9 by taking the FFT of the Bartlett-window tapered slip deficit at constant times and averaging over 100 timesteps. Distributions for six different experiments with $\kappa'_{cpl} = 0.05$, 0.1, 0.2, 0.5, 1, 1.5, and $\kappa'_{cpl} = 4$ are shown. Also indicated are the center of mass, P, values as defined in eq. (24) for $\kappa'_{cpl} = 1$ and $\kappa'_{cpl} = 0.5$. (c) Quantitative analysis of the power spectra of part b). Circle symbols denote estimated exponents, a, of a power-law frequency decay $\propto 1/f^a$, obtained by linear regression to get the slope of the spectra in part b) within the range $0.06 \leq f \leq 0.2$. Errorbars are formal uncertainties based on χ^2 (linear scale for a). Square symbols indicate the inverse of the first moment, $1/P$ (log scale for $1/P$). The solid line is a fit for $1/P$ in the range $0.05 \leq \kappa'_{cpl} \leq 1$, indicating that $1/P \propto {\kappa'_{cpl}}^{0.55\pm0.02}$ for small values of κ'_{cpl}.

history are almost entirely suppressed for $\kappa'_{cpl} \gtrsim 1$.) If we assume a constant spatial modulation (or interaction) frequency f_m for lateral cycle perturbations at constant κ'_{cpl} and constant slider number, an average dominant wavelength of the lateral slip irregularity, $\bar{\lambda}$, should scale with u when $u = f_m\bar{\lambda}$. This means that higher velocities will tend to organize the slip pattern over larger length scales. Higher coupling should, therefore, not only bring about a higher perturbation velocity u as was demonstrated in Figure 10a, but also longer wavelength asperities, as observed in the Σ-patterns of Figure 9.

While a spectral analysis of the sliding heterogeneity with time was already presented in section 3.2.1, I will now proceed to analyze the spatial frequency content of the asperities that are shown in Figure 9. Figure 10b is a plot of the average spectral power of the slip deficit asperities, $\bar{F}(\Sigma)$, versus frequency, f ($0 \leq f \leq f_{\text{Nyquist}} = 0.5$), for seven different models with varying coupling stiffness. The graphs were obtained by averaging the spectral density estimate for the last 100 timesteps in each experiment. Figure 10b confirms that high κ'_{cpl} seismicity is equivalent to a concentration of power in the lower frequencies, and weaker coupling results in an emphasis of the short spatial periods. The $\kappa'_{cpl} = 4$ spectrum corresponds to the synchronous end-member case with almost no variations along strike, similar to the $\kappa'_{cpl} = 1$ experiment for a ten slider model that was shown in Figure 8a.

The frequency distributions are roughly linear in the log-log-plot for spatial frequencies in the range $0.06 \lesssim f \lesssim 0.2$, and the rate of decay (the negative slope of the power-law part of the distributions) increases with increasing coupling. It is also found that there is a transition from a rough power spectrum at low κ'_{cpl} (e.g., solid line in Figure 10b for $\kappa'_{cpl} = 0.05$) to a smoother distribution for high coupling experiments (e.g., dash-dotted line for $\kappa'_{cpl} = 1$); the character of the spectral power distributions changes at $\kappa'_{cpl} \sim 0.7$. Also, the $\kappa'_{cpl} = 1$ spectrum is somewhat of an exception in that is does not follow the general trend for $f \gtrsim 0.2$ where we can observe more spectral power for $\kappa'_{cpl} = 1$ than for $\kappa'_{cpl} = 0.5$.

To quantify the observation that the slope of the power spectra decays as a function of κ'_{cpl}, I fitted a power law $1/f^a$ to the data in the interval $0.06 \leq f \leq 0.2$. The circle symbols in Figure 10c indicate the values for that best-fit exponent, a, versus κ'_{cpl}. We see that the slope changes from a very slow, $1/f$-type decay for $\kappa'_{cpl} = 0.05$ with $a \sim 0.5$ to more rapid power cut-off with exponents ~ 6.5 at $\kappa'_{cpl} = 0.7$. For larger κ'_{cpl}, Figure 10c indicates a saturation at large scale organization, corresponding to $\sim 1/f^6$. However, we observe considerable scatter and a decrease in the slope toward synchronous sliders at $\kappa'_{cpl} = 4$.

Finally, the first moment, P, of the $\bar{F}(\Sigma)$ curves,

$$P = \frac{1}{M}\int_0^{0.5} f\bar{F}(\Sigma)df \quad \text{with} \tag{24}$$

$$M = \int_0^{0.5} \bar{F}(\Sigma)df, \tag{25}$$

can be calculated as a measure for the "center of mass" of the spectral power distributions (see also Figure 10b). Square symbols in Figure 10c denote the inverse of P so that higher values of $1/P$ correspond to more power in the lower spatial frequencies. We can observe that $1/P$ increases with κ'_{cpl}, for $\kappa'_{cpl} \lesssim 1$ roughly as a power law with ${\kappa'_{cpl}}^{0.6}$. For larger κ'_{cpl} the plot again indicates some scatter and super-power law increase of $1/P$ when the systems gets close to synchrony. The spectral analysis has therefore demonstrated that we can interpret the increase in the length scale of slip organization for increasing κ'_{cpl} as a result of longer perturbation wavelengths. $1/P$ scales roughly in the same way with κ'_{cpl} as the perturbation velocity u. For $\kappa'_{cpl} \gtrsim 1$ we observed some qualitative differences between high and low κ'_{cpl} seismicity which result from the coherent sliding events of the asymptotic end-member state without variations in the seismic cycle.

Summing up, coupling modifies the single slider dynamics. The extreme cases of weak coupling with small wavelength cycle perturbations and very strong coupling where sliders move in synchrony border a range in which irregular sliding histories form as a result of sustained modulation waves.

4. LIMITATIONS AND IMPLICATIONS OF THE MODEL

Investigations of the single quasistatic slider system demonstrate that two state-variable friction laws can lead to deterministic chaos in a homogeneous system. Laboratory rock friction can therefore serve as yet another example of the peculiarities of nonlinear dynamics: while some aspects show the underlying determinism (e.g., map-predictions of the seismic period, Figure 4), any irregularity will be locally amplified by the sensitive dependence on initial conditions. If for nothing else, the system can be viewed as a tutorial for low-dimensional chaos from the Earth sciences, to be compared with other irregular systems as reviewed, for example, by *Turcotte* [1997].

In the following, I will discuss the possible effect of inertia and elaborate on conclusions one can draw from the results of interacting sliders.

4.1. Inertia

The influence of inertia on the dynamics of the slider block system might be important, yet it has been neglected in all the above models. We can estimate that the effect of mass will be twofold. First, inertia will tend to destabilize the system. As discussed by *Rice and Ruina* [1983], including a non-zero mass in the linear stability analysis of any single slider rate and state friction type system results in a decrease in the critical stiffness. A material with a constant κ' analog modulus will thus be more unstable with inertia than without. However, as calculations of *Rice and Tse* [1986] for single state-variable sliders show, the overall stick-slip characteristics of systems with inertia are similar to the quasistatic case, although the stress drop events are modified by the induced dynamic overshoot. Further, *Gu and Wong* [1994] demonstrated that period doubling cascades are also observed in inertial systems with two state-variables. In their models, the irregular parameter range as in Figure 3 toward lower values of κ' was not terminated by unstable sliding but rather by quasiperiodic system behavior.

Second, mass introduces another degree of freedom in the slider equations. Since three mathematical dimensions are a necessary condition for chaos, one state-variable sliders with inertia would also be possible candidates for a microscopic source of irregularity. I am not aware of any studies that show chaotic behavior for less than two state-variables for single slider rate and state friction systems. Yet, two dynamical sliders with asymmetric coupling were demonstrated to undergo period doubling cascades [*Huang and Turcotte*, 1990]. Based on the results of the studies mentioned above, we can state that the inclusion of inertia leads to modified system behavior, including shifted stability bounds. However, since inertia alone apparently does not change the overall characteristics, studying the simpler quasistatic system should be a good start, especially given the complexity that is already unraveled at this level of simplification.

As the comparison with the work of *Huang and Turcotte* [1990] further shows, even for simple slider systems two possible origins of irregularity are found which might lead to the seismicity that is observed in nature: either a complicated friction law working at the microscopic level might be the cause (as in the studies of *Gu et al.* [1984], *Gu and Wong* [1994], *Zhiren and Chen* [1994], and in this work), or irregularity might arise with simpler microscopic laws but heterogeneous interactions. This leads to the discussion of effects that arise when sliders are coupled.

4.2. Interaction

Seismicity in the Earth shows some regular features of almost periodic earthquake recurrence [e.g., *Bakun and McEvilly*, 1984], many more examples of irregular seismic cycles that are only quasiperiodic (large scale heterogeneity) [e.g., *Sieh*, 1981], and power-law magnitude-frequency distributions of events. The latter *Gutenberg and Richter* [1949]-type (GR type) size distributions might indicate criticality in the sense of *Bak et al.* [1988], with possible consequences for earthquake interaction ranges and predictability [e.g., *Sornette and Sornette*, 1989].

Assuming that the complex, possibly chaotic laboratory friction laws we studied have relevance for the behavior of fault zones in nature, we can try to evaluate whether a homogeneous fault in which these laws apply still produces regular seismic cycles in a continuum. If it does, then other mechanisms such as spatial heterogeneity of material parameters (noise input), geometrical complexity of fault traces [fractal grounds to start from, e.g., *King*, 1983], or mechanical fault interactions [e.g., *Harris*, 1998, and references therein] might be more important in leading to the observed irregularities in nature.

Previous studies have incorporated simpler friction laws in slider blocks [see, e.g, *Elmer*, 1996; and *Turcotte*, 1997, chap. 17, for reviews], or more sophisticated continuum models [e.g., *Horowitz and Ruina*, 1989; *Rice*, 1993; *Shaw*, 1995; *Cochard and Madariaga*, 1996]. However, the conditions under which fully dynamical models of faults produce periodic seismic cycles, large scale irregularity, or GR-type characteristics are still debated [e.g., *Rice and Ben-Zion*, 1996]. It appears that simple, one state-variable friction laws generically produce larger scale irregularity from a homogeneous model, but GR-statistics seem to be only the outcome of a small parameter range that might not be realized on Earth [*Shaw and Rice*, 1999].

The slider models from the previous section are an attempt to contribute to this discussion on a very simplified level. Coupled sliders show slip histories which do not result from the individual friction law's characteristics but are dominated by interaction effects. Similar findings were reported by *Horowitz and Ruina* [1989], and *Espanol* [1994] discussed the transition from periodicity to soliton-like solutions as in my Figure 9 for a velocity weakening friction law. More recently, *de Sousa Vieira* [1996] examined changes in the magnitude-frequency distributions of sliding events as a function of the coupling stiffness for a standard spring-block model with inertia and simple friction. In my quasistatic models, the variations in stick-slip cycle amplitude are a coupled oscillator effect. They can be described by a lateral interaction mechanism whose wavelength increases with the strength of coupling. Higher coupling was therefore observed to have a regularizing effect, leading to larger wavelength asperities and more regular slip patterns. The observed mechanism of regularization might have general relevance for seismicity in nature since a variety of models with different friction laws, with and without inertia, appear to show similar features.

My slider experiments have various shortcomings, such as the limitation to a single degree of freedom for motion, discrete blocks rather than a continuum model, and the limited parameter range that could be explored without damping. It is hence not possible to conclusively quantify the extent to which two state-variable friction can act as a source of chaos in coupled systems at this point. Given the small parameter range in which chaos is observed for a single slider and the observed regularizing effect of interactions, my conjecture is that elastic interactions on different scales dominate in the Earth, reducing the effect of microscopic chaos in friction, and leading to larger scale irregularity.

5. CONCLUSION

Two state-variable rate and state dependent friction laws that are derived from laboratory rock sliding experiments were shown to result in deterministic chaotic behavior for a simplified quasistatic model. Since a unimodal mapping can be derived from aperiodic time series, universal period doubling cascades are observed as a route to chaos. The single slider model is thus a good example with which to demonstrate bounds on the predictability of model rupture events. Interaction models imply that the microscopic friction law is not as important in coupled sliders. Other sources of irregularity such as mechanical interaction between faults might be more important in nature.

Acknowledgments. The author thanks the anonymous reviewer for his suggestions as well as Sascha Hilgenfeldt, Richard Kronauer, Richard O'Connell, Svetlana Panasyuk, James Rice, and especially Simone Bethge and James Kellogg for discussion and encouragement. Funding for this study was provided by the German Academic Exchange Service (DAAD) under a "Doktorandenstipendium HSP–III".

REFERENCES

Bak, P., Tang, C., and Wiesenfeld, K., Self-organized criticality, *Phys. Rev. A*, *38*, 364–373, 1988.

Bakun, W. and McEvilly, T., Recurrence models and Parkfield, California earthquakes, *J. Geophys. Res.*, *89*, 3051–3058, 1984.

Benettin, G., Galgani, A., Giorgilli, A., and Strelcyn, J.-M., Lyapunov characteristic exponents for smooth dynamical systems and for Hamiltonian systems: a method for computing all of them. Part 2: Numerical application, *Meccanica*, *15*, 21, 1980.

Blanpied, M. L. and Tullis, T. E., The stability and behavior of a frictional system with a two state variable constitutive law, *Pure Appl. Geophys.*, *124*, 415–444, 1986.

Burridge, R. and Knopoff, L., Model and theoretical seismicity, *Bull. Seism. Soc. Am.*, *57*, 341–371, 1967.

Carlson, J. M. and Langer, J. S., Properties of earthquakes generated by fault dynamics, *Phys. Rev. Lett.*, *62*, 2632–2635, 1989.

Cochard, A. and Madariaga, R., Complexity of seismicity due to highly rate-dependent friction, *J. Geophys. Res.*, *101*, 25,321–25,336, 1996.

de Sousa Vieira, M., Exponential distributions in a mechanical model for earthquakes, *Phys. Rev. E*, *54*, 5925–5928, 1996.

Dieterich, J. H., Modeling of rock friction 1. Experimental results and constitutive equations, *J. Geophys. Res.*, *84*, 2161–2168, 1979.

Dieterich, J. H., Mechanical behavior of crustal rocks, in *Mechanical behavior of crustal rocks: the Handin volume*, vol. 24, *Geophysical Monograph*, edited by N. L. Carter, M. Friedman, J. M. Logan, and D. W. Stearns, pp. 103–120, AGU, Washington, D. C., 1981.

Dieterich, J. H., A constitutive law for rate of earthquake production and its application to earthquake clustering, *J. Geophys. Res.*, *99*, 2601–2618, 1994.

Elmer, F.-J., Is Self-Organized Criticality possible in dry friction?, in *Physics of sliding friction*, edited by B. N. J.

Persson and E. Tosatti, pp. 433–447, Kluwer Academic, Dordrecht, The Netherlands, 1996.

Espanol, P., Propagative slipping modes in a spring-block model, *Phys. Rev. E.*, *50*, 227–235, 1994.

Feigenbaum, M. J., Quantitative universality for a class of nonlinear transformations, *J. Stat. Phys.*, *19*, 25, 1978.

Gomberg, J., Beeler, N. M., Blanpied, M. L., and Bodin, P., Earthquake triggering by transient and static deformations, *J. Geophys. Res.*, *103*, 24,411–24,426, 1998.

Grassberger, P. and Procaccia, I., Measuring the strangeness of strange attractors, *Physica D*, *9*, 189, 1983.

Gu, J.-C., Rice, J. R., Ruina, A. L., and Tse, S. T., Slip motion and stability of a single degree of freedom elastic system with rate and state dependent friction, *J. Mech. Phys. Solids*, *32*, 167–196, 1984.

Gu, Y. and Wong, T.-F., Nonlinear dynamics of the transition from stable sliding to cyclic stick-slip in rock, in *Nonlinear dynamics and predictability of geophysical phenomena*, vol. 83, *Geophysical Monograph*, edited by W. I. Newman, A. Gabrielov, and D. L. Turcotte, pp. 15–35, AGU, Washington, D. C., 1994.

Gutenberg, B. and Richter, C., *Seismicity of the Earth and Associated Phenomena*, Princeton Univ. Press, Princeton, 1949.

Harris, R. A., Introduction to special section: Stress triggers, stress shadows, and implications for seismic hazard, *J. Geophys. Res.*, *103*, 24,347–24,358, 1998.

Horowitz, F. G. and Ruina, A. L., Slip patterns in a spatially homogeneous fault model, *J. Geophys. Res.*, *94*, 10,279–10,298, 1989.

Huang, J. and Turcotte, D. L., Are earthquakes an example of deterministic chaos?, *Geophys. Res. Lett.*, *17*, 223–226, 1990.

Kaplan, J. L. and Yorke, J. A., Chaotic behavior of multidimensional difference equations, in *Functional Differential Equations and Approximations of Fixed Points*, vol. 730, *Lecture Notes in Mathematics*, edited by H.-O. Peitgen and H.-O. Walter, pp. 204–227, Springer, Berlin, 1979.

King, G. C. P., The accomodation of large strains in the upper lithosphere of the earth and other solids by self-similar fault systems: the geometric origin of b-value, *Pure Appl. Geophys.*, *121*, 761–815, 1983.

Libchaber, A., Laroche, C., and Fauve, S., Period doubling cascade in mercury, a quantitative measurement, *J. Physique. Lett.*, *43*, L211–L216, 1982.

Lorenz, E. N., Deterministic nonperiodic flow, *J. Atmos. Sci.*, *20*, 130, 1963.

Marone, C., Laboratory-derived friction laws and their application to seismic faulting, *Annu. Rev. Earth Planet. Sci.*, *26*, 643–696, 1998.

May, R. M., Simple mathematical models with very complicated dynamics, *Nature*, *261*, 459, 1976.

Niu, Z.-R. and Chen, D.-M., Lyapunov exponent and dimension of the strange attractor of elastic frictional system, *Acta Seimol. Sinica*, *8*, 575–584, 1995.

Ott, E., *Chaos in dynamical systems*, Cambridge Univ. Press, Cambridge, UK, 1993.

Press, W. H., Teukolsky, S. A., Vetterling, W. T., and Flannery, B. P., *Numerical Recipes in C: The Art of Scientific Computing*, Cambridge Univ. Press, Cambridge, UK, 2nd ed., 1993.

Ranjith, K. and Rice, J. R., Stability of quasi-static slip in a single degree of freedom elastic system with rate and state dependent friction, *J. Mech. Phys. Solids*, *47*, 1207–1218, 1999.

Rice, J. R., Spatio-temporal complexity of slip on a fault, *J. Geophys. Res.*, *98*, 9885–9907, 1993.

Rice, J. R. and Ben-Zion, Y., Slip complexity in earthquake fault models, *Proc. Natl. Acad. Sci. USA*, *93*, 3811–3818, 1996.

Rice, J. R. and Ruina, A. L., Stability of steady frictional slipping, *J. Appl. Mech.*, *50*, 343–349, 1983.

Rice, J. R. and Tse, S. T., Dynamic motion of a single degree of freedom system following a rate and state dependent friction law, *J. Geophys. Res.*, *91*, 521–530, 1986.

Ruina, A. L., Slip instability and state variable friction laws, *J. Geophys. Res.*, *88*, 10,359–10,370, 1983.

Ruina, A. L., *Friction Laws and Instabilities: A Quasistatic Analysis of some Dry Frictional Behavior*, Ph. D. thesis, Brown University, Providence, 1980.

Schmittbuhl, J., Vilotte, J.-P., and Roux, S., Velocity weakening friction: A renormalization approach, *J. Geophys. Res.*, *101*, 13,911–13,917, 1996.

Scholz, C. H., Earthquakes and friction laws, *Nature*, *391*, 37–42, 1998.

Shaw, B. E., Frictional weakening and slip complexity in earthquake faults, *J. Geophys. Res.*, *100*, 18,239–18,251, 1995.

Shaw, B. E. and Rice, J. R., Existence of continuum complexity in the elastodynamics of repeated fault ruptures (abstract), *EOS, Trans. AGU*, *80*(46), Fall Meet. Suppl., F688, 1999.

Sieh, K., A review of geological evidence for recurrence times of large earthquakes, in *Earthquake Prediction, an International Review*, vol. 4, *M. Ewing series*, edited by D. Simpson and P. Richards, pp. 209–216, AGU, Washington D. C., 1981.

Sornette, A. and Sornette, D., Self-organized criticality and earthquakes, *Europhys. Lett.*, *9*, 197–202, 1989.

Strogatz, S. H., *Nonlinear Dynamics and Chaos*, Addison-Wesley, Reading, MA, 1994.

Turcotte, D. L., *Fractals and chaos in geology and geophysics*, Cambridge Univ. Press, Cambridge, UK, 2nd ed., 1997.

Weiss, M. and Elmer, F.-J., A simple model for wearless friction: the Frenkel-Kontorova-Tomlinson model, in *Physics of sliding friction*, edited by B. N. J. Persson and E. Tosatti, pp. 153–178, Kluwer Academic, Dordrecht, The Netherlands, 1996.

Zhiren, N. and Chen, D., Period-doubling bifurcation and chaotic phenomena in a single degree of freedom elastic system with a two-state variable friction law, in *Nonlinear dynamics and predictability of geophysical phenomena*, vol. 83, *Geophysical Monograph*, edited by W. I. Newman, A. Gabrielov, and D. L. Turcotte, pp. 75–80, AGU, Washington, D. C., 1994.

Thorsten W. Becker, Department of Earth and Planetary Sciences, Harvard University, 20 Oxford St., Cambridge, MA 02138, USA. (e-mail: becker@eps.harvard.edu)

Note added in proof. In a work of which I was unaware until after the preparation of the final copy of this article, *Shkoller and Minster* [Shkoller, S. and Minster, J.-B., Reduction of Dietrich-Ruina attractors to unimodal maps, *Nonlin. Process. Geophys.*, *4*, 63–69, 1997] constructed unimodal mappings directly from the attractor of the two state-variable system. This alternative to my time-series approach yields comparable results, confirming that chaos can be an intrinsic feature in dry friction.

Spring-block models of seismicity: Review and analysis of a structurally heterogeneous model coupled to a viscous asthenosphere

Jon D. Pelletier

Department of Geosciences, University of Arizona, Tucson, Arizona

Since the introduction of the Burridge-Knopoff model for fault mechanics many spring-block models have been proposed which purportedly generate a Gutenberg-Richter (GR) distribution of earthquake moments. Upon close examination, not all of these models generate distributions consistent with GR. Of those that do, few are able to generate other realistic aspects of distributed seismicity such as numbers of foreshocks and aftershocks consistent with Omori's Law and both characteristic and noncharacteristic seismic cycle behavior. We have systematically constructed a structurally-heterogeneous spring-block model coupled to a viscous asthenosphere which generates many realistic features of distributed seismicity. Versions of the model without viscous coupling were found to be inadequate to generate realistic model seismicity. A structurally heterogeneous model with viscous coupling is compared in detail to previous spring-block models and data for observed seismicity in terms of the cumulative frequency-size distribution of events, the occurrence of foreshocks and aftershocks, the increase of cumulative Benioff strain release prior to large earthquakes, the dependence of seismicity on depth, fault length, and structural heterogeneity, the relationship between the average slip and moment of an earthquake, and the heterogeneity of the rupture process. These comparisons demonstrate the effectiveness of the model in simulating seismicity and suggest that heterogeneity and viscous coupling are necessary conditions for a realistic spring-block model of seismicity.

INTRODUCTION

Seismicity exhibits a complex variety of behavior including a broad distribution of energy release and spatial and temporal clustering of earthquakes over a wide range of scales. It has been recognized since the establishment of the Gutenburg-Richter (GR) Law [*Gutenburg and Richter*, 1944] that seismicity is statistically self-similar over a wide range of scales. This is a consequence of the power-law form of the GR Law (which states that the number of earthquakes in a region with moment greater that M_o is a power-law function of moment with exponent -2/3) since a power law has no characteristic scale. More recently, *Kagan* [1991a] has also documented the self-similar spatial clustering of hypocenters over a wide range of scales. Self-similarity has attracted the attention of many researchers who have proposed variations of the original spring-block model of *Burridge and Knopoff* [1967] in order to model seismicity on a single fault and/or on a collection of faults. Many of these models have been

GeoComplexity and the Physics of Earthquakes
Geophysical Monograph 120

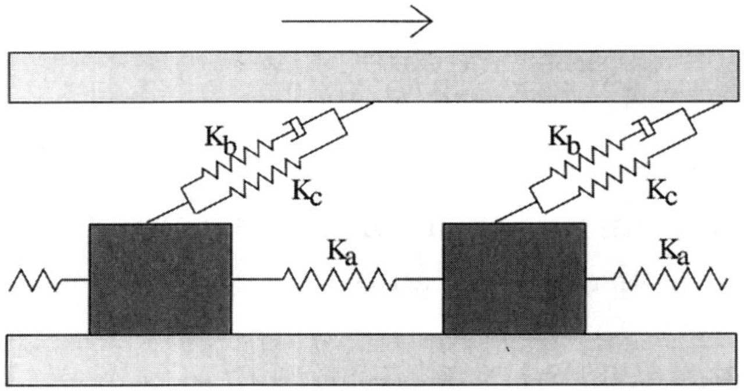

Figure 1a. Model geometry for the basic heterogeneous spring-block model. Blocks are coupled to one another by springs with spring constant K_a and are coupled to the driver plate with springs of spring constant K_b. Two other parameters define the model: the mean and standard deviation of the static coefficients of friction of the blocks.

analyzed and compared to observed seismicity in terms of the frequency-size distribution of events. In some cases the occurrence of foreshocks and aftershocks has also been examined. Based on the multiplicity of models that apparently exhibit GR statistics, however, the comparison of only one or two measures is clearly inadequate to validate spring-block models of seismicity. Instead, spring-block models must be compared against all available phenomenological data such as frequency-size statistics, spatiotemporal clustering of the number of earthquakes and stress release, the scaling of average slip with seismic moment, and the well-documented dependence of seismicity on depth, length, and structural heterogeneity of the fault surface or system of faults. The purpose of this paper is to review the previous work on spring-block models and to document the systematic construction of a spring-block model which generates model seismicity very similar to observed seismicity. The model was constructed systematically by starting with as basic a model as possible which reproduces GR statistics and then improving the model by trial and error until a model consistent with many observations of seismicity was obtained. This procedure not only identified which elements of our model were sufficient for generating realistic behavior but also served to rule out alternative possibilities. We have concluded that a model which includes structural heterogeneity as represented by variable static coefficients of friction and a coupling of blocks to viscous dashpots are necessary elements for a realistic spring-block model of seismicity.

PREVIOUS WORK ON SPRING-BLOCK MODELS

Spring-block models were originally proposed by *Burridge and Knopoff* [1967] (BK) who considered the relative motion of two rigid plates with elastic rebound of discrete blocks between the plates. Blocks which are frictionally coupled to one of the plates and elastically coupled to the other plate and to each other (Fig. 1a). The equations of motion for this system were integrated numerically in BK and it was shown that stick-slip behavior and a power-law distribution of event sizes resulted if the blocks obeyed a velocity-weakening friction law. A heavy-tailed distribution resulted because although most earthquakes involved the movement of only one or two blocks, occasionally the movement of one block would trigger the movement of adjacent blocks through their elastic coupling and these blocks would trigger still more in a cascade of slippage over a large area.

The late 1980s saw a renewed interest in this model since critical systems which generate power-law statistics and have no characteristic scale had been studied so successfully with techniques of statistical physics in the 1970s and early 1980s. Additionally, spring-block models gained interest because they were shown to exhibit deterministic chaos by *Huang and Turcotte* [1990] who considered a simple system of just two coupled blocks described by the logistic difference equation. With different block masses, the authors showed that their system exhibited deterministic chaos and patterns of seismicity similar to those on real fault segments. A similar coupled two-block system was compared to the behavior of subduction zones by *Ruff* [1992]. The blocks in his model were associated with large asperities. *Carlson and Langer* [1989] studied a linear array similar to the original Burridge-Knopoff model. Although the system was deterministic and completely homogeneous, the system behavior was extremely complex and chaotic. Their model seemed to show that it is possible to generate many aspects of seismicity with a homogeneous model provided that the system had nonlinearity. However, the behavior that they observed does not hold down to the continuum limit [*Ben-Zion and Rice*, 1995; *Ben-Zion and Rice*, 1997]. Therefore, it seems necessary to associate the blocks with structural heterogeneities in order to motivate the discreteness of the spring-block model.

Nakanishi [1990,1991] considered a version of the spring-block model in which only one block is allowed to move at any time. The slippage of one block can lead to the instability of one or more neighboring blocks, which can then slip in a subsequent time step or steps until all blocks are stable again. This modification made simulation of spring-block models far less computationally intensive and allowed for the investigation of larger system sizes and two-dimensional arrays of blocks. *Nakanishi* [1990,1991] obtained similar results to those of *Carlson and Langer* [1989], including their power-law distribution of event sizes. He found that the equations of motion could be nondimensionalized such that the system behavior is fully described by the stiffness, the ratio of the spring constants connecting blocks to those connecting each block to the driver plate. For a large stiffness, events which move the whole system at once are common. *Wang* [1995] also emphasized the role of the stiffness (or seismic coupling) on the frequency-size distribution of the resultant model seismicity. The behavior for large stiffness appears to be similar to the frequency-size distribution observed for a single fault, with frequent characteristic earthquakes which rupture the entire fault superimposed on a background seismicity with a power-law distribution. For smaller stiffnesses, an exponential rolloff in the power-law frequency-size distribution is observed. There is a critical stiffness for which a power-law distribution is observed up to the largest events. Distributed seismicity has been claimed to be an example of self-organized criticality [*Bak and Tang*, 1989; *Ito and Matsuzaki*, 1990; *Barriere and Turcotte*, 1994]. However, since Nakanishi's model exhibits the power-law distribution characteristic of critical phenomena for a particular value of the stiffness, his model is self-organized critical. Since a robust model of seismicity should generate both characteristic and noncharacteristic behavior, as Nakanishi's model does, a self-organized critical model which only produces a single frequency-size distribution may not be robust enough for a general model of seismicity. *Grasso and Sornette* [1998] have reached the same conclusion.

Although the models of *Carlson and Langer* [1989], *Carlson* [1991], *Bak and Tang* [1989], *Brown at al.* [1991], *Rundle and Brown* [1991], and *Huang et al.* [1992] exhibit power law statistics, the seismicity generated by these models is not consistent with GR. These studies computed non-cumulative distributions of event sizes with exponents close to -1: $N(A) \propto A^{-1}$. To compare this distribution with GR the distribution must be integrated giving $N(>A) \propto \log(A)$ which is not a power law. Thus, although these models purportedly generated frequency-size distributions consistent with those of observed seismicity, they were actually significantly different from GR.

Another, more subtle, problem with some spring-block models has to do with how the model moves the driver plate relative to the movement of slipped blocks. In spring-block models the relative motion of the plates represents the buildup of stress between two tectonic plates. This buildup happens at a rate much lower than that of the elastic rebound of the slipped blocks. In all computer models, however, motion takes place in discrete jumps. In order to ensure that the rate of tectonic loading is always much less than the rate of elastic rebound, tectonic loading must occur only as long as all blocks are stable. A naive implementation of a spring-block model moves the driver plate forward by equal discrete amounts in successive time steps and relaxes all blocks whose elastic stress exceeds the frictional restoring force *after* the driver plate has been moved. If the driver plate is incremented by equal, discrete amounts without relaxing unstable blocks before the driver plate completes its motion, it is possible for two or more unconnected patches of the fault surface to become unstable during the same time step and be counted as one earthquake. The models of *Brown et al.* [1991] and *Rundle and Klein* [1993] suffer from this artifact. To correct this, the driver plate must be advanced only until the next block in the model becomes unstable.

A third weakness of many previous analyses is the dependence of model behavior on initial conditions. If the spring-block model of Fig. 1a is started with uniform spacing between blocks, a single earthquake in which all of the blocks in the system move the same amount will occur repeatedly. If a small amount of randomness is included in the initial block positions, most of the stress will be released in a single large earthquake but there will be smaller events as well. If the variation in the initial block positions is increased, the elastic rebound of the blocks will be further decoupled and the ratio of the strain released in the largest event to the strain released in background seismicity will be decreased. The models of *Nakanishi* [1990,1991], for example, include this initial randomness in the block positions. It is unlikely that a model which depends so heavily on initial conditions can be a viable model for seismicity since seismicity displays very similar statistical behavior under a wide range of geometries and fault histories.

Randomness has also been introduced into some spring block models by incrementing the stress randomly over time until some uniform threshold has been reached.

For example, in the models of *Bak and Tang* [1989], *Barriere and Turcotte* [1990], and *Ito and Matsizaki* [1990] a site is chosen at random during each time step and a unit of stress is added to that site during that time step. When a site has four units of stress accumulated on it, the site becomes unstable and redistributes stress to its nearest neighbors. While this approach may be one way to model structural heterogeneity of the fault surface, it does not allow the model heterogeneity to be characterized by a parameter which can be varied to study the model dependence on the degree of heterogeneity. It is hard to imagine any other motivation for the randomness in those models besides structural heterogeneity.

One way in which the simple repeating earthquake behavior of spring-block models with uniform friction can be rectified is by letting the blocks have a distribution of static coefficients of friction representing structural heterogenity of the fault surface. If the discrete elements of the spring-block model are associated with fault asperities, it is most realistic to consider nonuniform coefficients of static friction to represent the intermittent contact between fault surfaces and/or bends in the fault which concentrate stress in particular areas. The importance of structural heterogeneity on seismicity has been emphasized by many studies. For example, elementary considerations based on fracture mechanics theory indicate that strong fault heterogeneities, acting as barriers, determine the frequency-magnitude relations, non-uniform fault slip, clustering of events, and random ground accelerations observed [*Hanks*, 1979; *Hanks and McGuire*, 1981; *Aki*, 1984; *Aki*, 1992; *Dmowska and Lovison*, 1992; *Rice*, 1993; *Ben-Zion and Rice*, 1993; *Ben-Zion and Rice*, 1995; *Ben-Zion and Rice*, 1997; *Dmowska, Zhang, and Rice*, 1996]. Several studies have modeled the topography and/or stress distribution of the fault surface as a stochastic function and compared the resulting frequency-size distribution of events with that of observed seismicity [*Andrews*, 1980; *Andrews*, 1981; *von Seggern*, 1980; *Kagan and Knopoff*, 1987; *Yin and Ranalli*, 1995; *de Rubeis et al.*, 1996]. In this approach, the stress difference along a fault is modeled as a Brownian walk as inferred by *Hanks* [1979]. The distribution of intervals between zero-crossings of a Brownian walk is associated with the frequency-size distribution of earthquakes. In addition, many studies have included heterogeneities in mechanical models of earthquakes and found the heterogeneities to be of central importance in the model behavior [*Das and Aki*, 1977; *Rundle and Jackson*, 1977; *Nur and Israel*, 1980; *Papageorgiou and Aki*, 1983a,b; *Rundle, Kanamori, and McNally*, 1984; *Lomintz-Adler and Lemus-Diaz*, 1989; *Knopoff, Landoni, and Abinante*, 1992; *Rundle and Klein*, 1993; *Ben-Zion and Rice*, 1995]. These models are often classified as asperity models. They include heterogeneities that break upon rupture and drive slip on weaker regions. In contrast, barrier models include heterogeneities which do not break during an earthquake. *Mikumo and Miyatake* [1983] studied a threshold model of rupture with spatially inhomogeneous breaking strengths. They found that a broad distribution of breaking strengths led to a rich buildup of activity prior to the principal rupture in direct analogy with experimental work on rock fracture by *Mogi* [1962]. A number of seismic observations suggest that heterogeneity in both individual fault planes and in systems of faults exerts control on the history of seismicity. *Lay et al.* [1980] investigated the clustering of seismicity in subduction zones. They found evidence for richer foreshock activity on the Kurile Island Arc than in the Aleutian Island Arc and greater still than that of the Chilean subduction zone. They attributed this difference to the heterogeneity of seismic coupling in these areas. Greater clustering and foreshock activity was associated with the larger variance in asperity size of the Kurile Island Arc than that of the Aleutians or the Chilean subduction zone. *Foxall, Michelini, and McEvilly* [1993] and *Lees and Nicholson* [1993] have obtained a three-dimensional tomographic image of seismic wave speed anomalies for southern California. They attribute anomalies to lithospheric heterogeneities and argue that these heterogeneities exert control on the recent history of seismicity in the San Andreas fault system. *Malin et al.* [1989] has performed a similar imaging of heterogeneities in the Parkfield section of the San Andreas fault and reached similar conclusions. *Dodge, Beroza, and Ellsworth* [1996] have interpreted the prevalence of foreshocks prior to several major southern California earthquakes in terms of the degree of structural heterogeneity on the fault plane. They plot the number of immediate foreshocks on the fault plane of each mainshock versus the width of the foreshock zone, a simple measure of the jaggedness or structural heterogenity of the fault plane. They found that the number of foreshocks increased with the heterogenity of the fault plane, similar to the dependence on heterogeneity observed in acoustic emission [*Scholz*, 1968a,b; *Hirata*, 1987]. This is consistent with the conclusions of *Jones* [1984] and *Abercrombie and Mori* [1995] who investigated the dependence of foreshock activity as a function of depth on the San Andreas fault system. Both studies found that the number of foreshocks decreases rapidly with increasing depth in the crust. *Frohlich* [1987] obtained similar results for other fault systems. *Kisslinger and Jones* [1991] also identified a similar depth dependence in the number of aftershocks in Southern California. This decrease in foreshock and aftershock activity may be related to a decrease in the degree of structural heteroegenity since heterogeneity decreases with increasing depth (as inferred from strong ground motion studies) [*Anderson and Hough*, 1984]. *Novelo-Casanova et al.* [1985] and *Jin and Aki* [1989] presented evidence for a correlation between Gutenberg-Richter b-values and spatial and temporal variations in heterogeneity as measured by attenuation.

A particularly important quantitative study of the dependence of seismicity on structural heterogeneity and fault length for strike-slip faults was performed by *Stirling et al.* [1996]. These authors quantified the difference in the observed frequency of intermediate and large-magnitude events compared to the number predicted based on an extrapolation of the GR distribution for small events. They found that short, smooth faults tend to display characteristic earthquake behavior while large, rough faults tend to release a larger portion of their seismic moments in background events. Although *Stirling et al.* [1996] did not fit exponents to their data, a visual inspection indicates that the fault zone complexity is inversely proportional to the strike slip offset. The frequency ratio of observed to extrapolated large-magnitude events increased, in turn, approximately as the square root of the strike slip offset. Combining these two observations, the frequency ratio varies as one over the square root of the fault complexity, quantified as the number of steps per unit fault length.

Studies of slider block models which include the effects of structural heterogeneity have been performed. *Rundle and Klein* [1993], *Steacy et al.* [1996], *Gross* [1996] and *Steacy and McCloskey* [1999] studied a cellular automaton version of the two-dimensional slider block model with a statistical distribution of failure thresholds. For a large variance in the failure threshold, they found an exponential rolloff in the power law frequency-size distribution similar to that observed in the model of *Nakanishi* [1990,1991] for a small value of the stiffness. For a small variance in failure thresholds, they found many events which ruptured the whole system. *Knopoff, Landoni, and Abinante* [1992] considered a one-dimensional slider block model with heterogeneous friction. They also observed Gutenberg-Richter cumulative frequency-size statistics with an exponential rolloff at large energies.

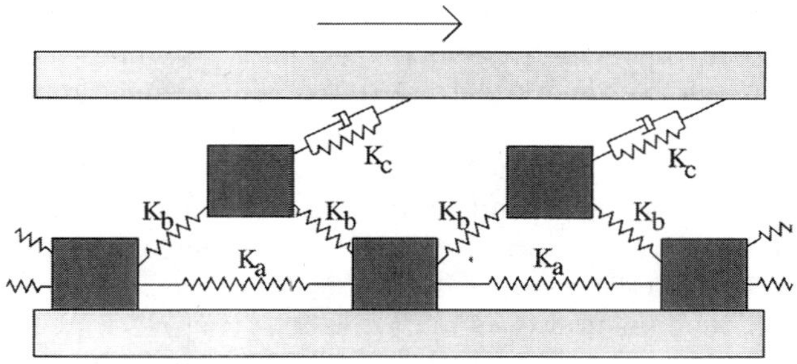

Figure 1b. Model geometry for the heterogeneous spring-block model with viscous coupling (from *Hainzl et al.* [1999]).

Hainzl at al. [1999] constructed one of the few models to generate realistic foreschocks and aftershocks in a seismic cycle. They achieved this result by coupling the blocks of a cellular-automaton spring-block model to an intermediate set of blocks coupled to dashpots (Figure 1b). This intermediate layer represents the viscous asthenosphere. Their model generated a cumulative frequency-size distribution consistent with GR as well as aftershocks and foreshocks consistent with Omori's Law for particular model parameters with a realistic ratio of aftershocks to foreshocks. However, their model generates only one frequency-size distribution (corresponding to noncharacteristic behavior). The behavior of their model depends on an initially random distribution of stresses on model blocks. One additional drawback of their model was that aftershocks and foreshocks were only consistent with Omori's law for a particular ratio of the aftershock decay time to the characteristic earthquake recurrence time.

In this paper, we extend the model of *Hainzl et al.* [1999] to include a distribution of static coefficients of friction in order to model structural heterogeneity of the fault surface or network. The realistic aftershock and foreshock behavior of the model of *Hainzl et al.* [1999] is preserved in our model and the statistics are even more robust since the number of foreshocks and aftershocks obey Omori's Law for nearly all combinations of the model parameters. In addition, the model exhibits a number of other realistic features of seismicity.

Besides the viscous coupling mechanism of *Hainzl et al.* [1999], rate and state-dependent friction has also been proposed as a mechanism for aftershocks [*Dieterich*, 1972;1994]. In order to test this hypothesis within the context of a spring-block model, we have studied a version of our model with a stress-dependent time-to-failure instead of viscous coupling. This version generates realistic seismicity for a short time, but just as for a model with no viscous coupling and no stress-dependent friction, the model eventually settles into a state which generates unrealistic seismicity with frequency-size statistics inconsistent with GR and no foreshocks or aftershocks. Therefore, we conclude that rate and state-dependent friction does not work as a model for aftershocks, at least within the context of a spring-block model.

A fundamental question for each spring-block model is whether they represent a single, isolated fault plane or distributed seismicity on a collection of faults. In this paper we compare the output of a two-dimensional spring-block model to seismicity along individual segments of subduction zones as well as to complex fault networks such as Southern California. This may be reasonable since spring-block models do not depend on dimensionality [*Nakanishi*, 1998]. Spring-block models in a linear (1D) or planar (2D) geometry exhibit qualitatively and quantitatively very similar behavior. A block in a planar (2D) spring-block model has twice as many neighboring blocks as in a linear (1D) system, resulting in a stiffness twice as great. However, if the stiffness of the two models is made comparable by lowering the spring constant of interblock springs in the 2D model, identical statistics are obtained. For a collection of faults, a fractal dimension between 1 and 2 is often observed [*Barton and Larson*, 1985; *Turcotte*, 1986; *Aviles et al.*, 1987; *Hirata et al.*, 1987; *Okubo and Aki*, 1987; *Hirata*, 1989; *Matsumoto et al.*, 1992; *Davy*, 1993; *Scholz*, 1997], indicating that a collection of faults differs from a single fault only in its dimensionality. Since spring-block models exhibit very similar behavior for one and two-dimensional systems, this provides a rational basis for applying spring-block models to a variety of fault geometries.

SYSTEMATIC DEVELOPMENT OF A STRUCTURALLY HETEROGENEOUS SPRING-BLOCK MODEL COUPLED TO A VISCOUS ASTHENOSPHERE

Heterogeneity

We first consider a two-dimensional version of the spring-block model of Figure 1a. Each block is coupled to four neighboring blocks by springs with spring constant K_a and to the driver plate with a single spring of spring constant K_b. In addition, each block is frictionally coupled with the lower plate and has a dynamic coefficient of friction lower than the static coefficient of friction by a constant factor. Many other friction laws are possible but we have chosen this one for simplicity. *Huang and Turcotte* [1992] have shown that more complex velocity-dependent friction laws do not significantly influence the behavior of a system of two blocks. *Narkouskaia and Turcotte* [1992] showed that a linear transformation of the block motions exists between a system with one ratio of dynamic to static coefficient of friction and another. Thus, changing the ratio does not affect the behavior of the model. To model the structural heterogeneity of the fault system, the coefficients of static friction are sampled from a Gaussian distribution with a given mean and variance. If a negative coefficient of friction is sampled initially from the distribution, a new coefficient of friction is sampled until a positive coefficient is obtained. Border blocks have no force exerted on them from outside ("open" boundary conditions; similar results were obtained for closed and periodic boundary conditions) and for which the initial forces between blocks are all zero. *Huang et al.* [1992] found that a version of this spring-block model with homogeneous friction could be nondimensionalized by the seismic coupling K_a/K_b. The analog of the seismic coupling for a heterogeneous model is obtained by multiplying K_a/K_b by the ratio of the mean static coefficient of friction $\overline{\mu}$ to the standard deviation $\overline{\Delta\mu}$ and dividing by the fault length L:

$$\frac{K_a\overline{\mu}}{K_b\overline{\Delta\mu}L} \tag{1}$$

The fault length L enters into the above relation because a larger system will require stiffer interblock springs to rupture the entire system. This system size dependence scales with the linear dimension of the system.

The heterogeneous spring-block model described above was integrated forward in time. An example of the seismic history is shown in Figure 2. Earthquake sizes are commonly estimated by the moment, the product of average slip and slip area, and magnitude, defined here in terms of the seismic moment as 1.5 times the logarithm (base 10) of the seismic moment [*Kanamori and Anderson*, 1975]. In Figure 2 the magnitude is plotted as a function of time in model units. Initially the model displays a buildup of seismicity before a large event and a decline of activity before the next cycle. However, after several cycles, the model becomes completely random and no longer exhibits a broad range of event sizes. We calculated the cumulative frequency-size distribution of seismic moments for the early portion exhibiting cyclic behavior. In Figure 3 the cumulative frequency-size distribution of seismic moments is plotted for models of system size 64x64 in which all of the model parameters were kept fixed except for the value of K_b which was varied from 0.6 to 0.8 to 1.0. The small events for each

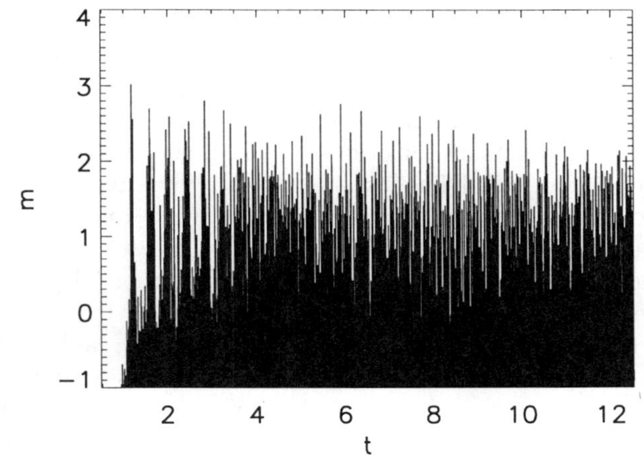

Figure 2. Magnitude versus time in model units for the basic heterogeneous model. Although the initial behavior is consistent with realistic seismic cycle behavior, this is only a transient feature. After several identifiable cycles, the seismicity becomes entirely random with no foreshocks or aftershocks and a frequency-size distribution inconsistent with GR.

case were well approximated by a power-law with an exponent of -2/3, in close agreement with that of observed seismicity for both global [*Gutenberg and Richter*, 1944; *Hanks and Kanamori*, 1979] and regional [*Evernden*, 1970; *Fiedler*, 1974] values. The large events are either overabundant (for $K_b = 0.6$) relative to the power-law trend with many events which rupture nearly the entire system or they are underabundant (for $K_b = 1.0$) with a rolloff in the power-law distribution with an exponential tail. This variation in large-event behavior is similar to that observed for single faults for which a characteristic earthquake distribution often results [*Schwartz and Coppersmith*, 1984; *Davison and Scholz*, 1985; *Kagan*, 1993; *Wesnousky*, 1994] while a Gutenburg-Richter distribution of events with an exponential rolloff often exists for larger areas including the global distribution [*Kagan*, 1994].

It should be noted that considerable care must be taken in any analysis of the tail of the cumulative frequency-size distribution since there are few events in this region and the uncertainty in the statistics of such small numbers can be large. Due to this uncertainty, there have been several different functional forms proposed for the tail of the GR distribution. It has been proposed that the tail of the distribution is a power law with an exponent of -1 rather than -2/3 [*Pacheco et al.*, 1992; *Romanowicz and Rundle*, 1993; *Okal and Romanowicz*. 1994]. However, *Kagan* [1997] and Sornette and Sornette (Sornette, D., and A. Sornette, General theory of the modified Gutenberg-Richter law for large seismic moments, unpublished manuscript, 1998) have carefully tested the likelihood of various functional forms for the frequency-size distribution of large-magnitude events and found that an exponential rolloff was the most consistent with available data.

Although the cumulative frequency-size relationships obtained from the early time behavior of this model are encouraging, another element needs to be added to the system so that this is not simply a transient feature. We have incorporated both stress-dependent friction and viscous coupling in an attempt to obtain realistic seimicity in steady-state model behavior.

Stress-dependent friction

Dieterich [1972,1994] has proposed that aftershocks are related to rate and state-dependent friction. Our first attempt to improve the basic heterogeneous spring-block model was to introduce stress-dependent friction. We introduced stress-dependence by decreasing the time-to-failure of a block by a multiplicative factor dependent on the magnitude of the stress step and that decayed with time according to Dieterich's model for seismic rate increase following a stress step [*Dieterich*, 1994]. Representative model results are shown in Figure 4. As with the basic heterogeneous model, although there is initally seismic cycle behavior, it is only tran-

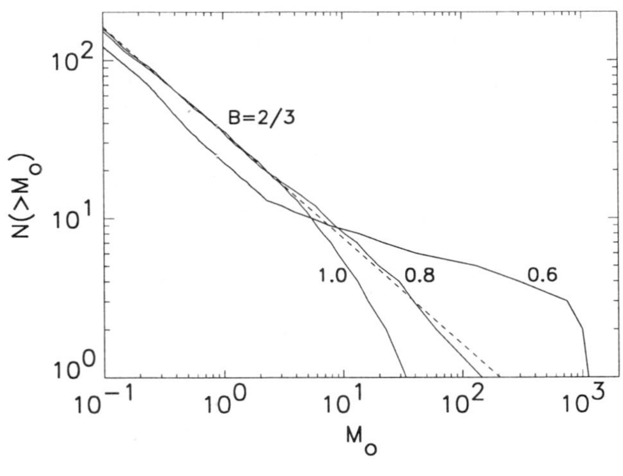

Figure 3. Cumulative frequency-size distribution, the number of events with moments larger than M_o, for three different values of K_b and all other constants fixed. The distribution of small events for all three model runs is consistent with the power-law observed in GR. The frequency of large events varies from characteristic earthquake behavior for small values of K_b to an exponential rolloff for larger values of K_b.

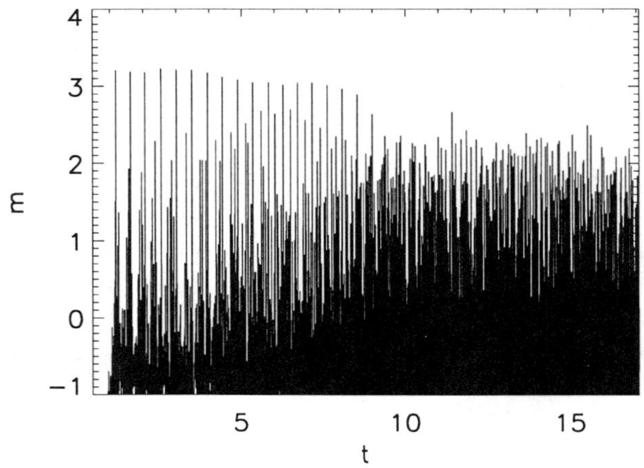

Figure 4. Magnitude versus time in model units for the heterogeneous model with stress-dependent time-to-failure. This model exhibits cyclical early-time behavior that eventually disappears into unrealistic seismic behavior with no foreshocks or aftershocks and a frequency-size distribution inconsistent with GR.

sient behavior. Once the model achieves a dynamic steady-state condition the seismicity is completely random and the cumulative frequency-size distribution is inconsistent with GR. This result suggests that stress-dependent friction is not sufficient as a model for aftershocks or foreshocks, at least within the context of a spring-block model.

Coupling to viscous asthenosphere

One additional way in which spatiotemporal clustering can be introduced into a spring-block model is to couple the blocks to an additional layer viscously coupled to the upper plate. The viscous layer introduces a time-delay to the transfer of stress within the fault.

Figure 1b illustrates the model of *Hainzl et al.* [1999]. When a block slips there is an instantaneous stress transfer from one block to it's neighboring blocks through the elastic coupling of the springs with spring constant K_a. There is also, however, an overdamped transfer of stress between neighboring blocks through blocks coupled both elastically and viscously to the driver plate. These blocks do not change position immediately after the slippage of a coupled block in the bottom layer but rather creep to a new equilibrium position with an exponentially decreasing rate through time.

Cumulative frequency-size distribution. We have implemented the model of Figure 1b with heterogeneous friction. Representative model results are given in Figures 5a-c for three different values of K_a with all other constants held fixed. Figure 5a-c presents the seismic history after each model has achieved a dynamic steady-state following the initial transient related to the uniform initial block positions. This model successfully generates well-defined seismic cycles for large seismic coupling (Figure 5a). The cumulative frequency-size distribution for the models of Figure 5 are given in Figure 6 with the power-law trend representing GR plotted as the dashed line. As the value of K_a is lowered from 0.4 to 0.025, the seismic coupling decreases and the model behavior switches from characteristic to non-characteristic earthquake behavior. Thus, the model exhibits the same range of frequency-size distributions observed in real faults. There is a rolloff in the GR distribution at small magnitudes as well The model also exhibits a systematic decrease in B value for larger seismic couplings. This is consistent with the observed decrease in B value with increasing depth [*Mori and Abercrombie*, 1997] assuming that seismic coupling decreases with depth due decreasing heterogeneity with increasing depth.

Stress diffusion. A strong correlation exists in observed seismicity between mainshock precursor time intervals and the magnitude of the mainshock [*Scholz*, 1990]. This observation has motivated the dilitancy-diffusion model [*Scholz et al.*, 1973] and models of seismicity based on the diffusion of pore fluids [*Nur and Booker*, 1972; *Miller et al.*, 1999] because it implies stress diffusion with a diffusivity of 1 m^2/s. Stress diffusion is also, however, consistent with a spring-block model coupled to a viscous layer. A basic calculation in *Turcotte and Schubert* [1982] shows that the displacement of the lithosphere coupled to a viscous asthenosphere is governed by the diffusion equation with a diffusivity D given by

$$D = \frac{h_a h_L G}{\mu} \tag{2}$$

where h_a is the thickness of the asthenosphere, h_L is the thickness of the elastic lithosphere, G is the shear modulus of the lithosphere, and μ is the viscosity of the asthenosphere. Using h_a=100 km, h_L=30 km, G=30 GPa, and μ=4x10^{19} Pa s we obtain D=0.75 m^2/s, in close agreement with the value obtained from the precursor-magnitude relation.

Omori's Law and relative frequency of foreshocks and aftershocks. Omori's Law states that the number of aftershocks following a main shock rupture decays as a power-law function of the time since the mainshock:

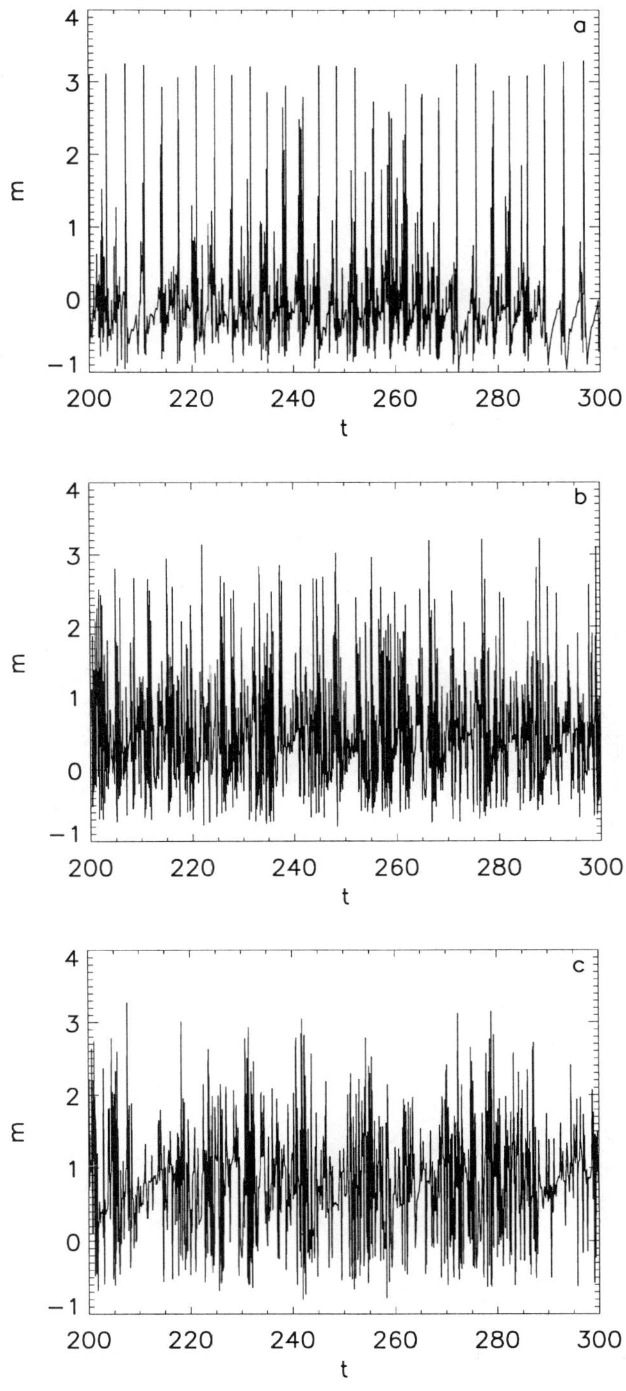

Figure 5. Magnitude versus time for seismicity of a model fault with (a) large seismic coupling exhibiting characteristic earthquake behavior (K_a=0.4), (b) intermediate seismic coupling (K_a=0.1), and (c) relatively low seismic coupling exhibiting no large events (K_a=0.025).

$r(t_o - t) \propto (t_o - t)^{-p}$ where $p \approx 1$. The number of aftershocks is defined by excess number of earthquakes above the background rate of seismicity following a mainshock rupture. The same function has also been found to describe the increase in the number of foreshocks before a mainshock [*Kagan and Knopoff*, 1980; *Jones and Molnar*, 1979]. There are many more aftershocks than foreshocks. Stacked plots of model seismicity as a function of time before and after mainshock ruptures are presented in Figures 7a-c for the models of Figure 5a-c respectively. The plots have been stacked by adding up the seismicity before and after many mainshocks in equally-spaced bins. The ratio of foreshocks to aftershocks is found to decrease with decreasing seismic coupling from a value near one for a system exhibiting characteristic earthquake behavior to a value much less than one for a system exhibiting noncharacteristic earthquake behavior. Thus, the model suggests a very specific pattern to look for in seismic data: the relative occurrence of foreshocks to aftershocks is related to whether the fault displays characteristic or noncharacteristic earthquake behavior.

The stacked number of earthquakes per unit time above the background rate before (Figure 8a) and after (Figure 8b) mainshocks are plotted in Figure 8. These plots indicate that although the number of foreshocks and aftershocks vary for models with different seismic coupling, the exponent in Omori's Law remains the same and is in agreement with the observation $p \approx 1$ for real aftershock data. This result contrasts with that of *Hainzl et al.* [1999] who found that their model only produced an exponent consistent with Omori's Law for particular values of their model parameters.

Kagan's Law for spatiotemporal clustering. *Kagan* [1970] was the first to use the pair correlation function to study the spatiotemporal clustering of hypocenters. The results he obtained are a fundamental observation of seismicity. Kagan [1991a, 1991b] calculated the number of pairs of hypocenters per unit area and time as a function of the distance and time between hypocenters. He found that the frequency of pairs decreased as a power-law function of the distance between pairs over a wide range of distances: $c(|r_1 - r_2|) = |r_1 - r_2|^{-q}$ with values of q that decreased with increasing time between hypocenters. He obtained the same result when he exchanged time for space: he found that the frequency of pairs decreased as a power-law function of the time between pairs with exponents that decreased in magnitude with increasing distance between hypocenters. The corresponding results for the output of the het-

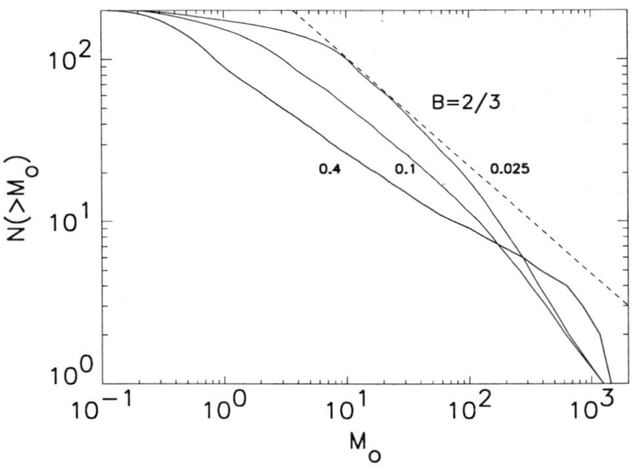

Figure 6. Cumulative frequency-size distribution for the seismicity of Figure 5. The model generates seismicity which is consistent with GR for small events but varies from characteristic earthquake behavior for models with large seismic coupling to a GR distribution with exponential rolloff for models with a low seismic coupling.

erogeneous spring-block model with viscous coupling is given in Figure 9. For small separations (1-2 lattice sites) the exponent of the pair correlation function in time is close to -1 while for larger separations the exponent q decreases in close agreement with Kagan's result. The time in Figure 9 is normalized by the viscous damping time T_o in our model.

Cumulative Benioff strain increase before large earthquakes. *Bufe and Varnes* [1993] identified a new quantitative precursor by observing that the cumulative Benioff strain released prior to large earthquakes increased as a power-law function of time before the mainshock: $B(t_o - t) \propto (t_o - t)^{-m}$ where B is the cumulative Benioff strain release. *Bufe and Varnes* [1993] documented this pattern for seismicity prior to Loma Prieta and *Bufe et al.* [1994] identified the same pattern in seismicity preceding several large earthquakes in the Aleutian Arc. Their work follows that of *Keilis-Borok et al.* [1988] and *Keilis-Borok and Rotwain* [1990] who have used the increase in intermediate-magnitude earthquakes to predict times of increased probabilities for large earthquakes. The average values of m for both San Francisco and segments of the Aleutian Arc were close to $\frac{1}{3}$. These authors chose to analyze the Benioff strain because it is a measure intermediate between earthquake number and seismic moment. If one computes the cumulative seismic moment before large events the curve is dominated by the seismic moment release of the few largest events and it is impossible to see any

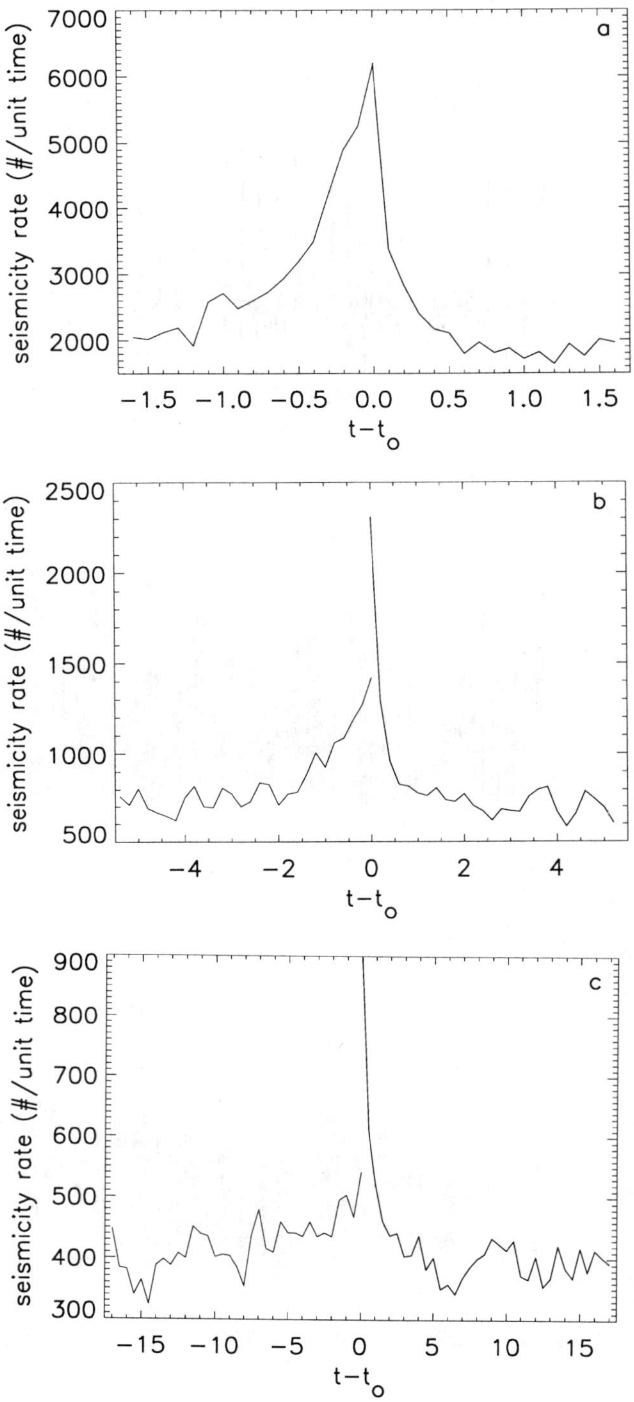

Figure 7. Rate of foreshocks and aftershocks preceeding and following the mainshock rupture for the model with (a) large seismic coupling (K_a=0.4), (b) intermediate seismic coupling (K_a=0.1), and (c) relatively low seismic coupling (K_a=0.025). The ratio of foreshocks to aftershocks decreases with decreasing seismic coupling.

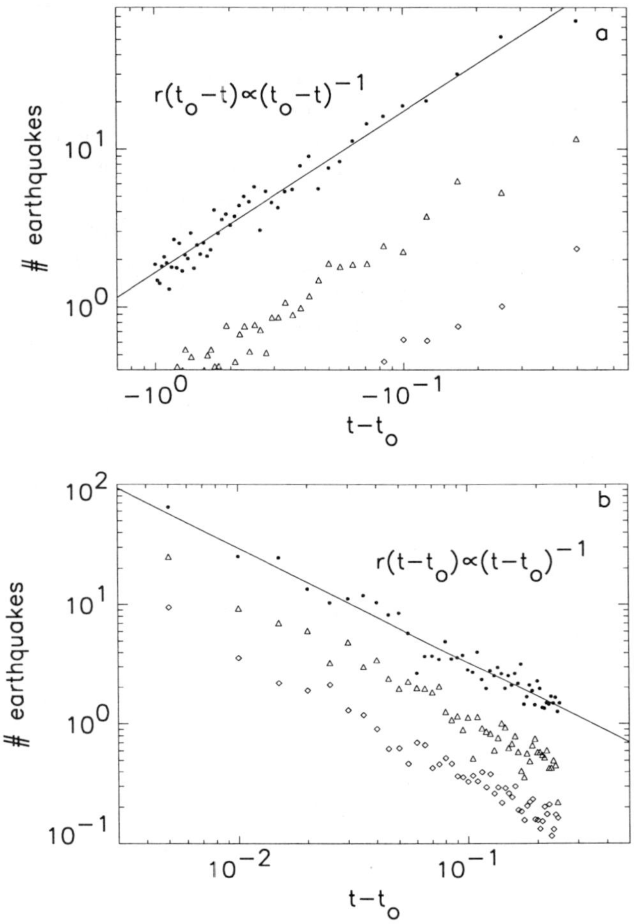

Figure 8. Number of earthquakes per unit time (a) leading up to and (b) following the main shock rupture for the model of Figure 5a (filled circles), 5b (unfilled triangles), and 5c (unfilled diamonds). The exponent of the power-law increase and decrease of earthquake number is consistent with Omori's Law and is independent of the seismic coupling.

pattern in a statistic dominated by such as small number of events. Earthquake number, on the other hand, weighs each earthquake equally and so it is a poor measure of increase in the number or moment release of intermediate-magnitude events. The stacked cumulative Benioff strain prior to mainshock rupture for the three model calculations of Figure 5 are plotted in Figure 10. The cumulative Benioff strain increases as a power-law with an exponent of $m = \frac{1}{3}$, independent of seismic coupling. This is entirely consistent with the results of *Bufe and Varnes* [1993] and *Bufe et al.* [1994].

Rupture histories. *Hartzell and Heaton* [1985] have computed far-field displacement functions for large earthquakes from subduction zones around the Pacific. They observed rough, episodic rupture histories in structurally heterogeneous subduction zones such as the Kurile Islands and broad, smooth ruptures for more structurally uniform subduction zones such as in Alaska. Although there are no rupture dynamics in our cellular-automaton spring-block model, it is possible to obtain approximate dynamics by assuming a constant rupture velocity. The rupture begins at the first unstable block and propagates radially outward at a constant velocity in each direction in which the neighboring block also becomes unstable. This process is iterated at each time step until the earthquake rupture stops. Far-field displacement functions are then calculated by binning the slip distances which take space at equal distances from the hypothetical receiver. Far-field displacement functions for a fault with a relatively large (a) and a relatively small (b) seismic coupling are shown in Figure 11. The fault with large seismic coupling exhibits the broad, smooth rupture that *Hartzell and Heaton* [1985] observed in structurally homogeneous subduction zones while Figure 11b is a heterogeneous rupture with a very rapid increase in stress release in the beginning and a slow falloff towards the end of the rupture. Grayscale plots of the total slip of each portion of the fault surface for the earthquakes of Figure 11a and 11b are shown in Figure 12 and 13, respectively. Dark areas indicate greater amounts of slip. Note that earthquakes do not neces-

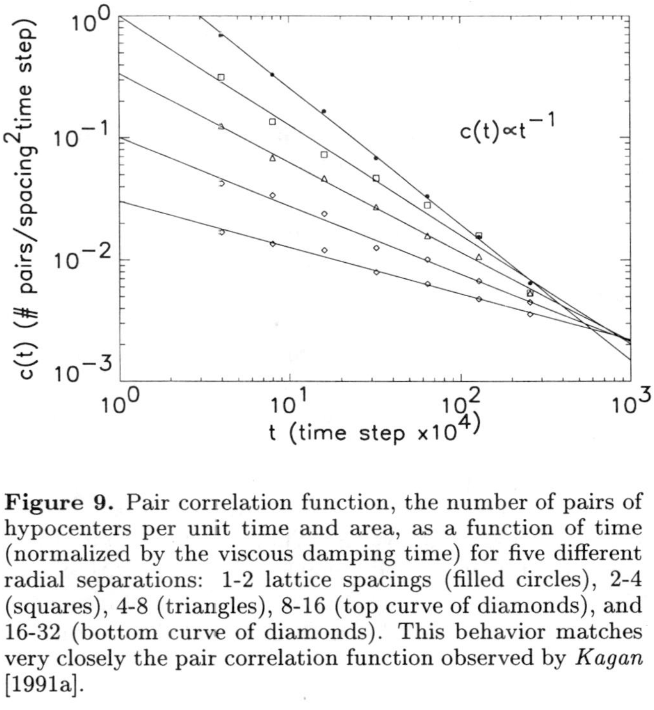

Figure 9. Pair correlation function, the number of pairs of hypocenters per unit time and area, as a function of time (normalized by the viscous damping time) for five different radial separations: 1-2 lattice spacings (filled circles), 2-4 (squares), 4-8 (triangles), 8-16 (top curve of diamonds), and 16-32 (bottom curve of diamonds). This behavior matches very closely the pair correlation function observed by *Kagan* [1991a].

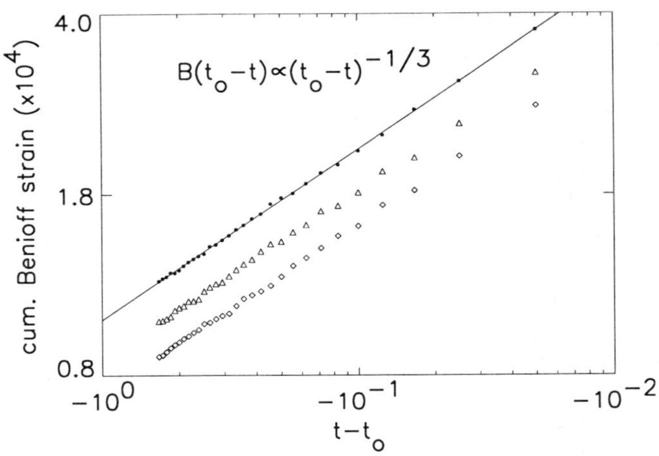

Figure 10. Cumulative Benioff strain preceeding large earthquakes for a model with relatively large seismic coupling (filled circles), intermediate seismic coupling (open triangles), and small seismic coupling (open triangles). A least-square fit to the data yield a power-law relationship with exponent -1/3: $B(t_o - t) \propto (t_o - t)^{-1/3}$ independent of the seismic coupling, consistent with the observations of *Bufe and Varnes* [1993].

sarily rupture simply connected regions: the rupture of Figure 12 has three patches of fault surface which do not slip even though they are completely surrounded by regions of slip.

Scaling of average slip versus moment. The average slip versus moment is plotted in Figure 14. A least-squares fit to the data indicate a relationship of $d \propto M_o^{0.17}$. This same relationship was obtained for different values of the seismic coupling. This relationship matches exactly that of earthquakes in Parkfield, California, for which the source geometries are well known [*Sammis et al.*, 1999].

Dependence of large-magnitude seismicity on fault length and structural complexity. *Stirling et al.* [1996] have performed the most complete quantitative analysis of the dependence of the frequency-size distribution of earthquakes on fault length and structural complexity for strike-slip faults. These authors quantified the large-magnitude behavior of strike-slip faults with the ratio of the frequency or moment of observed large events to the frequency or moment of large events based on an extrapolation of the best-fit power-law distribution of small-magnitude events. They observed that small, smooth faults displayed behavior consistent with a relatively large seismic coupling (i.e. the ratio of frequency or moment of observed to extrapolated large-magnitude events was large) compared to large, rough faults. We have performed a similar analysis for fault length in the Aleutian Arc. Figure 15 plots the ratio of the largest observed seismic moment to that predicted based on an extrapolation of the power-law fit to smaller events for each segment of the Aleutian Arc. The data source for the moments is *Kagan* [1997]. We observe the same dependence on fault length that *Stirling et al.* [1996] observed for strike-slip faults: shorter faults have a larger seismic coupling and a larger fraction of their total seismic moment released in large events.

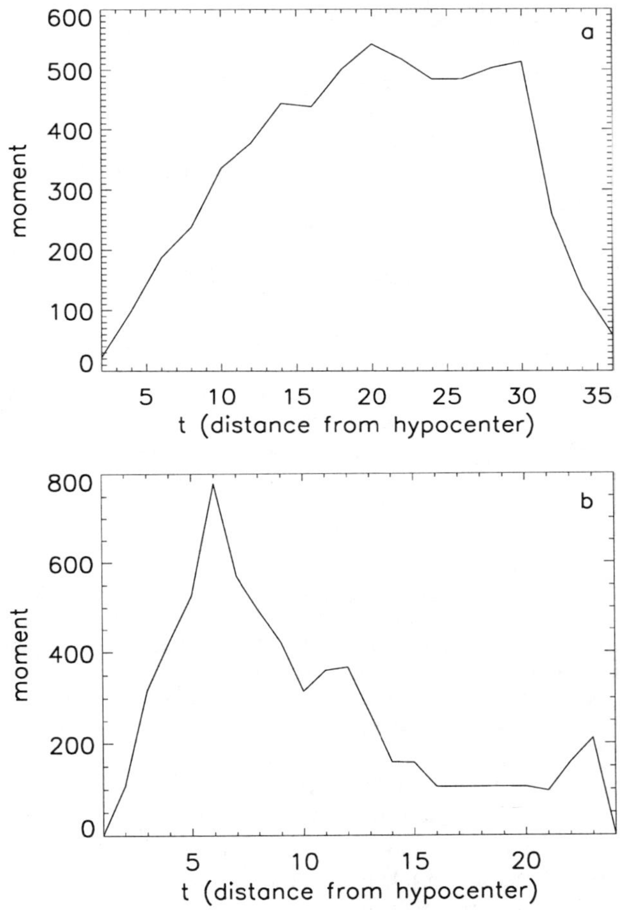

Figure 11. Source-time function for a large rupture of a model fault with (a) large seismic coupling and a broad, smooth rupture and (b) small seismic coupling and a rough, episodic rupture history. These source-time functions can be compared to the broad, smooth rupture histories of large earthquakes of Alaska and the rough ruptures of the heterogeneous Kurile Islands subduction zone observed by *Hartzell and Heaton* [1985].

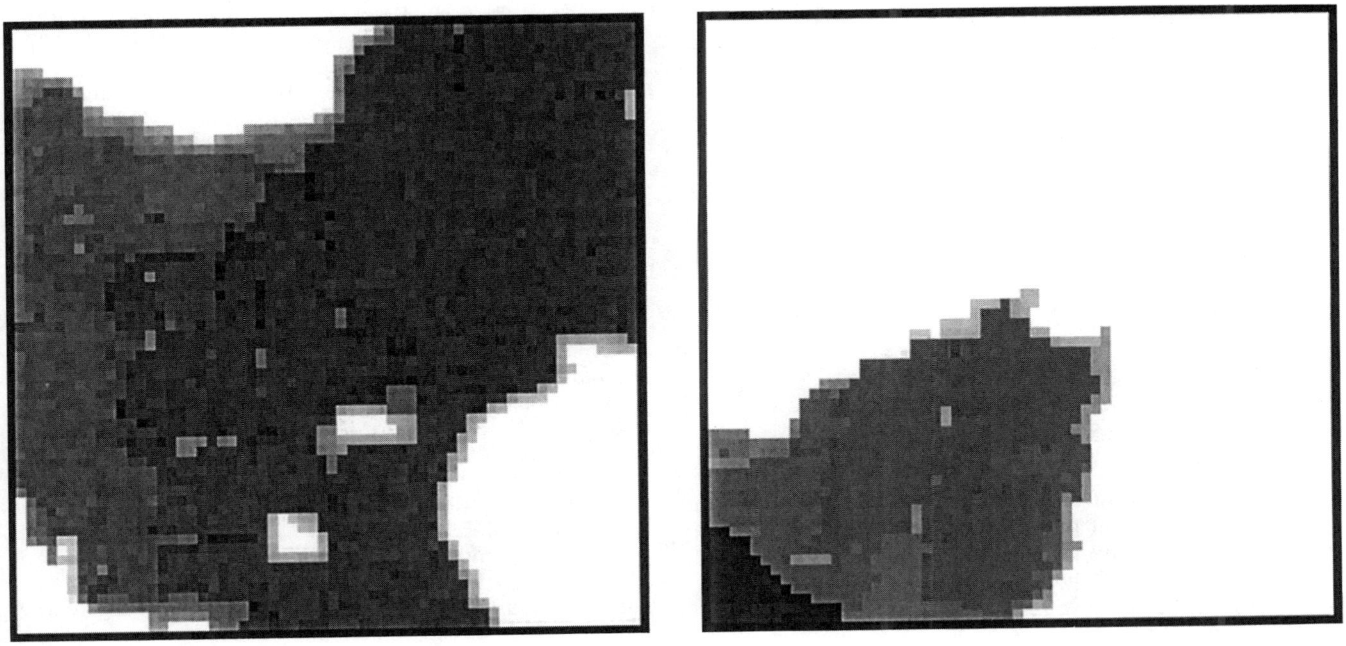

Figure 12. Grayscale plot of fault slip for a large rupture of a model fault with (a) large seismic coupling and (b) relatively small seismic coupling.

CONCLUSION

Although many spring-block models have been proposed, on the whole they have failed to add significantly to our basic understanding of fault mechanics. Part of the reason for this may be that so many different models have been proposed, all of which purportedly reproduce GR statistics. Another reason may be that too few studies investigate alternative models to see if their model is unique in reproducing observed patterns of seismicity. In this study we have attempted to set a new standard for spring-block models by presenting a model which generates realistic seismicity with respect to many different measures of seismicity. We have also constructed the model in such a way that we are able to rule out stress-dependent friction as a significant generating mechanism for foreshocks and aftershocks. Lastly, we have tried to provide a brief review of the last decade of work to put our work and the recent work of others into context.

Our model suggests that the roles of fault length, depth, and structural heterogeneity can be understood in terms of their influence on the seismic coupling of fault systems. Seismic coupling, in turn, influences the relative frequency of large and small-magnitude events, the relative frequency of foreshocks and aftershocks, the spatiotemporal clustering of events, the buildup of cumulative Benioff strain before large earthquakes, and the rupture histories of individual earthquakes. The statistical similarity of seismicity on strike-slip faults and in subduction zones suggests that seismicity in sub-

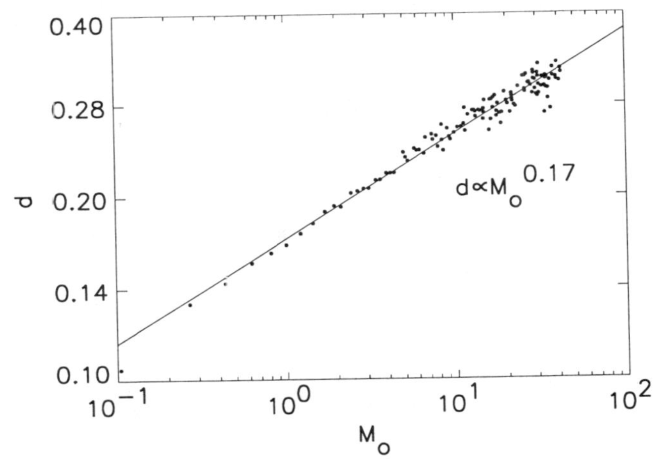

Figure 13. Average displacement d versus seismic moment M_o. A least-square fit to the data yield a power-law relationship $d \propto M_o^{0.17}$, consistent with the value obtained for earthquakes at Parkfield, California by *Sammis et al.* [1999].

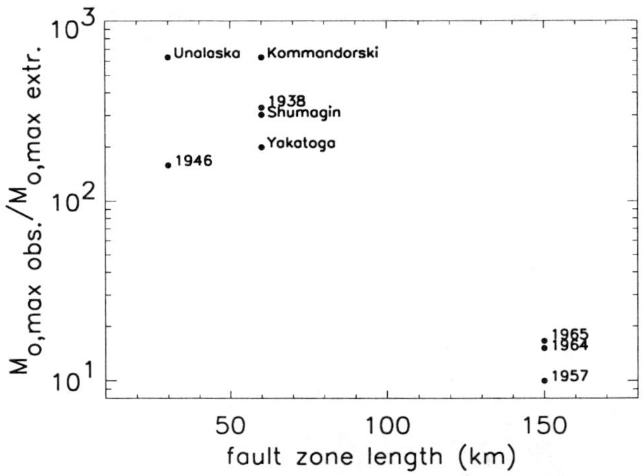

Figure 14. Ratio of the maximum moment observed to the maximum moment predicted based on an extrapolation of the GR fit to small events for different segments of the Aleutian Arc, as a function of segment length. Moment data from *Kagan* [1997].

duction zones may be able to be quantified in a similar way to that of strike-slip faults in the work of *Stirling et al.* [1996]. *Lay et al.* [1980] have qualitatively described the seismicity on different subduction zones in terms of their structural complexity. The model in this paper, together with data analysis of the patterns it suggests to look for in real sesimicity, may provide a starting point for a better quantification of patterns of seismicity in subduction zones. In addition, the model may help us to better understand the depth dependence of seismicity in both subduction zone and strike-slip fault systems.

REFERENCES

Abercrombie, R., and J. Mori, Foreshock occurrence as a function of mainshock depth and mechanism, *Seismol. Res. Lett., 66*, 23, 1995.

Aki, K., Asperities, barriers, characteristic earthquakes, and strong motion prediction, *J. Geophys. Res., 89*, 5867-5872, 1984.

Aki, K., Higher-order interrelations between seismogenic structures and earthquake processes, *Tectonophysics, 211*, 1-12, 1992.

Anderson, J.G., and S.E. Hough, A model for the shape of the Fourier amplitude spectrum of acceleration at high frequencies, *Bull. Seismol. Soc. Am., 74*, 1969-1993, 1984.

Andrews, D., A stochastic fault model. 1. Static case, *J. Geophys. Res., 85*, 3867-3887, 1980.

Andrews, D., A stochastic fault model. 2. Time-dependent case, *J. Geophys. Res., 86*, 10,831-10834, 1981.

Aviles, C.A., C.H. Scholz, and J. Boatwright, Fractal analysis applied to characteristic segments of the San Andreas fault, *J. Geophys. Res., 92*, 331-344, 1987.

Bak, P., and C. Tang, Earthquakes as a self-organized critical phenomenon, *J. Geophys. Res., 94*, 15,636-15,637, 1989.

Barriere, B., and D.L. Turcotte, Seismicity and self-organized criticality, *Phys. Rev. E, 49*, 1151-1160, 1994.

Barton, C.C., and E. Larson, Fractal geometry of two-dimensional fracture networks at Yucca Mountain, Southwestern Nevada, in *Fundamentals of Rock Joints*, edited by Ove Stephansson, pp. 77-84, Proc. Int. Symp. on Fundamentals of Rock Joints, 1985.

Ben-Zion, Y., and J.R. Rice, Earthquake failure sequences along a cellular fault zone in a three-dimensional elastic solid containing asperity and non-asperity regions, *J. Geophys. Res., 98*, 14,109-14,131, 1993.

Ben-Zion, Y., and J.R. Rice, Slip patterns and earthquake populations along different classes of faults in elastic solids, *J. Geophys. Res., 100*, 12,959-12,983, 1995.

Ben-Zion, Y., and J.R. Rice, Dynamic simulation of slip on a smooth fault in an elastic solid, *J. Geophys. Res., 102*, 17,771-17,784, 1997.

Brown, S.R., C.H. Scholz, and J.B. Rundle, A simplified spring-block model of earthquakes, *Geophys. Res. Lett., 18*, 215-218, 1991.

Bufe, C.G., and D.J. Varnes, Predictive modeling of the seismic cycle of the greater San Francisco Bay region, *J. Geophys. Res., 98*, 9,871-9,883, 1993.

Burridge, R., and L. Knopoff, Model and theoretical seismicity, *Bull. Seismol. Soc. Am., 57*, 341-371, 1967.

Carlson, J.M., Two-dimensional model of a fault, *Phys. Rev. A, 44*, 6226-6232, 1991.

Carlson, J.M., and J.S. Langer, Mechanical model of an earthquake fault, *Phys. Rev. A, 40*, 6470-6484, 1989.

Das, S., and K. Aki, Fault planes with barriers: A versatile earthquake model, *J. Geophys. Res., 82*, 5658-5670, 1977.

Davison Jr., F.C. and C.H. Scholz, Frequency-moment distribution of earthquakes in the Aleutian arc: A test of the characteristic earthquake model, *Bull. Seismol. Soc. Am., 75*, 1349-1361, 1985.

Davy, P., On the frequency-length distribution of the San Andreas fault system, *J. Geophys. Res., 98*, 12,141-12,151, 1993.

Dieterich, J.H., Time-dependent friction as a possible mechanism for aftershocks, *J. Geophys. Res., 77*, 3771-3781, 1972.

Dieterich, J., A constitutive law for rate of earthquake production and its application to earthquake clustering, *J. Geophys. Res., 99*, 2601-2618, 1994.

Dmowska, R., and L.C. Lovison, Influence of asperities along subduction interfaces on the stressing and seismicity of adjacent areas, *Tectonophysics, 211*, 23-43, 1992.

Dmowska, R., G. Zhang, and J.R. Rice, Seismicity and deformation at convergent margins due to heterogeneous coupling, *J. Geophys. Res., 101*, 3015-3029, 1996.

Dodge, D.A., G.C. Beroza, and W.L. Ellsworth, Detailed observations of California foreshock sequences: Implications for the earthquake initiation process, *J. Geophys. Res., 101*, 22,371-22,392, 1996.

Evernden, J.F., Study of regional seismicity and associated problems, *Bull. Seismol. Soc. Am., 60*, 393-446, 1970.

Fiedler, G., Local b value related to seismicity, *Tectonophysics, 23*, 277-282, 1974.

Foxall, W., A. Michelini, and T.V. McEvilly, Earthquake travel time tomography of the suthern Santa Cruz Mountains: Control of fault rupture by lithological heterogeneity of the San Andreas fault zone, *J. Geophys. Res., 98*, 17,691-17,710, 1993.

Frohlich, C., Aftershocks and temporal clustering of deep earthquakes, *J. Geophys. Res., 92*, 13,944-13,956, 1987.

Grasso, J.-R., and D. Sornette, Testing self-organized criticality by induced seismicity, *J. Geophys. Res., 103*, 29,965-29,987, 1998.

Gross, S., Magnitude distributions and slip scaling of heterogeneous seismic sources, *Bull. Seismol. Soc. Am., 86*, 498-504, 1996.

Gutenberg, B., and C.F. Richter, Frequency of earthquakes in California, *Bull. Seismol. Soc. Am., 34*, 185-188, 1944.

Hainzl, S., G. Zoller, and J. Kurths, Similar power laws for foreshock and aftershock sequences in a spring-block model for earthquakes, *J. Geophys. Res., 104*, 7243-7253, 1999.

Hanks, T.C., b values and $\omega^{-\gamma}$ seismic source models: Implications for tectonic stress variations along active crustal fault zones and the estimation of high-frequency ground motion, *J. Geophys. Res., 84*, 2235-2242, 1979.

Hanks, T., and H. Kanamori, A magnitude-moment scale, *J. Geophys. Res., 84*, 2348-2352, 1979.

Hanks, T., and R. McGuire, The character of high-frequency strong ground motion, *Bull. Seismo. Soc. Am., 71*, 2071-2095, 1981.

Hartzell, S.H., and T.H. Heaton, Teleseismic time functions for large, shallow subduction zone earthquakes, *Bull. Seismol. Soc. Am., 75*, 965-1004, 1985.

Hirata, T., Omori's power-law aftershock sequences of microfracturing in rock fracture experiment, *J. Geophys. Res., 92*, 6215-6221, 1987.

Hirata, T., Fractal dimension of fault systems in Japan: Fractal structure in rock fracture geometry at various scales, *Pure Appl. Geophys., 131*, 157-170, 1989.

Hirata, T., T. Satoh, and K. Ito, Fractal structure of spatial distribution of microfracturing in rock, *Geophys. J. R. Astr. Soc., 90*, 369-374, 1987.

Huang, J., and D.L. Turcotte, Evidence for chaotic fault interactions in the seismicity of the San Andreas fault and Nankai trough, *Nature, 348*, 234-236, 1990.

Huang, J., G. Narkounskaia, and D.L. Turcotte, A cellular-automata slider-block model of earthquakes II. Demonstration of self-organized criticality for a 2-D system, *Geophys. J. Int., 111*, 259-269, 1992.

Ito, K., and M. Matsizaki, Earthquakes as a self-organized critical phenomenon, *J. Geophys. Res., 95*, 6853-6860, 1990.

Jin, A., and K. Aki, Spatial and temporal correlation between coda Q^{-1} and seismicity and its physical mechanism, *J. Geophys. Res., 94*, 14,041-14,059, 1989.

Jones, L.M., Foreshocks (1966-1980) in the San Andreas system, California, *Bull. Seismol. Soc. Am., 74*, 1361-1380, 1984.

Jones, L.M., and P. Molnar, Some characteristics of foreshocks and their possible relationship to earthquake prediction and premonitory slip on faults, *J. Geophys. Res., 84*, 3596-3608, 1979.

Kagan, Y.Y., Fractal dimension of brittle fracture, *J. Nonlin. Sci., 1*, 1-17, 1991a.

Kagan, Y.Y., Long-term earthquake clustering, *Geophys. J. Int., 104*, 117-133, 1991b.

Kagan, Y.Y., Statistics of characteristic earthquakes, *Bull. Seismol. Soc. Am., 83*, 7-24, 1993.

Kagan, Y.Y., Observational evidence for earthquakes as a nonlinear dynamic process, *Physica D, 77*, 160-192, 1994.

Kagan, Y.Y., Seismic moment-frequency relation for shallow earthquakes: regional comparison, *J. Geophys. Res., 102*, 2835-2852, 1997.

Kagan, Y.Y., and L. Knopoff, Dependence of seismicity on depth, *Bull. Seismol. Soc. Am., 70*, 1811-1822, 1980.

Kagan, Y.Y., and L. Knopoff, Random stress and earthquake statistics: Time dependence, *Geophys. J. R. Astr. Soc., 88*, 723-731, 1987.

Kanamori, H., and D.L. Anderson, Theoretical basis of some empirical relations in seismology, *Bull. Seismol. Soc. Am., 65*, 1073-1096, 1975.

Keilis-Borok, V.I., L. Knopoff, I.M. Rotwain, and C.R. Allen, Intermediate-term prediction of occurrence times of strong earthquakes, *Nature, 335*, 690-694, 1988.

Keilis-Borok, V.I., and I.M. Rotwain, Diagnosis of time of increased probability of strong earthquakes in different regions of the world: Algorithm CN, *Phys. Earth Plan. Int., 61*, 57-72, 1990.

Kisslinger, C., and L.M. Jones, Properties of aftershock sequences in Southern California, *J. Geophys. Res., 96*, 11,947-11,958, 1991.

Knopoff, L., J.A. Landoni, and M.S. Abinante, Dynamical model of an earthquake fault with localization, *Phys. Rev. A, 46*, 7445-7449, 1992.

Lay, T., L. Ruff, and H., Kanamori, The asperity model and the nature of large earthquakes in subduction zones, *Earthquake Pred. Res., 1*, 3-71, 1980.

Lees, J.M., and C. Nicholson, Three-dimensional tomography of the 1992 southern California earthquake sequence: Constraints on dynamic earthquake rupture, *Geology, 21*, 387-390, 1993.

Malin, P.E., S.N. Blakeslee, M.G. Alvarez, and A.J. Martin, Microearthquake imaging of the Parkfield asperity, *Science, 244*, 557-559, 1989.

Matsumoto, N., K. Yomogida, and S. Honda, Fractal analysis of fault systems in Japan and in the Phillipines, *Geophys. Res. Lett., 19*, 357-360, 1992.

Mikumo, T., and T. Miyatake, Numerical modeling of space and time variations of seismic activity before large earthquakes, *Geophys. J. R. Astron. Soc., 74*, 559-583, 1983.

Miller, S.A., Y. Ben-Zion, and J.P. Burg, A three-dimensional fluid-controlled earthquake model: Behavior and implications, *J. Geophys. Res., 104*, 10621-10638, 1999.

Mogi, K., Study of elastic shocks caused by the fracture of heterogeneous materials and their relation to earthquake phenomena, *Bull. Earthquake Res. Inst. Univ. Tokyo, 40*, 125-173, 1962.

Mori, J., and R. Abercrombie, Depth dependence of earthquake frquency-magnitude distributions in California: Implications for rupture initiation, *J. Geophys. Res., 102*, 15,081-15,090, 1997.

Nakanishi, H., Cellular automaton model of earthquakes with deterministic dynamics, *Phys. Rev. A, 41*, 7086-7089, 1990.

Nakanishi, H., Statistical properties of the cellular automata model for earthquakes, *Phys. Rev. A, 43*, 6613-6621, 1991.

Nakanishi, H., Complex behavior in earthquake dynamics, *Inter. J. Mod. Phys. B, 12*, 273-284, 1998.

Narkounskaia, G., and D.L. Turcotte, A cellular-automata, slider-block model for earthquakes I. Demonstration of chaotic behavior for a low-order system, *Geophys. J. Int., 111*, 250-258, 1992.

Novelo-Casanova, D.A., E. Berg, V. Hsu, and C.E. Helsley, Time-space variation of seismic S-wave attenuation (Q^{-1}) and magnitude distribution (b-values) for the Petatlan earthquake, *Geophys. Res. Lett., 12*, 789-792, 1985.

Nur, A., and J.R. Booker, Aftershocks caused by pore fluid flow?, *Science, 175*, 885-887, 1972.

Nur, A., and M. Isreal, The role of heterogeneities in faulting, *Phys. Earth Plan. Int., 21*, 225-236, 1980.

Okal, E.A., and B.A. Romanowicz, On the variation of b-values with earthquake size, *Phys. Earth Planet. Int., 87*, 55-76, 1994.

Okubo, P.G., and K. Aki, Fractal geometry in the San Andreas fault system, *J. Geophys. Res., 92*, 345-355, 1987.

Omori, F., On the aftershocks of earthquakes, *J. Coll. Sci., Tokyo Imp. Univ., 7*, 111-200, 1895.

Pacheco, J., C. Scholz, and L. Sykes, Changes in frequency-magnitude relationship from small to large earthquakes, *Nature, 355*, 71-73, 1992.

Papageorgiou, A.S., and K. Aki, A specific barrier model for the quantitative description of heterogeneous faulting and the prediction of strong ground motion. I. Description of the model, *Bull. Seismol. Soc. Am., 73*, 693-722, 1983a.

Papageorgiou, A.S., and K. Aki, A specific barrier model for the quantitative description of heterogeneous faulting and the prediction of strong ground motion. II. Applications of the model, *Bull. Seismol. Soc. Am., 73*, 953-978, 1983b.

Romanowicz, B., and J. Rundle, On scaling relations for large earthquakes, *Bull. Seismol. Soc. Am., 83*, 1294-1297, 1993.

de Rubeis, V., R. Hallgrass, V. Loreto, G. Paladin, L. Pietronero, and P. Tosi, Self-affine asperity model for earthquakes, *Phys. Rev. Lett., 76*, 2599-2602, 1996.

Ruff, L.J., Asperity distribution and large earthquakes in subduction zones, *Tectonophysics, 211*, 61-84, 1992.

Rundle, J.B., and D.D. Jackson, Numerical simulation of earthquake sequences, *Bull. Seismol. Soc. Am., 67*, 1363-1377, 1977.

Rundle, J.B., H. Kanamori, and K.C. McNally, An inhomogeneous fault model for gaps, asperities, barriers, and seismic migration, *J. Geophys. Res., 89*, 10,219-10,231, 1984.

Rundle, J.B., and W. Klein, Scaling and critical phenomena in a cellular automatom slider-block model for earthquakes, *J. Stat. Phys., 72*, 405-412, 1993.

Sammis, C.G., R.M. Nadeau, and L.R. Johnson, How strong is an asperity?, *J. Geophys. Res., 104*, 10,609-10,619, 1999.

Scholz, C., The frequency-magnitude relation of microfracturing in rock and its relation to earthquakes, *Bull. Seismol. Soc. Am, 58*, 399-415, 1968a.

Scholz, C., Microfractures, aftershocks, and seimicity, *Bull. Seismol. Soc. Am., 58*, 1117-1130, 1968b.

Scholz, C.H., Size distributions for large and small earthquakes, *Bull. Seismol. Soc. Am., 87*, 1074-1077, 1997.

Scholz, C.H., *The Mechanics of Earthquakes and Faulting*, Cambridge Univ. Press, New York, 1990.

Scholz, C.H., L.R. Sykes, and Y.P. Aggarwal, Earthquake prediction: A physical basis, *Science, 181*, 803-809, 1973.

Schwartz, D.P., and K.J. Coppersmith, Fault behavior and characteristic earthquakes: Examples from the Wasatch and San Andreas fault zones, *J. Geophys. Res., 89*, 5681-5698, 1984.

Steacy, S.J., J. McCloskey, C.J. Bean, and J. Ren, Heterogeneity in a self-organized critical earthquake model, *Geophys. Res. Lett., 23*, 383-386, 1996.

Steacy, S.J., and J. McCloskey, Heterogeneity and the earthquake magnitude-frequency distribution, *Geophys. Res. Lett., 26*, 899-902, 1999.

Stirling, M.W., S.G. Wesnousky, and K. Shimayaki, Fault trace complexity, cumulative slip, and the shape of the magnitude-frequency distribution for strike-slip faults: a global survey, *Geophys. J. Int., 124*, 833-868, 1996.

Turcotte, D.L., A fractal model for crustal deformation, *Tectonophysics, 132*, 261-269, 1986.

Wang, J.-H., Effect of seismic coupling on the scaling of seismicity, *Geophys. J. Int., 121*, 475-488, 1995.

Wesnousky, S.G., The Gutenberg-Richter or characteristic earthquake distribution, which is it?, *Bull. Seismol. Soc. Am., 84*, 1940-1959, 1994.

Department of Geosciences, University of Arizona, Gould-Simpson Building, 1040 E. Fourth St., Tucson, AZ 85721-0077, phone: 520-621-6024, email: jon@geo.arizona.edu

Statistical Analysis of a Model for Earthquake Faults With Long-Range Stress Transfer

W. Klein, M. Anghel and C. D. Ferguson[1]

Physics Department, Center for Polymer Physics, and Center for Computational Science, Boston University, Boston, Massachussets

J. B. Rundle[2] and J. S. Sá Martins

Colorado Center for Chaos and Complexity and CIRES, University of Colorado, Boulder, Colorado

We present both theoretical and numerical analysis of a cellular automaton version of a slider-block model for earthquake faults that includes long-range stress transfer. Theoretically we develop a coarse-grained description in the mean-field (infinite range) limit and discuss the relevance of the meta-stable state, limit of stability (spinodal) and nucleation to the phenomenology of the model. We also simulate the model and confirm the relevance of the theory for systems with long, but finite range, interactions. Results of particular interest include the existence of Gutenberg-Richter like scaling consistent with that found on real earthquake fault systems, the association of large events with nucleation near the spinodal, the existence of more than one scaling regime, and the result that such systems can be described, in the mean-field limit, with techniques appropriate to systems in equilibrium.

1. INTRODUCTION

It is well known that the frequency-size statistics of earthquakes obey Gutenberg-Richter (GR) scaling relations [*Gutenberg and Richter*, 1954]. However, it is not known whether critical phenomena produce these empirical relations. To see if this connection exists, geophysicists [*Gu et al*, 1984; *Huang and Turcotte*, 1990; *Rundle and Jackson*, 1977; *Rundle and Brown*, 1991] and condensed matter physicists [*Carlson and Langer*, 1989; *Carlson et al*, 1991, 1994; *Bak and Tang*, 1989; *Feder and Feder*, 1991; *Olami et al*, 1992; *Rundle and Klein*, 1993] have been examining models of earthquake faults which hopefully contain the minimal, but essential, physics. In the first of these models, *Burridge and Knopoff* [1967] (BK) constructed a one-dimensional network of slider-blocks connected to their nearest-neighbors and a loader plate via linear elastic springs. The BK model and other models [*Gu et al*, 1984; *Huang and Turcotte*, 1990; *Rundle and Jackson*, 1977; *Rundle and Brown*, 1991; *Carlson and Langer*, 1989; *Carlson et al*, 1991, 1994; *Bak and Tang*, 1989; *Feder and Feder*, 1991; *Olami et al*, 1992; *Rundle and Klein*, 1993] influenced by it can exhibit frequency-size statistics similar to GR scaling. However, until recently no theoretical

[1]Now at Federation of American Scientists, Washington, DC

[2]Also at Department of Physics and Department of Geological Sciences, University of Colorado at Boulder, Boulder, Colorado

GeoComplexity and the Physics of Earthquakes
Geophysical Monograph 120

analysis has demonstrated a connection between this scaling and critical phenomena. On the contrary, after analyzing the behavior of a single BK slider-block, *Vasconcelos* [1996] argued that a first-order phase transition occurs but that no continuous phase transition, giving rise to critical phenomena, exists. In this work we present a theoretical and numerical analysis of a cellular automaton(CA) long-range interaction version of the BK model introduced by Rundle, Jackson, and Brown [*Rundle and Jackson*, 1977; *Rundle and Brown*, 1991] (RJB). Our work indicates the presence of a spinodal critical point, which can give rise to the GR scaling observed in these models.

In addition to describing a coarse-grained theory we have developed for this model [*Klein et al*, 1997], and discussing the assumptions that were made, we also present the results of simulations that are consistent with this theory [*Ferguson et al*, 1999]. In particular, we find critical slowing down as the spinodal critical point is approached, frequency-size statistics for a wide range of realistic fault parameters that are consistent with the values predicted by the theory and a strong relationship between earthquakes in the model and behavior near a mean-field spinodal including nucleation events [*Anghel et al*, 2000].

We have also investigated further a previous result [*Rundle et al*, 1995] that the mean-field (infinite range interactions) version of this model can be described by an equilibrium theory. We find that for slow tectonic driving and long-range interactions the CA model demonstrates what we call local ergodic behavior. We demonstrate this using a measure of effective ergodicity developed by *Thirumalai and Mountain* [1990, 1993]. This result, as well as the theoretical analysis, gives further weight to the claim [*Klein et al*, 1997; *Rundle et al*, 1995] that various aspects of this model can be described by the techniques of equilibrium statistical mechanics.

Finally we have done a careful study of the scaling of this long range CA model and found that there is more than one scaling region and that this substructure can be predicted from analysis of the behavior of clusters near the spinodal critical point [*Anghel et al*, 2000].

The remainder of this paper is structured as follows: In Section 2 we briefly describe the RJB version of the BK model. In Section 3 we discuss the applicability of long-range interactions to models of earthquake faults and develop a coarse-grained theory for the long-range CA model. In Section 4 we discuss the assumptions made in the development of this theory and present numerical evidence supporting their validity. In Section 5 we review spinodal nucleation and present the predictions of the theory as well as the corresponding simulation results. In Section 6 we apply a measure of effective ergodicity, the energy-fluctuation metric, to the CA model and find that the system tends toward ergodic behavior as the tectonic velocity approaches zero and as the interaction range increases. In Section 7 we discuss the arresting of nucleation events and return to the question of coarse graining. In Section 8 we investigate the different scaling regions and the implication of this type of scaling for the structure of earthquake events. Finally, in Section 9, we discuss our results.

2. MODEL

In this section we describe the model introduced by Rundle, Jackson, and Brown. This model consists of a two-dimensional array of massless blocks that interact with their neighbors and a tectonic loader plate via linear springs with constants $T(r)$ and K_L, respectively, where r is the distance between blocks. Initially, each block i receives a random position U_i from a uniform distribution. The loader plate contribution to the stress is set equal to zero initially, and the stress σ_i on each block is measured and compared to a threshold value σ_i^F. If $\sigma_i < \sigma_i^F$ then the block is not moved. If, however, $\sigma_i \geq \sigma_i^F$ the block is moved according to the rule

$$U_i(t+1) = U_i(t) + \left[\frac{\sigma_i(t) - \sigma_i^R}{K}\right](1 - \eta_i(t))\Theta\left(\sigma_i(t) - \sigma_i^F\right). \quad (1)$$

Here the step function $\Theta(x) = 0 \,\forall\, x \leq 0$ and $= 1 \,\forall\, x > 0$, and the effective spring constant $K = K_L + K_C$ where $K_C = \sum_{j,i\neq j} T_{ij}$, and T_{ij} is the matrix of interaction coefficients. The sum over j includes all of the neighbors of block i. The residual stress, σ_i^R, is a parameter that specifies the stress on a block immediately after failure. All blocks are tested and moved until no block has a stress greater than σ_i^F. At this point the plate is moved. Two procedures are used. The first, which we will refer to as the zero velocity limit, requires that we find the block that has the highest σ_i. We then move the plate so that each block receives the stress that just brings the block with the highest stress to its failure threshold. This guarantees that in the vast majority of plate updates there will be only one initial failure site and hence one earthquake per plate update. The second method we will employ is to move the plate a fixed distance $V\Delta T$, thereby increasing the stress σ_i on each block i by $K_L V \Delta T$. The quantity ΔT determines the "tectonic" time scale which is taken to be $\Delta T = 1$ for the moment. We will return to the question of this time scale in Section 7. After the plate is moved the stress on each block is calculated via

$$\sigma_i(t) = \sum_j T_{ij} U_j(t) + K_L V \sum_n \Theta(n - t), \qquad (2)$$

and the process outlined above is repeated.

The step function mathematically expresses the Mohr-Coulomb friction law, which has the useful property that it is always scale invariant [*de Sousa Vieira and Herrmann*, 1994]. In this paper, we use the stochastic CA model [*Rundle et al*, 1995], in which $\eta_i(t)$ is a random noise equal to ρW, where ρ is a uniformly distributed random number $\in [0, 1]$ and the predetermined noise amplitude W is in the range $0 \leq W \leq 1$.

After block i slips, each neighbor j of the failed block receives an amount of stress equal to $T_{ij}(\sigma_i(t) - \sigma_i^R)/K$, while the system dissipates the amount $K_L(\sigma_i(t) - \sigma_i^R)/K$. This stress transfer may cause these neighbors to slip and so on, thus initiating an earthquake, or avalanche, which continues until every block has $\sigma_j < \sigma_j^F$. The size of an earthquake equals the number of slipped blocks after each plate update. We count each slip as a block failure even if a block slips more than once. In most cases of interest in this work multiple slips of a block during one event are rare. The count begins anew after each update of the loader plate. It is the properties of the earthquakes or avalanches in this model that we have studied theoretically and via simulations.

3. LONG-RANGE INTERACTIONS AND COARSE-GRAINING OF MODEL

In this work we will concentrate on the long range interaction version of the CA model. We do so for the following reasons. Linear elasticity theory yields long-range stress tensors for a variety of geophysical applications [*Rybicki*, 1986], including idealized viscoelastic faults. For a two-dimensional dislocation in a three-dimensional homogeneous elastic medium, the magnitude T of the static stress tensor goes as $T \sim 1/r^3$ [*Steketee*, 1958]. While geophysicists do not know the actual stress tensors for real faults, they expect that long-range stress tensors, which are similar to the $1/r^3$ interaction, apply to faults. It is suspected that microcracks in a fault, as well as other "defects" such as water, screen the $1/r^3$ interaction, leading to a proposed $e^{-\alpha r}/r^3$ interaction, where $\alpha \ll 1$, implying a slow decay to the long-range interaction over the fault's extent [*Rundle and Klein*, 1995].

In addition, a fault's interaction strength varies with and extends over a fault's depth, which is on the order of a kilometer or more. Because our CA slider-block models a macroscopic fault asperity, which has a linear dimension on the order of ten meters, the interaction range should span on the order of one hundred blocks in length. To approximately account for the screening and depth effects, we can truncate the $1/r^3$ interaction or any appropriate long-range interaction.

Moreover, recent observations [*Hill et al*, 1993; *Gomberg*, 1996] of seismic activity following some large magnitude ($\gtrsim 7$ on the Richter scale) events indicate long-range correlations of subsequent activity with the main quake, occurring within a few minutes after and extending up to 1200 km (much further than the typical aftershock zone) from the main shock. *Hill et al.* [1993] proposed models with long-range stress tensors, including $1/r^3$, to explain these spatial correlations.

In nature, the average earthquake stress drop $\Delta\sigma = \sigma^F - \sigma^R$ is approximately 0.01 to 1 MPa, which is small compared to the failure threshold ($\sigma^F \approx 10$ MPa) or the breaking strength of rock [*Scholz*, 1990]. Consequently, the ratio $\Delta\sigma/\sigma^F \approx 0.001$ to 0.1. For a fixed σ^F in the long-range CA model, $\Delta\sigma$ decreases as the interaction range increases because the mean interaction strength or inter-block spring constant decreases as the interaction range increases [*Rundle et al*, 1996]. To obtain a consistent ratio $\Delta\sigma/\sigma^F$ with geological faults, the CA model's interaction region must include several hundred blocks.

Finally, in addition to questions of interest in earthquake physics we want to investigate mean-field effects in driven dissipative systems which appear as the CA model becomes more long-range. In particular, our coarse-grained theory (developed below) indicates that a spinodal critical point influences earthquake scaling, the structure of an earthquake, and the amount of time required to form an earthquake during a tectonic plate update. In addition, earlier simulations [*Rundle et al*, 1995] and theoretical analysis [*Klein et al*, 1997] indicate that these systems are locally ergodic in the limit of infinite interaction range. The concept of locally ergodic will be addressed in section 6. We know that long-range systems [*Monette and Klein*, 1992; *Rundle and Klein*, 1989; *Gross et al*, 1994; *Rundle et al*, 1995] can display different physics than short-range systems. This model, due to the fact that it is driven and dissipative, is a particularly interesting one on which to study this phenomenon. We begin with the theoretical analysis.

In Ref. [*Klein et al*, 1997], we derived a coarse-grained theory for the CA model. In this paper, we present a more detailed description of this theory. We begin by rewriting Eqs. (2) and (1) to eliminate the position variable $U_j(t)$ and to develop a stress evolution equation. Multiply Eq. (1) by T_{ij}, sum over j, and use Eq. (2) to obtain

$$\begin{aligned}\sigma_i(t+1) - \sigma_i(t) = \\ \frac{1}{K}\sum_j T_{ij}\left(\sigma_j(t) - \sigma_j^R\right)\Theta\left(\sigma_j(t) - \sigma_j^F\right) \\ + \eta_j'(t) + K_L V, \qquad (3)\end{aligned}$$

where the noise $\eta_j'(t) = \sum_j T_{ij}\left(\sigma_j(t) - \sigma_j^F\right)\eta_j(t)$ remains spatially random as long as T_{ij} is a radially symmetric interaction, which it is assumed to be. Eq. (3) gives the new stress on block i at time $t+1$ in terms of the previous stresses from block i, the neighboring j blocks, the loader plate, and the noise. We have assumed that blocks fail only once per plate update, which will be discussed later in this section and in Section 4.

The $K_L V$ term was obtained by assuming that the stress loading at time $t+1$ in Eq.(2), i.e.

$$K_L V \sum_n \Theta(n - t - 1), \qquad (4)$$

minus the stress loading at time t was simply the difference between the two sums at $t+1$ and t. This point will require a much more careful discussion, and we will return to it in Section 7.

For long-range interactions, the number of neighbors $q \gg 1$. Each block at lattice site i interacts via springs with all blocks at lattice sites j contained within a square interaction region with area $(2R_c+1)^2$ so that $q = (2R_c+1)^2 - 1$. We consider two stress Green's functions: (1) $T_{ij} \sim NK_C/|\vec{i} - \vec{j}|^3$, which is truncated at an interaction range R_c (an infrared cutoff), where $|\vec{i} - \vec{j}|$ is the lattice distance between cells i and j, K_C is the effective spring constant due to block interactions, and N is a finite normalization constant that insures the condition $K_C = \sum_{j,i\neq j} T_{ij}$, and (2) $T_{ij} = 0$ for $|i-j| > R_c$, where i and j in $|i-j|$ are the x and y components of $\vec{i}$ and $\vec{j}$, and $T_{ij} = K_C/q$ for $|i-j| \leq R_c$ and $= 0$, otherwise. Note that these Green's functions are similar in that both weaken the nearest-neighbor interaction. We also assume a short range (ultraviolet) cutoff in T_{ij}, which arises due to the natural short range cutoff of geological and condensed matter systems, so that $T_{ij} \leq A < \infty \;\forall\; i$ and j. The limit $q \to \infty$ produces mean-field behavior, which is qualitatively different than the nearest-neighbor, $q = 4$, model. For both choices of the Green's function the effective interaction range, $R^2 = \sum_{j,i\neq j} |\vec{i} - \vec{j}|^2 T_{ij}$, grows with q as $R \sim q^{1/4}$ in $d=2$ for interaction (1) and as $R \sim q^{1/d}$ for interaction (2).

The only physics of interest in our CA model is on length scales greater than R. By developing a coarse-grained description [*Ma*, 1976] of Eq. (3), with a coarse graining size of R^2, we can sum over fluctuations on length scales smaller than R and retain the physics at larger length scales. We proceed therefore from a microscopic description of the stresses on the individual blocks to a macroscopic description of the stresses on a coarse-grained block in which the stresses become continuous classical variables or fields. In the remainder of this paper we will assume that σ_i^F and σ_i^R are spatial constants σ^F and σ^R. We will also use interaction (2) which we will call the equivalent neighbor model.

To accomplish the coarse-graining, we define a coarse-grained cell with a volume q centered at block i, specify a coarse-grained time τ, and we average the stress over both the coarse-grained volume and time. Therefore, $\sigma_i(t)$ becomes $\bar{\sigma}_i(\tau)$ on the left hand side of Eq. (3), where the bar denotes coarse-grained average.

On the right-hand-side of Eq. (3), we convert the summation over the individual j blocks covering a large region into a summation between interacting coarse-grained blocks restricted to local interactions. Because the first term in Eq. (3) is written as a discrete convolution, we can use Fourier transform techniques to convert Eq. (3) into a coarse-grained form. First, expand the Fourier transform of T_{ij} in a power series using the transform variable $|\vec{k}| = k$. Note that with the short and long distance cutoffs, we have a bounded function on a finite support and hence are assured that the power series in k exists. Second, we truncate the power series at k^2 in order to express the stress tensor as a local interaction between coarse-grained blocks and invert the Fourier transform to obtain for the first term in Eq. (3)

$$\begin{aligned}\sum_j T_{ij}\left(\sigma_j(t) - \sigma^R\right)\Theta\left(\sigma_j(t) - \sigma^F\right) \sim \\ -\, qK_C \sum_j \Delta_{ij}\left(\sigma_j(t) - \sigma^R\right)\Theta\left(\sigma_j(t) - \sigma^F\right) \\ -\, K_L\left(\sigma_i(t) - \sigma^R\right)\Theta\left(\sigma_i(t) - \sigma^F\right) \qquad (5)\end{aligned}$$

where $-K_L$ and qK_C are the zeroth and second moments of T_{ij}, respectively, Δ_{ij} is the matrix (discrete) representation of the Laplacian, and the sum preceding the Laplacian is over the coarse-grained blocks with a length scale set by the coarse graining scale $R \sim q^{1/2}$.

Next, we convert the discrete summation in Eq. (5) into a continuous integral which transforms σ_j into a continuous variable. Notice that the step function in Eq. (5) specifies that only blocks whose stress equals or exceeds σ^F will fail in a coarse-grained time interval τ and thus contribute to the summation. Consequently, we can compute the partial sum of only those blocks that fail within τ.

Before we compute the partial sum, we consider what happens to the time averaged stress on a block due to the inter-block springs in the mean-field limit. From the symmetry of the interaction, this stress becomes very small as $q \to \infty$. We also numerically verified this condition by measuring the mean stress on the blocks and comparing it to the mean stress solely due to the loader plate. Recall that the loader plate stress equals K_L times the mean distance between the actual positions of the blocks and the positions where the loader plate exerts zero force on the blocks. Within a coarse-graining time as q increases, the mean stress on the blocks approaches the mean loader plate stress on the blocks. Using standard mean-field arguments [*Gross et al*, 1994], it is expected that the stress fluctuations away from the mean value approach zero as $q^{-1/2}$.

The above discussion implies that the blocks are weakly interacting with each other in the mean-field limit. Thus, from the central limit theorem, within a coarse-grained volume on a time scale shorter than the coarse-grained time τ, the blocks' stress distribution will equilibrate to a Gaussian centered about $\bar{\sigma}(\vec{x}, \tau)$, where $\vec{x}$ labels the coarse-grained volume. In the next section, we discuss the numerical evidence that the stress distribution equilibrates to a Gaussian on time scales less than τ.

Also, within the coarse-grained time, only blocks with stress lying between σ_o and σ^F will fail, where $0 < \sigma_o < \sigma^F$, because the loader plate will add only enough stress to bring blocks with stress σ_o or greater to failure. Here σ_o is a parameter to be determined later. The above considerations lead to the following Gaussian representation of the partial sum

$$\frac{1}{q}\sum_j{}' \left(\sigma_j(t) - \sigma^R\right) \Theta\left(\sigma_j(t) - \sigma^F\right) = (\sigma^F - \sigma^R)\times \sqrt{\frac{\beta}{\pi}} \int_{\sigma_o}^{\sigma^F} d\sigma \exp\left[-\beta(\sigma - \bar{\sigma}(\vec{x}, \tau))^2\right] \quad (6)$$

where the prime on the summation means that the sum includes only the blocks that fail inside the coarse grained volume in a coarse grained time. The factor $\sigma^F - \sigma^R$ arises from the fact that if all q blocks failed then the left hand side of Eq.(6) is exactly $\sigma^F - \sigma^R$ in the $\beta \to \infty$ limit.

In developing this Gaussian approximation, we assume that a block fails at most once during a coarse grain time τ. These assumptions are correct in the $q \to \infty$ limit for $V < V_c = (\sigma^F - \sigma^R)/K$, as demonstrated in the next section. The velocity V_c is a good approximation to the spinodal value for the ranges of β we investigate in this work. Beyond this value of the velocity the low stress phase is unstable. A detailed discussion of this last point will be presented in a future publication [*Goldstein et al*, 2000]. In this work we will restrict ourselves to discussion of the low stress phase. However, we will return to this point in Section 7. In Eq. (6), β specifies the level of noise in the system and plays the role of an inverse temperature so that $\beta \gg 1$ implies low noise compared to $\beta \ll 1$ for high noise. The quantity β also determines the width of the Gaussian and the weight assigned to the "entropy" term, developed below. Because of β's dual roles, we assume that the Gaussian and "entropy" terms have the same noise dependence. Using different noise factors would not significantly change the derivation of the coarse-grained equation. (Because the noise determines the parameter β, it sets the scale for the width of the Gaussian as well as the weight given to the "entropy" term. Although we use only one β for these two functions, using different β's would not substantially alter our conclusions.) The quantity $\sqrt{\beta/\pi}$ approximates the normalization for $\beta \gg 1$, which follows because a large β produces a narrow Gaussian distribution and allows σ^F to be replaced by infinity in the upper integration limit, causing negligible error in the normalization. In this work we always assume $\beta \gg 1$ since small amplitude noise is what is expected on earthquake faults.

Finally, we derive an "entropy" term that counts the number of ways $N(\bar{\sigma}(\vec{x}, \tau))$ to distribute the available stress $q\bar{\sigma}(\vec{x}, \tau)$ among the q blocks within a coarse-grained volume, where the stress values range from σ^R to σ^F. Over the coarse-graining time, the stress inside a coarse-grained volume will tend to cluster around the mean value of this stress range, i.e., $\bar{\sigma} = (\sigma^F + \sigma^R)/2$. The entropy equals minus the logarithm of the distribution $N(\bar{\sigma}(\vec{x}, \tau))$. Since the long-range CA model behaves as an non-equilibrium thermodynamical systems near equilibrium [*Rundle et al*, 1995], the system's entropy can be expressed as the potential of a generalized force (see, e.g., *Nicolis and Prigogine*, [1977]), which gives the following contribution

$$
\begin{aligned}
\frac{\delta S(\bar{\sigma}(\vec{x},\tau))}{\delta\bar{\sigma}(\vec{x},\tau)} &= -\frac{\delta \ln N(\bar{\sigma}((\vec{x},\tau))}{\delta\bar{\sigma}(\vec{x},\tau)} \\
&= \frac{\beta^{-1}}{\sigma^F-\sigma^R}\ln\left[\frac{\bar{\sigma}(\vec{x},\tau)-\sigma^R}{\sigma^F-\bar{\sigma}(\vec{x},\tau)}\right] - \frac{\beta^{-1}}{\sigma^F-\sigma^R}\times \\
&\int_{\sigma^R}^{\sigma^F} d\sigma \ln\left[\frac{\sigma-\sigma^R}{\sigma^F-\sigma}\right]\exp\left[-\beta(\sigma-\bar{\sigma}(\vec{x},\tau))^2\right] \quad (7)
\end{aligned}
$$

to the $\bar{\sigma}(\vec{x},\tau)$ equation of motion.

To obtain the coarse-grained equation of motion, we combine Eqs. (3), (5), (6), and (7), take the temporal and spatial continuum limits, and obtain

$$
\begin{aligned}
\frac{\partial\bar{\sigma}(\vec{x},\tau)}{\partial\tau} &= K_L V + \bar{\eta}(\vec{x},\tau) + \frac{[qK_C\nabla^2 - K_L]}{K}\times \\
&\frac{(\sigma^F-\sigma^R)}{2}\Bigg[\operatorname{erf}\left[-\sqrt{\beta}(\sigma^F-\bar{\sigma}(\vec{x},\tau))\right] \\
&- \operatorname{erf}\left[\sqrt{\beta}(\sigma_o-\bar{\sigma}(\vec{x},\tau))\right]\Bigg] - \frac{\beta^{-1}}{\sigma^F-\sigma^R}\times \\
&\Bigg[\ln\left[\frac{\bar{\sigma}(\vec{x},\tau)-\sigma^R}{\sigma^F-\bar{\sigma}(\vec{x},\tau)}\right] - \left[\frac{\beta}{\pi}\right]^2\times \\
&\int_{\sigma^R}^{\sigma^F} d\sigma \ln\left[\frac{\sigma-\sigma^R}{\sigma^F-\sigma}\right]\exp\left[-\beta(\sigma-\bar{\sigma}(\vec{x},\tau))^2\right]\Bigg] \quad (8)
\end{aligned}
$$

where $\operatorname{erf}(z)$ is the error function, which came from the Gaussian in Eq. (6), and $\bar{\eta}(\vec{x},\tau)$ is the coarse-grained noise. We can obtain the equation for the time independent spatially homogeneous solutions to Eq. (8) by setting the noise and the derivatives to zero to obtain

$$
\begin{aligned}
K_L V &= \frac{K_L}{K}\frac{(\sigma^F-\sigma^R)}{2}\Bigg[\operatorname{erf}\left[\sqrt{\beta}(\sigma^F-\bar{\sigma})\right] \\
&- \operatorname{erf}\left[\sqrt{\beta}(\sigma_o-\bar{\sigma})\right]\Bigg] + \frac{\beta^{-1}}{\sigma^F-\sigma^R}\times \\
&\Bigg[\ln\left[\frac{\bar{\sigma}-\sigma^R}{\sigma^F-\bar{\sigma}}\right] - \left[\frac{\beta}{\pi}\right]^{1/2}\times \\
&\int_{\sigma^R}^{\sigma^F} d\sigma \ln\left[\frac{\sigma-\sigma^R}{\sigma^F-\sigma}\right]\exp\left[-\beta(\sigma-\bar{\sigma})^2\right]\Bigg]. \quad (9)
\end{aligned}
$$

The right-hand-side of Eq. (9) represents the rate of stress dissipation over the coarse-grained time and system volume, and the left-hand-side of Eq. (9) equals the rate of stress input into the system.

The initial conditions specify all the parameters except σ_o. We can determine σ_o by considering the $q \to \infty$ limit in which blocks within the interaction range are essentially non-interacting. This non-interacting effect occurs because if every block interacts with all other blocks there can be no spatial scale for fluctuations so that the interactions can be combined into an effective or mean-field. By assuming that every block interacts with all other blocks, we can still calculate mean-field thermodynamics [*Ma*, 1976], and we have that the spatial and temporal average of $\bar{\sigma}(\vec{x},\tau)$ must be $\bar{\sigma} = (\sigma^F + \sigma^R)/2$. Substituting $\bar{\sigma}$ into Eq. (9) and noting that the "entropy" term equals zero for this $\bar{\sigma}$, we get the following equation

$$
\begin{aligned}
\frac{\sigma^F-\sigma^R}{2K}&\Bigg[\operatorname{erf}\left[\sqrt{\beta}\frac{\sigma^F-\sigma^R}{2}\right] \\
&- \operatorname{erf}\left[\sqrt{\beta}(\sigma_o - \frac{\sigma^F+\sigma^R}{2})\right]\Bigg] = V, \quad (10)
\end{aligned}
$$

which we can solve for σ_o.

Now that we have derived the coarse-grained equation we can obtain the physics of the CA model after specifying the four parameters $\beta, K/K_L, VK_L$, and $(\sigma^F-\sigma^R)/K$, which determine the behavior of the equation and the model. Using the values of these parameters representative of those for real earthquake faults, such that $\Delta\sigma/\sigma^F \approx 0.001$ to 0.1 and $V \sim 0$, we numerically solve the time independent spatially homogeneous equation Eq. (9) to illustrate some important aspects of the theory. As noted in Section 2, a typical CA slider-block dissipates an amount of stress $\Delta\sigma = (K_L/K)(\sigma^F - \sigma^R)$. We choose values of K_L, K, σ^F, and σ^R that give $\Delta\sigma/\sigma^F$ values consistent with real faults. In Figure 1 $K_L = 1, K = 100, \sigma^F = 50$, and $\sigma^R = 10$ so that $\Delta\sigma/\sigma^F = 8 \times 10^{-3}$, and in Figure 2 we increase K to 1000 so that $\Delta\sigma/\sigma^F = 8 \times 10^{-4}$. For both figures we use $\beta = 5$, assuming that noise does not strongly influence the fault, and we set $V = 0.005$ to slowly drive the system. To obtain these curves, we also need to compute σ_o, which we numerically calculated from Eq. (10). In Figs. 1 and 2 the solid curve comes from the right-hand-side of Eq. (9) and equals the mean rate of stress dissipation, while the dashed straight line comes from the left-hand-side of Eq. (9) and equals the mean rate of stress input. Therefore, the intersection of the two curves determines the spatially and temporally averaged stress $\bar{\sigma}$ at which the rates of stress dissipation and input balance each other.

Analyzing the solutions of Eq.(8), we make the following observations. In Figs. 1 and 2 the intersection of the horizontal line representing the mean velocity V and the curve representing the right hand side of Eq.(8)

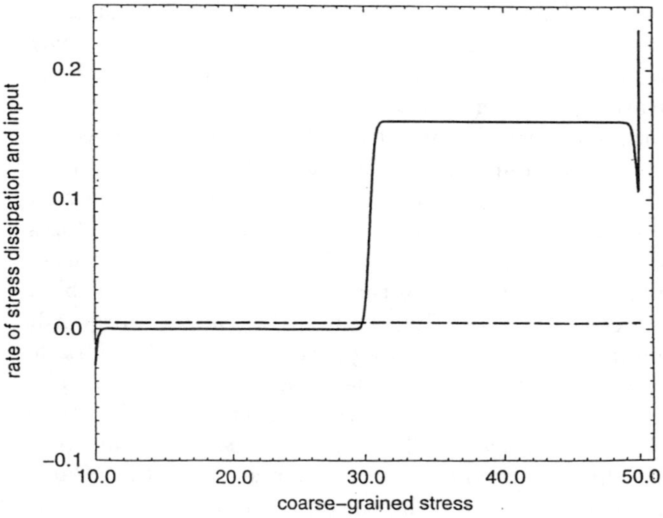

Figure 1. Solution to the time independent spatially homogeneous Eq. (9), using parameters: $K = 100, \sigma^F = 50, \sigma^R = 10, \beta = 5$, and $V = 0.005$. The solid line represents the rate of stress dissipation from the right-hand-side of Eq. (9), and the dashed line represents the rate of stress input from the left-hand-side of Eq. (9).

occurs in the low stress regime. In simulations of the CA model with approximately the same parameters, $\overline{\sigma}$ is approximately equal to the values in these figures obtained from the coarse grained equation (10). Because the steady-state solutions in Figs. 1 and 2 come from a mean-field theory we do not expect these solutions to agree exactly with the simulations, which are limited to finite, but long, range interactions. What happens if either $\Delta\sigma$ is decreased or V is increased, or both. The curves would then intersect at two additional positive values near $\overline{\sigma} = 50$. Although both of these intersection points belong to high stress regimes, the left intersection (lower stress) is unstable, and the right one (higher stress) is meta-stable. In stress space, a very small distance separates these meta-stable and unstable states. Observe also in the figures that the well depth decreases as K increases, thus decreasing the range of driving velocities over which the system can explore the higher stress states. For even higher values of V the high stress phase is stable while the low stress phase is meta-stable. If V is set higher still the low stress phase ceases to exist.

Even though the top of the rate of stress dissipation curve appears flat, it is slightly curved with a curvature that depends on β. For smaller β the top becomes more curved. In contrast, for larger β, the curve steepens and approaches a step function, and the well becomes deeper and tends toward the half-way point between the top and bottom of the curve. As discussed in Ref. [*Klein et al*, 1997], the top of the curve and the bottom of the well delineate spinodal critical points, which separate meta-stable and unstable states. By increasing K we can bring the spinodal critical points closer to the rate of stress input line. Consequently, as the intersection point nears the spinodal critical points, the system is more likely to experience critical behavior such as scaling induced by the spinodal. In Section 5, we present simulation results that demonstrate the effects of the spinodal.

4. COARSE-GRAINING ASSUMPTIONS

In this section we present simulation results relevant to clarifying the coarse-graining assumptions used in the previous section. The three assumptions we examine are: 1) The number of times a block fails per coarse graining time is one. 2) The stress at which the blocks fail is σ^F and not greater. 3) The time-averaged stress distribution within a coarse-grained volume is a Gaussian. These assumptions are all made in the mean-field ($q \to \infty$) limit.

The assumption that a block should fail no more than once during a plate update as $q \to \infty$ for $V < (\sigma^F - \sigma^R)/K$ (the spinodal velocity) is tested in Fig-

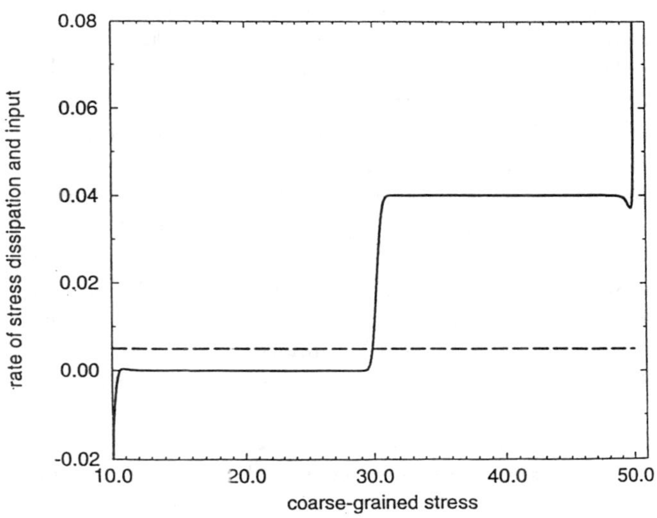

Figure 2. Solution to the time independent spatially homogeneous Eq. (9), using the same parameters as Fig. 1, except $K = 1000$. The solid line represents the rate of stress dissipation from the right-hand-side of Eq. (9), and the dashed line represents the rate of stress input from the left-hand-side of Eq. (9).

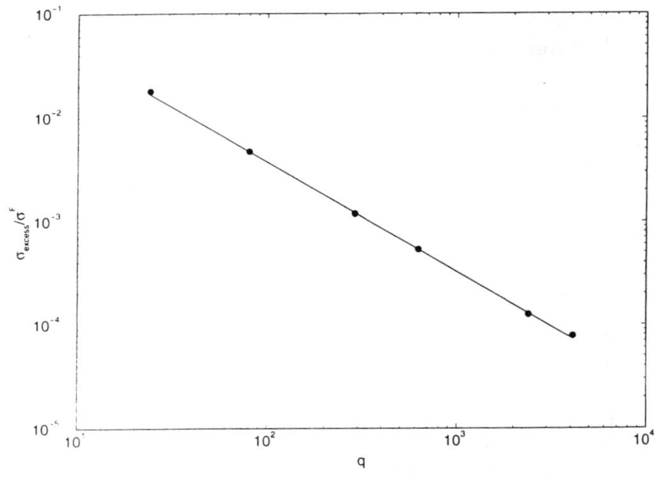

Figure 3. The number of multiple failures normalized by the total number of failures versus q, where the number of plate updates $n = 35000$, using parameters: $K_L = 1, K_C = 200, \sigma^F = 50, V \sim 0, 256 \times 256$ system size, $W = 0.3$ and the K_C/q interaction.

ure 3. We show a log-log plot of the number of multiple failures normalized by the total number of failures. Since the value of K used in the simulations is large we test the assumption for $V \sim 0$ to satisfy $V < V_c$. The data show that this ratio decreases as q^{-2}. Since a coarse-graining time is greater than a plate update the simulation results validate the assumption that the multiple failures can be ignored as q becomes large.

We also assumed that a block's stress before failure does not exceed σ^F in the $q \to \infty$ limit for $V < (\sigma^F - \sigma^R)/K$. Figure 4 shows that as q becomes large, the time-averaged value of the failure stress that exceeds σ^F decreases almost linearly with q. Thus, the simulations again validate the assumption. Note that for the first two assumptions tested with data shown in Figs. 3 and 4 we used a very large value of $K_C = 200$. Because this value typically produces a great number of multiple failures and high values of failure stress relative to σ^F for small q, these data signify a stringent test of the theory.

Next we present evidence that the stress within a coarse-grained volume becomes Gaussian distributed within a short time scale compared to the time required for all the blocks within that volume to fail. Figure 5 shows binned temporally averaged stress within a coarse-grained volume of size 256×256. The sequence of plots depicts the evolution of this time-averaged stress from $\approx 20\%$ to 100% of the blocks failing (the caption shows the exact percentages). Before computing the temporal average of each block's stress, the blocks failed several times each to allow the system to move away from the transient state due to the initial conditions. Notice that after about $\approx 20\%$ failures the stress distribution clearly deviates from a flat distribution, which is the distribution after a single plate update. After $\approx 40\%$ failures the stress distribution appears Gaussian distributed, and it becomes more so as more failures occur. The solid curves in Fig. 5 come from Gaussian fits to the data, where the results are listed in Table . To compute these fits, we used Levenberg-Marquardt method of nonlinear fitting [*Press et al*, 1992]. In Table observe that the amplitude A increases while the mean $\bar{x}$ and the standard deviation σ decrease as the stress distributions evolve. The σ and $\tilde{\chi}^2$ decrease means that the fits are becoming narrower and closer to a Gaussian distribution. The mean $\bar{x}$ approaches the value predicted by the steady-state solution to the mean-field theory. These fits show that the stress distribution equilibrates to a Gaussian on a time scale shorter than the time required for all blocks to fail.

5. THEORETICAL PREDICTIONS AND SIMULATION RESULTS

In this section we develop scaling relations relevant to the long-range CA model and present simulation results that are consistent with these relations. Our basic premise is that the scaling relations that are seen in the model are caused by the spinodal critical point. However, unlike most attempts to associate the scaling of the earthquake events in slider block or avalanche

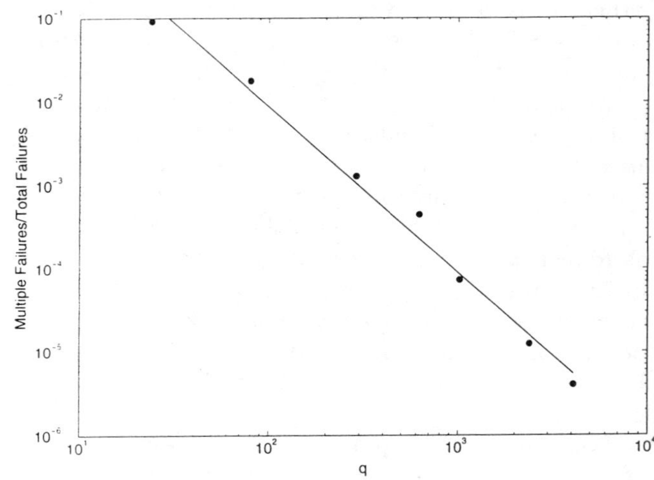

Figure 4. Time-averaged excess stress normalized by σ^F versus q, using the same system parameters as Fig. 3.

models with critical fluctuations [*Bak and Tang*, 1989; *Feder and Feder*, 1991; *Olami et al*, 1992], we associate the large earthquake events with arrested spinodal nucleation droplets and the small events with critical phenomena fluctuations. In addition there are system size events which are not in the scaling regime. We will return to this point at the end of this section. In order to explain our point we will briefly review the spinodal nucleation idea and explain the concept of arrested nucleation which we will return to in Section 7.

Spinodal nucleation is concerned with nucleation near the spinodal critical point in systems with long-range (but not infinite) interactions [*Heermann et al*, 1982]. The reason that long range is essential is that spinodal effects are not seen in models with short range interactions, as will be seen below. However, as a practical matter, in models with long range interactions, such as the CA model we are considering, spinodal nucleation is the dominant event. The classical events will be strongly suppressed as we will see below [*Unger and Klein*, 1984].

We begin our discussion of spinodal nucleation with the partition function. This would be the functional integral of the right hand side of Eq.(8). To simplify the discussion we will use a simple "ϕ^4" theory. Since we are only interested in the scaling dependence of various physical quantities this model will suffice. The partition function Z is

$$Z = \int \delta\phi \exp(-\beta f(\phi)) \tag{11}$$

where

$$f(\phi) = \int d\,\vec{x} \left[\frac{R^2}{2}(\nabla\phi(\vec{x}))^2 + \epsilon\phi^2(\vec{x}) + \phi^4(\vec{x}) - h\phi(\vec{x})\right]. \tag{12}$$

Here ϵ is $T - T_c$, T_c is the critical temperature, h can be thought of as an applied magnetic field or a chemical potential, and $f(\phi)$ is the Ginzburg-Landau free energy functional. We can scale all lengths with the interaction range R and in the $R \gg 1$ limit obtain the free energy per unit volume, $F = -\lim_{V\to\infty} \ln Z/(\beta V)$, where V is the volume of the system, from a saddle point evaluation of Eq.(11). We will use the relation [*Gunton et al*, 1983]

$$\frac{\partial\phi(\vec{x},t)}{\partial t} = -M\frac{\delta f(\phi)}{\delta\phi}. \tag{13}$$

Here M is a mobility which we will take to be a constant and we now assume a time dependence for $\phi(\vec{x},t)$ given by the Langevin equation 13. One could add a noise to

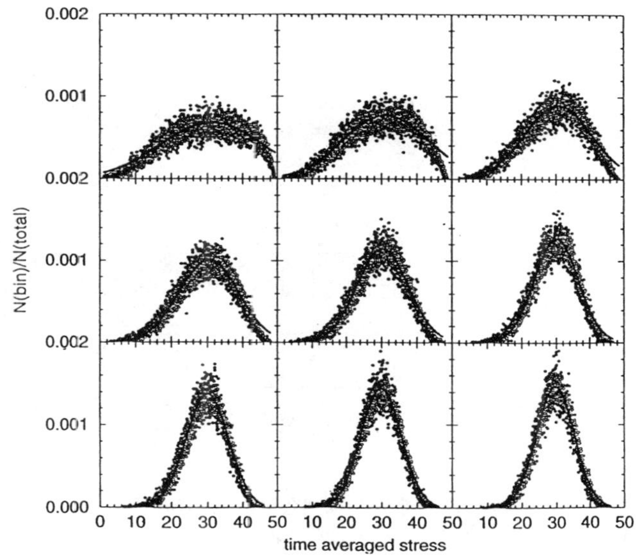

Figure 5. Histograms of the time averaged stress in a coarse-grained volume containing 256 × 256 blocks with bin size equal to 0.02 in units of stress, showing the temporal evolution of the stress in the system, where the earliest to latest times go from left to right and top to bottom. These plots correspond to an increasing number of cumulative failures in terms of the percentage of the total number of blocks that have failed, where the percentages are 20.4, 33.0, 43.2, 55.5, 65.3, 76.8, 86.3, 95.6, and 105.8. The data were collected only after each block has failed several times. The binned simulation data are represented by circles, while the solid curve is a Gaussian fit to the data. The other system parameters are $\sigma^F = 50$, $K_L = 1$, $V \sim 0$, $W = 0.3$, and $K_C = 1$.

the right hand side of Eq.(13) but it will be irrelevant for our considerations. From Eqs. (11) and (13), and scaling all lengths with R, we obtain

$$\frac{\partial\phi(\vec{x},t)}{\partial t} = -MR^d(-\nabla^2\phi(\vec{x},t) - 2|\epsilon|\phi(\vec{x},t) + 4\phi^3(\vec{x},t) - h), \tag{14}$$

where we have taken $T < T_c$. The equilibrium values of the order parameter $\phi(\vec{x},t)$ can be obtained from Eq.(14) by setting all derivatives equal to zero and assuming a solution that is a temporal and spatial constant or, equivalently, from Eq. (11) in the $R \to \infty$ limit. The free energy per unit volume f is then

$$f = -|\epsilon|\phi^2 + \phi^4 - h\phi. \tag{15}$$

The minima of the free energy corresponding to phases are then the solutions to

$$-2|\epsilon|\phi + 4\phi^3 - h = 0. \tag{16}$$

Table 1. Gaussian fit parameters corresponding to the curves in Fig. 5, where the fitting form is $y = A\exp\left[-((x-\bar{x})/\sigma)^2\right]$. The two leftmost columns contain the percentage of the number of failed blocks compared to the total number of blocks in the coarse-grained volume and the cumulative number of failures that determined the % Failure column. The rightmost column lists the reduced Chi Squared of the fit.

% Failure	slips	$A[\times 10^{-3}]$	$\bar{x}$	σ	$\tilde{\chi}^2[\times 10^{-5}]$
20.4	13,349	0.678 ± 0.007	31.526 ± 0.233	20.455 ± 0.479	2.747
33.0	21,624	0.768 ± 0.007	31.271 ± 0.171	17.181 ± 0.334	2.712
43.2	28,281	0.894 ± 0.008	30.909 ± 0.118	13.826 ± 0.215	2.295
55.5	36,357	1.014 ± 0.008	30.573 ± 0.089	11.867 ± 0.160	2.116
65.3	42,795	1.147 ± 0.008	30.373 ± 0.069	10.313 ± 0.121	2.163
76.8	50,355	1.272 ± 0.008	30.225 ± 0.055	9.210 ± 0.097	2.013
86.3	56,557	1.375 ± 0.008	30.063 ± 0.047	8.485 ± 0.083	1.788
95.6	62,673	1.440 ± 0.008	29.911 ± 0.043	8.094 ± 0.075	1.529
105.8	69,365	1.473 ± 0.008	29.826 ± 0.041	7.920 ± 0.072	1.377

As is easily seen, for $\epsilon < 0$ there are two minima: $\phi_+ > 0$ and $\phi_- < 0$. For $h > 0$ the ϕ_+ minimum has a lower free energy per unit volume and represents the stable phase. The minimum ϕ_-, with a negative value of ϕ, represents the meta-stable phase and has a higher value of the free energy. As h is increased the meta-stable minimum becomes shallower, and finally at $h = h_s$ the meta-stable minimum disappears. This value of h is referred to as the spinodal value of the field and $\phi_- = \phi_s$, the value of the meta-stable minimum when $h = h_s$, is the spinodal value of the order parameter ϕ. We are interested in the mechanism of nucleation out of the meta-stable well for $\Delta h = h_s - h \ll 1$, $R \gg 1$, $\epsilon < 0$ and fixed, and hence $\phi \sim \phi_s$. This problem has been addressed both theoretically [*Unger and Klein*, 1984; *Lee et al*, 1995; *Klein and Unger*, 1983] and via simulations [*Monette and Klein*, 1992; *Ray and Klein*, 1990; *Heermann and Klein*, 1983a, 1983b].

The initial step in the theoretical treatment is to find the spinodal values of ϕ and h (ϕ_s and h_s). These values mark the place where the meta-stable well vanishes. Hence, ϕ_s and h_s must be solutions of Eq. (16) and

$$-2|\epsilon| + 12\phi^2 = 0, \tag{17}$$

since ϕ_s is an inflection point. We now define a new field, $\psi(\vec{x},t) = \phi(\vec{x},t) - \phi_s$, and rewrite Eq. (14)

$$\begin{aligned}\frac{\partial\psi(\vec{x},t)}{\partial t} &= -MR^d(-\nabla^2\psi(\vec{x},t) - 12\left[\frac{|\epsilon|}{6}\right]^{1/2}\psi^2(\vec{x},t) \\ &\quad + 4\psi^3(\vec{x},t) + \Delta h),\end{aligned} \tag{18}$$

where we have used Eqs. (16) and (17) and the fact that ϕ_s is a solution of Eq. (16) when $h = h_s$. We now assume, and will demonstrate below, that $\psi(\vec{x},t) \ll 1$ so that we can ignore $\psi^3(\vec{x},t)$ relative to $\psi^2(\vec{x},t)$. We stress that $|\epsilon| > 0$ and does not approach zero. It is Δh that approaches zero as we approach the spinodal in this analysis. The fundamental equation for nucleation near the spinodal is then

$$\frac{\partial\psi(\vec{x},t)}{\partial t} = -MR^d\left(-\nabla^2\psi(\vec{x},t) + \Delta h - C\psi^2(\vec{x},t)\right), \tag{19}$$

where we have set $C = 12|\epsilon|^{1/2}/6^{1/2}$.

The nucleation or critical droplet is assumed to be an equilibrium fluctuation (see [*Unger and Klein*, 1984; *Gunton et al*, 1983; *Cahn and Hilliard*, 1958, 1959; *Langer*, 1967]) that lies on the boundary between those fluctuations which lower their free energy by growing and those that lower their free energy by shrinking. Thus, the nucleating droplet is a stationary point of the free-energy functional and therefore should be a solution to the time independent form of Eq. (19)

$$-\nabla^2\bar{\psi}(\vec{x}) + \Delta h - C\bar{\psi}^2(\vec{x}) = 0. \tag{20}$$

To understand how the droplet scales we don't have to solve Eq. (20). We need only note that

$$\bar{\psi}(\vec{x}) \sim (\Delta h)^{1/2}\tilde{\psi}(\vec{x}(\Delta h)^{1/4}), \tag{21}$$

where $\tilde{\psi}(\vec{x}(\Delta h)^{1/4})$ is of order 1 and a solution to

$$-\nabla^2\tilde{\psi}(\vec{x}) + 1 - C\tilde{\psi}^2(\vec{x}) = 0. \tag{22}$$

In order to see that this is indeed a critical droplet we can specialize to one dimension (d=1) [*Klein and Unger*, 1983]. In one dimension Eq. (20) reduces to

$$-\frac{d^2\bar{\psi}(x)}{dx^2} + \Delta h - C\bar{\psi}^2(x) = 0. \tag{23}$$

There are two boundary conditions required by the equation and those are fixed by physical considerations associated with the idea of an isolated droplet. First, as $x \to \infty$ we expect $\psi(x)$ to approach its value in the meta-stable state, ψ_{ms}. This can easily be seen from Eq. (18), once all derivatives are set equal to zero, to be $\psi_{ms} = -[\Delta h/C]^{1/2}$. The second boundary condition is that the droplet should be smooth so that we expect $\frac{d\,\psi(x)}{d\,x} = 0$ at $x = 0$. With these conditions the solution of Eq. (23) is

$$\bar{\psi}(x) = -\left[\frac{\Delta h}{C}\right]^{1/2} + 3\left[\frac{\Delta h}{C}\right]^{1/2} \frac{1}{\cosh^2\left(\frac{C^{1/4}\Delta h^{1/4}}{\sqrt{2}}x\right)}, \tag{24}$$

which has all of the scaling properties derived above. We should also note that for $\Delta h \neq 0$ there is a well defined length scale for nucleation, $\xi \sim \Delta h^{-1/4}$, or in scaled units $\xi \sim R\Delta h^{-1/4}$, which is also a measure of the correlation length.

We now test the stability of this solution to Eq. (23) by performing a linear stability analysis. We insert

$$\psi(x,t) = \bar{\psi}(x) + \eta(x,t), \tag{25}$$

where $\bar{\psi}(x)$ is given in Eq.(24), into

$$\frac{\partial\psi(x,t)}{\partial t} = -MR^d\left[-\frac{\partial^2\psi(x,t)}{\partial x^2} + \Delta h - C\psi^2(x,t)\right] \tag{26}$$

and linearize with respect to $\eta(x,t)$, which is assumed to be a small perturbation. We obtain

$$\frac{\partial\eta(x,t)}{\partial t} = -MR^d\left[\frac{\partial^2\eta(x,t)}{\partial x^2} + 2C\left(\left[\frac{\Delta h}{C}\right]^{1/2} - 3\left[\frac{\Delta h}{C}\right]^{1/2} \frac{1}{\cosh^2\left(\frac{C^{1/4}\Delta h^{1/4}}{\sqrt{2}}x\right)}\right)\eta(x,t)\right]. \tag{27}$$

This equation has solutions of the form $\eta(x,t) = e^{-\lambda_i t}\eta_i(x)$ where the λ_i are the eigenvalues associated with the eigenvectors $\eta_i(x)$ of the operator

$$\lambda_i\eta_i(x) = -MR^d\left[-\frac{d^2\eta_i(x)}{dx^2} + 2C\left(\left[\frac{\Delta h}{C}\right]^{1/2} - 3\left[\frac{\Delta h}{C}\right]^{1/2} \frac{1}{\cosh^2\left(\frac{C^{1/4}\Delta h^{1/4}}{\sqrt{2}}x\right)}\right)\eta_i(x)\right]. \tag{28}$$

Equation (28) has the form of a time independent Schrödinger equation with a shallow (since $\Delta h \ll 1$) well. One expects that there will be both positive and negative eigenvalues corresponding to bound and free states. This is indeed the case. One can show [*Unger and Klein*, 1984; *Klein and Unger*, 1983] that there is one negative eigenvalue and a continuum of positive eigenvalues. The eigenvector with negative eigenvalue, $\eta_1(x)$, is then unstable since its amplitude grows exponentially with time. One can easily see [*Unger and Klein*, 1984] that the eigenvector with negative eigenvalue has the form

$$\eta_1(x) \sim \frac{1}{\cosh^3\left(\frac{C^{1/4}\Delta h^{1/4}}{\sqrt{2}}x\right)}. \tag{29}$$

It is also easy to see that the negative eigenvalue $-\lambda_1$, where $\lambda_1 > 0$, scales as $\Delta h^{1/2}$. This implies that the droplet whose profile $\psi(x)$ is given in Eq.(24) is a saddle point solution to Eq. (23) and that the system can move away from the meta-stable state via the growth of this saddle point droplet. Moreover, the scaling of λ_1 with $\Delta h^{1/2}$ implies that the growth of the droplet slows as $\Delta h \to 0$. That is the characteristic growth time t_g scales as $\Delta h^{-1/2}$. Before discussing the simulation test of this prediction there are other quantities that we can predict from the theoretical treatment of the CA model which can also be tested via simulation. In order to obtain these predictions it is necessary to go a bit deeper into nucleation theory.

The main contribution to the partition function for the meta-stable state comes from integrating in the neighborhood of the meta-stable minimum ($h > 0$ and $\phi_- < 0$). However, as stated above, one of the main assumptions of nucleation theory is that the critical, or nucleating, droplet can be viewed as an equilibrium fluctuation that samples the configurations that are close to the saddle point separating the meta-stable minimum from the stable one ($h > 0$ and $\phi_+ > 0$) [*Langer*, 1967]. As we approach the spinodal point, the saddle point fluctuations and the fluctuations close to the meta-stable minimum are both important. Including these two contributions to the partition function we get

$$Z \simeq Z_0 + Z_1 = Z_0\left(1 + \frac{Z_1}{Z_0}\right) \simeq Z_0 \exp\left(Z_1/Z_0\right), \tag{30}$$

where Z_0 comes from the integration near the meta-stable minimum, and Z_1 is due to the integrations near the saddle point. Including contributions to Z from configurations close to the saddle point we integrate along one direction in which the integrand increases as we move away from the saddle point. If we perform a simple Gaussian approximation this direction will lead to a diverging result. We remove this divergence by an analytic continuation of the Gaussian integral [*Langer*, 1967; *Unger and Klein*, 1984]. Upon removing the spurious divergence by deforming the contour of the integral, we find that the contribution to Z from the region near the saddle point becomes complex. The imaginary part of the meta-stable free energy is related to the lifetime, τ, of the meta-stable state

$$\begin{aligned} I_n &= \tau^{-1} = \lim_{V\to\infty} \left(\frac{Im(\ln Z)}{V}\right) f_d \\ &= \lim_{V\to\infty} \left(\frac{Im(Z_1/Z_0)}{V}\right) f_d, \end{aligned} \tag{31}$$

where f_d is a kinetic prefactor that describes the initial exponential growth rate of the unstable mode at the saddle point. The probability of finding a critical droplet is inversely proportional to the average time time, τ, spent in the meta-stable state before one appears. Implicitly, we assume that the critical droplets are noninteracting and that there is roughly a volume V inside which there is at most one droplet with probability τ^{-1}. Therefore, τ^{-1} measures the nucleation rate, I_n, which is the number of critical droplets per unit volume per unit time.

The coefficient in front of the free-energy functional is proportional to R^d. Therefore, for large R, the partition function can be approximated by the steepest-descent integral in which we performing quadratic expansions of $f(\phi)$ in the neighborhood of the two stationary points and evaluate the resulting Gaussian integrals. For the meta-stable minimum we find

$$Z_0 = \exp\left(-\beta f(\phi_-)\right) \prod_i \left(\frac{\pi}{\lambda_i^{(0)}}\right)^{1/2}, \tag{32}$$

where the $\lambda_i^{(0)}$ are the eigenvalues of the free-particle Schrödinger equation

$$\left(-\nabla^2 - 2|\epsilon| + 12\phi_-^2\right) \eta_i^{(0)}(\vec{x}) = \lambda_i^{(0)} \eta_i^{(0)}(\vec{x}), \tag{33}$$

or, equivalently, are the curvatures of the free energy functional $f(\phi)$ in the directions of the principal-axis coordinates with their origin at the meta-stable point ϕ_-. The spectrum has a continuum of states with $\lambda_i^{(0)} > 0$. The saddle point integration gives a similar result

$$Z_1 = \exp\left(-\beta f(\bar{\psi})\right) \prod_j \left(\frac{\pi}{\lambda_j}\right)^{1/2}, \tag{34}$$

where λ_j are the eigenvalues of Eq. (28), which is equivalent to the previous Schrödinger equation with a potential described by the critical droplet profile $\bar{\psi}(\vec{x})$ — in $1d$ this is given by Eq. (24).

Thus, we can write

$$\frac{Z_1}{Z_0} = \exp(-\beta\Delta f) \prod_j \left(\frac{\pi}{\lambda_j}\right)^{1/2} / \prod_i \left(\frac{\pi}{\lambda_i^{(0)}}\right)^{1/2} \tag{35}$$

where

$$\Delta f = f(\bar{\psi}) - f(\phi_-) \tag{36}$$

is the free-energy cost of the nucleating droplet. Inserting the critical droplet profile $\bar{\psi}(\vec{x})$ — which in $1d$ is described by Eq. (24) — into the free energy functional we obtain for the free energy cost of the droplet

$$\Delta f = -BR^d \Delta h^{3/2-d/4}, \tag{37}$$

where B is a constant independent of Δh.

The final result for the nucleation rate can be written in the standard form,

$$I_n = I_0 \exp(-\beta\Delta f), \tag{38}$$

in which the fundamental fluctuation rate I_0, whose expression is given by

$$I_0 = \lim_{V\to\infty} \frac{f_d}{V} Im \left(\prod_j \left(\frac{\pi}{\lambda_j}\right)^{1/2} / \prod_i \left(\frac{\pi}{\lambda_i^{(0)}}\right)^{1/2} \right), \tag{39}$$

has two parts [*Gunton et al*, 1983; *Langer*, 1969]. We recognize besides the kinetic prefactor f_d a statistical prefactor that comes from Gaussian fluctuations associated with the steepest descent part of the saddle point evaluation of the partition function [*Unger and Klein*, 1984; *Lee et al*, 1995; *Langer*, 1967].

The statistical prefactor has three components. First, we note that the attractive potential in Eq. (28) creates one bound state with a negative eigenvalue, d independent translational modes of that state with 0 eigenvalue, and a continuum of positive eigenvalues. As we already know, the negative eigenvalue λ_1 scales as $\Delta h^{1/2}$, and can be simply inserted into the square root; this is justified through the analytic continuation argument. Its contribution to I_0 is proportional to $i(\pi/\lambda_1)^{1/2} \simeq i\Delta h^{-1/4}$.

The d translational modes must be handled separately since the integrand does not decay as we move away from the stationary point, and thus we cannot perturb around that point. Their contribution is equal to the volume of the subspace in the ϕ space which is spanned by the set of critical droplet configurations that leave $f(\phi)$ invariant. We get a contribution to I_0 [*Unger and Klein*, 1984] which is proportional to $\Delta h^{3d/8}V$, where V is the volume of the system and it simply reflects the fact that the nucleating droplet can appear anywhere inside the system. This will cancel out the V dependence of the nucleation rate and will make the nucleation rate an intensive quantity as it should be. As a remark, the translational modes are the 0 energy Goldstone modes generated by having the critical droplet spontaneously breaking the translational symmetry of the uniform meta-stable state.

We are now left in evaluating I_0 with just the continuum of states with $\lambda_i^{(0)} > 0$ and $\lambda_j > 0$ which can be handled by a density functional [*Langer*, 1967]. The result is proportional to $(\Delta h^{1/2}/\pi)^{(1+d)/2}$. Putting all these contributions together, the form of the static part to the prefactor is $\xi^{-d} \sim R^{-d}\Delta h^{d/4}$.

The kinetic part of the prefactor depends on the form of the dynamics driving the system. In particular it depends on the conservation laws and how conserved quantities couple to the variable $\psi(x)$. In the Langevin dynamics we are discussing in this section, as well as in the dynamics for the earthquake model, the order parameter is not conserved and the kinetic prefactor scales as [*Unger and Klein*, 1984; *Langer*, 1967]:

$$f_d \sim \lambda_1/\pi. \tag{40}$$

Combining these factors we obtain the probability per unit volume and per unit time of a critical or nucleating droplet, $I_n(\Delta h)$, to be

$$I_n(\Delta h) = \frac{A\Delta h^{1/2}\exp\left[-B\beta R^d\Delta h^{3/2-d/4}\right]}{\xi^d}. \tag{41}$$

The constants A and B are independent [*Klein and Unger*, 1983] of Δh and are not relevant to our discussion. The factor $\Delta h^{-1/2}$ sets the time scale. Since the droplets are assumed to be independent of each other then $I_n(\Delta h)$ is proportional to the number of droplets per unit volume. From Eq. (41) it can be seen that for a fixed Δh, $I_n(\Delta h)$ is exponentially damped in R. In the mean-field limit, $R \to \infty$, nucleation is completely suppressed. However, for fixed R, which is more in line with simulations, the exponential part of $I_n(\Delta h)$, usually known as the nucleation barrier, can be made smaller by decreasing Δh. Due to the exponential dependence on Δh the probability of a critical droplet remains quite small until the argument of the exponential becomes of order 2-3. This is usually referred to in the literature as the Becker-Döring limit or limit of meta-stability [*Gunton et al*, 1983]. Nucleation will occur from the meta-stable state on some time scale no matter what the size of the nucleation barrier. For practical purposes, however, in simulations such as the ones we are performing, nucleation will occur only in the vicinity of the Becker-Döring limit where the time scales become attainable. For this discussion we do not need to know the precise value of this limit. We only need to note that the condition that nucleation occurs only in the neighborhood of this limit implies that

$$R^d\Delta h^{-d/4} = \xi^d = D\Delta h^{-3/2}, \tag{42}$$

where D is a constant independent of Δh.

We are now in a position to derive one of the three scaling relations we have tested with simulations on the CA model. The first is a scaling relation between the mean equilibration time $t_{\rm equil}$ and K. The equilibration time is defined as the number of time steps required for the system to relax after a plate update. It is also a measure of the earthquake duration. Taking the time average of each equilibration time over all plate updates determines the mean equilibration time, which is the mean time to form an earthquake. Our prediction is that this equilibration time will scale with the correlation length in analogy with the "ϕ^4" model. To obtain this scaling in variables appropriate to the CA model we note that from Eqs. (8) and (9) K behaves as a field, similar to the external magnetic field Δh in the ϕ^4 model discussed above. That is, varying K either brings the system closer to the spinodal (increasing K) or moves it further away. The location of the spinodal for $V \sim 0$ will be $K^{-1} \sim 0$. The correlation length, or mean radius of gyration, of the earthquake clusters will scale as

$$\xi \sim K^{1/4}, \tag{43}$$

in analogy to $\xi \sim \Delta h^{-1/4}$ in the ϕ^4 model. Direct comparison of Eq. (7) with Eq. (19) leads to the conclusion that the characteristic, or growth, time is

$$t_{equil} \sim \xi^2 \sim \Delta h^{-1/2} \sim K^{1/2}. \tag{44}$$

This relation means that as K increases and brings the system closer to the spinodal, the mean time to form an earthquake increases, corresponding to critical slowing down as observed in condensed matter systems near critical points. Figure 6 shows that for large $q = 1088$ (circles), but much smaller than the system size, t_{equil} follows a power law form of $t_{equil} \sim K^{0.35\pm0.1}$. However, as the interaction region includes more blocks $q \approx 10000$ (diamonds), or becomes more mean-field, the data points approach the predicted mean-field result given in Eq. (44). The alert reader will notice the iden-

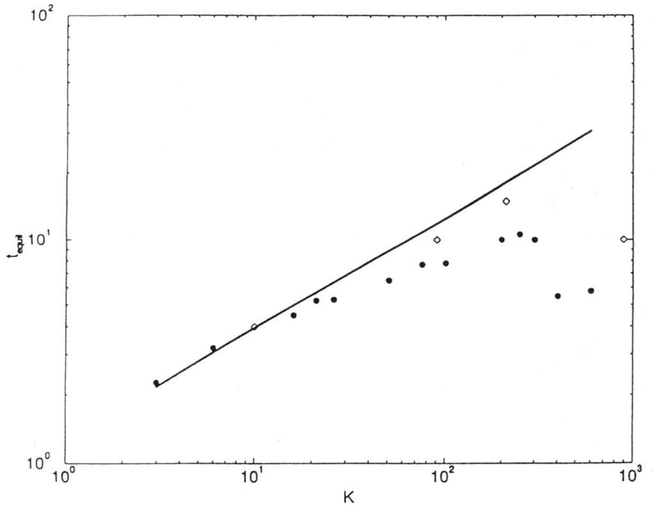

Figure 6. Log-log plot of the mean equilibration time to form a cluster per plate update versus the effective spring constant K. The circles are for a 256 × 256 system with $V \sim 0, \sigma^F = 50, K_L = 1, W = 0.3$, and $q = 1088$ using the K_C/q interaction. The diamonds are for a 128 × 128 system with $V \sim 0, \sigma^F = 50 \pm 10$ (random uniform distribution of failure thresholds), $\sigma^R = 0, K_L = 1$, and $q \approx 10000$ using the $1/r^3$ truncated interaction. The straight line has slope $= 1/2$, the predicted mean-field exponent.

tical scaling behavior of the equilibration time, t_{equil}, and the growth time, t_g, that defines the exponential growth of the critical droplet. This is a direct consequence of the Gaussian (symmetric) expansion around the saddle point droplet, in which the growth (climbing) direction of fluctuations toward the top of the saddle point has identical scaling properties with the evolution of the critical droplet as it moves away from the top of the saddle point. As we have already mentioned, the critical droplet has equal probabilities to grow, descend from the top of the saddle point, and nucleate the stable phase or to shrink and fall back into the metastable state. Although they scale in the same way with K, the equilibration time and the growth time differ by many orders of magnitude. The time scale to grow a critical earthquake is measured in years or tens of years while the equilibration time (earthquake duration) is measured in minutes.

A very interesting and important aspect of these data which not only impacts on our understanding of this class of models, but also has implications for understanding spinodal nucleation in condensed matter systems, is the large value of q required to see the critical slowing down predicted by the mean-field theory. Large values of q allow the spinodal to be more closely approached by raising the nucleation barrier (see Eq. 41)) and hence decreasing the value of Δh needed to reach the Becker-Döring limit.

In order to derive the next two scaling relations we need to introduce the idea of arrested nucleation. We begin by noting that from Eq. (9) the low velocity phase is a low stress phase. As the system is brought closer to the spinodal by raising the velocity or the coupling constant K, the low stress phase becomes meta-stable, and the high stress phase becomes stable. One would expect then that nucleation would take the system from the low stress meta-stable phase into the high stress stable phase. However, the high stress phase is not seen in the simulations performed in this work. Our hypothesis is that the nucleation process is stopped, or arrested. We will discuss the mechanism responsible for stopping the nucleation events in Section 7. In this section we will pursue the consequences of this idea.

Our working hypothesis is that large stress saddle point objects occur as in the theory outlined above. These objects are stopped or arrested and decay. The release of the stress from the high stress region during the decay is the earthquake. We have found these high stress regions in simulations and that data will be presented in a future publication [*Goldstein et al*, 2000].

Since the earthquake is associated with the release of stress the number of failed blocks should be proportional to the number of high stress blocks. That is, we are claiming that the high stress region defines the range of the earthquakes that scale in this model as it is presently simulated and that the Gutenberg-Richter scaling, as well as other scaling laws, can be obtained from understanding the statistical distribution of the high stress regions.

This raises the question as to why we believe that the only blocks that fail in the earthquakes that are in the scaling regime, are those originally in the high stress region. There are very large break-out events that span the system but they do not scale. Here, however, we are only considering the events that scale. First, that is essentially what the simulations show in this work and in other versions of the model we have studied [*Goldstein et al*, 2000]. The number of break-out events is a small fraction of the total number and that fraction decreases as we increase q. In addition, the time scale between break-out events and the time scale over which the fluctuation metric analysis indicates that the system is ergodic are of the same order [*Ferguson*, 1996]. One should note that there are some small break-out pieces of scaling events but the effect is quite small and does not affect the scaling. We will return to break-out events in Section 8.

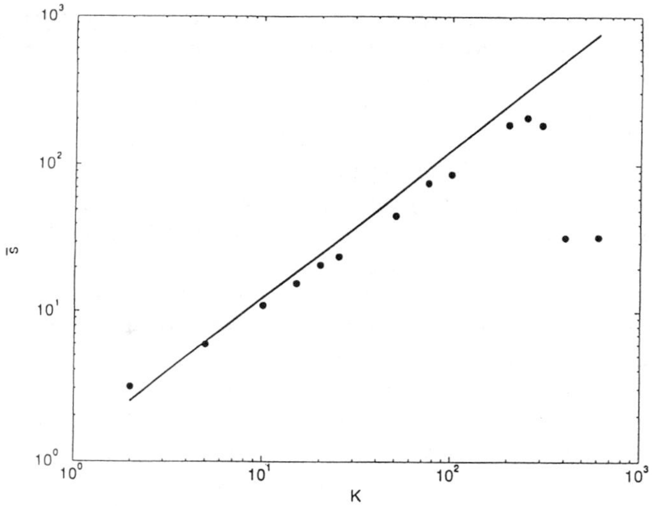

Figure 7. Log-log plot of the mean cluster size $\bar{s}$ versus the effective spring constant K. The 256 x 256 system has $V \sim 0, \sigma^F = 50, K_L = 1, W = 0.3$, and $q = 1088$ using the K_C/q interaction. The straight line has slope = 1, the predicted mean-field exponent.

Second, we have done an analysis of what conditions would favor such a containment and what would favor break-out from the high stress region resulting in a characteristic event in which approximately all blocks in the system fail [*Rundle et al*, 1998]. That study indicated that the stress profile in the model, as it is run in these studies, is too rough for a significant number of break-out events to occur and that the statistics will be dominated by scaling events that are contained within the original high stress region. The stress profile is defined as the field $\sigma^F - \bar{\sigma}(\vec{x}, t)$. By rough we mean a surface defined by the stress profile in $d = 2$ with a fractal dimension greater than 2.5.

With these considerations the mean mass of the clusters $\bar{s}$ will scale as the mean mass of the nucleation or critical droplet. From Eqs. (21) and (24) the density of the critical droplet scales as $\Delta h^{1/2}$ and its volume as ξ^d. This implies that the mass of the critical droplet $\bar{s}$ scales as

$$\bar{s} \sim \xi^d \Delta h^{1/2} = R^d \Delta h^{-d/4+1/2}. \tag{45}$$

From the condition that nucleation takes place near the Becker-Döring limit and Eq. (42) we have

$$\bar{s} \sim \Delta h^{-1}. \tag{46}$$

As in the above discussion we can express this relation in terms of K. That is $K \sim \Delta h^{-1}$ so that $\bar{s} \sim K$. The data in Figure 7 are clearly consistent with this prediction. However, this is not the final word on this subject and we will return to this discussion in Section 7.

Finally, in this section we derive the exponent for cluster scaling. Since critical slowing down is only cleanly attainable for extremely large q we will initially remove it from the calculation of nucleation rates. We expect then that the number of critical droplets over a time scale proportional to $\Delta h^{-1/2}$ is proportional to ξ^{-d} where we have used Eq.(41) multiplied by the time (proportional to $\Delta h^{-1/2}$) and assumed, as usual, that $R^d \Delta h^{3/2-d/4}$ is a constant since nucleation takes place only near the Becker-Döring limit, as expressed by the Eq. (42). From our assumption that the mass of the earthquakes scales like the critical droplet we have that the number of earthquakes or clusters per unit volume scales as ξ^{-d}. But ξ^d is related to $\bar{s}$ through

$$\bar{s} \sim \xi^{2d/3}. \tag{47}$$

This follows from Eqs. (42) and (45). Clearly this implies from the above discussion and Eq.(41) that

$$n_c(\bar{s}) \sim \frac{1}{\bar{s}^{3/2}}. \tag{48}$$

Figure 8 and Figure 9 demonstrate that the simulations confirm this analytic result. Note that even with the two different forms of the long-range interaction, as shown in Figs. 8 and 9, the model produces the same mean-field exponent.

This scaling exponent for clusters, together with assumptions of how the slip scales with the number of failed blocks, translates into a Gutenburg-Richter b value of 3/4. If critical slowing down is included in the nucleation rate, the same arguments used above give a cluster scaling exponent of 2 and a b value of 1. The reader interested in the details of the relation between b values and cluster scaling exponents is referred to reference [*Rundle et al*, 1997b].

As we mentioned above, the arrested nucleation droplets are associated with the large events. The smaller events we associate with the critical phenomena fluctuations near the spinodal. Since the clusters associated with the spinodal critical point have the same scaling as the arrested nucleation droplets without critical slowing down [*Klein*, 1990; *Stauffer and Aharony*, 1992] we would expect the same scaling for both regions of the cluster scaling curves in Figs. 6 and 7. There should be a relative displacement in the straight lines but the normalization with the total number of clusters makes this displacement impossible to see without significantly more events.

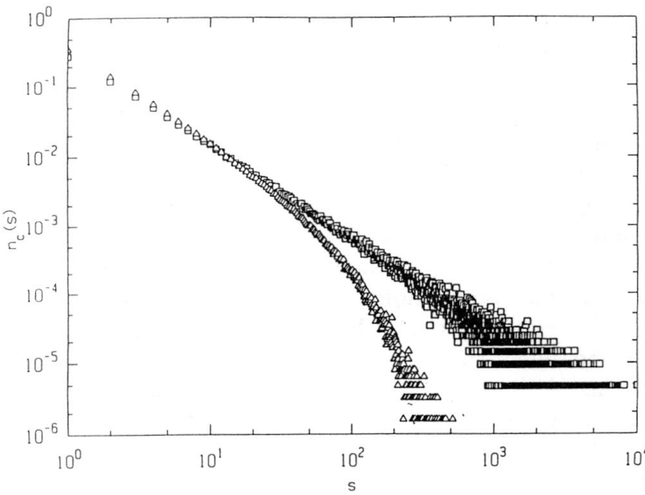

Figure 8. Log-log plot of the number of clusters $n_c(s)$ with s failed blocks divided by the total number of clusters. The 128×128 system was run with $V \sim 0, \sigma^F = 50, \sigma^R = 0$, and $K_L = 1$, using the truncated $1/r^3$ interaction. The triangles are for $K = 9.95$, and the squares correspond to $K = 90.45$. The slope of the straight part of the curves is 1.5 ± 0.1 for the triangles and 1.38 ± 0.15 for the squares. Both slopes are consistent with 3/2 as predicted by the theory.

The scaling curve for the mean cluster size versus K will not be affected by the fact that the smaller clusters associated with the critical phenomena fluctuations scale differently than arrested nucleation droplets since the mean cluster size is dominated by the much larger clusters associated with arrested spinodal nucleation. The approximate size of these two types of objects is estimated below.

Finally we point out that, while the clusters associated with the critical phenomena fluctuations can explain the smaller cluster end of the cluster scaling curves in Figs. 6 and 7, they cannot explain the upper end of theses scaling plots. Conversely, arrested nucleation droplets have the right scale for the larger end of the cluster scaling but not the lower end.

To see this we return first to the mean size $\bar{s}$ of the arrested nucleation droplets given in Eq. (45), i.e.

$$\bar{s} = \xi^d \Delta h^{1/2}. \tag{49}$$

Since, from Eq. (42), $\xi^d \Delta h^{1/2} = R^d$ in $d = 2$ we have that the arrested nucleation droplets have a mean size $\bar{s} \sim 10^3$. Clearly this is only an estimate, but $\bar{s}$ is consistent with the upper end of the scaling plot in Fig. 7. We expect some spread in the size of the arrested nucleation droplets but if the mean size is of order 10^3 it is difficult to imagine nucleation or activation events at a scale of order 10, which is the lower end of the scaling plots in Figs. 6 and 7.

The solution to this problem we believe is to consider another mechanism for events at the smaller mass scales. Since we are near a mean-field spinodal, or more precisely a pseudo spinodal, we should expect critical phenomena fluctuations as well as arrested nucleation droplets. This is in analogy with the standard "ϕ^4" models [*Unger and Klein*, 1984]. We can estimate the size of the clusters associated with the critical phenomena fluctuations by first noting that the distance from the spinodal of the simulations that gave rise to the scaling plots in Figs. 6 and 7 is of order 10^{-2}. We can see that first by noting that if K^{-1} is used as a scaling field as it is in Fig. 6, the straightest line is for $K \sim 10^2$. In Fig. 7, where a slightly different form of the potential is used, the value of K is .5x10^2. Note that Eq. (10) implies that for $V \sim 0$ the spinodal value of $K^{-1} \sim 0$.

For the clusters associated with critical phenomena fluctuations near spinodals the mass or mean size scales as

$$\bar{s} \sim \rho_{fc}\xi^d = \left(\frac{\Delta h^{1/2}}{R^d \Delta h^{3/2-d/4}}\right) \times \left(R^d \Delta h^{-d/4}\right) = \Delta h^{-1}, \tag{50}$$

where ρ_{fc} is the density of critical phenomena fluctu-

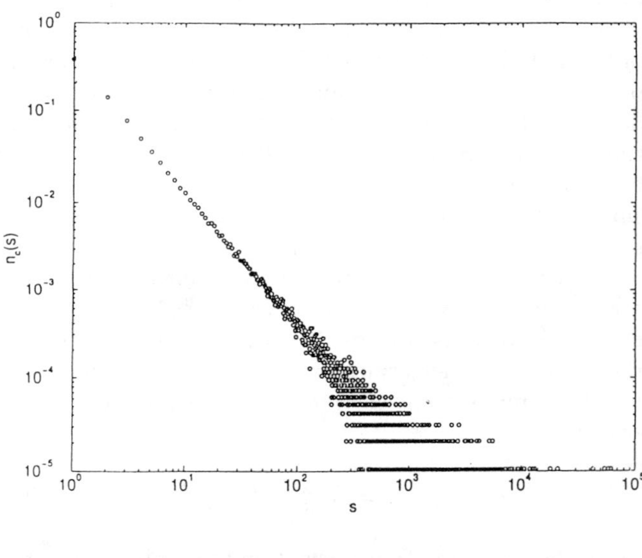

Figure 9. Log-log plot of the number of clusters $n_c(s)$ with s failed blocks divided by the total number of clusters. The 256×256 system was run with $V \sim 0, \sigma^F = 50, W = 0.3, K_L = 1, K_C = 50$, and $q = 1088$ using the K_C/q interaction. The slope of the straight part of the curve is 1.5 ± 0.1, which is consistent with the theory's prediction of 3/2.

ations while their volume is ξ^d. The derivation and details of this scaling can be found in Section 8 of this manuscript. Also see references [*Monette and Klein*, 1992] and [*Ray and Klein*, 1988]. With $\Delta h \sim 10^{-2}$ and $d = 2$ we have that the size of the clusters associated with critical phenomena fluctuations is on the order of 100. Note that the mean cluster size is an average over all clusters and hence is dominated by the larger clusters in the scaling plot. The mean size of 100 gives us an estimate of the region of the upper end of the cluster scaling associated with the critical phenomena clusters. Larger clusters would be in the exponential decay. This indicates, as stated earlier, that associating the larger end of the scaling plot for the "earthquakes" with critical phenomena fluctuations is untenable.

As a footnote to this section we want to point out that the derivation of these scaling laws implicitly makes use of the fact that there is only one divergent length in the system that controls all singular behavior. This is the so called hyper-scaling assumption. In general mean-field systems do not have hyper-scaling except in the upper critical dimension which in this case is six [*Klein and Unger*, 1983]. However, the condition that nucleation takes place only in the vicinity of the Becker-Döring limit fixes $R^d \Delta h^{3/2-d/4}$ and restores hyper-scaling. The discussion of this point is subtle. The interested reader can find the details in [*Ray and Klein*, 1988].

Before ending this section we need to make three points that we will return to in Section 8. First: When we refer to the clusters associated with the critical phenomena fluctuations we are not referring to the fluctuations themselves but to the building blocks of these fluctuations. Second: We have tentatively identified a third type of cluster that contributes to the GR scaling which also scale as $1/\bar{s}^{3/2}$. We will return to this point in Section 8. Third: We will need a considerably more sophisticated analysis of the scaling region to see these different cluster types.

6. ENERGY-FLUCTUATION METRIC AND ERGODICITY

We have been applying equilibrium statistical mechanics to the CA model. Before we can use these methods, we must establish that this model displays the characteristics of an equilibrium system under the conditions we have used in the simulations. Recently, *Rundle et al.* [1995] demonstrated that the CA model exhibits Boltzmann fluctuations in its internal energy field for $K \gg 1$, i.e. near mean-field. In this section, we present evidence that this model tends toward an ergodic system as the interactions become long-range by using a numerical test, called the energy-fluctuation metric $\Omega(t)$ [*Thirumalai and Mountain*, 1990], which determines if a simulated system is effectively ergodic over an observational time scale. The quantity $\Omega(t)$ measures the difference between the time average and the ensemble average of a system's energy and is defined as

$$\Omega(t) = \frac{1}{N} \sum_{i=1}^{N} \left[\epsilon_i(t) - \bar{\epsilon}(t)\right]^2 , \qquad (51)$$

where the sum runs over the N blocks of the system, the running time average of block i's energy $\epsilon_i(t)$ from $t' = 0$ to t is

$$\epsilon_i(t) = \frac{1}{t} \int_0^t \epsilon_i(t') dt', \qquad (52)$$

and the spatial average of the energy is

$$\bar{\epsilon}(t) = \frac{1}{N} \sum_{i=1}^{N} \epsilon_i(t). \qquad (53)$$

For ergodic systems, $\Omega(t) \sim 1/t$ [*Thirumalai and Mountain*, 1990]. Consequently, plots of $1/\Omega(t)$ versus t increase linearly if a system is effectively ergodic.

We will illustrate the effects of the loader plate velocity V, interaction range R, and spring constant K_C on the energy-fluctuation metric $1/\Omega$. Figure 10 shows $1/\Omega$ versus loader plate update n for constant R as V increases from 0.01 to 1 from top left to bottom right. Each plot uses the same abscissa scale for ease of comparison. The top two plots do not depict the initial transient due to the initial conditions. As V decreases, $1/\Omega$ becomes straighter and increases in a more linear fashion over a longer time.

Figure 11 shows $1/\Omega$ versus n for varying R. As in Fig. 10, each plot uses the same abscissa scale and does not depict the initial transient. Comparing the top right and bottom two plots, observe that as the interaction range increases from nearest-neighbors to $R = 4$, $1/\Omega$ becomes straighter and increasingly linear over a longer time. Note that in these plots, we are moving the loader plate at what we call a "critical" velocity V_c in which frequency-size plots produce power law scaling over a narrow range of velocities centered about V_c. We emphasize that this scaling results from the coalescence of typically many initial growth sites during each plate update. While we recognize that this coalescence violates the separation of time scales inherent in most earthquake processes, we have driven the system in this manner in order to have a reference velocity V_c that

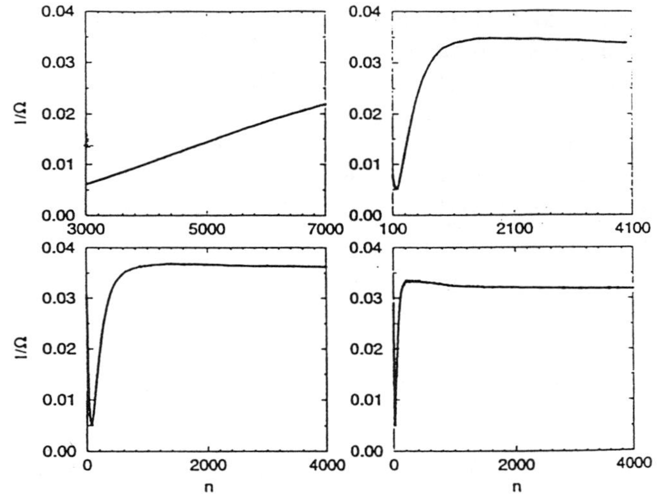

Figure 10. Inverse of the energy-fluctuation metric $1/\Omega$ versus loader plate update n for a 256×256 lattice with closed boundaries and $R = 4(q = 80)$, using parameters: $\sigma^F = 50 \pm 10$ (random uniform distribution of failure thresholds), $\sigma^R = 0$, and $K_C = K_L = 1$. From left to right and top to bottom, $V = 0.01, 0.15, 0.3086$, and 1.

gives similar "critical" behavior for different interaction ranges. This allows us to examine the ergodicity differences for different interaction ranges, R. The top left and bottom right plots in show that for a constant V but different R, $1/\Omega$ shows a more linear increase over a longer time as R goes from nearest-neighbors to $R = 4$.

In comparison to Fig. 11, using realistic fault parameters of $V = 0.001$, which produces about one initial growth site per plate update, and of $K_C = 100$ and $K_L = 1$, which correspond to strong coupling among the blocks and relatively weak coupling to the loader plate, Figure 12 shows that $1/\Omega$ displays linear behavior over a long time. If we remove the coupling among the blocks by setting $K_C = 0$, the system does not exhibit ergodic behavior, as shown in Figure 13. Here the sinusoidal character of $1/\Omega$ [*Thirumalai and Mountain*, 1993] arises from the quasi-periodic movement of the blocks as the loader plate continually drives them to σ^F and they relax to σ^R. The energy-fluctuation results indicate that as $V \to 0$ and as R and K_C increase, the CA system becomes more ergodic. This observation in conjunction with the previous findings demonstrating Boltzmann statistical fluctuations [*Rundle et al*, 1995] indicates that the long-range, slowly driven CA system tends toward an equilibrium system.

Our interpretation of this result is that the free energy surface obtained by functionally integrating Eq. (8) has multiple minima within the low stress phase. These minima correspond to different energies. Small events, which are contained within the high stress region, as discussed in the previous section, appear to leave the system in the same free energy minimum. Break-out events, which are relatively rare in our simulations and become less frequent as q is increased, seem to push the system into a new minimum. This interpretation, and its confirmation, is the subject of current investigation.

7. COARSE GRAINING REVISITED AND ARRESTED NUCLEATION

In this section we return to the discussion of arrested nucleation. The question we need to address is what causes the nucleation of the high stress phase to be halted so that it is never seen in these simulations. The answer to this question requires that we reexamine the coarse graining procedure and the derivation of Eq. (8).

As discussed in Section 4, there is a coarse graining time which is the fundamental time unit in our coarse grained description. That is, the fundamental time unit in Eq. (8) is the time scale required to have the stress distribution in a coarse grained volume become Gaussian. This will, as seen in Section 4, require a finite fraction of the number of blocks in the coarse grained volume to fail. This time is not constant however in

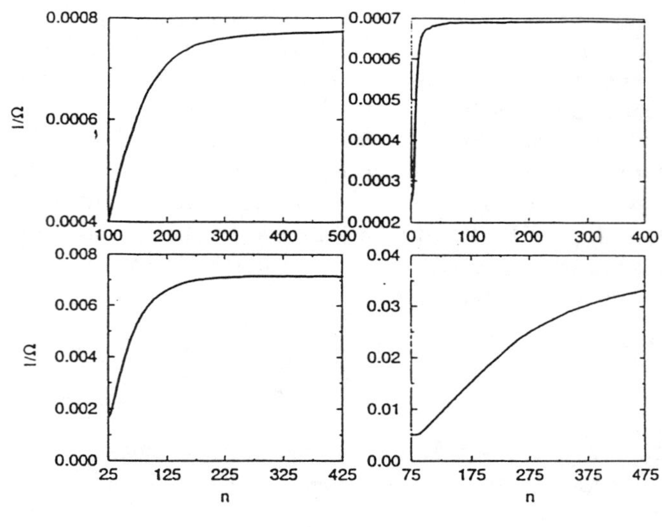

Figure 11. Inverse of the energy-fluctuation metric $1/\Omega$ versus loader plate update n using the same parameters as Fig. 10, except for R and V. From left to right and top to bottom, $R =$ nearest-neighbors and $V = 0.3086$; $R =$ nearest-neighbors and $V = V_c = 5.0$; $R = 2(q = 24)$ and $V = V_c = 1.0$; and $R = 4(q = 80)$ and $V = V_c = 0.3086$.

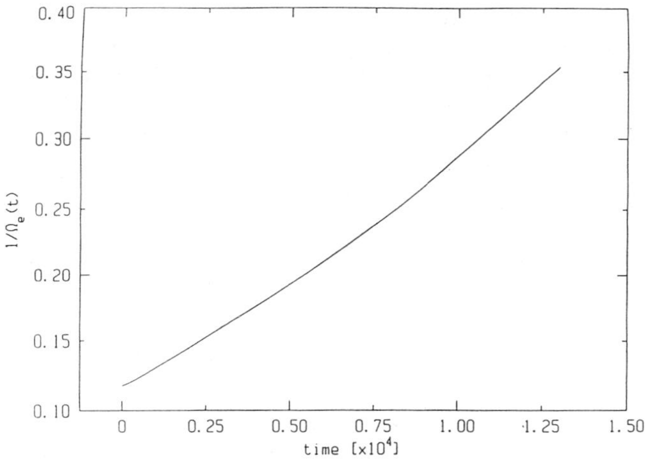

Figure 12. Inverse of the energy fluctuation metric $1/\Omega$ versus loader plate update n for a 256×256 lattice with closed boundaries and $R = 4 (q = 80)$, using parameters: $\sigma^F = 50, K_L = 1, K_C = 100, V = 0.001$, and $W = 0.1$.

units of plate update time. Since the coarse graining time is the basic unit of time that we are forced to have once we adopt a coarse grained description, this implies that the unit of plate update time and hence, the plate velocity is not constant but is in fact a fluctuating variable.

Clearly, the plate update number required to fail a finite but fixed number of blocks depends on the size of the "earthquake" events that occur as that fraction of blocks fails. The larger the events the smaller number of plate updates, and hence, the slower the velocity of the plate in units of coarse graining time. This implies that the coarse grained loading rate depends on the "earthquake" activity during the coarse grained time unit. We expect then that the velocity of stress loading, or the plate velocity, will be lowered by increased "earthquake" activity and increased by decreased activity. In the mean-field model all sites fail at the failure threshold as discussed in Section 4. This allows us to relate the number of failures in an event to the stress change. With these considerations we can express the time dependent velocity $V(\tau)$ as

$$V(\tau) = V + \frac{\alpha}{L^2} \int d\vec{x} \frac{\partial \sigma_D(\vec{x}, \tau)}{\partial \tau}. \qquad (54)$$

where L^2 is the volume of the system in $d = 2$, α is a positive constant and the time derivative $\sigma_D(\vec{x}, \tau)$ is the rate at which stress is dissipated at $\vec{x}$ at time τ. Note that for static considerations the time derivative vanishes and we return to the situation discussed in Section 4 and Eq. (9). However, when we are considering time dependent quantities the $\int d\vec{x} \frac{\partial \sigma_D(\vec{x}, \tau)}{\partial \tau}$ term in the plate velocity will now depend on τ. Also note that when the net rate of stress dissipation is negative, as would occur during times of large earthquake activity, the time dependent velocity $V(\tau)$ is lower than the bare velocity V which is now the time average, over infinite time, of $V(\tau)$ in a steady state. The idea of arrested nucleation is that large events are preceded by an increased activity in this model and this activity drives the plate velocity down effectively raising the nucleation barrier.

There are three points that need to be made about this velocity modification. The first is that we can relate the time dependence of the velocity to the rate of stress dissipation only in the mean-field limit where all blocks fail at the failure threshold as discussed in Section 4. Second, the term

$$\frac{\partial \sigma_D(\vec{x}, \tau)}{\partial \tau} \propto \sqrt{\frac{\beta}{\pi}} \int_{\sigma_o}^{\sigma^F} d\sigma \exp\left[-\beta(\sigma - \bar{\sigma}(\vec{x}, \tau))^2\right]. \qquad (55)$$

as can be seen from Eqs. (6) and (8). Finally these considerations predict that there should be increased activity before any large events that could be described as arrested nucleation where "before" refers to the coarse graining time interval preceding the arrested nucleation event. This has been seen to be correct. The data and a detailed discussion will be presented in a future publication [*Goldstein et al*, 2000]. We have also tested this idea by performing simulations [*Goldstein et al*, 2000] in

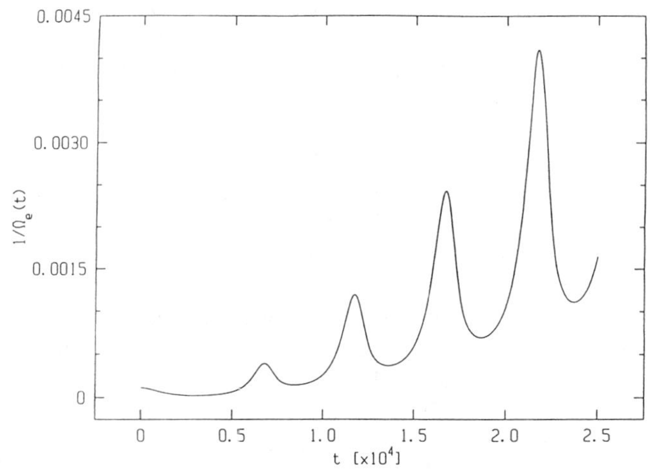

Figure 13. Inverse of the energy-fluctuation metric $1/\Omega$ versus loader plate update n using the same parameters as Fig. 12, except $K_C = 0$.

which the plate updates were done so that the amount of stress put into the system was proportional to the number of blocks that failed after the preceding plate update. In the mean-field limit this insures that the plate velocity is a constant in units of the coarse graining time. With this form of update the high stress phase was observed [*Goldstein et al*, 2000], consistent with the ideas presented in this section.

8. CLUSTERS AND SCALING IN NEAR MEAN-FIELD MODELS

In this section we examine the GR scaling of this model in greater detail. In order to do this we make use, as we did in Section 5, of the analogy between the $R \to \infty$ version of the "ϕ^4" model and the RJB model. This is possible because both of these models have a dynamics which is given by a Langevin equation [*Klein et al*, 1997; *Ferguson et al*, 1999].

We begin this discussion by noting that the thermal model with Boltzmann statistics described by a Langevin equation can be mapped onto a percolation model. The derivation of this result is in Ref. [*Klein*, 1990] so we will not repeat it here. However, the physical meaning of this result is important for our discussion. It is probably simplest to understand the physics in the language of Ising models which can also be described by Langevin equations [*Gunton et al*, 1983].

Suppose an Ising model with interaction range R, in equilibrium, is simulated with a Monte Carlo Algorithm. In the neighborhood of a critical point (in the case $R >> 1$ we also consider spinodals) we take a snapshot of the spins at time t. Between each pair of spins that are in the same direction and are in each others interaction range we toss a bond with probability $p_b = 1 - \exp(-J(1-\rho)/K_BT)$. Here T is the temperature, J is the coupling constant, K_B is Boltzmann's constant and ρ is the density of the up spins. The quantity ρ is related to the magnetization per spin m by $\rho = (1+m)/2$. For the critical point $\rho = 1/2$ and $p_b = 1 - \exp(-J/2K_BT)$ [*Coniglio and Klein*, 1980]. With this bond probability the Ising critical point is a percolation threshold for both long and short range interactions. For long range interactions the same statement is true for spinodals when the density is included in the bond probability as above. In this form the metastable phase has a negative magnetization.

In addition, with this percolation definition, thermal quantities are isomorphic to their counterparts in percolation. For example, the isothermal compressibility is isomorphic to the mean cluster size. The magnetization per spin is isomorphic to the probability that a site selected at random belongs to the infinite cluster and the free energy per spin is isomorphic to the mean number of clusters per site [*Stauffer and Aharony*, 1992; *Coniglio and Klein*, 1980]. Consider once more the partition function in Eq. (11). In terms of the variable $\psi(\vec{x}) = \phi(\vec{x}) - \phi_s$, which is the time independent analog of $\psi(\vec{x}, t)$ defined in Eq. (18) the partition function is

$$Z = \int \delta\psi \exp\left[-\beta \int d\vec{x}\frac{R^2}{2}\left(\nabla\psi(\vec{x})^2\right) + \Delta h\psi(\vec{x}) - 4\left(\frac{|\epsilon|}{6}\right)^{1/2}\psi^3(\vec{x}) + \psi^4(\vec{x})\right], \quad (56)$$

where again we take $\epsilon < 0$. We now define a new field $\tilde{\psi}(\vec{x}) = \psi(\vec{x}) - A$ and determine A such that the linear term,$\Delta h\psi(\vec{x})$, is replaced by a quadratic one of the form $\Delta h^{1/2}\tilde{\psi}^2$. That is, we perform a shift so that the partition function becomes

$$Z = \int \delta\tilde{\psi} \exp\left[-\beta \int d\vec{x}\frac{R^2}{2}\left(\nabla\tilde{\psi}(\vec{x})\right)^2 + c_1\Delta h^{1/2}\tilde{\psi}^2(\vec{x}) - c_2\tilde{\psi}^3(\vec{x}) + c_3\tilde{\psi}^4(\vec{x})\right]. \quad (57)$$

The constants c_1, c_2 and c_3 are easily obtained [*Unger and Klein*, 1984] but are not of interest here.

We scale all lengths with respect to $R\Delta h^{-1/4}$ and assume $\tilde{\psi}(\vec{x}) \propto \Delta h^{1/2}\bar{\psi}(\vec{x})$ and obtain

$$Z = \int \delta\bar{\psi} \exp\left[-\beta R^d \Delta h^{3/2-d/4} \int d\vec{x}' \left(\nabla\bar{\psi}(\vec{x}')\right)^2 + c_1\bar{\psi}^2(\vec{x}') - c_2\bar{\psi}^3(\vec{x}') + c_3\Delta h^{1/2}\bar{\psi}^4(\vec{x}')\right]. \quad (58)$$

We now neglect the $\bar{\psi}^4$ term due to the $\Delta h^{1/2}$ coefficient and the fact that we are near the spinodal so that $\Delta h << 1$. Hence

$$Z = \int \delta\bar{\psi} \exp\left[-\beta R^d \Delta h^{3/2-d/4} \int d\vec{x}' (\nabla\bar{\psi}(\vec{x}'))^2 + c_1\bar{\psi}^2(\vec{x}') - c_2\bar{\psi}^3(\vec{x}')\right]. \quad (59)$$

If $\beta R^d \Delta h^{3/2-d/4} >> 1$, which will be true if R is large enough as we will see below, we can evaluate this integral with saddle point techniques and find that the free energy F and, hence, the mean number of clusters is

$$F = CR^d \Delta h^{3/2-d/4}, \quad (60)$$

where C is an unimportant constant of order one. The length scale $R^d \Delta h^{-1/4}$ is the correlation length and, by the cluster mapping, the mean cluster radius [*Unger and Klein*, 1984].

To see that in MF systems the quantity F in Eq. (60) must be large we derive the relation known as the Ginsburg criterion [*Unger and Klein*, 1984]. If a system is MF then the fluctuations can be ignored compared to mean values. The magnitude of the fluctuations per unit volume are measured in these systems by the isothermal susceptibility χ_T. which near MF spinodals diverges as $\chi_T \sim \Delta h^{-1/2}$. The fluctuation magnitude in a region the size of the correlation length ξ is then $\xi^d \Delta h^{-1/2}$ where $\xi = R\Delta h^{-1/4}$ [*Unger and Klein*, 1984]. The order parameter ψ scales as $\psi \propto \Delta h^{1/2}\xi^d$ and we need to compare the fluctuations to the square of the order parameter. That is for MF to be a good approximation we must have

$$\frac{\xi^d \Delta h^{1/2}}{\xi^{2d}(\Delta h^{1/2})^2} = \frac{\Delta h^{-1/2}}{R^d \Delta h^{-d/4}\Delta h} << 1, \tag{61}$$

or

$$R^d \Delta h^{3/2-d/4} >> 1. \tag{62}$$

This implies that the number of clusters in a region with a radius of the correlation length is large in the MF limit. This is also true in the neighborhood of the MF critical point and is in contrast to systems with short range interactions where the number of clusters in a region the size of ξ^d is one [*Coniglio and Klein*, 1980]. Note that this is the number of finite clusters. There is also, in the vicinity of the spinodal, an "infinite" cluster which is the size of the system with a finite (non-fractal) density [*Stauffer and Aharony*, 1992] that represents the ordered phase or the phase with a finite magnetization per spin. Near the spinodal this would be the meta-stable phase. If we subtract off this cluster then the remaining system has only finite clusters and a residual magnetization of zero. Hence, by symmetry, there must be of order $R^d \Delta h^{3/2-d/4}$ fundamental clusters of each sign whose net contribution to the remaining magnetization is zero.

What is the density of these clusters? In the usual percolation models (R small) the density of sites in a region the size of the connectedness length, the quantity that plays the role of the correlation length, scales as Δp^β where Δp is the parameter that controls how far the system is from the percolation threshold [*Stauffer and Aharony*, 1992]. For the spinodal the connectedness length is isomorphic to the correlation length [*Klein*, 1990; *Coniglio and Klein*, 1980] and one might have expected that the density of the clusters would be proportional to $\Delta h^\beta = \Delta h^{1/2}$. However, there are $R^d \Delta h^{3/2-d/4}$ clusters in this region as described above. Any structure that is that numerous must have a low free energy cost, basically of order one. The free energy F in Eq. (60) is much greater than 1 from Eq. (62). However if we interpret F as the free energy cost of all of the fundamental clusters then each of them has a free energy cost of order one. This in turn implies that the MF scaling of thermodynamic quantities such as the free energy and the density refer to the entire collection of fundamental clusters. By this reasoning the density of the fundamental clusters near the spinodal ρ_{fc} should be

$$\rho_{fc} \propto \frac{\Delta h^{1/2}}{R^d \Delta h^{3/2-d/4}}. \tag{63}$$

This result was confirmed for random percolation in the MF limit in Ref.([*Ray and Klein*, 1988]) and for the spinodal in Refs.([*Ray and Klein*, 1990; *Alexander et al*, 2000]).

It is important to understand that these clusters, unlike the situation in short range models, are *not* the critical phenomena fluctuations. The density of the critical phenomena fluctuations can be obtained from the equation for the partition function. Returning to Eq. (58) we have in the Gaussian approximation

$$Z = \int \delta\bar{\psi} \exp\Big[-\beta R^d \Delta h^{3/2-d/4} \int d\vec{x}' (\nabla\bar{\psi}(\vec{x}'))^2 + \bar{\psi}^2(\vec{x}')\Big], \tag{64}$$

where we have ignored the $\bar{\psi}^3(\vec{x}')$ term. We will see that this is justified self consistently. Since we are only interested in scaling properties we can assume that $\bar{\psi}(\vec{x}')$ is a spatial constant for $|\vec{x}'| < 1$, or in unscaled variables for lengths less than the correlation length, and zero on larger scales. The partition function becomes

$$Z = \int d\bar{\psi} \exp\left[-\beta R^d \Delta h^{3/2-d/4} \bar{\psi}^2\right], \tag{65}$$

where $\bar{\psi}$ is the magnitude of $\bar{\psi}(\vec{x}')$. Since the integral is damped exponentially when $\beta R^d \Delta h^{3/2-d/4}\bar{\psi}^2 > 1$ we have that

$$\bar{\psi} = \frac{1}{\left(\beta R^d \Delta h^{3/2-d/4}\right)^{1/2}}. \tag{66}$$

Since $\tilde{\psi}(\vec{x}') = \Delta h^{1/2}\bar{\psi}(\vec{x}')$ the density of the critical phenomena fluctuations scale as

$$\tilde{\psi}(\vec{x}') \sim \frac{\Delta h^{1/2}}{\left(R^d \Delta h^{3/2-d/4}\right)^{1/2}}. \tag{67}$$

This result justifies neglecting $\bar{\psi}^3(\vec{x}')$ since in MF systems $R^d\Delta h^{3/2-d/4} >> 1$ (see Eq. (62)). Also note that the critical phenomena fluctuations are denser than the fundamental clusters whose density is given in Eq. (63).

Are there clusters the size of critical phenomena fluctuations? In Ising models this is the subject of current research. If such clusters do exist they are a sign that the system is not truly mean-field as they denote a probing of the non-Gaussian nature of the free energy. In the CA model we are studying such clusters have been found and they will be discussed later in this section.

Since the Langevin equation derived in Section 3 will guarantee a Boltzmann statistics quite similar to those of the ϕ^4 model we expect that the same cluster mapping and existence of different cluster types will hold for the CA model of RJB. The question we address next is what is the bond probability in the CA model. At present we do not have a theoretical treatment of the bond probability. However, we can argue as follows: The bond probability that we seek will tell us what fraction of the sites that are in the high stress state will actually fail in the evolution of the model. This is similar to the idea that in Ising models not all spins in the same direction within the range of interaction of a chosen spin are connected to that spin. There are, in fact spins that are in the same direction by random chance and not due to correlation. The imposition of the bond probability corrects for this over counting [*Coniglio and Klein*, 1980]. In the CA model we know what sites fail so that we do not need the bond probability. It is of interest to obtain it theoretically as a check on this idea. However, we should expect that if the bond probability is used on high stress sites after a plate update, but before the sites are allowed to fail, then we will get a cluster that is statistically the same as the clusters of failed sites that we have been using. Note that all of the properties that we would expect from clusters was found for the clusters of failed sites in Section 5. This relation between the clusters of high stress sites and the clusters of failed sites is a subject for further research.

The relation between the fundamental clusters and the critical phenomena fluctuations can be understood by noting that the fundamental clusters are constructed as independent objects [*Klein*, 1990; *Coniglio and Klein*, 1980]. As such the fluctuations in the number of fundamental clusters should be of the order of $\left(R^d\Delta h^{3/2-d/4}\right)^{1/2}$. Multiplying this by the density of the fundamental clusters in Eq. (63) we obtain the density of the critical phenomena fluctuations in Eq. (67). This result implies that the critical phenomena fluctuations are an incoherent superposition of the fundamental clusters. This has been verified computationally and the results will appear in another publication [*Alexander et al*, 2000].

The analysis in this section and the one in Section 5 have yielded three types of clusters: fundamental clusters, arrested nucleation droplets that are not affected by critical slowing down and arrested nucleation droplets that are affected by critical slowing down. A fourth cluster type also arises when the clusters that make up a critical phenomena fluctuation via incoherent superposition coalesce into one cluster which we will call coalescence clusters. Finally, we can have break-out events where the various forms of clusters discussed above can trigger a system wide event. These break-out events are not described by the scaling theory and are outside the theoretical framework presented in Section 3. For a discussion of break-out events see Ref. [*Rundle et al*, 1998].

We now derive the scaling law for the two new classes of clusters within the scaling region. For the arrested nucleation clusters, both with and without critical slowing down, the derivation was presented in Section 5. A key element in this derivation is that the system is always simulated for a fixed *average* value of

$$R^d\Delta h^{3/2-d/4} = C. \tag{68}$$

From Eq.(63) we have that the number of sites in the fundamental clusters $\bar{s}_{fc}$ scale as

$$\bar{s}_{fc} \propto \frac{\Delta h^{1/2} R^d \Delta h^{-d/4}}{R^d\Delta h^{3/2-d/4}}, \tag{69}$$

or $\bar{s}_{fc} \propto \Delta h^{-1}$. The number of fundamental clusters in a region the size of the correlation length is proportional to $R^d\Delta h^{3/2-d/4}$ as we argued above (see Eq. (60)). Hence, the number of fundamental clusters per unit volume N_{fc} is

$$N_{fc} \propto \frac{R^d\Delta h^{3/2-d/4}}{R^d\Delta h^{-d/4}} = \Delta h^{3/2}. \tag{70}$$

From Eqs.(69) and (70) we obtain

$$N_{fc} \propto \frac{1}{\bar{s}_{fc}^{3/2}}. \tag{71}$$

For the number of coalescence clusters N_{cc} we have

$$N_{cc} \propto \frac{(R^d\Delta h^{3/2-d/4})^{1/2}}{R^d\Delta h^{-d/4}} = \frac{\Delta h^{3/2}}{C^{1/2}}. \tag{72}$$

The mass, or number of sites, of a coalescence cluster S_{cc} is

$$\bar{s}_{cc} \propto \frac{\Delta h^{1/2} R^d \Delta h^{-d/4}}{(R^d\Delta h^{3/2-d/4})^{1/2}} = \Delta h^{-1} C^{1/2}, \tag{73}$$

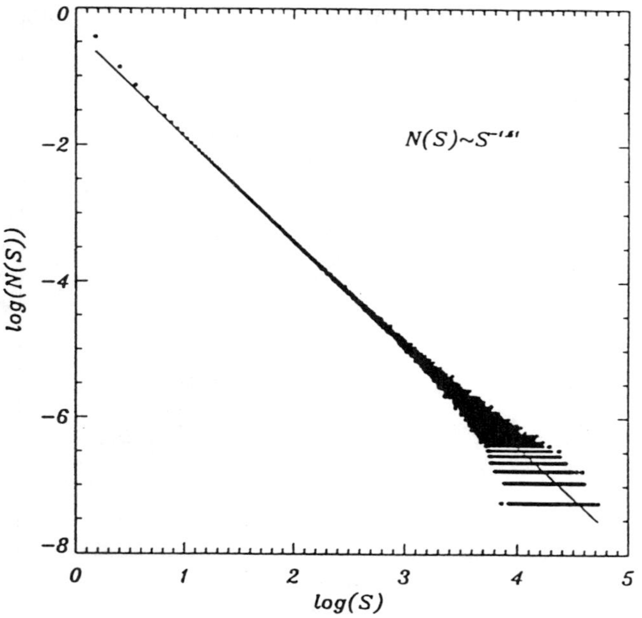

Figure 14. Log-log plot of the number of clusters versus the number of sites in the cluster. The system size is 256x256, R=30, there are 18x10[6] events and the bin size is one. Other parameters are given in Section 8 of the text.

or

$$N_{cc} \propto \frac{C^{1/4}}{\bar{s}_{cc}^{3/2}}. \tag{74}$$

To summarize: There are four different cluster scaling categories: fundamental clusters, coalescence clusters and arrested nucleation without critical slowing down, all of which have a predicted cluster scaling exponent τ of $\tau = 1/2$, and arrested nucleation with critical slowing down with a cluster scaling exponent of 2.

In Figure 14 we plot the number of clusters of all types as a function of their size $\bar{s}$. The interaction range is such that one site transfers stress to, or interacts with, 3720 other sites. There are 18 million clusters and all the sites have a failure threshold of $\sigma^F = 1$ and an average residual stress of $\sigma^R = 0.25$ We have used a noise in the stress drop chosen from a flat random distribution between ± 0.25, i. e. $\sigma^F - \sigma^R = 0.75 \pm 0.25$.

From the previous discussion and the discussion of critical slowing down in Section 5, we would expect to see clusters scaling with at least two exponents (1.5 and 2) and three regions with different amplitudes. As can be seen from the figure we see only one slope with an exponent consistent with the value 1.5. There is no sign of a scaling region with a slope of -2 or of different amplitudes. However a closer examination of Fig. 14 indicates a significant spread of the data at the large cluster end. The first thought would be that the explanation of this spread is poor statistics. Even with the 18 million events shown in Fig. 14 there are of the order of a few hundred clusters in each stress bin at the large cluster end. However, if we compare the width of the spread of the data for several different number of events we see that the spread, or fan, stops shrinking once we have obtained about 7 million events (see Figure 15). This is an indication that the data spread could be signalling the presence of the other scaling regimes.

From the analysis in this section of the different cluster types in ϕ^4 systems and the derivation of the Langevin equation that describes the model of RJB in Section 3 it seems reasonable to analyze our cluster scaling data in terms of these different scaling catagories. To accomplish this we need to translate the scaling results derived above into the language appropriate to the simulations of the model of RJB.

As in Section 5 the parameter $K^{-1} = (K_L + qK_c)^{-1}$ (see Eq.(1)) acts as the field Δh. For fundamental clusters we make use of the fact that within the scaling regime a block fails at most once during an "earthquake" (see Section 4) to calculate the stress drop of the cluster. From Eq. (63) we have that the number of sites,N_s, in a fundamental cluster in a spatial region the size of the stress transfer, R, is

$$N_s = \rho_{fc} \times R^d \propto \frac{R^d K^{-1/2}}{R^d K^{-3/2+d/4}} \propto K^{1/2}. \tag{75}$$

Since each site that fails distributes $1/R^d$ units of stress, where we have taken the stress drop per site to be one for the purposes of this argument, we have that the total stress drop for a fundamental cluster, $\Delta\sigma_{fc}$, is of the order

$$\Delta\sigma_{fc} \propto K^{1/2} R^{-d} \sim 10^{-2} \tag{76}$$

in two dimensions and for $R = 30$.

Since we do not know the amplitudes associated with the scaling of the free energy/mean number of clusters $R^d \Delta h^{3/2-d/4}$ or the order parameter $\Delta h^{1/2}$ this value of 0.01 for $\Delta\sigma_{fc}$ is approximate. The point is that the fundamental clusters are made up of sites whose minimum stress after the "plate update" prior to their failure is greater than a number the order of

$$\sigma_m = \sigma^F - \Delta\sigma_{fc} = 1 - 0.01 = 0.99. \tag{77}$$

Figure 16 is a log-log plot of N_{fc} versus $\bar{s}_{fc}$ for the fundamental clusters as defined above. Note that the data spread or fan is gone and the cluster scaling exponent is consistent with the theoretical prediction of

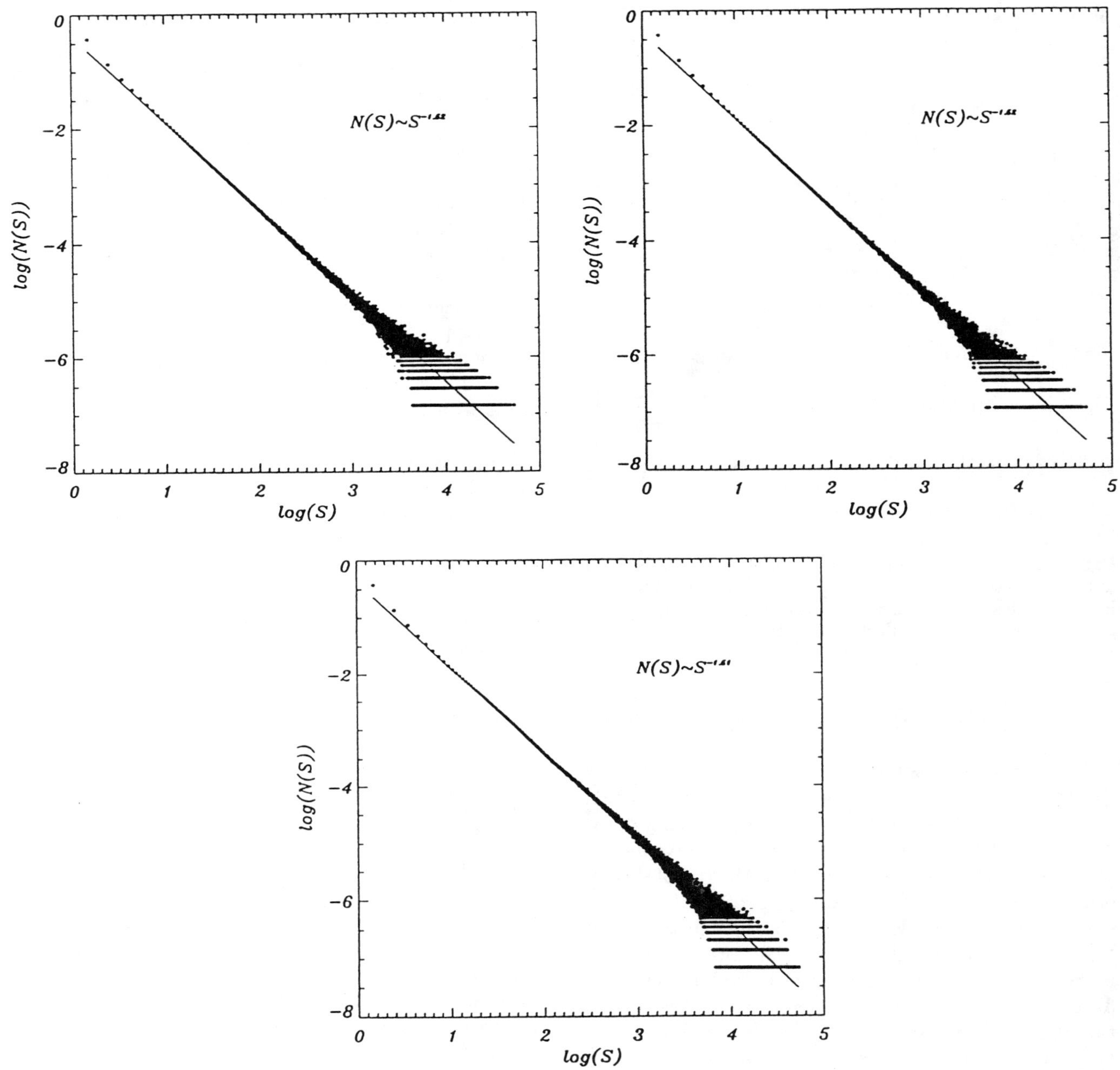

Figure 15. Log-log plots of cluster of all sizes for the same parameters as in Fig. 14 except that the number of clusters for these plots is a) 7 million, b) 9 million, c) 15 million. The number of clusters of each size is normalized by the total number of clusters in that figure. As in the previous figure, the bin size is one.

-1.5. The fundamental clusters are by far the most prevalent. There are slightly more than 17.1 million fundamental clusters out of 18 million events.

We turn now to the next size cluster. The ϕ^4 system indicates that the next scale of correlated objects are the critical phenomena fluctuations. The analysis of the density of these fluctuations (see Eq. (67)) led to the conclusion that they were caused by a variation in the number of fundamental clusters of the order of $C = (R^d \Delta h^{3/2-d/4})^{1/2}$. If this correlated object is a

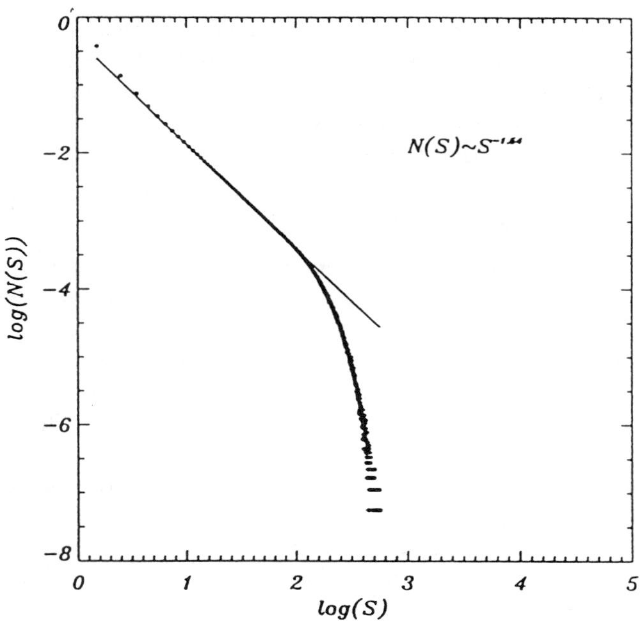

Figure 16. The number of fundamental clusters versus cluster size. The number of events is approximately 17.1 million out of 18 million and the bin size is one.

cluster then the cluster is made up of C fundamental clusters. With the values of R and K we use, $C \sim 4-5$. From the stress drop of a fundamental cluster $\Delta\sigma_{fc}$ given above we have that the stress range of the coalescence clusters is between 0.99 and 0.95. That is, the minimum stress on a coalescence cluster site after the "plate update " and before the start of the failure sequence is between 0.99 and 0.95. In Figure 17 we present a log-log plot of the number of coalescence clusters versus cluster size. Note the scaling with exponent ~ -1.5 and the absence of a fan. We want to emphasize here that these stress cutoffs we have used are approximate in that we can only derive them from scaling laws hence there are constants, or amplitudes, that we do not know that can alter these values. We also wish to stress the fact that the scaling estimates for cluster densities and stress drops are mean values and that the scaling region extend on both sides of these estimates.

Finally we turn to the arrested nucleation clusters. As discussed in Section 5 the mean mass or size of an arrested nucleation event is of the order $\xi^d \Delta h^{1/2}$. However, here as in the above discussion there are constants or prefactors which we do not know (for example see the critical droplet density profile in Eq. (24)). Because of this uncertainty we will look at all clusters with minimum stresses prior to failure which are less than 0.9. The number 0.9 was chosen due to the fact that the coalescence clusters had a minimum stress between 0.95 and 0.99 however, we expect some gap in stress drop due to the fact that there will be clusters bigger than the coalescence clusters (which correspond to a melting of the fundamental clusters that make up a critical phenomena fluctuation) and saddle point objects. In Figure 18 we show a log-log plot of the clusters with a minimum stress less than or equal to 0.9. The slope is -2 in agreement with the assumption of arrested nucleation with no critical slowing down. If we had taken all clusters with a minimum stress prior to failure less than 0.95 then the slope of the scaling plot would be approximately 1.85. Note that there is no arrested nucleation events that scale with the exponent -1.5. It appears that the interaction range we use is large enough to see critical slowing down (see the discussion in Section 5). We checked the stress drop range between 0.9 and 0.95, which is the estimate of the range of arrested nucleation objects with no critical slowing down, and found no evidence of scaling.

We have begun to probe the physical consequences of the existence of different cluster sizes with different physical mechanisms responsible for their presence. As

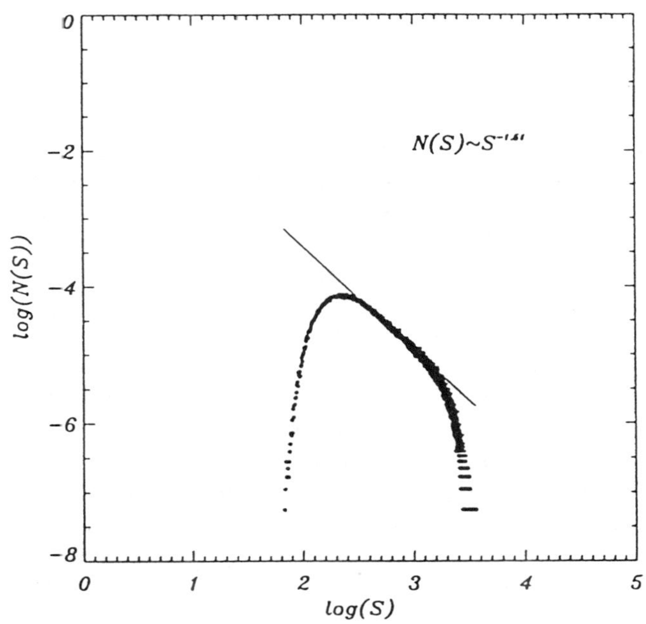

Figure 17. Log-log plot of the number of coalescence clusters versus cluster size. The parameters are the same as the previous figure. The total number of events is approximately 700,000.

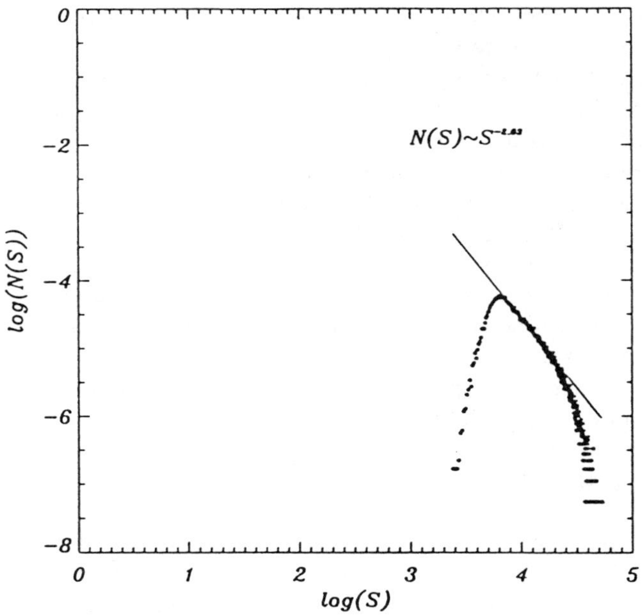

Figure 18. Log-log plot of the number of clusters with minimum stress prior to failure less than or equal to 0.9. The slope is consistent with the prediction of −2. There are approximately 200,000 events plotted. The bin size is one hundred due to the small number of clusters of each size in the large cluster scaling region.

an example we looked at the the number of fundamental, coalescence, and arrested nucleation clusters which precede large events, those which follow large events, and those which are between large events. Large events are those that exhibit break-out and are bigger than any of the events that are described by scaling. We plot the relative number of fundamental clusters, in Figure 19, coalescence clusters, in Figure 20, and arrested nucleation clusters, in Figure 21, that occur in a window of 500 updates both before and after large events, as well as in a window of 500 updates between consecutive large events, as a function of the large event size. The data clearly indicates that the number of coalescence and arrested nucleation events decreases both before and after a large event and stays relatively constant between large events. Note, however, the fluctuations in the number of clusters between large events grows as the event size increases. This is almost certainly due to the small number of very big events leading to bad statistics. On the other hand the fundamental clusters seem to increase in number before and after large events. For the system size used in our simulations the large events span the entire fault and this pattern of activity is characteristic for earthquake activity *inside* the area of the large shock. The analysis of the activity pattern outside the location of the large shocks, using a much larger system size, as well as a systematic comparison with real seismic activity patterns are the subject of current investigations.

9. SUMMARY AND CONCLUSION

In this paper we have presented theoretical arguments and data that support our assertion in a previous publication [*Klein et al*, 1997] that slider block models of the type described in Refs. [*Rundle and Jackson*, 1977] and [*Rundle and Brown*, 1991] obey an Ito-Langevin equation in the limit of long range stress transfer. This derivation and the energy-fluctuation metric studies in Section 6 also support our conclusion, based on data analysis [*Rundle et al*, 1995], that the slider block system can be described with equilibrium techniques in the mean-field limit. Also see Refs. [*Xu and Sornette*, 1997] and [*Rundle et al*, 1997] for additional discussion of this point.

The energy-fluctuation metric study in Section 6 also shows that the time scale over which the slider block

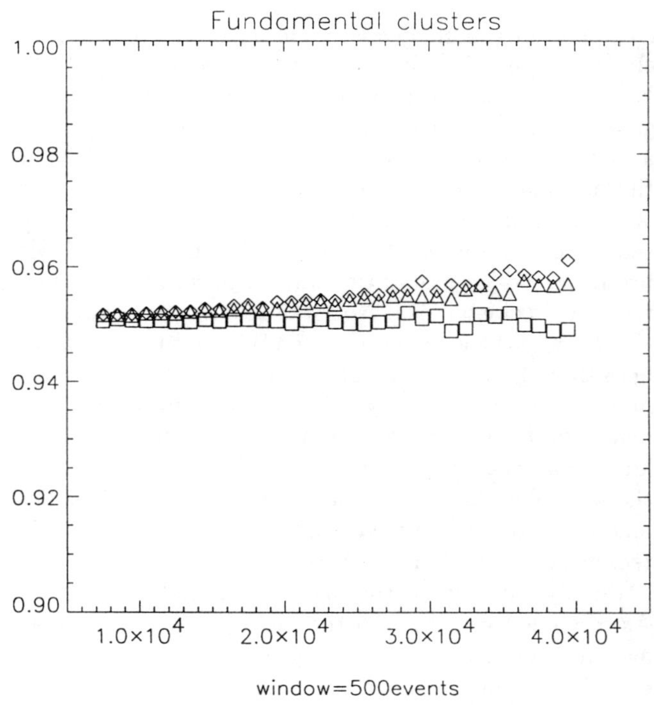

Figure 19. The relative number of fundamental clusters before (diamonds), after (triangles), and between (squares) large events as a function of the large event size. The parameters are the same as in Fig. 14.

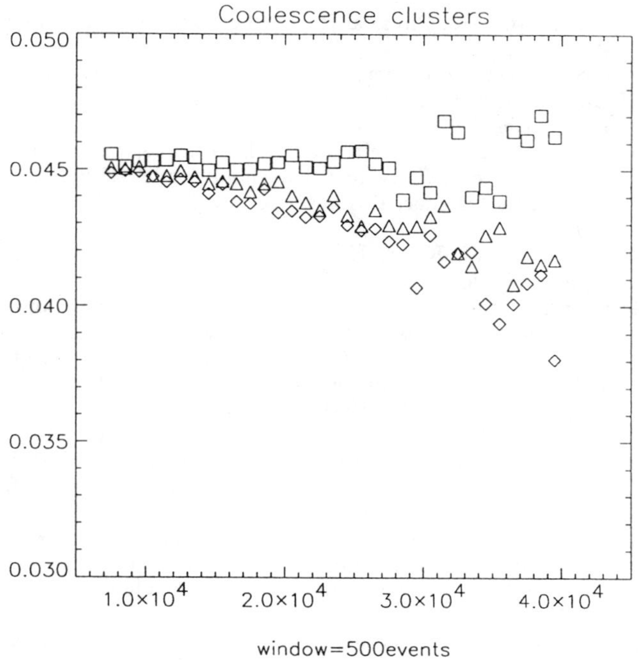

Figure 20. The relative number of coalescence clusters before (diamonds), after (triangles), and between (squares) large events as a function of the large event size. The parameters are the same as in Fig. 14.

system can be described with equilibrium techniques for systems with finite range stress transfer is finite and grows longer as the range of the stress transfer increases.

The picture that emerges from these considerations is of a multiple minima, high dimensional free energy surface (in the system with noise) where the system remains trapped in a particular well for a time scale that diverges as the stress transfer range goes to infinity, that is, in the mean-field limit.

Another result that emerges from this work is the association of GR scaling with a spinodal, or pseudo-spinodal. We have argued that the lower or smaller end of the cluster scaling plots in Figs. 6 and 7 can be explained by associating "earthquake" events with critical phenomena fluctuations. However the upper or larger end of the scaling plots have clusters that are too large to be explained by critical phenomena scaling at the distance the data was taken from the spinodal critical point. We have explained these larger events with the concept of arrested nucleation and provided a theoretical derivation of cluster number scaling, critical slowing down, and cluster size scaling that are consistent with data from simulations of the model. We have also discussed the mechanism that we believe is responsible for the arresting of the nucleation: namely that the plate velocity is not constant in the right time frame but must be treated as a fluctuating variable. We have also mentioned the results of two additional tests that support this idea [*Goldstein et al*, 2000]. This work together with the study in Ref. [*Rundle et al*, 1998] indicates that in this model there are at least four separate "earthquake" orders of magnitude that are generated by different mechanisms. The smallest is associated with the fundamental clusters that are the building blocks of the critical phenomena fluctuations near the spinodal critical point. The second largest events are those associated with the coalescence of fundamental clusters into an object the size of a critical phenomenon cluster. The third largest is associated with arrested spinodal nucleation, including critical slowing down, and the fourth, which is not on the scaling plot (i. e. these events do not scale on a single fault) is associated with break-out events.

Finally, these results raise several interesting questions. Among them are the relation between the events of different sizes that we have seen and the different size events seen in the work of Carlson, Langer and Shaw

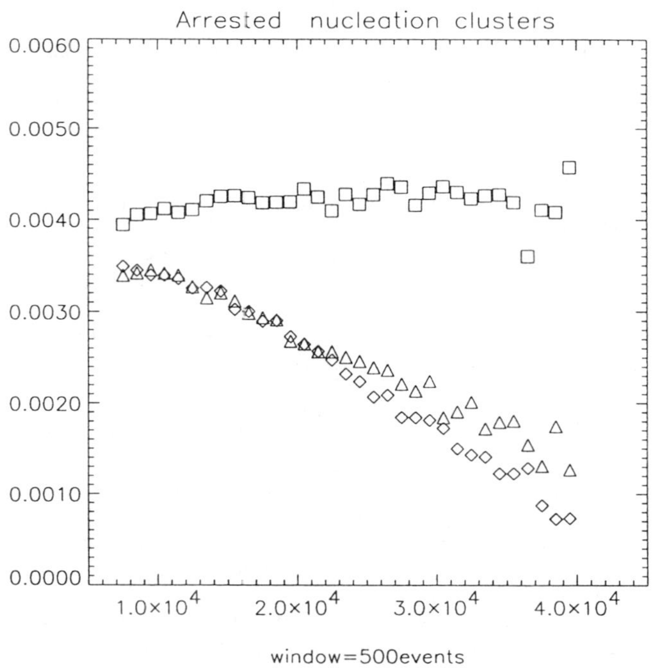

Figure 21. The relative number of arrested nucleation clusters before (diamonds), after (triangles), and between (squares) large events as a function of the large event size. The parameters are the same as in Fig. 14.

[*Carlson and Langer*, 1989; *Carlson et al*, 1991, 1994], the possible existence of additional event size scales, the possible connection between break-out events and the breaking of ergodicity in the energy-fluctuation metric and the relation between the events on the various size scales. For example can critical phenomena size events trigger a break-out event or can that only be done by an arrested nucleation event? Finally, what is the relation between these events and the nucleation phase seen by *Ellsworth and Beroza* [1995] and will other updating mechanisms, CA rules, or other model modifications alter the picture emerging from the study of this model? These and other questions raised by this work are currently being pursued.

Acknowledgments.
W. K. and C. D. F. received support from DOE grant DE-FG02-95ER14498, and J. B. R. and J. S. S. M. received support from DOE grant DE-FG03-95ER14499. We acknowledge useful conversations with F. Alexander, Y. Ben Zion, R. Brower, H. Gould and J. Goldstein. We also acknowledge assistance from the Center for Information Technology of Boston University.

REFERENCES

Alexander, F. J., W. Klein, and M. Anghel, in preparation, 2000.

Anghel, M., W. Klein, J. B. Rundle, and F. Alexander, 2000, Physical Review Letters, submitted, 2000.

Bak, P., and C. Tang, Earthquakes as a self-organized critical phenomenon, Journal of Geophysical Research, *94*, 15635-15637, 1989.

Burridge, R., and L. Knopoff, Model and theoretical seismicity, Bulletin of the Seismological Society of America, *57*, 341-371, 1967.

Cahn, J. W., and J. E. Hilliard, Free energy of a nonuniform system. I. Interfacial free energy, Journal of Chemical Physics, *28*, 258-267, 1958.

Cahn, J. W., and J. E. Hilliard, Free energy of a nonuniform system. III. Nucleation in a two-component incompressible fluid, Journal of Chemical Physics, *31*, 688-699, 1959.

Carlson, J. M., and J. S. Langer, Mechanical model of an earthquake fault, Physical Review A, *40*, 6470-6484, 1989.

Carlson, J. M., J. S. Langer, B. E. Shaw, and C. Tang, Intrinsic properties of a Burridge-Knopoff model of an earthquake fault, Physical Review A, *44*, 884-897, 1991.

Carlson, J. M., J. S. Langer, and B. E. Shaw, Dynamics of earthquake faults, Reviews of Modern Physics, *66*, 657-670, 1994.

Coniglio, A., and W. Klein, Clusters and Ising critical droplets: a renormalization group approach, Journal of Physics A, *13*, 2775-2780, 1980.

de Sousa Vieira, M., and H. J. Herrmann, Self-similarity of friction laws, Physical Review E, *49*, 4534-4541, 1994.

Ellsworth, W. L., and G. C. Beroza, Seismic evidence for an earthquake nucleation phase, Science, *268*, 851-855, 1995.

Feder, H. J. S., and J. Feder, Self-organized criticality in a stick-slip process, Physical Review Letters, *66*, 2669-2672, 1991.

Ferguson, C. D., Numerical investigations of an earthquake fault based on a cellular automaton, slider-block model, Ph.D. thesis, Boston University, 1996.

Ferguson, C. D., W. Klein and J. B. Rundle, Spinodals, scaling, and ergodicity in a threshold model with long-range stress transfer, Physical Review E, *60*, 1359-1373, 1999.

Goldstein, J., W. Klein, H. Gould, and J. B. Rundle, in preparation, 2000.

Gomberg, J., talk presented at the 21st International Conference on Mathematical Geophysics, 1996.

Gross, N., W. Klein, and K. Ludwig, Structure and failure of the linear theory of continuous ordering, Physical Review Letters, *73*, 2639-2642, 1994.

Gu, J. C., J. C. Rice, A. L. Ruina, and S. T. Tse, Slip motion and stability of a single degree of freedom elastic system with rate and state dependent friction, Journal of Mechanics and Physics of Solids, *32*, 167-196, 1984.

Gunton, J. D., P. Sahni, and M. San Miguel, in *Phase Transitions and Critical Phenomena*, volume 8, edited by C. Domb and J. Lebowitz, pp. 269-466, Academic Press, New York, 1983.

Gutenberg, B., and Richter, C. F. *Seismicity of the Earth and Associated Phenomena*, Princeton University Press, Princeton, NJ, 1954.

Heermann, D. W., W. Klein, and D. Stauffer, Spinodals in a long-range interaction system, Physical Review Letters, *49*, 1262-1264, 1982.

Heermann, D. W., and W. Klein, Nucleation and growth of nonclassical droplets, Physical Review Letters, *50*, 1062-1065, 1983.

Heermann, D. W., and W. Klein, Percolation and droplets in a medium-range three-dimensional Ising model, Physical Review B, *27*, 1732-1735, 1983.

Herz, A. V. M., and J. J. Hopfield, Physical Review Letters, *75*, 1222-1225, 1995.

Hill, D. P., P. A. Reasenberg, A. Michael, W. J. Arabaz, G. Beroza, D. Brumbaugh, J. N. Brune, R. Castro, S. Davis, D. dePolo, W. L. Ellsworth, J. Gomberg, S. Harmsen, L. House, S. M. Jackson, M. J. S. Johnston, L. Jones, R. Keller, S. Malone, L. Munguia, S. Nava, J. C. Pechmann, A. Sanford, R. W. Simpson, R. B. Smith, M. Stark, M. Stickney, A. Vidal, S. Walter, V. Wong, and J. Zollweg, Seismicity remotely triggered by the magnitude 7.3 Landers, California, earthquake, Science, *260*, 1617-1623, 1993.

Huang, J., and D. Turcotte, Evidence for chaotic fault interactions in the seismicity of the San Andreas fault and Nonkai trough, Nature, *348*, 234-236, 1990.

Klein, W., and C. Unger, Pseudospinodals, spinodals, and nucleation, Physical Review B, *28*, 445-448, 1983.

Klein, W., Fractals and multifractals in early stage spinodal decomposition and continuous ordering, Physical Review Letters, *65*, 1462-1465, 1990.

Klein, W., J. B. Rundle, and C. D. Ferguson, Scaling and nucleation in models of earthquakes faults, Physical Review Letters, *78*, 3793-3796, 1997.

Langer, J. S., Theory of the condensation point, Annals of Physics, *41*, 108-157, 1967.

Langer, J. S., Statistical theory of the decay of the metastable states, Annals of Physics, *54*, 258-275, 1969.

Lebowitz, J. L., and O. Penrose, Journal of Mathematical Physics, *7*, 98, 1966.

Lee, J., M. A. Novotny, and P. A. Rikvold, Method to study relaxation of metastable phases: macroscopic mean-field dynamics, Physical Review E, *52*, 356-372, 1995.

Ma, S. K., *Modern Theory of Critical Phenomena*, Benjamin, Reading, Mass., 1976

Monette, L., and W. Klein, Spinodal nucleation as a coalescence process, Physical Review Letters, *68*, 2336-2339, 1992.

Nicolis, G., and I. Prigogine, *Self Organization in Non-Equilibrium Systems*, J. Wiley, New York, 1977.

Olami, Z., H. J. S. Feder, and K. Christensen, Self-organized criticality in a continuous, nonconservative cellular automaton modeling earthquakes, Physical Review Letters, *68*, 1244-1247, 1992.

Press, W. H., S. A. Teukolsky, W. T. Vetterling, and B. P. Flannery, *Numerical Recipes: The Art of Scientific Computing*, 2nd edition, Cambridge University Press, Cambridge and New York, 1992.

Ray, T., and W. Klein, Crossover and the breakdown of hyperscaling in long-range bond percolation, Journal of Statistical Physics, *53*, 773-794, 1988.

Ray, T., and W. Klein, Nucleation near the spinodal in long-range Ising models, Journal of Statistical Physics, *61*, 891-902, 1990.

Rundle, J. B., and D. D. Jackson, Numerical simulation of earthquake sequences, Bulletin of the Seismological Society of America, *67*, 1363-1377, 1977.

Rundle, J. B., and W. Klein, Nonclassical nucleation and growth of cohesive tensile cracks, Physical Review Letters, *63*, 171-174, 1989.

Rundle, J. B., and S. R. Brown, Origin of rate dependence in frictional sliding, Journal of Statistical Physics, *65*, 403-412, 1991.

Rundle, J. B., and W. Klein, Proceedings of the 33rd Symposium on Rock Mechanics, edited by J. R. Tillerson and W. R. Wawwersik, A. A. Balema, Rotterdam, 1992.

Rundle, J. B., and W. Klein, Scaling and critical phenomena in a cellular automaton slider-block model for earthquakes, Journal of Statistical Physics, *72*, 405-412, 1993.

Rundle, J. B., and W.Klein, Dynamical segmentation and rupture patterns in a 'toy' slider block model for Earthquakes, Nonlinear Proc. in GeoPhysics, *2*, 61-81, 1995.

Rundle, J. B., W. Klein, S. Gross and D. Turcotte, Boltzmann fluctuations in numerical simulations of nonequilibrium lattice threshold systems, Physical Review Letters, *75*, 1658-1661, 1995.

Rundle, J. B., W. Klein, and S. Gross, Rupture characteristics, Recurrence, and predictability in a slider-block model for earthquakes, in *Reduction and Predictability of Natural Disasters*, Santa Fe Institute Studies in the Sciences of Complexity, volume XXV, edited by J.B. Rundle, D.L. Turcotte, and W. Klein, pp. 167-203, Addison-Wesley, Reading, Mass., 1996.

Rundle, J. B., W. Klein, S. Gross, and D. L. Turcotte, Reply, Physical Review Letters, *78*, 3798, 1997.

Rundle, J. B., W. Klein, S. Gross, and C. D. Ferguson, Traveling density wave models for earthquakes and driven threshold systems, Physical Review E, *56*, 293-307, 1997.

Rundle, J. B., E. Preston, S. McGinnis, and W. Klein, Why earthquakes stop: growth and arrest in stochastic fields, Physical Review Letters, *80*, 5698-5701, 1998.

Rybicki, K., in *Continuum Theories in Solid Earth Physics*, edited by R. Teisseyre, Elsevier, Amsterdam, 1986.

Scholz, C. H., The Mechanics of Earthquakes and Faulting, Cambridge University Press, Cambridge, 1990.

Stauffer, D., and A. Aharony, *Introduction to Percolation Theory*, 2nd edition, Taylor and Francis, Washington, D. C., 1992.

Steketee, J. A., On Volterra's dislocations in a semi-infinite elastic medium, Canadian Journal of Physics, *36*, 192-205, 1958.

Thirumalai, D., and R. D. Mountain, Ergodic convergence properties of supercooled liquids and glasses, Physical Review A, *42*, 4574-4587, 1990.

Thirumalai, D., and R. D. Mountain, Activated dynamics, loss of ergodicity, and transport in supercooled liquids, Physical Review E, *47*, 479-489, 1993.

Unger, C., and W. Klein, Nucleation theory near the classical spinodal, Physical Review B, *29*, 2968-2708, 1984.

Van Kampen, N. G., and B. U. Felderhof, *Theoretical Methods in Plasma Physics*, North-Holland, Amsterdam, 1967, and references therein.

Vasconcelos, G.L., First order phase transition in a model for earthquakes, Physical Review Letters, *76*, 4865-4868, 1996.

Xu, H. J., and D. Sornette, Non-Boltzmann fluctuations in numerical simulation of nonequilibrium lattice threshold systems, Physical Review Letters, *78*, 3797, 1997.

W. Klein and M.Anghel, Physics Department, Center for Polymer Physics, and Center for Computational Science, Boston University, 590 Commonwealth Ave., Boston, MA 02215

C. D. Ferguson, Federation of American Scientists, Washington, DC 20002

J. B. Rundle and J. S. Sá Martins, Colorado Center for Chaos and Complexity and CIRES, University of Colorado, Boulder, CO 80309

Traveling Wave and Rough Fault Earthquake Models: Illuminating the Relationship Between Slip Deficit and Event Frequency Statistics

Susanna J. Gross

CIRES, University of Colorado, Boulder, Colorado

We discuss and interpret new simulation results from a recently proposed, physically-based model of earthquakes called the traveling wave model. This model produces a mixture of scaling and characteristic event ruptures. Stresses are transferred well beyond nearest neighbors in the two-dimensional lattice which represents the fault in the model. Cohesive forces due to small scale fault topography produce large scale friction, showing how friction is a function of length scale. Two variants of this new earthquake model, called 1) the basic traveling wave and 2) the rough fault model, illustrate the significant role the loading process has upon the statistics of the simulated events produced. Scale invariant rupture area distributions arise in these models in the same manner that they do in cellular automata slider block models, as a result of the force balance applied to the location of each site. Event frequency statistics are in the range of those observed for natural seismicity. Results from these models are compared to classical slider block models with long range springs.

1. INTRODUCTION

Models of earthquake sources can be broadly classified into three categories, observational, complex synthetic, and simplified synthetic. Models of specific earthquake observations generally focus on description of the patterns of moment release, not the source preparation process. [i.e. *Ruff*, 1992, *Mendoza et al.*, 1994.] By contrast, source preparation models have generally not been fit to observations of specific earthquake sequences. They are instead concerned with reproducing the generic properties of seismicity. Complex synthetic source preparation and rupture models include most known and empirically derived physical processes that operate in seismic source zones, and require strong material heterogeneity to produce populations of small earthquakes [*Rundle and Klein*, 1993, *Ben-Zion and Rice*, 1995]. Trade-offs and interactions between different physical processes can make interpretation of the results of the more complicated models difficult. These models often include long range stress transfer, dynamical formulations of the rupture process, explicit material heterogeneity and empirical friction relations, [*Mikumo and Miyatake*, 1979, *Ward*, 1992, *Ward and Goes*, 1993, *Rice* 1993, *Ben-Zion and Rice*, 1993].

Simplified synthetic models of source processes often neglect known physics in favor of easier interpretation and are better suited to finding the answers to thought experiments such as: Is long range stress transfer necessary for long range clustering? Is material heterogeneity necessary for slip complexity and the Gutenberg-Richter distribution? A large number of authors have constructed slider block [*Burridge and Knopoff* 1967] and other simplified source models. Examples include *Rundle and Klein* [1993], *Ding and Lu* [1993], *de Sousa Vieira et al.* [1993], *Lomitz-Adler* [1993], *Pepke et al.*, [1994] and *Rydelek and Sacks*, [1996].

GeoComplexity and the Physics of Earthquakes
Geophysical Monograph 120

One reason for the continued interest in slider block models is their ability to reproduce the power law distribution of earthquake magnitudes remarkably well, even though many of these models lack any effects of momentum, fault populations, or even explicit material heterogeneity. The dynamics of these models are expressed in the form of simple rules representing Coulomb friction and short range stress transfer. *Rundle and Kanamori* [1987] have shown that the healing process can produce heterogeneous stress distributions on faults that have failed in a large earthquake, and it is this healing process that perhaps most strongly distinguishes slider block models from more realistic models of earthquakes. The rapid healing causes slip to be concentrated in a pulse just behind the rupture front, another respect in which these models unexpectedly resemble the phenomenology of real earthquakes [*Heaton*, 1990, *Zheng et al.*, 1996].

The traveling wave model was first presented by *Rundle et al.* [1997] who also carried out an analysis of the model based upon statistical mechanics. The rough fault model presented below is a new variant of the traveling wave model. The present paper also presents some new interpretations of size distribution scalings. Both models are built on the tradition of simple models, but they represent an attempt to rationalize the evolution rules in terms of quantitative physical principles, so they are a step closer to more realistic source models. Rather than including detailed empirical descriptions of fault friction, we have instead taken the approach of seeing what friction properties our model can produce from the geometry of the rough fault in the model alone. Notice that we use the term "friction" here in the sense of a macroscopic resistance to sliding, in contrast to the cohesive forces over microscopic length scales on the surface. In the course of the work, we have gained a greater understanding of the mechanisms that produce the power law distribution of simulated earthquakes in all slider block models, and how these mechanisms are related to the configuration of the underlying slip deficit field. The strength of the coupling springs linking adjacent blocks has a strong influence upon the geometry of the slip deficit field, and also the size distribution of synthetic earthquakes. We have found two distinct scaling regimes corresponding to strongly and weakly coupled cases.

2. SIMULATION TECHNIQUES

The traveling wave and rough fault models presented here simulate earthquakes on a single two-dimensional fault with a square lattice of 10,000 sites distributed

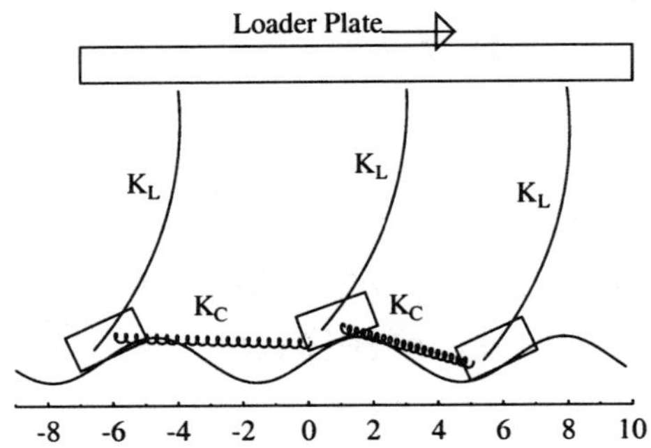

Figure 1. This sketch represents a one dimensional slice through the rough fault or traveling wave model, with the fault surface roughness exaggerated for visibility. Loader springs with constants K_L drive blocks. The blocks are coupled together with springs having constants K_C/d^3. Since there is normal stress exerted by the loader plate, the low spots on the fault are more energetically favored than the high spots.

upon it. The dynamics are quasi-static, and there is no explicit material heterogeneity. As figure 1 shows, the sites rest upon a fault surface with a sinusoidal irregularity that moves relative to the loader plate, causing the loading springs to stretch as the sites move to follow the more energetically favored locations on the irregular fault surface. The same sinusoidal fault topography is experienced by each site, apart from variations in phase, so there is no spatial inhomogeneity of fault properties. Coupling springs connect not only the four nearest neighbors, as in most cellular automata slider block models, but also all other sites within r lattice units of each other, with coupling spring constants that decay with the distance cubed. The inverse-cube decay is characteristic of static stress transfer in three-dimensional elastic solids. The cases with $r = 2$ connect each site with 24 neighbors, and $r = 5$ connects each site with 120 neighbors. A model with range of interaction r connects each block to $(2r+1)^2 - 1$ neighbors. The equations describing temporal evolution of the basic traveling wave and rough fault models come from the minimization of the potential energy, which has a term that represents the effects of a rough surface, and that reproduces many of the effects of friction.

The potential energy E is:

$$
\begin{aligned}
E &= \sum_{i=1}^{N} \frac{K_L}{2}\phi_i^2 \qquad (1)\\
&+ \sum_{j=-r}^{r} \frac{K_C}{2d(j)^3}(\phi_i - \phi_{i+j})^2\\
&+ 2\gamma\cos(\kappa[\phi_i - vt + \epsilon_i]) + h\phi_i,
\end{aligned}
$$

in which the spring constant K_c controls the coupling between sites. Coupling stresses are transferred between sites a distance $d(j)$ apart following an inverse cube law out to the range of interaction r. The slip deficit $\phi = s_i(t) - vt$ is a measure of the displacement of each site behind the load point, whose loading spring has spring constant $K_L = 1$. The cosine term in the expression above can be thought of as the leading term in a Fourier expansion representing fault topography, with wavelength $2\pi/\kappa$, amplitude γ and phase that advances with velocity v as a function of time t. This cohesion term is a result of irregularities in the surface that create energetically favored locations for the sites. The variable ϵ_i is a phase scatter dependent upon location that is redistributed every time there is a major jump. Conceptually, every time a site moves past a feature on the fault surface, the next feature will appear at a location having a phase ϵ_i relative to the average position of features. The term $h\phi_i$ gives the potential well the simplest form of asymmetry, so it is included for generality. This expression can be solved to find the position of each site, ϕ_i, which minimizes the potential energy E as a function of the locations of the other sites and the coupling and loader spring constants, K_C and K_L.

The sketch in figure 1 shows the basic geometry of a one-dimensional slice through either the traveling wave or rough fault model. The cohesion term affects the force balance in the direction of site movement because normal stresses encourage the movement of the sites into the low spots, which are potential energy minima.

We find the net force on the i'th site by taking a negative derivative of the potential energy with respect to the slip deficit ϕ_i.

$$\begin{aligned} F_i &= K_L\phi_i \\ &+ \sum_{j=-r}^{r} \frac{K_C}{d(j)^3}(\phi_i - \phi_{i+j}) \\ &- 2\gamma\kappa\sin(\kappa[\phi_i - vt + \epsilon_i]) + h \end{aligned} \qquad (2)$$

At each time t, the site locations ϕ_i are found that reduce the residual force F to a small value η. The level of accuracy, or threshold stress η can be thought of as representing the frictional effects of features smaller than the dominant wavelength $2\pi/\kappa$. The introduction of this threshold regularizes the problem and renders it much more tractable. Solutions have a tolerance or noise level η, because the locations of sites are only updated if their net forces are greater than η. Reducing the size of η allows us to approach the solutions which

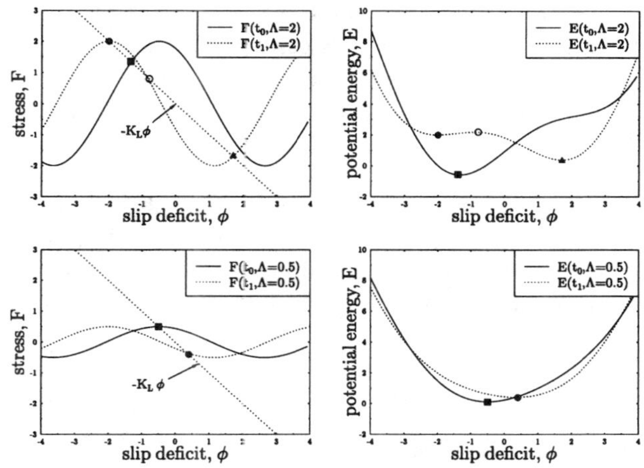

Figure 2. Curves on these plots represent the evolution of the friction term with time, showing the impact of the friction upon the forces and potential energy experienced by a single site in the traveling wave model. The slanted lines on the left-hand panels represent forces exerted by the springs. Intersections of the two curves are solutions to the force balance equation.

would be obtained in the case of a fault whose only friction arises from irregularities described by the potential in equation 1.

As an example of how the traveling wave model represents the earthquake process, consider figure 2, in which both the energy (1) and the force (2), are plotted for a single site and the noise $\epsilon_i = \eta_i = 0$. In the panels on the left, the term $-K_L\phi$ is the bold dotted line, and the solid and light dotted lines are the cosine terms at the two times t_0, t_1. The point(s) of intersection represent solutions to the equation $F(\phi) = 0$. The right panels show the corresponding points on the plot of $E(\phi(t))$. In the top panel, the quantity $\Lambda = 2\gamma\kappa^2/K_L$ is $\Lambda = 2$, whereas in the bottom panels, $\Lambda = 0.5$. As *Rundle et al.* [1996, 1998] have shown, when $\Lambda < 1$ (bottom), there is only one point of intersection, corresponding to the one stable solution. However, for $\Lambda > 1$, there can be one, two, or three points of intersection. A single solution point occurs when there exists only one minimum in the energy curve. Three solutions occur when there are two minima, separated by a maximum. Two solutions occur when the meta-stable minimum and the unstable maximum merge into one point.

The distribution of the values ϵ_i of the phase scatter provides information about the spatial scale of the fault surface irregularities as compared to the lattice spacing. If the phase scatter is small, there is a strong correlation between the fault shape seen by all sites. Models of this kind might be suitable for studies of the preparation process, rupture and recovery of a large asperity, because these models contain events in which all sites jump simultaneously and quasi-periodically. The smaller events that occur in the adjustment process that follows a nucleation event of this type may resemble aftershocks.

The techniques needed to implement this model on a computer are different from conventional slider block models, because the usual "update rules" are replaced with a nonlinear routine based upon Newton's method for locating roots.

3. TEMPORAL EVOLUTION ALGORITHMS

3.1. Basic Traveling Wave Model

As described in *Rundle et al.* [1998], the basic equation to be solved for the slip deficit field $\phi_i(t)$ is based upon a gradient decent method subject to noise :

$$\frac{d\phi_i}{dt} = F_i + \eta. \tag{3}$$

However, due to the nonlinear nature of the sine term in F_i, this algorithm is computationally expensive to implement for large lattices. We have therefore been motivated to develop alternative approximate solution schemes that nonetheless yield the correct general temporal evolution and scaling behavior. These algorithms are described below.

1. Site locations ϕ and starting phases $\kappa\epsilon$ are either generated or read in from a prior model. Other parameters like the coupling spring constant K_C, the range of interaction r and the solution accuracy η are defined at this time as well.

2. The stresses F_i on all of the sites due to the cohesion, coupling and loading springs are computed.

3. The stresses on all the sites are compared with the solution accuracy η and the site closest to failure is selected. The failure time $t_f = t + \Delta t$ is found that makes the stress on that site exceed the threshold η. This time must be in the future, and is found by projecting the rate of change of F_i with slip deficit into the future,

$$\frac{\Delta F_i}{\delta \phi_i} \approx \frac{\partial F_i}{\partial \phi_i} = K_L - 2\gamma\kappa^2 \cos\kappa[\phi_i + vt + \epsilon_i]. \tag{4}$$

To estimate ΔF_i, we compare F_i to the small threshold η, hence

$$\Delta F_i \approx F_i - \eta. \tag{5}$$

Using $\delta\phi_i = -v\Delta t$, we finally have

$$\Delta t \approx \frac{\eta - F_i}{v\{K_L - 2\gamma\kappa^2 \cos\kappa(\phi_i + vt + \epsilon_i)\}}. \tag{6}$$

4. The time is set to the estimated failure time, t_f. If the stress at that time, F_i is not greater than η, steps 2 and 3 are repeated. We increment the time only enough to bring the next site to failure and run the rupture without increasing time because of the tremendous difference between plate velocities, which move a few centimeters per year, and rupture velocities, which move a few kilometers per second.

5. Now the stress on the least-stable block is greater than the threshold, η. Before slip adjustment, the region of potential slip is set to the entire lattice.

6. Stresses are recomputed for the region of potential slip, and then compared with the threshold η. All those that exceed the threshold are flagged for adjustment. If there are no additional adjustments needed, statistics for this event are computed and saved, and the algorithm returns to step 3.

7. Using Newton's method, a new location for each site is calculated that satisfies

$$F_i(\phi_i) < \frac{\eta}{10}, \tag{7}$$

No sites are moved at this step.

8. The locations of the sites are then updated together. If the number of updates exceeds 20, the new location is averaged with the old one so that oscillations are damped. The region of potential slip is adjusted to contain all sites that slipped plus a buffer of width r, and then step 6 begins again.

3.2. Rough Fault Model

The evolution algorithm for the rough fault version of the traveling wave model is like the basic traveling wave model except for the loading process. During loading in the basic traveling wave model the sites change position relative to the cohesive forces. Loading in the rough fault model is different, because during loading the sites move to follow the features of the cohesion term, accumulating stress from the loading springs until they jump. Only steps 3 and 4 are different. To estimate the failure time in this case, we calculate Δt using the relation:

$$\frac{dF_i}{dt} \approx \frac{\Delta F_i}{\Delta t}. \tag{8}$$

We again estimate ΔF_i from (5). To calculate $\frac{dF_i}{dt}$, we differentiate (2):

$$\begin{aligned}\frac{dF_i}{dt} &= \frac{d}{dt}F_i(\phi_j(t),t) = \Sigma_j \frac{\partial F}{\partial \phi_j}\frac{\partial \phi_j}{\partial t} + \partial F_i/\partial t \\ &= \{K_L - 2\gamma\kappa^2 \cos\kappa[\phi_i + vt + \epsilon_i]\}(-v) \\ &- \left(2\gamma\kappa^2 \cos\kappa[\phi_i + vt + \epsilon_i]\right) v \\ &= -K_L v \end{aligned} \tag{9}$$

Combining (5), (8), and (9), we obtain

$$\Delta t \approx (\eta - F_i)/vK_L. \tag{10}$$

Step four computes the failure time from: $t_f = t + \Delta t$, which is then used to update the locations:

$$\phi(t + \Delta t) = \phi(t) + v\Delta t \tag{11}$$

and the stresses $F_i(t + \Delta t) = F_i(t) + vK_L\Delta t$.

When sites move in the rough fault model, they always release elastic energy. In the traveling wave model, sites must move to store elastic energy as well as to release it. The rules for slip adjustment are the same. For the rough fault model, the solution accuracy η may be interpreted as a friction resulting from irregularities small compared to the microscopic length scale that characterizes the sites.

4. MODEL PHENOMENA

4.1. Friction Results

One of the most interesting properties of the traveling wave and rough fault models is their ability to model bulk, macroscopic friction, arising from "microscopic" cosine bumps. The cohesion term in the potential representing fault topography produces a bulk frictional force when it is averaged over the fault surface. This synthetic bulk frictional force resists the motion of the loader plate in whatever direction the loading force is applied, just as natural frictional forces do. Unlike conventional slider block models, the failure stress η does not dominate resistance to loader plate movement. Physically, the frictional force in the traveling wave and rough fault models arises as a result of the normal stress σ_N pushing the fault together, acting on the fault topography, that gives rise to the cohesive energy 2γ and h. This normal stress renders some displacements more energetically favored than others. The amplitude of the cohesion term γ is proportional to the normal stress σ_N. The displacement of a site thus represents a balance between coupling forces, loading shear stresses, small scale friction, and the cohesive energy (potential field) created by compressive stresses normal to the fault.

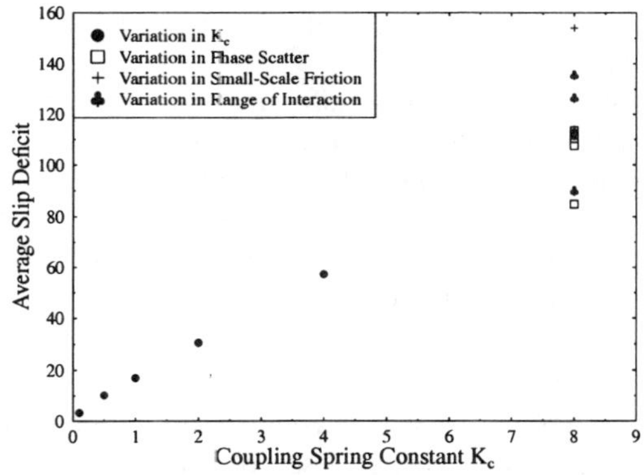

Figure 3. Variations in coupling spring constant have a large effect upon the magnitude of the cohesive forces in the rough fault model, which are measured in terms of average slip deficit. The small-scale friction also contributes, but is not necessary in order to produce significant cohesion. These models all have square lattices 100 elements on a side.

The small-scale friction term, or error term η is the frictional slip threshold that applies to single sites. The cohesive bulk frictional force is much larger scale, because it is an average property of the entire model, representing the average resistance to sliding of the whole system of 10,000 sites. Even when the small scale friction is made exceedingly small, the large scale friction remains essentially unchanged, because the cohesion comes from the balance between fault topography and normal stress. The topography given to the model fault in this case is a simple sinusoid, but it is still capable of producing cohesive forces which oppose the motion of the loading plate. The cohesive forces do not average to zero because the most energetically favored locations for sites are not symmetrically distributed about the minima in cohesive energy. Stable locations for sites are offset toward the loading direction from the cohesive energy minimum, which causes the cohesive force to oppose the motion of the loading plate. As figure 3 illustrates, cohesive forces that give rise to bulk friction in the model are proportional to the coupling spring constant. That is, there is a significant dependence of the bulk friction on microscopic interactions between sites.

If the fault model is re-scaled, so that all the sites together represent a single block of a larger model, then bulk friction of the smaller scale becomes the sliding stress η of the large scale. This picture justifies the introduction of significant microscopic friction into models which are supposed to represent macroscopic fault scales. Faults possess irregularities on a broad range of scales, so it is reasonable to suppose that realistic models of faults might have significant frictional forces on multiple scales, as the models with nonzero η and cohesion do.

There are significant differences between the cohesive friction of the traveling wave models and classic Coulomb friction. The cohesive models have no explicit difference between static and dynamic macroscopic friction, because there are no processes modeled which are inherently time dependent.

4.2. Scaling Properties

The traveling wave and rough fault models produce sequences of failures that resemble natural earthquakes in their power-law size distributions and in some cases show increased proportions of large magnitude "characteristic" events. The mechanisms that produce the power-law distribution of small events are the same in the traveling wave and rough fault models as in classic slider block models, so much of the discussion that follows applies to all three types of models.

Sizes of simulated earthquakes in a slider block model are governed by the slip deficit distribution, also called the displacement field. Each site is displaced from its equilibrium position by an amount ϕ which is a function of location across the lattice. If the rupture front encounters a set of sites which are already displaced in the direction in which the rupture is causing sites to slip, the rupture may stop because the sites on the edge cannot be induced to fail by the stresses transfered from the rupture. Discontinuities in the displacement field correspond to bands of strong positive and negative stresses, features unfavorable for the propagation of ruptures. This means that discontinuities in displacement field are likely to arrest ruptures. If we call each region of similar displacement in the displacement field a segment, we expect the population of segments in the displacement field will at least in part control the population of simulated earthquakes generated. Relatively smooth fields with typically large distances between displacement discontinuities should produce simulations rich in large earthquakes. Rough distributions with short distances between discontinuities should produce more small area ruptures.

The examples shown in figure 4 are statistics from earthquake models of three different types and a few choices of model parameters. At small values of the noise threshold η, the cohesion term is very important, and the phase scatter ϵ strongly influences the scaling. Large amounts of phase scatter produce greater numbers of small area ruptures, and small phase scatter reduces the influence of the cohesion term significantly, producing distributions more similar to the original slider block model and to natural earthquakes. In

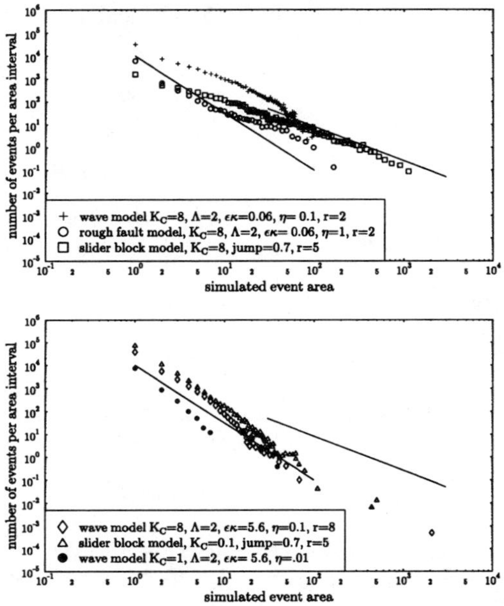

Figure 4. The parameters for three types of models are shown in the figure keys on these plots. The top panel shows cases with size distributions having a power-law slope close to -3/2, and the lower panel shows cases with slopes nearer -5/2. The high noise case plotted with diamonds exhibits scaling more similar to weakly coupled models, since the noise disrupts large scale structure in the slip deficit field. The traveling wave case plotted with filled circles has a characteristic earthquake distribution. The maximum event area possible is 10,000 elements.

effect, larger phase scatter (annealed disorder) serves to reduce the correlation length of the slip deficit field, thus producing smaller events. When η is very large, the cohesion term is only a small perturbation upon it, and all three types of models exhibit similar scaling. This scaling is dependent only upon the degree of coupling, K_c/K_L in the models.

Variations in the range of interaction r do not produce any obvious systematic trends in the event-size statistics. Longer ranges of interaction do enhance the edge effects in the models though, and they increase the initial transient, meaning that many more events need to be simulated before the statistical characteristics stabilize. Longer ranges of interaction are also more realistic, since the range of interaction in the earth is effectively infinite, but the strongest stress transfer is at close ranges, and so we have usually used a range of interaction near 5, which includes interactions only 1% the size of the nearest neighbor stress transfer.

4.2.1. Power Spectra. In recent work *Rundle et al.*, 1998], John Rundle proposed a stochastic fracture mechanics framework that extends the usual Gibbs-Griffith

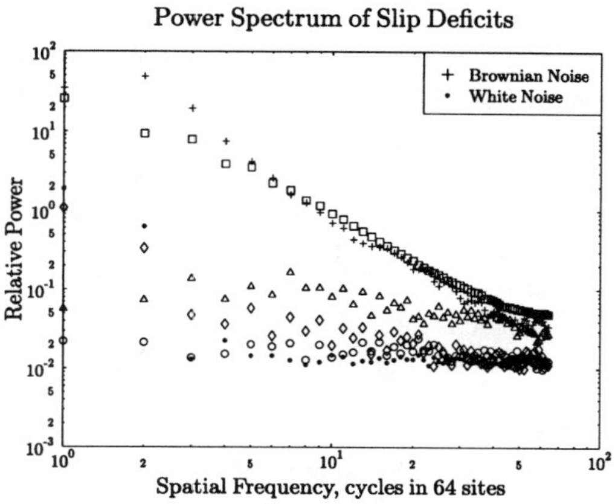

Figure 5. Power spectra of slider block slip deficit fields for four models are compared with those for brownian and white noise. Symbols represent the same models as in figure 4.

approach to brittle shear fracture. This framework is an attempt to quantify the physical idea that a shear fracture will only extend through that region of a material in which the average difference between fracture strength $\sigma^F(x)$ and shear stress $\sigma(x)$ is sufficiently small. Thus we are interested in the statistical correlation properties of the field $\Sigma(x) = \sigma^F(x) - \sigma(x)$. In the slider block models of figure 4, $\sigma^F(x)$ is a spatial constant. For the traveling wave models, the analogous quantities, 2γ and h, are constants. Therefore, correlations in $\sigma(x)$ determine the correlations in $\Sigma(x)$, which result from curvature in $\phi(x,t)$, the slip deficit field (see equation 2).

Correlations in the slip deficit field can be analyzed by studying the power spectrum of slip deficit, which is the Fourier transform of the pair correlation function of the slip deficit between sites. Examples of power spectra of slip deficit fields for the models shown previously in figure 4 are given in figure 5 using methods that have been discussed extensively in the literature [*Voss*, 1988]. These power spectra were measured from simulation results by taking the slip deficit field $\phi(x,y,t)$, and extracting strips $\phi(x,t)$. These one dimensional data were then Fourier transformed and squared to produce the power spectra. The slopes of these power spectra at the high frequency end (large wavenumber k) yield values for the power spectrum exponent $-\beta$.

Exponents β range from roughly zero to about 2. The latter value is also characteristic of brownian noise, as can be seen from the crosses (+). The theory of stochastic fracture [*Rundle et al.*, 1998] predicts that the eventual size of shear fractures should depend on the value of the Hurst exponent H characterizing correlations in slip deficit field. For $H = 0.5$, events of all sizes are expected. For values of H less than 0.5, system-wide nucleation ("characteristic") events are possible. Finally, for values of H greater than 0.5, a maximum size event is predicted.

We determined the value of H by measuring the slope of the power spectrum and using the relationship (*Voss*, [1988], and Appendix A):

$$D = E + 1 - H = E + \frac{3-\beta}{2} \qquad (12)$$

In (12), E is the Euclidean dimension of the independent variable over which the power spectrum is computed. Because we compute the power spectrum for one dimensional curves, here $E = 1$. D is the associated fractal dimension of the curve, H is the Hurst exponent and β is the exponent characterizing the power spectrum. Note that equation (12) assumes that the curves are self similar over all scales, and thus, for example, that there is a well defined value for β, over all spatial frequencies, leading to well defined values of H and D. The power spectra for the slip deficits in figure 5 can be seen to have fixed slopes ($-\beta$) only at the high frequency (shorter wavelength) end of the spectrum in general. This implies the existence of a finite correlation length.

With these considerations in mind, we turn to an examination of figures 4 and 5. To check the predictions of stochastic fracture mechanics, we compare models that a) have a large enough value of r that stress concentrates at the rupture tip as the rupture grows, and b) preferably have the same value for K_C. The two models that best fit these criteria are the wave model with $K_C = 8$ and $r = 8$ (diamonds), and the slider block model with $K_C = 8$ and $r = 5$ (squares). Also plotted on figure 5 are power spectra for a brownian walk (crosses, $\beta = 2$, $H = 0.5$) and for white noise (dots, $\beta = 0$, $H = -0.5$). Figure 5 shows that the slider block model has a power spectrum that is very similar to the brown walk, except for the lowest frequencies, indicating a finite correlation length of perhaps 40 lattice units. By contrast, the wave model has a spectrum that corresponds to white noise out to the longest wavelengths, where the simulation data are less reliable. Comparing figures 4 and 5, we see that the slider block model with $H = 0.5$ has a frequency of occurrence curve (figure 4) that is linear out to the largest events as predicted. By contrast, the wave model with $H = -0.5$ has small and intermediate events, but an excess of large events.

4.2.2. Fractal Dimension. The fractal dimension D of the power spectrum is a function of the slope β (12). Setting $\beta = 2$ for the brownian case gives $D = 3/2$.

The value of $\beta = 0$ for white noise yields a fractal dimension $D = 5/2$, also in rough agreement with the exponent from the distribution of areas. Notice that these fractal dimensions are the same as the values for the Fisher exponents τ in the frequency size distributions shown in figure 4. Our simulations suggest that physically, the coincidence between the values of D and τ arises because the rupture process detects features of the slip deficit field in much the same way as a box-counting algorithm determines fractal dimension. In fact, this physical idea can be regarded as an alternative statement of the fundamental principle underlying stochastic fracture mechanics.

Evidently, rupture propagation as defined in slider block models is sensitive to the fractal dimension of the underlying displacement field. Since ruptures are more likely to stop at discontinuities in displacement, the population of rupture areas is a measure of the population of displacement features in the displacement field on a fault. Thus, similarities in values of D and τ should not be completely unexpected.

4.2.3. Discussion. We interpret the scaling limits in terms of interactions between the model evolution and the displacement distribution. In the limit of weak coupling, the sites move independently. Each one has an equal chance to be in each stress state, because they independently cycle through all stress states, and the initial stress distribution is random. The uniform distribution of stresses maps to an approximately uniform distribution of displacements in simulations with weak coupling, because there is little stress transfer. A slider block model with weak coupling naturally produces a displacement field similar to white noise because the positions of neighboring blocks are nearly independent of one another. White-noise scaling is also found in cases of the traveling wave and rough fault model with very large phase scatter, because the site locations become uncorrelated.

For strongly coupled slider block models, stress transfer becomes the dominant contributor to the total stress exerted on each block. Blocks still have a uniform probability of being at any stress level over time, but this no longer maps to a uniform probability of being at any displacement level. Instead, it is gradients of the displacement field that generate stresses. The displacement accumulating across a lattice takes a step, positive or negative, depending upon the value of the stress on a particular block. The accumulation of slip deficit across a slider block lattice is similar to the accumulation of steps in a random walk, and so it is not surprising that it produces a population of ruptures having the same scaling properties as brownian noise.

If the slider blocks are assumed to be representative of parts of a flat fault, elasticity theory may be used to define what coupling spring constant and range of interaction are most physical. For adjacent cracks connected to one another by a three-dimensional continuous elastic medium, the displacement on one crack changes the stress on a neighboring crack 40% as much as it changes the stress on itself. This corresponds to a K_c/K_L ratio of 0.4. If the cracks nearest a slipped crack do not move, cracks two crack dimensions distant will still be loaded to some extent, but only 0.6% as strongly as the crack unloads itself. This argument provides some physical basis for neglecting the longest range stress transfer in a slider block model. If the cracks are not adjacent, being separated by unbroken material, the coupling will be reduced, and K_c/K_L ratios less than 0.4 become appropriate. Physically realistic cases with $K_c/K_L = 0.4$ exhibit scaling very similar to the cases with $K_c/K_L = 8$ shown in figure 4. Thus, the most physically reasonable choice for the coupling spring constants leads to physically reasonable distributions of event areas, similar to those observed for natural earthquakes. The models produce magnitude distributions that approximate observed Gutenberg-Richter magnitude distributions, with simulated b-values ≈ 0.85, slightly lower than the most typical values for real earthquakes. Since these models represent single homogeneous faults, it is not surprising that their statistics are not identical to b-values observed in entire networks of natural faults. Smaller events should be more numerous in networks of faults, where geometric and frictional irregularities produce effectively lower coupling and introduce barriers to rupture.

5. CONCLUSIONS

We have discussed some new results arising from the recently proposed traveling wave model for earthquakes, together with a new variation, the rough fault model. We presented evidence illustrating the relationship between the evolving slip deficit field and the frequency-area statistics for event occurrence. The basic result is that the strength of coupling between sites affects the statistics of the slip deficit field, which in turn determines the frequency-area relationship. For strong coupling (large K_c), one obtains brown noise statistics in the slip deficit, and for weak coupling (small K_c), the slip deficit field has statistics similar to white noise. These statistics determine both the power spectral exponent β through the relationship (12), and evidently also the value of the Fisher exponent τ, and ultimately the b-value of simulated events.

In addition to this basic result, we also find: (1) The large scale friction caused by the cohesion term in these models is proportional to K_c, the parameter defining the strength of coupling on the microscopic scale. The bulk friction is also proportional to normal stresses and oppose the direction of applied loading, like real friction. (2) Both the rough fault and basic traveling wave model produce characteristic events in the limit of low phase scatter and friction. (3) Power-law event statistics similar to earthquake size statistics arise in the traveling wave and rough fault models when frictional forces and stress transfer are significant compared to cohesion, as in a traditional slider block model.

APPENDIX A: FRACTAL RELATIONSHIPS

To derive the relationship between the correlation of a random walk $V(x)$ and its fractal dimension D, [*Voss*, 1988, section 1.6.5] we first need to consider the statistical definition of a random walk with Hurst parameter H,

$$< |V(x+dx) - V(x)|^2 > \propto |dx|^{2H}. \tag{A1}$$

The definition of the autocorrelation function G is,

$$< |V(x+dx) - V(x)|^2 >= 2[< V^2 > - G_V(dx)], \tag{A2}$$

and the definition of the autocorrelation slope is

$$G_V \propto dx^{\beta-1} \tag{A3}$$

from which we can see that

$$2H = \beta - 1. \tag{A4}$$

Now it is necessary to relate the random walk parameter H to the fractal dimension D of the surface that includes the walk. Two dimensional slider block models are concerned with the displacement of blocks which occupy a two dimensional lattice. The slip deficit is the analog of V in the above equations, and the displacements accumulate in the spatial dimensions x and y instead of the more frequently encountered time. The three dimensions that describe the state of the lattice are two of space and one of slip deficit, but we are concerned with a single profile of slip deficit across the lattice, so only one of the spatial dimensions need be considered. Counting the number N of boxes needed to cover the set as a function of the length of a box L,

$$N \propto L^{-D} \tag{A5}$$

defines the fractal dimension D.

The boxes fill the spatial dimension x completely, because there is some slip deficit defined for every slider block. The only complication comes when we consider how many boxes are occupied by the lattice slice in the slip deficit dimension V. If the slice has the statistical properties of a random walk, it will occupy a number of boxes in the slip deficit dimension N_v which is related to the size of the boxes in the spatial dimension, L_x,

$$N_V = \frac{L_x^H}{L_V}. \tag{A6}$$

And the total number of occupied boxes in a slice of length X will scale like

$$N = \frac{X}{L_x} \cdot \frac{L_x^H}{L_V}. \tag{A7}$$

So if we reduce the two dimensions of the boxes together, making $L_x = L_V$, then the total number of boxes is

$$N = \frac{X}{L^{2-H}}, \tag{A8}$$

which implies

$$D = 2 - H. \tag{A9}$$

Combining equation 16 and 21 gives

$$D = 2 - \frac{\beta - 1}{2}, \tag{A10}$$

which may be solved for β,

$$\beta = 5 - 2D. \tag{A11}$$

Acknowledgments. I appreciate the constructive comments made by anonymous reviewers. Financial support of S.J.G. was provided by NSF grant EAR-9526814. This work would not have been possible without extensive conversations and advice from John Rundle.

REFERENCES

Ben-Zion Y. and J. R. Rice, 1993. Earthquake failure sequences along a cellular fault zone in a three-dimensional elastic solid containing asperity and non-asperity regions, *J. Geophys. Res.*, 98, 14109-14131.

Ben-Zion Y. and J. R. Rice, 1995. Slip patterns and earthquake populations along different classes of faults in elastic solids, *J. Geophys. Res.*, 100, 12959-12983.

Burridge, R. and L. Knopoff, 1967. Model and theoretical seismicity, *Bull. Seism. Soc. Am.*, 57, 341-371.

de Sousa Vieria, M., G. L. Vasconcelos, and S. R. Nagel, 1993. Dynamics of spring-block models: Tuning to criticality, *Phys. Rev. E*, 47, 2221-2224.

Ding E. J. and Y. N. Lu, 1993. Analytical treatment for a spring-block model, *Phys. Rev. Lett.*, 70, 3627-3630.

Heaton, T. H., 1990. Evidence for and implications of self-healing pulses on slip in earthquake rupture, *Phys. Earth Planet. Int.*, 64, 1-20.

Lomitz-Adler, J., 1993. Automaton models of seismic fracture: Constraints imposed by the magnitude-frequency relation, *J. Geophys. Res.*, 98, 17745-17756.

Mendoza, C., S. Hartzell, and T. Monfret, 1994. Wide-band analysis of the 3 March, 1985 central Chile earthquake: Over-all source process and rupture history, *Bull. Seism. Soc. Am.*, 84, 269-283.

Mikumo, T., and T. Miyatake, 1979. Earthquake sequences on a frictional fault model with non-uniform strengths and relaxation times, *Geophys. J. R. astr. Soc.*, 59, 497-522.

Pepke, S. L., J. M. Carlson, and B. E. Shaw, 1994. Prediction of large events on a dynamical model of a fault, *J. Geophys. Res.*, 99, 6769-6788.

Rice, J. R., 1993. Spatio-temporal complexity of slip on a fault, *J. Geophys. Res.*, 98, 9885-9907.

Ruff, L. J., 1992. Asperity distributions and large earthquake occurrence in subduction zones, *Tectonophysics*, 211, 61-83.

Rundle, John B. and H. Kanamori, 1987. Application of an inhomogeneous stress (patch) model to complex subduction zone earthquakes: A discrete interaction matrix approach, *J. Geophys. Res.*, 92, 2606-2616.

Rundle, J. B. and W. Klein, 1993. Scaling and critical phenomena in a cellular-automaton slider-block model for earthquakes, *J. Stat. Phys.*, 72, 405-412.

Rundle, J. B., W. Klein, and S. Gross, 1996. Dynamics of a traveling density wave model for earthquakes, *Phys. Rev. Lett.*, 76, 4285-4288.

Rundle, John, Eric Preston, Seth McGinnis and William Klein, 1998, Why earthquakes stop: growth and arrest in stochastic fields, *Phys. Rev. Lett.*, 80, 5698-701.

Rydelek Paul A. and I. Selwyn Sacks, 1996. Earthquake slip rise time and rupture propagation: Numerical results of the quantum earthquake model, *Bull. Seism. Soc. Am.*, 86, 567-574.

Voss, R., 1988. Fractals in nature, from characterization to simulation, in Peitgen, H., and D. Saupe, editors, *The Science of Fractal Images*, Springer-Verlag, New York, 312pp.

Ward, S. N., 1992. Synthetic earthquake models for long-term prediction, *Geotimes*, 37, 18-20.

Ward, S. N. and S. D. B. Goes, 1993. How regularly do earthquakes recur? A synthetic seismicity model for the San Andreas fault, *Geophys. Res. Lett.*, 20, 2131-2134.

Zheng, G., R. Dmowska, and J. R. Rice, 1996. Modeling earthquake cycles in the Shumagin subduction segment, Alaska, with seismic and geodetic constraints, *J. Geophys. Res.*, 101, 8383-8392.

S. Gross, CIRES, Campus Box 216, University of Colorado, Boulder, CO, 80303 (e-mail: sjg@colorado.edu)

A Statistical Physics Approach to Earthquakes

D.L. Turcotte

Department of Geological Sciences, Cornell University
Ithaca, New York 14853

W.I. Newman

Departments of Earth & Space Sciences, Physics & Astronomy, and Mathematics, University of California
Los Angeles, California 90095

A. Gabrielov

Departments of Mathematics and Earth & Atmospheric Sciences, Purdue University
West Lafayette, Indiana 47907

Earthquakes occur on a hierarchy of faults in response to tectonic stresses. Earthquakes satisfy a variety of scaling relations; the most important is the Guttenberg-Richter frequency magnitude scaling. A variety of simple models have been shown to exhibit similar scaling. These models are said to exhibit self-organized criticality and one of the models, the slider-block model, was developed as a simple analog to earthquakes. We show that an inverse cascade model can explain the behavior of these models. Metastable clusters coalesce in a manner analogous to the development of river networks. This cascade gives the observed scaling for the models and its implications for earthquakes are discussed.

1. INTRODUCTION

Earthquakes are a pervasive feature of the deformation of the earth's brittle crust. The hypotheses of stick-slip behavior and elastic rebound explains the physics of earthquakes in a general way. Earthquakes occur on preexisting faults and stick-slip events occur if the friction on the faults is velocity weakening. When the slip-event (earthquake) occurs the elastic strains in the adjacent rock are relieved generating seismic waves and heat. The laws of elasticity can be solved in a self-consistent way to obtain periodic earthquake cycles, [*Turcotte and Spence, 1974*].

But earthquakes do not occur in periodic cycles. The earth's crust is extremely complex and faults and earthquakes in a region occur on a wide range of scales. There is considerable evidence that faults and earthquakes interact on a range of scales from thousands of kilometers to millimeters or less. Evidence in support of this hypothesis comes from the universal validity of scaling relations. The most famous of these is the Gutenberg-Richter frequency-magnitude relation [*Gutenberg and Richter, 1954*]

$$\log N = -bm + \log a \tag{1}$$

GeoComplexity and the Physics of Earthquakes
Geophysical Monograph 120

where N is the number of earthquakes per unit time with a magnitude greater than m occurring in a specified area and b and a are constants. This relation is valid for earthquakes both regionally and globally. The constant "*b*" or "*b-value*" varies from region to region, but is generally in the range of $0.8 < b < 1.2$ [*Frohlich and Davis, 1993*]. The constant "*a*" is a measure of the regional level of seismicity.

As we will discuss, complex phenomena often exhibit power-law (fractal) scaling [*Mandelbrot 1967, 1982; Turcotte, 1997*]. For earthquakes, power-law (fractal) scaling implies the validity of the relation

$$N = CA^{-\gamma} \tag{2}$$

where N is the number of earthquakes per unit time with rupture area greater than A occurring in a specified area; C and γ are constants with $D = 2\gamma$ the fractal dimension. *Aki* [*1981*] showed that (1) and (2) are entire equivalent with

$$\gamma = b = \frac{D}{2} \tag{3}$$

Thus, the universal applicability of the Gutenberg-Richter relation implies universal fractal behavior of earthquakes. This in turn implies that earthquakes in the earth's crust are a classic example of complexity and chaotic behavior.

The transition from a deterministic, periodic behavior to a chaotic behavior is clearly illustrated by slider block models. These models are considered to be analogs for the behavior of faults in the earth's crust. The simplest example is a single slider block of mass m pulled over a surface by a spring attached to a constant velocity driver plate. The interaction of the block with the surface is controlled by friction. Many friction laws have been proposed, the simplest is the static-dynamic friction law. If the block is stationary the static frictional force is F_s, if the block is slipping the dynamic frictional force is F_d. If $F_s > F_d$ stick-slip behavior is obtained, the motion of the block is made up of periodic slip events.

The behavior of a pair of slider blocks pulled over a surface and connected by a connector spring was studied in detail by *Huang and Turcotte* [*1990*]. The equations of motion for the two blocks were solved simultaneously. Solutions were governed by two parameters, the stiffness of the system $\alpha = k_c/k_p$ (k_c the spring constant of the connector spring and k_p the spring constant of the puller springs) and the ratio of static to dynamic friction $\Phi = F_s/F_d$. For some values of these parameters deterministic chaos was found. The period doubling route to chaos was observed with positive values of the Lyapunov exponent in the chaotic regions. The behavior of the pair of slider blocks is very similar to the behavior of the logistic map [*May 1976*] which is considered to be a classic example of chaotic behavior.

The chaotic behavior of the low-dimensional Lorenz equations [*Lorenz 1963*] is now accepted as evidence that the behavior of the atmosphere and oceans is chaotic. Similarly, the chaotic behavior of a pair of slider blocks is evidence that earthquakes exhibit chaotic behavior. It is instructive to make comparisons between the behavior of the Earth's atmosphere and the behavior of the Earth's crust. Both are extremely complex on a very wide range of scales. From a practical point of view it is impossible, in either case, to make a sufficient number of measurements to fully specify the problem. Both have significant random components. Both are undoubtedly chaotic. But what are the relative rolls of random versus chaotic behavior?

Massive numerical simulations are routinely used to forecast the weather. In many cases, they are quite accurate on time scales of 24 to 48 hours, but on the scale of weeks they are of little value. The motions of the storm systems are relatively stable considering the complexity involved. In many cases the paths of major storms such as hurricanes can be predicted with considerable accuracy, but in other cases there are major uncertainties. One approach to establishing whether a particular predicted storm path is stable or not is to apply alternative numerical models. The alternative models may use different discretizations or may differ in other ways. If the models all predict essentially the same path then the path is taken to be stable. If the models predict radically different paths then any forecast is considered to be suspect. This is also the way classical chaotic systems such as the Lorenz attractor behave. At many points on the evolving trajectory there is relatively little sensitivity to small perturbations and at other points there is extreme sensitivity.

But what about the Earth's crust? Forecasting or predicting an earthquake is quite different than forecasting the path and intensity of a hurricane. The hurricane exists but the earthquake does not exist until it happens. *Geller et al.* [*1997*] have argued, based on the chaotic behavior of the Earth's crust, that "earthquakes cannot be predicted". This is certainly true in the sense that the exact time of occurrence of an earthquake cannot be predicted. But this is also true of hurricanes, the exact path of a hurricane cannot be predicted. But probabilistic forecasts of hurricane paths with a most probable path are routinely made and there use is of great value in terms of requiring evacuations and in other preparations.

An essential question concerning earthquakes is whether similar useful probabilistic forecasts can be made. In fact, this is already being done in terms of hazard assessments. Certainly earthquakes do not occur randomly on the surface of the Earth. Also, the occurrence of large earthquakes can be associated with the occurrence of small earthquakes. A systematic approach to the quantitative assessment of the earthquake hazard will be given in Section 3 of this paper. A more intriguing question is whether useful forecasts of the temporal occurrence of earthquakes can be made and, in particular, whether useful forecasts can be made based on concepts of complexity. This will also be discussed in Section 6 of this paper.

2. SELF-ORGANIZED CRITICALITY

One explanation for fractal statistics is scale invariance. The power-law distribution is the only statistical distribution that does not result in a characteristic length scale. Thus, natural phenomena that do not inherently have a natural length scale would be expected to obey power-law (fractal) statistics. However, there may be a more fundamental basis for the applicability of fractal statistics. In the past ten years, a variety of numerical models have been found to exhibit a universal behavior that has been called self-organized criticality. In self-organized criticality the "input" to a complex system is slow and steady; whereas the output is a series of events or "avalanches" that follow power-law (fractal) frequency-size statistics. Regional seismicity is often taken as a naturally occurring example of self-organized criticality. The input is the motion of the tectonic plates and the output is the earthquakes.

The concept of self-organized criticality was introduced by *Bak et al.* [*1988*] as an explanation for the behavior of the "sandpile" model. In this model, a square array of boxes is considered and at each time step a particle is dropped into a randomly selected box. When a box accumulates four particles, they are redistributed to the four adjacent boxes, or in the case of edge boxes lost from the grid. Because the redistributions involve only nearest neighbor boxes, it is known as a cellular-automata model. Redistributions of particles can lead to further instabilities with, at each step, the possibility of avalanches of particles being lost from the grid. Each of the multiple redistributions during a time step contribute to the size of the model "avalanche." The size of an avalanche can be associated with the number of particles lost from the grid during the sequence of redistributions or by the number of boxes that participate in redistributions.

This model was called a "sandpile" model because of the resemblance to an actual sandpile on a table. The randomly dropped particles in the model are analogous to the addition of particles to the actual sandpile and the model avalanches are analogous to sand avalanches down the sides of the sandpile. In some cases the sand avalanches lead to the loss of particles off the table. Extensive numerical studies of the "sandpile" model were carried out by *Kadanoff et al.* [*1989*]. They found that the noncumulative frequency-size distribution of avalanches satisfies (2) with $\gamma \approx 1$.

A second example of "self-organized criticality" is the behavior of the large arrays of slider blocks. The slider-block model with a pair of slider-blocks considered above can be extended to include large numbers of slider-blocks. Multiple slider-block simulations were first considered by *Burridge and Knopoff* [*1967*]. *Otsuka* [*1972*] considered a two-dimensional array of slider-blocks and obtained power-law distributions for the size of slip events.

Carlson and Langer [*1989*] considered long linear arrays of slider blocks with each block connected by springs to the two neighboring blocks and to a constant-velocity driver. They used a velocity-weakening friction law and considered up to 400 blocks. Slip events involving large numbers of blocks were observed, the motions of all blocks involved in a slip event were coupled, and the applicable equations of motion had to be solved simultaneously. Because of certain similarities, these are sometimes known as molecular-dynamics simulations. Although the system is completely deterministic, the behavior was apparently chaotic. Frequency-size statistics were obtained for slip events. The events fell into two groups: smaller events obeyed a power-law (fractal) relationship with a slope near unity, but there was an anomalously large number of large events that included large numbers of slider blocks. The observed behavior was said to be characteristic of self-organized criticality. The motion of the driver plate is the steady input. The slip events are the avalanches with a fractal distribution.

Rundle and Jackson [*1977*] and *Nakanishi* [*1991*] studied multiple sliderblock models using a cellular automata approach. A linear array of slider blocks was considered but only one block was allowed to move in a slip event. The slip of one block could lead to the instability of either or both of the adjacent blocks, which would then be allowed to slip in a subsequent step or steps, until all blocks were again stable. *Brown et al.* [*1991*] proposed a modification of this model involving a two-dimensional array of blocks. The use of the cellular automata approach greatly reduces the complexity of the calculations and the results using the two approaches are generally very similar. A wide variety of slider-block models have been proposed and studied; these have been reviewed by *Carlson et al.* [*1994*] and *Turcotte* [*1997*].

The standard multiple slider-block model consists of a square array of slider-blocks as illustrated in Figure 1. Each block with mass m is attached to the driver plate with a driver spring, spring constant k_p. Adjacent blocks are attached to each other with connector springs, spring constant k_c. A simulation requires the specification of the stiffness $\alpha = k_c/k_p$, the ratio of static to dynamic friction $\Phi = F_s/F_d$, and the area of the square array.

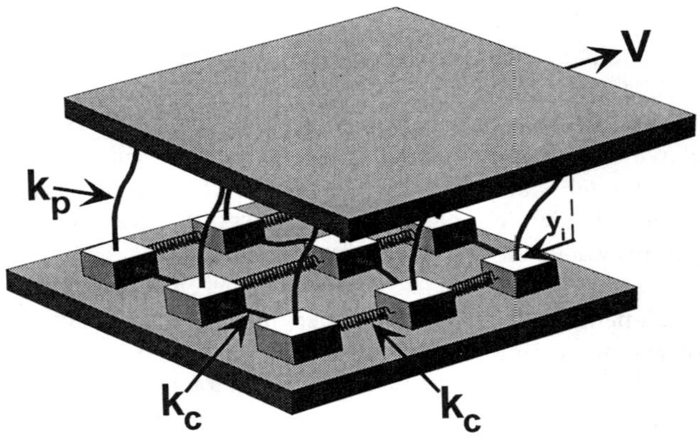

Figure 1. Illustration of the two-dimensional slider-block model. An array of blocks, each with mass m, is pulled across a surface by a driver plate at a constant velocity, V. Each block is coupled to adjacent blocks with either leaf or coil springs (spring constant k_c) and to the driver plate with leaf springs (spring contant k_p).

Huang et al. [*1992*] carried out a large number of simulations on a square array of blocks using static-dynamic friction and the cellular automata approach described above. Their noncumulative frequency-area distributions for model slip events are given in Figure 2. The number of slip events per time step with area A_e, N_e/N_o is given as a function of A_e. Results are given for a stiffness α = 30, a friction ratio Φ = 1.5, and four grid sizes, 20 x 20, 30 x 30, 40 x 40, and 50 x 50. For the smaller events there is general agreement with the power-law relation (2) taking $\gamma \approx 1$. For stiff systems, α large, slip events including all blocks occur regularly. These large events give peaks in the distributions at A_e = 400, 900, and 1,600 as illustrated in Figure 2. For soft systems, α relatively small, large events occur infrequently and the power-law distribution is terminated smoothly.

There are strong similarities between the behavior of the sandpile model and the slider-block model. In both cases smaller slip events have a noncumulative power-law frequency-size distribution with a slope near unity. Whereas the sandpile model is random in the selection of boxes, the slider-block model is fully deterministic. The slider-block model provides a bridge between chaotic behavior (two slider-blocks) and self-organized critical behavior (large numbers of slider-blocks). Adjacent solutions for the chaotic behavior of a pair of slider-blocks diverge exponentially. The divergence of adjacent solutions for large numbers of slider-blocks have a power-law divergence.

Since the concept of self-organized criticality was first introduced, earthquakes have been identified as an example of this phenomena in nature [*Bak and Tang, 1989*]. These authors also pointed out the similarity in slopes between self-organized critical behavior and earthquakes. However, since the model data are noncummulative, the agreement with the cumulative earthquake data must be considered fortuitous. The power-law exponents are significantly higher for the application than for the theoretical model.

In order to further consider the applicability of the Gutenberg-Richter relation to seismicity we consider the frequency-magnitude distribution of earthquakes in southern California on a yearly basis using data from the Southern California Seismographic Network. In Figure 3 the number of earthquakes N_{CE} in each year between 1980 and 1994 with body-wave magnitudes greater than m is given as a function of m. In general, there is good agreement with (1) taking b = 1.05 and $a = 2.06 \times 10^5$ yr^{-1}. The exceptions can be attributed to the aftershock sequences of the Whittier (1987), Landers (1992), and Northridge (1994) earthquakes.

With aftershocks removed, the near uniformity of the background seismicity in southern California illustrated in Figure 3 is clearly striking. Small earthquakes behave like a thermal background noise. This is observational evidence that the earth's crust is continuously on the brink of failure. Further evidence for this comes from induced seismicity. Whenever the crust is loaded, whether in a tectonically active area or not, earthquakes are induced. Examples of loading include the filling of a reservoir behind a newly completed dam or the high pressure injection of fluids in a deep well.

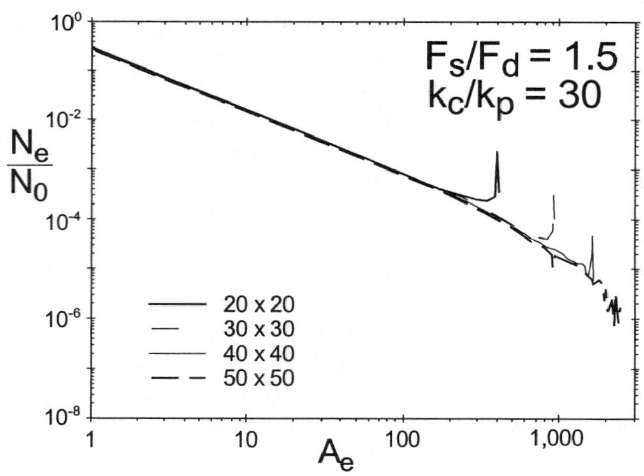

Figure 2. Results for a two-dimensional slider-block model with multiple blocks. The ratio of the number of events, N_e, with size, A_e, to the total number of events, N_o, is plotted against A_e [*Huang et al., 1992*]. Results are given for systems with stiffness $\alpha = k_c/k_p = 30$, $\Phi = 1.5$, and grid sizes 20 x 20, 30 x 30, 40 x 40, and 50 x 50. The peaks at A_e = 400, 900, and 1,600, correspond to catastrophic events involving the entire system.

While there are important similarities between the slider-block models and earthquakes there are also important differences. Slider-block models would be representative of a distribution of earthquakes on a single fault. However, the Gutenberg-Richter distribution of earthquakes is not associated with a single fault but with a hierarchy of faults. The earthquakes included in the Southern California data given in Figure 3 occur over a broad zone with a width of about 200 km on a wide variety of faults associated with the San Andreas system.

3. HAZARD ASSESSMENT

An important aspect of self-organized criticality relative to earthquakes is the implications for earthquake forecasting and prediction. The repeated observational confirmation of the validity of the Gutenberg-Richter relation (1) as well as its independence of geology, fault-geometries and other physical properties implies that the observed frequency of occurrence of small earthquakes can be extrapolated to estimate the recurrence frequencies of larger earthquakes. This is routinely done and is a primary basis for published maps of the earthquake hazard.

Probabilistic seismic hazard assessment is based on extrapolations of available data. This technique follows directly from the acceptance of the methods of statistical physics for regional seismicity, in particular the acceptance of the Guttenberg-Richter relation (1). *Kossobokov* and *Turcotte [1996]* proposed a systematic global assessment of the seismic hazard based on the extrapolation of the occurrence of small earthquakes to larger earthquakes. The method was based on the epocenters of earthquakes with body-wave moment magnitudes $m > 4$ taken from the Global Hypocenter Database of the National Earthquake Information Center for the period of 1964-1995. The magnitude 4 cutoff was taken because this is the minimum magnitude for which the global catalog is complete for this period.

The surface of the earth was divided into 1° *x* 1° regions and the number of earthquakes per year with body-wave magnitudes greater than $m = 4$ in each region was determined The seismic intensity factor, I_4 was defined to be the number of magnitude $m = 4$ and larger earthquakes that have occurred in a given 1° x 1° region per year. Each 1° x 1° region varied in area with respect to changing latitude. These variations were scaled out by using the cosine of the latitude as a normalizing factor. A global map of the seismic intensity factor is given in Figure 4. The boundaries of plate tectonics are clearly defined. Seismicity is particularly intense in subduction zones (i.e., the ring of fire around the Pacific) as expected. A broad band of seismicity extends from southern Europe to southeast Europe to southeast Asia, this is associated with the continent-continent collision zone between the

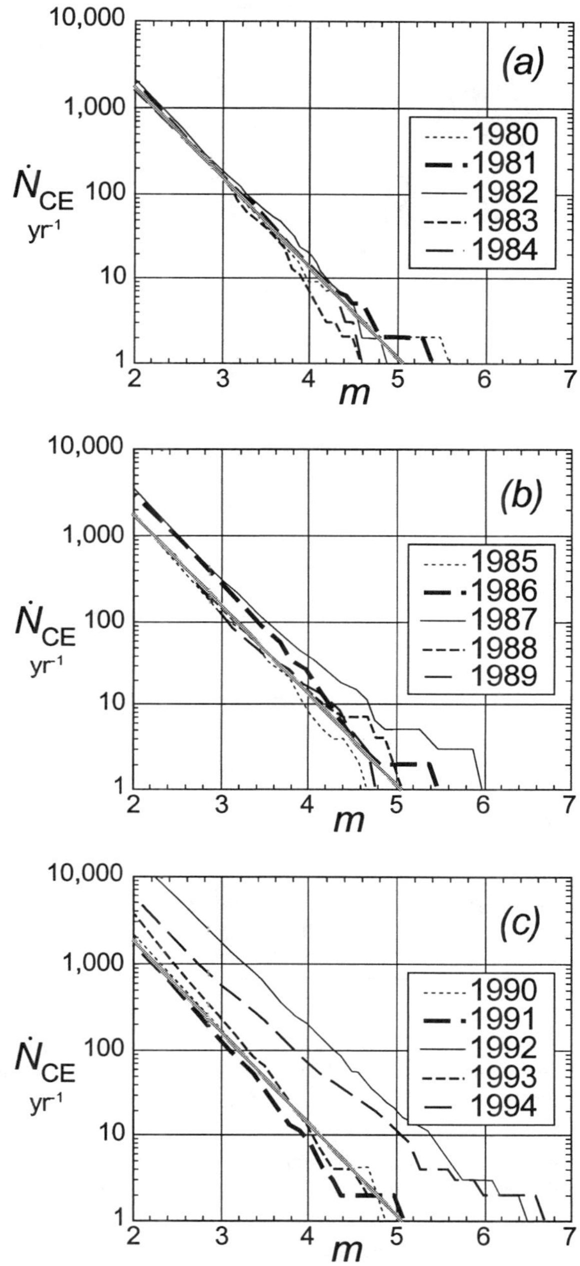

Figure 3. Cumulative number of earthquakes per year, $\dot{N}_{CE}$, occurring in southern California with magnitudes greater than m as a function of m. Fifteen individual years are considered (*SCSN Catalog, 1995*): (a) 1980-1984; (b) 1985-1989; (c) 1990-1994. The solid straight line in (a) to (c) is the Gutenberg-Richter relation (1) with $b = 1.05$ and $a = 2.06 \times 10^5$ yr^{-1}. The larger numbers of earthquakes in 1987, 1992, and 1994 can be attributed to the aftershocks of the Whittier, Landers, and Northridge earthquakes, respectively. If aftershocks are excluded, the background seismicity in southern California is nearly uniform in time, this implies a thermodynamic behavior.

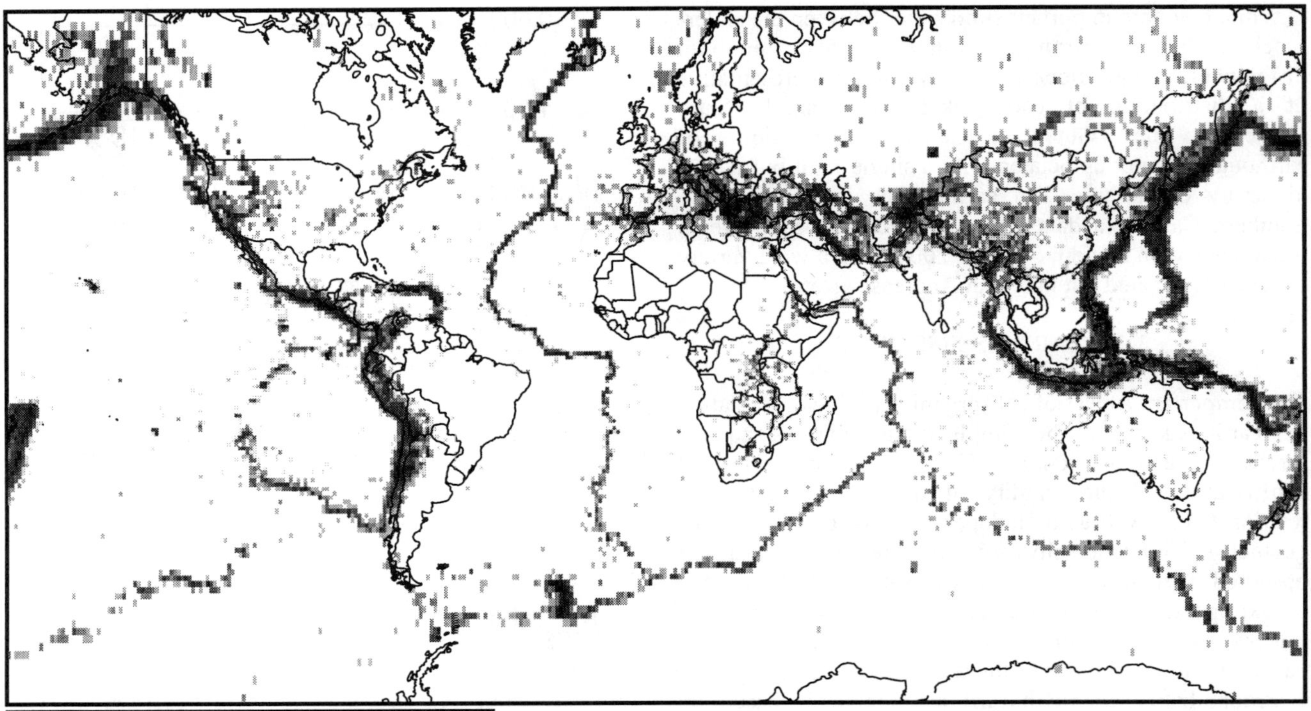

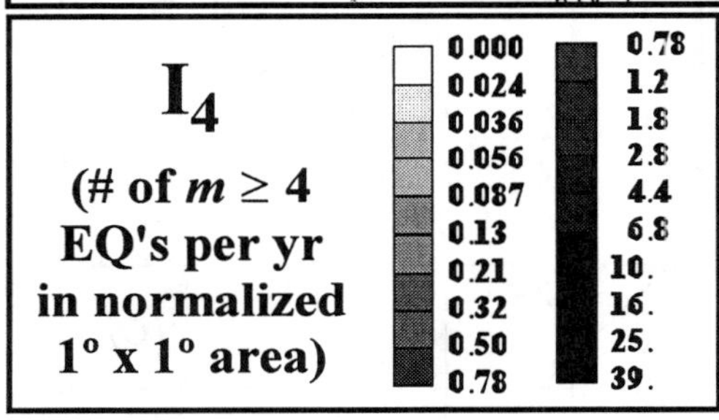

Figure 4. Global map of the seismic intensity factor, I_4, the average annual number of earthquakes per year during 1964-1995 with magnitudes $m \geq 4$ in each normalized 1° x 1° cell [*Kossobokov and Turcotte, 1996*]. Data from the NEIC Global Hypocenter Database.

Eurasian plate and the African, Arabian, and Indian plates. The minimum value of the seismic intensity factor considered is $I_4 = 1/32$ yr^{-1}, one magnitude $m = 4$ or larger earthquakes in the 32 years for which records were used. The maximum value of the seismic intensity factor is about $I_4 = 40$ yr^{-1} (forty magnitude $m = 4$ or larger earthquakes per year).

The basis for using the seismic intensity factor I_4, to estimate the risk of having large earthquakes in a region is illustrated in Figure 5. Using the Guttenberg-Richter relation (1), the number of earthquakes per year in a 1° x 1° area with a body-wave magnitude m is directly related to the seismic intensity factor I_4 (the number of earthquakes per year in a 1° x 1° area with a magnitude greater than 4). As an example, consider a region where $I_4 = 1$ yr^{-1} (one magnitude 4 per year), the risk of having a magnitude $m = 6$ earthquakes is 0.0016 yr^{-1} (return period of 63 years) and the risk of having a magnitude $m = 8$ earthquakes is 2.5 x 10^{-4} yr^{-1} (return period of 4,000 yrs).

The primary advantage of this approach is that it is totally based on a generally accepted data set. There are no ambiguities with regards to the technique. This is not the case for other approaches to assessing the seismic hazard, which combine geophysical and geological observations in arbitrary ways. Different studies give different weights to historical and paleo-seismic data, and to the presence of "active" faults. Considering the many uncertainties regarding fault depth, seismic attenuation, available database, and the occurrence of an earthquake on a particular fault, we believe that this simple approach for assessing the seismic hazard is reasonable based on the present level of knowledge.

4. FOREST FIRE MODEL

Although the forest-fire model [*Bak et al., 1992; Drossel and Schwable, 1992*] was not the first model associated with self-organized criticality, it is probably the most

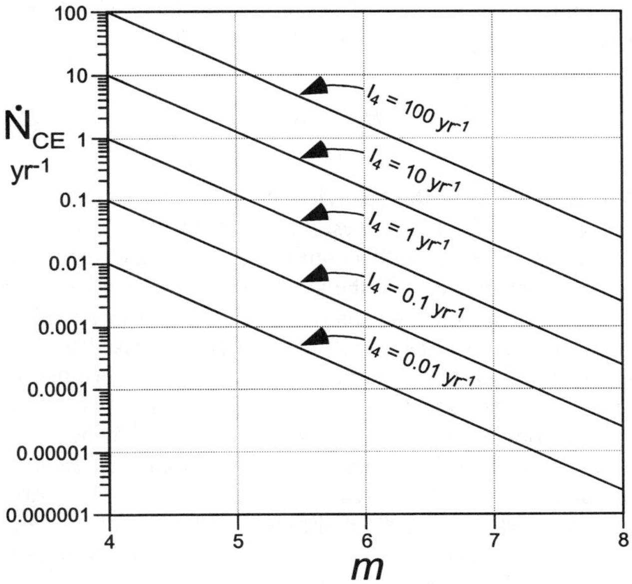

Figure 5. Similarity basis for extending the seismic intensity factor I_4 to higher earthquake magnitudes. For $I_4 = 0.01$, 0.1, 1, 10, and 100 yr^{-1}, the cumulative number of earthquakes per year in a 1° x 1° area, $\dot{N}_{CE}$, with magnitudes greater than m are plotted as a function of m. Each line is derived by using the Gutenberg-Richter relation (1) with $b = 0.9$, and calculating the constant $\dot{a}$ from $\log \dot{a} = 3.6 + \log I_4$.

illustrative. The forest-fire model we consider consists of a square grid of sites. At each time step, a model tree is dropped on a randomly chosen site, if the site is unoccupied, the tree is planted. The sparking frequency, f, is the inverse number of attempted tree drops on the square grid before a model match is dropped on a randomly chosen site. If $f = 1/100$, there have been 99 attempts to plant trees (some successful, some unsuccessful) before a match is dropped at the 100th time step. If the match is dropped on an empty site nothing happens. If it is dropped on a tree, the tree ignites and a model fire consumes that tree and all adjacent (nondiagonal) trees. In terms of self-organized criticality the steady input is the planting of trees, the "avalanches" are the fires, and the number of trees on the grid fluctuates about a quasi-equilibrium value. The frequency-size distribution of fires is found to be fractal.

Having specified the size of the square grid, N_g, and the sparking frequency, f, a simulation is run for N_s time steps and the number of fires N_F with area A_F is determined. The area, A_F, is the number of trees that burn in a fire. Examples of four typical fires during a run are given in Figure 6. In these examples the grid size is 128 x 128 (N_g = 16,384), $1/f = 2000$, and fires with A_F = 5, 51, 505 and 5327 trees are illustrated. Figure 6d is an example of a special class of forest-fires which spans the grid.

Noncumulative frequency-size statistics for the model forest-fires are given in Figure 7. The number of fires per time step with area A_F, N_F/N_S is given as a function of A_F, results are given for a grid size 128 x 128 and three sparking frequencies $1/f$ = 125, 500, and 2000. In all cases, the smaller fires correlate well with the power-law (fractal) relation (2) taking $\gamma \approx 1$. The results for the forest-fire model are very similar to the results for the slider-block model given in Figure 2. The sparking frequency f plays a role similar to the stiffness parameter α. If f is relatively large the power-law region is confined to small fires. If f is small, fires occur that span the grid and there is a peak in the frequency-magnitude distribution for large fires. The parameter f can be "tuned" so that fires cross the grid. It should be noted that the two models behave similarly even though the slider-block model is entirely deterministic whereas the forest-fire model is stochastic. It should also be noted that actual forest and wild fires behave in a similar manner (*Malamud et al. 1998*).

5. INVERSE CASCADE MODEL

We now introduce an inverse cascade model with substantial generality in order to explain self-organized

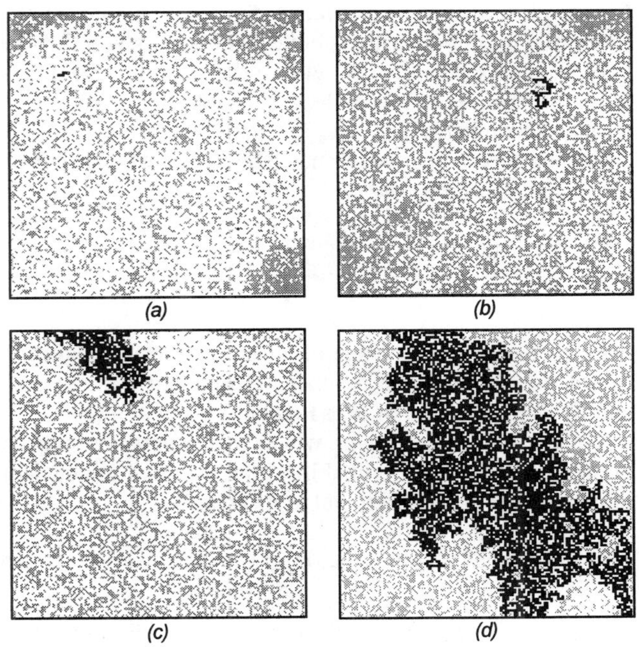

Figure 6. Four examples of typical model forest fires are given. This run was carried out on a 128 x 128 grid with $f = 1/2000$. The heavily shaded regions are the forest fires. The lightly shaded regions are trees. The white regions are unoccupied sites. The areas A_F of the four forest fires are (a) 5 trees, (b) 51 trees and (c) 505 trees and (d) 5327 trees. The largest forest fire spans the entire grid.

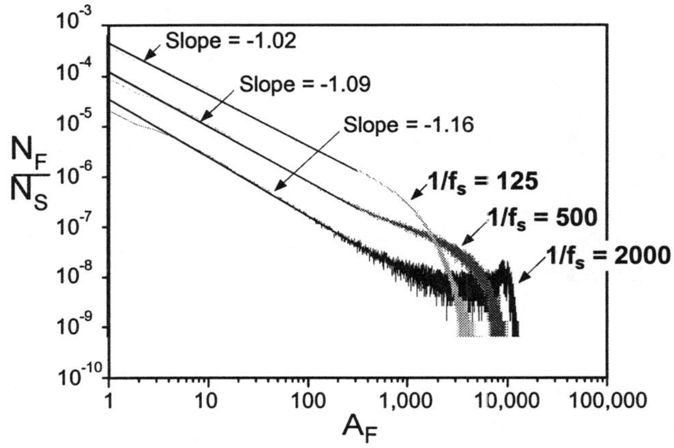

Figure 7. Frequency-size distributions of model fires with a constant grid size and several firing frequencies. The number of fires per time step with size A_F, N_F/N_s is given as a function of A_F where A_F is the number of trees burnt in each fire. Results are given for grid size 128 x 128 and three sparking frequencies, f = 1/125, 1/500, 1/2000. The small fires correlate well with the power-law relation (2) taking γ = 1.02 to 1.16. The finite grid size effect can be seen at the smallest firing frequency, f = 1/2000. At A_F = 2000 fires begin to span the entire grid.

critical behavior [*Turcotte et al., 1999*]. This model will be introduced as an explanation for the behavior of the forest-fire model described above. The results will then be extended to explain the behavior of the sandpile and slider-block models, and finally the results will be discussed in terms of earthquakes.

For the forest-fire model the frequency-area distribution of fires satisfies the relation

$$N_f = \frac{N_{f1}}{A} \tag{4}$$

over a relatively wide range of areas as shown in Figure 7. In writing (4) N_f is the number of fires that burn A trees and N_{f1} is the number of fires that burn a single tree.

We define the area over which a fire spreads to be a cluster of trees. The probability that a model match will ignite a cluster is proportional to its area so that

$$N_f \sim N_c A \tag{5}$$

where N_c is the number of clusters of area A. From (4) and (5) we have

$$N_c = \frac{N_{c1}}{A^2} \tag{6}$$

where N_{c1} s the number of single tree clusters on the grid. The cumulative number of clusters larger than area A, N_{cc} is obtained by integrating (6) with the result

$$N_{cc} = \frac{N_{c1}}{A} \tag{7}$$

Our hypothesis for the behavior of the forest-fire model is that clusters of trees coalesce to form larger clusters as trees are planted. And that significant numbers of trees are lost from the grid only in the largest fires that terminate the self-similar (power-law) distribution. We construct rules for assigning an order (rank) to each cluster based on the *Strahler* [*1957*] classification that was originally developed for branching in river networks. In this classification system, a stream with no upstream tributaries is defined to be of order one; when two order one streams combine, they form a stream of order two, and so forth. However, when streams of different orders combine, the order of the dominant stream prevails. The rules that define rank (order) for our cluster model are:

1. We define a single tree that is added to a system to be a cluster of rank 1.
2. If a new tree is added adjacent to an existing cluster, we say that it is added to the cluster without changing that cluster's rank, unless the cluster is a single tree. In that special case, we define the two trees as forming a cluster of rank 2.
3. If a new tree connects two existing clusters of ranks i and j, respectively, then the rank of this new cluster is defined as $i + 1$ when $i = j$ and as max $\{i, j\}$ when $i \neq j$. In words, this is equivalent to saying that when two clusters of equal rank coalesce, then the rank increases by one; however, if the two clusters are not of equal rank, then the rank of the larger cluster prevails.
4. In principle, a tree could instantaneously connect three or four clusters, but such an event is so rare that it can be neglected in our formulation.
5. We terminate the inverse cascade of elements from small to large clusters by eliminating clusters of a specified high rank.

The application of order (rank) to clusters is illustrated in Figure 8. In Figure 8a binary trees of rank 1, 2, 3, and 4 are illustrated. In Figure 8b a single tree is planted to create a cluster of rank 1. A tree planted between two single trees creates a cluster of rank 2. A tree planted between two 2nd rank clusters forms a cluster of rank 3. A tree planted between two 3rd rank clusters forms a cluster of rank 4.

The inverse cascade model assumes that metastable clusters grow by coalescence. The process is referred to as an inverse cascade because it proceeds from the smallest to the largest scales. Each cluster has a specified rank. The number of clusters of rank i is N_i. We give a set of dynamical equations governing the coalescence of clusters in our inverse cascade. Conservation of cluster numbers at each order requires

$$\frac{dN_1}{dt} = C - 2r_{11} - \sum_{j=2}^{\infty} r_{1j} - f_l, \; for \; i = 1 \tag{8}$$

$$\frac{dN_i}{dt} = r_{i-1,i-1} - 2r_{ii} - \sum_{j=i+1}^{\infty} r_{ij} - f_i, \; for \; i > 1 \tag{9}$$

with i = 1, 2, 3, ... The constant C is the rate at which new single tree (rank 1) clusters are planted. The rate of coalescence between clusters of ranks i and j is r_{ij}. A cluster of rank i is formed only be the coalescence of two clusters of order $i - 1$, the rate at which this occurs is $r_{i-1,\, i-1}$. Clusters of rank i are lost by coalescence with clusters of rank i at a rate r_{ii} and by coalescence with clusters of all higher ranks r_{ij}, $j > i$. Clusters of rank i are also destroyed in fires at a rate f_i.

In addition to providing equations for the conservation of cluster numbers N_i we must also provide equations for the conservation of cluster areas (numbers of trees). These equations can be written

$$\frac{d}{dt}(N_1) = C - 2r_{11} - \sum_{j=2}^{\infty} r_{1j} - f_1, \; for \; i = 1 \tag{10}$$

$$\frac{d}{dt}(N_i A_i) = 2r_{i-1,i-1}A_{i-1} + \sum_{k=1}^{i-1} r_{ki}A_k - 2r_{ii}A_i - \sum_{j=i+1}^{\infty} r_{ij}A_i - f_i A_i, \; for \; i > 1 \tag{11}$$

where A_i is the mean area of clusters of rank i. The total area of clusters of rank i, N_iA_i, is increased by the coalescence of two clusters of rank i-1 to form a cluster of order i and by the coalescence with all smaller clusters k = 1, 2, ⋯, i − 1. The total area is decreased by the coalescence of two clusters of rank i to form a cluster of rank i + 1 and by the coalescence with clusters of all higher ranks j = i+ 1, i + 2 ⋯⋯. Since A_1 = 1, (10) is identical to (8). The two equations (9) and (11) prescribe the number of clusters of rank i, N_i, and the mean area of clusters of rank i, A_i.

In order to proceed it is necessary to specify the rates r_{ij} and f_i included in (9) and (11). We assume that the coalescence rate is proportional to the product of the perimeters P_i and P_j of the two clusters

$$r_{ij} = \xi P_i P_j \tag{12}$$

and the perimeters are given by

$$P_i = \eta N_i A_i^{1/2} \tag{13}$$

This assumption can be generalized and it will be discussed further after a solution is found. The probability of a spark hitting a cluster is proportional to the area of the cluster so that

$$f_i = \beta N_i A_i \tag{14}$$

Substitution of (12), (13), and (14) into (8), (9), and (11) gives

$$\frac{dN_1}{dt} = C - 2\xi\eta^2 N_1^2 - \xi\eta^2 \sum_{j=2}^{\infty} N_1 A_j^{1/2} - \beta N_1 \tag{15}$$

$$\frac{dN_i}{dt} = \xi\eta^2 N_{i-1}^2 A_{i-1} - 2\xi\eta^2 N_i^2 A_i - \xi\eta^2 \sum_{j=i+1}^{\infty} N_i N_j A_i^{1/2} A_j^{1/2} - \beta N_i A_i \tag{16}$$

$$\frac{d(N_i A_i)}{dt} = 2\xi\eta^2 N_{i-1}^2 A_{i-1}^2 + \xi\eta^2 \sum_{k=l}^{i-1} N_k N_i A_k^{3/2} A_i^{1/2} - 2\xi\eta^2 N_i^2 A_i^2 - \xi\eta^2 \sum_{j=i+1}^{\infty} N_i N_j A_i^{3/2} A_j^{1/2} - \beta N_i A_i^2 \tag{17}$$

We first note that the loss terms associated with fires are proportional to N_i whereas all coalescence terms are proportional to the product $N_i\, N_j$. The conclusion is that the losses of trees associated with fires are important only in the very largest clusters where N_i is small and these fires terminate the inverse cascade. Thus we will neglect these terms and set $\beta = 0$ in order to determine the structure of the inverse cascade. We are interested in determining the steady-state structure of the inverse cascade so that we will set the time derivatives equal to zero.

In this steady state (15), (16), and (17) become

$$C' = 2N_1^2 + \sum_{j=2}^{\infty} N_1 N_j A_j^{1/2} \tag{18}$$

$$N_{i-1}^2 A_{i-1} = 2N_i^2 A_i + \sum_{j=i+1}^{\infty} N_i N_j A_j^{1/2} A_j^{1/2} \tag{19}$$

$$2N_{i-1}^2 A_{i-1}^2 + \sum_{k=1}^{i-1} N_k N_i A_k^{3/2} A_i^{1/2} = 2N_i^2 A_i + \sum_{j=i+1}^{\infty} N_i N_j A_i^{3/2} A_j^{1/2} \tag{20}$$

$$C' = \frac{C}{\xi\eta^2} \tag{21}$$

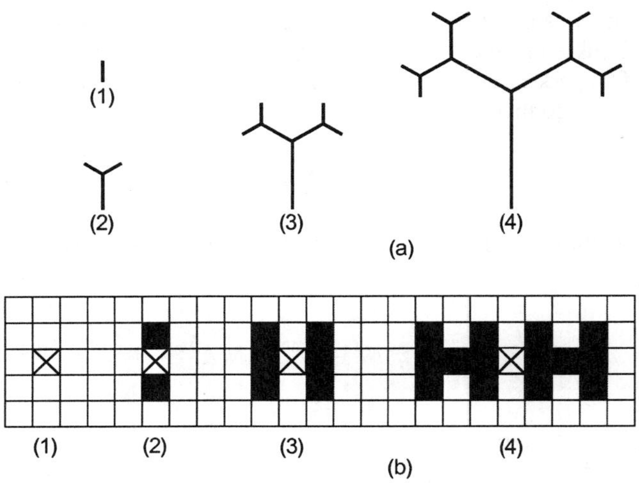

Figure 8. Illustration of the concept of rank (order) applied to cluster coalescence. (a) Binary fractal trees of order (rank) 1, 2, 3 and 4 are illustrated. (b) Clusters of rank 1, 2, 3 and 4 are illustrated. The crosses are newly planted trees that form the clusters.

Equation (19) has a self-similar solution since it is invariant under the transformation $i \to i + 1$ and depends only on $N_i A_i^{1/2}$. Thus we assume that

$$N_i A_i^{1/2} = N_1 x^{i-1} \tag{22}$$

where x is a constant that must be determined. Substitution of (22) into (19) gives

$$x^{2i-4} = 2x^{2i-2} + \sum_{j=i+1}^{\infty} x^{i+j-2} \tag{23}$$

However the geometrical series can be explicitly summed to give

$$\sum_{j=i+1}^{\infty} x^{i+j-2} = x^{2i-1}\left(1 + x + x^2 + \cdots\right) = \frac{x^{2i-1}}{1-x} \tag{24}$$

so that (23) can be written

$$x^3 - 2x^2 - x + 1 = 0 \tag{25}$$

This cubic equation has a single root in the allowed range $0 < x < 1$ so that we have

$$x = 0.55495813\cdots \tag{26}$$

Substitution of this result into (22) gives our first scaling relation for our cluster cascade

$$N_i A_i^{1/2} = N_1 (0.55495813)^{i-1} \tag{27}$$

From (18) and (22) we have with (24)

$$C' = N_1^2\left[2 + \frac{x}{1-x}\right] = \frac{N_1^2}{x^2} \tag{28}$$

which can be used to relate C' to N_i with x from (26).

We now turn to (20), substitution of (22) into (20) gives

$$2x^{i-1}A_{i-1} + \sum_{k=1}^{i-1} x^{k+1} A_k = 2x^{i+1}A_i + \frac{x^{i+2}}{1-x} A_i \tag{29}$$

And substituting (25) gives

$$2x^{i-1}A_{i-1} + \sum_{k=1}^{i-1} x^{k+1} A_k = x^{i-1} A_i \tag{30}$$

This equation does not have an exact self-similar solution since it is not invariant under the transformation $i \to i + 1$. In order to further study (30) we introduce

$$x^{i-1} A_i = y^{i-1} \tag{31}$$

assuming $y > 1$. Summing the finite series in (30) gives

$$2xy^{i-2} + x^2 \frac{\left(y^{i-1} - 1\right)}{\left(y-1\right)} = y^{i-1} \tag{32}$$

For large i we can make the approximation

$$y^{i-1} - 1 \approx y^{i-1} \tag{33}$$

In this limit (32) becomes

$$y^2 - \left(x+1\right)^2 y + 2x = 0 \tag{34}$$

This quadratic equation has one root with $y > 1$ so that we have

$$y = \frac{1}{x} = 1.8019377 \tag{35}$$

And from (31) we have approximately for large i

$$A_i \approx \frac{1}{x^{2(i-1)}} = (3.24697602)^{i-1} \qquad (36)$$

From (22), (27), and (36) we have

$$N_i \approx N_1 x^{2(i-1)} = N_1 (0.307978526)^{i-1} \qquad (37)$$

The ratio of numbers of clusters in adjacent ranks is given by

$$\frac{N_i}{N_{i+1}} = 3.24697602 \qquad (38)$$

This is known as the number-order ratio (it is also known as the bifurcation ratio in river networks). The average areas of clusters in adjacent ranks is given by

$$\frac{A_{i+1}}{A_i} = 3.24697602 \qquad (39)$$

This is known as the area-order ratio.

From (38) and (39) we have

$$N_i = \frac{N_1}{A_i} \qquad (40)$$

Since the association of numbers of clusters with cluster rank is equivalent to a logarithmic binning, this is equivalent to a cumulative distribution. This result is identical to the result given for the forest fire model in (7). The concept of clusters can also be extended to both sandpile and slider-block models. In these cases, the clusters are the metastable regions that will avalanche or slip when an event is triggered. In both cases, the cumulative distribution of cluster sizes satisfies (2) with $\gamma \approx 1$. These scaling relationships are typical of self-organized criticality. Remarkably, this scaling has been deduced using solely analytic means from our inverse-cascade hierarchical cluster model.

6. SEISMIC ACTIVATION

We have discussed how small earthquakes can be used to quantify the hazard associated with large earthquakes. An important question is whether small earthquakes can be used to forecast the temporal occurrence of large earthquakes. As shown in Figure 3 the occurrence of the smallest earthquakes seems to have very little temporal dependence. However, there is accumulating evidence that there may be an activation of intermediate sized earthquakes prior to a great earthquake. The occurrence of a relatively large number of intermediate sized earthquakes in northern California prior to the 1906 San Francisco earthquake has been noted [*Sykes and Jaume, 1990*]. It has also been proposed that there is a power-law increase in seismicity with respect to the time remaining until rupture prior to a major earthquake. This was first proposed by *Bufe and Varnes* [*1993*]. They considered the cumulative amount of Benioff strain (square root of seismic energy) in a specified region. They showed that an accurate retrospective prediction of the Loma Prieta earthquake could be made assuming a square root of time increase in Benioff strain prior to the earthquake.

Systematic increases in intermediate level seismicity prior to a large earthquake have been proposed by several authors [*Varnes, 1989; Bufe et al., 1994; Knopoff et al., 1996; Varnes and Bufe, 1996; Johansen et al., 1996; Brehm and Braile, 1998, 1999*]. A systematic study of the optimal spatial region and magnitude range to obtain the power-law seismic activation has been carried out by *Bowman et al.* [*1998*]. Four examples of their results are given in Figure 9. Clear increases in seismic activity prior to the Kern County, Loma Prieta, Landers, and Northridge earthquakes are illustrated. The optimal radius for precursory activation is given as a function of earthquake magnitude in Figure 10. The dependence on the square root of rupture area is also shown. The radius over which activation occurs is about ten times the length of rupture $r_c \approx 10 A_E^{1/2}$. *Dobrovolsky et al.* [*1979*] and *Keilis-Borok and Kossobokov* [*1990*] reported a similar scaling for the maximum distance between an earthquake and its precursors using pattern recognition techniques.

Based on pattern recognition techniques a number of intermediate range earthquake prediction algorithms were developed at the International Institute of Earthquake Prediction Theory and Mathematical Geophysics in Moscow [*Keilis-Borok, 1990; Keilis-Borok and Rotwain, 1990; Keilis-Borok and Kossobokov, 1990*]. The pattern recognition includes quiescence [*Schreider, 1990*], increase in the clustering of events, and changes in aftershock statistics [*Molchan et al., 1990*]. The algorithm *M8* was developed to make intermediate term predictions of the largest earthquakes ($M > 8$). This method utilizes overlapping circles of seismicity with diameters of 384, 560, 854, and 1333 km for earthquakes, with magnitudes 6.5, 7.0, 7.5, and 8.0 respectively. Within each circle four quantities are determined. The first three are measures of intermediate levels of seismicity and the fourth is a measure of aftershock activity. The first quantity that must be specified is the lower magnitude cutoff M_{min} for

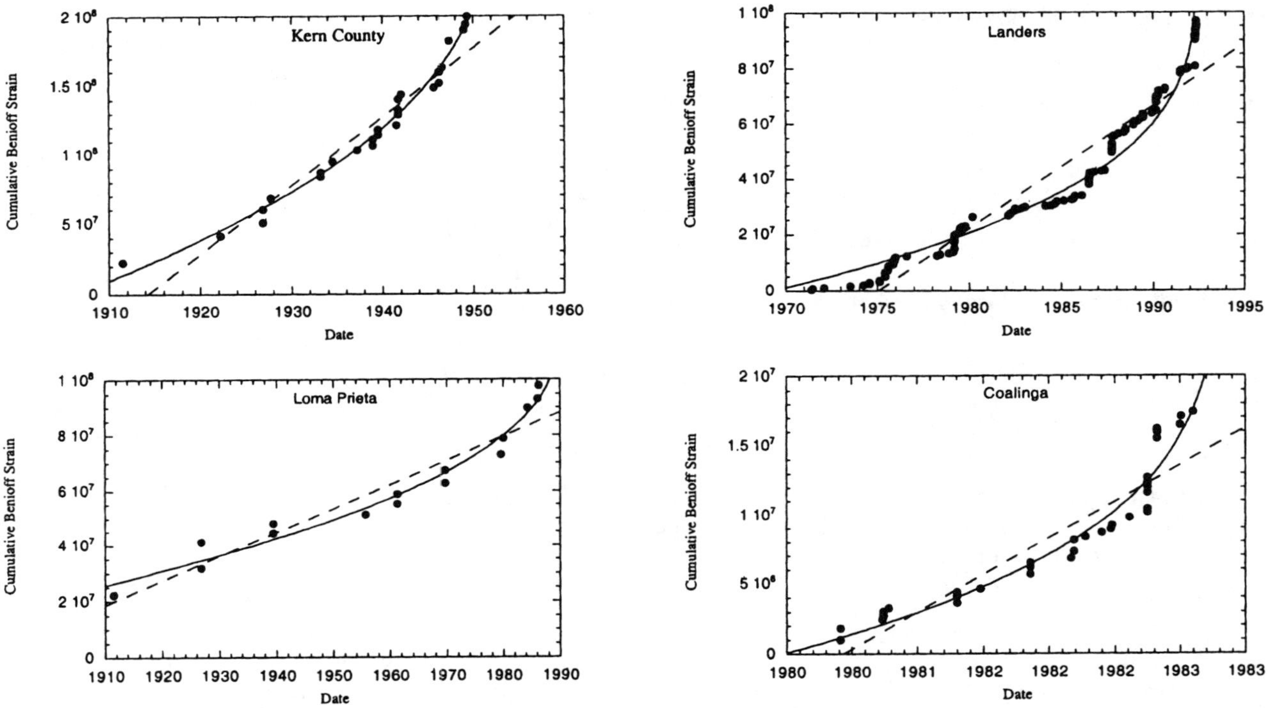

Figure 9. Power-law increases in the cumulative Benioff strains prior to four major earthquakes in California [*Bowman et al., 1998*].

earthquakes to be considered in the circle. Two magnitude cutoffs are considered for each circular region. The long term number of earthquakes per year N in the circle with magnitudes greater than M_{min} is determined, M_{min10} corresponds to $N = 10$ and M_{min20} corresponds $N = 20$. The first quantity $N_1(t)$ is the number of earthquakes per year in the circle with magnitude greater than M_{min10} and $N_2(t)$ is the number of earthquakes per year with magnitudes greater than M_{min20}. The second quantity is the five year trend in activity $L_1 = dN_1/dt$ and $L_2 = dN_2/dt$ for running five year windows. Clearly $N(t)$ and $L(t)$ are strongly correlated. The third quantity, $Z_1(t)$ or $Z_2(t)$, is the ratio of the average linear dimension of rupture to the average separation between earthquakes for a year in a circle. The final measure is the number of aftershocks in a specified magnitude range and time window following a main shock. This quantity $B(t)$ is a measure of aftershock activation.

Since N, L, and Z are determined both for $N = 10$ and 20 earthquakes per year there are seven time series to be considered. An earthquake alarm or time of increased probability (TIP) is issued if 6 of the 7 quantities, including B, exceed their average values by a specified value of 75% for B and 90% for the others. In order to issue an alarm these conditions must be satisfied for two successive six month time periods and the alarm lasts for five years. Details of this algorithm have been given by Keilis-Borok (1996). Two examples of the application of this algorithm are given in Figure 11 for the Loma Prieta (1989) and the Landers (1992) earthquakes.

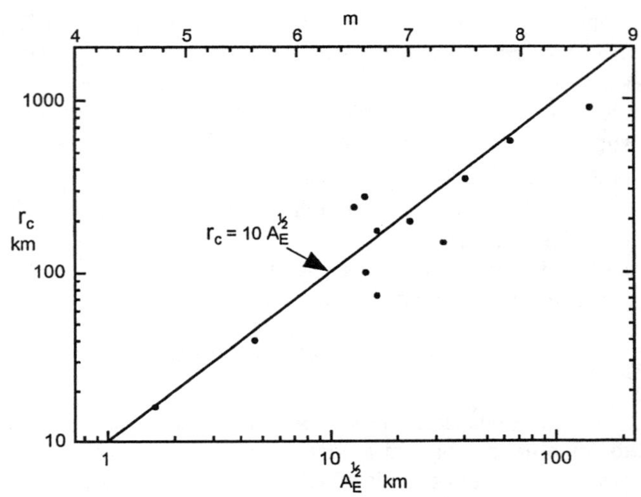

Figure 10. Optimum radius for precursory seismic activation r_c is given as a function of the square root of the rupture area $\sqrt{A_E}$ and the magnitude m for a number of earthquakes [*Bowman et al., 1998*].

The algorithms described above were developed using pattern recognition approaches. Although they have had demonstrated predictive successes [*Kossobokov et al., 1999; Rotwain and Novikova, 1999*], their use remains quite controversial. The main difficulty is that, although success to failure ratios of predictions is quite high, the time and spatial windows of alarms are also quite high.

There are clearly very strong similarities between the M8 and the seismic activation algorithms. Consider the Loma Prieta earthquake, the intermediate sized events that led to the M8 TIP (Figure 11) were the same events that give the increase in Benioff strain (Figure 9). Both these approaches are based on the concept that correlation length increases before major earthquakes [*Harris, 1998*].

7. DISCUSSION

The inverse cascade model introduced in Section 5 of this paper provides a quantitative explanation for the behavior of three models that are said to exhibit self-organized criticality, these are the forest-fire, slider-block, and sandpile models. We now discuss the implications of this model for earthquakes.

For simplicity we will discuss earthquakes in terms of static and dynamic friction coefficients. An earthquake occurs on a fault when the stress on the fault exceeds the static friction. To a first approximation the rupture on the fault spreads over the connected region on the fault where the applied stress exceeds the dynamic friction. After an earthquake occurs on a fault the stress on the rupture zone is everywhere less than the dynamic friction. Stress on the fault builds due to the tectonic motion of the plates. Regions (clusters) on the fault exceeds the dynamic friction. These clusters grow by coalescence as the stress increases. A few earthquakes may occur on the fault but they remove an insignificant fraction of the accumulating stress.

There is also a fractal distribution of fault sizes. The inverse cascade described above operates on every fault. In addition to the increased stress due to the tectonic motion, stress will also be transferred whenever an earthquake occurs. The mean regional stress will decrease when an earthquake occurs over a region which has a scale of the order of its rupture length, but there will also be stress shadows where the stress will increase.

There is observational evidence that the number of intermediate sized earthquakes increases prior to a large earthquake. This activation occurs over a length scale which is about ten times the characteristic rupture lengths. Stresses are correlated on this scale. There is a hierarchy of inverse cascades on the fractal distribution of faults. The cascades on the larger faults appear to be correlated but the cascades on the smaller faults remain uncorrelated.

Acknowledgements. We wish to acknowledge the support of NSF Grant EAR 9804859. We are grateful to Bill Klein, Bruce Malamud, Gleb Morein, John Rundle, Charlie Sammis, and Lennie Smith for many useful discussions.

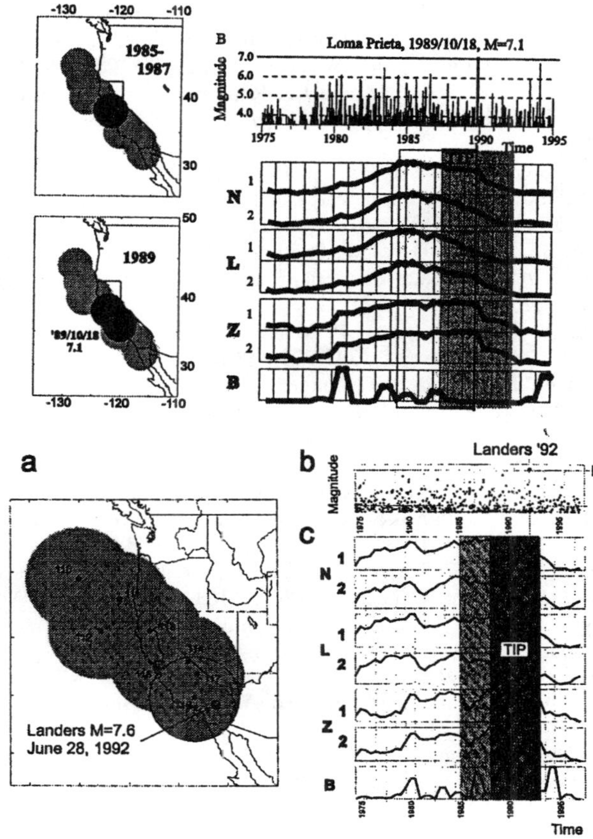

Figure 11. Applications of the M8 algorithm to (1) the $M - 7.1$ Loma Prieta earthquake and (2) the $M = 7.6$ Landers earthquake. In (a) the circular regions where TIP's have been declared are shown, in (b) the seismicity in the relevant region is shown, and in (c) the values of the seven parameters N_1, N_2, L_1, L_2, Z_1, Z_2, and B are shown along with the time interval of the TIP.

REFERENCES

Aki, K., A probabilistic synthesis of precursory phenomena, in *Earthquake Prediction*, D.W. Simpson & P.G. Richards, eds., pp. 566-574, American Geophysical Union, Washington D.C., 1981.

Bak, P., K. Chen, and C. Tang, A forest-fire model and some thoughts on turbulence, *Phys. Lett., A147*, 297-300, 1992.

Bak, P., and C.J. Tang, Earthquakes as a self-organized critical phenomenon, *J. Geophys. Res., 94*, 15,635-15,637, 1989.

Bak, P., and C.J. Tang, and K. Wiesenfeld, Self organized criticality, *Phys. Rev., A38*, 364-374, 1988.

Bowman, D.D., G. Ouillon, C.G. Sammis, A., Sornette, and D. Sornette, An observational test of the critical earthquake concept, *J. Geophys. Res., 103*, 24,359-24,372, 1998.

Brehm, D.J. and L.W. Braile, Intermediate-term earthquake prediction using precursory events in the New Madrid seismic zone, *Bull. Seis. Soc. Am., 88*, 564-580, 1998.

Brehm, D.J. and L.W. Braile, Intermediate-term earthquake prediction using the modified time-to-failure method in southern California, *Bull. Seis. Soc. Am., 89*, 275-293, 1999.

Brown, S.R., C.H. Scholz, and J.B. Rundle, A simplified spring-block model of earthquakes, *Geophys. Res. Let., 18,* 215-218, 1991.

Bufe, G.G., S.P. Nishenko, and D.J. Varnes, Seismicity trends and potential for large earthquakes in the Alaska-Aleutian region, *Pure. Ap. Geophys., 142*, 83-99, 1994.

Bufe, G.G., and D.J. Varnes, Predictive modeling of the seismic cycle of the greater San Francisco Bay region, *J. Geophys. Res., 98,* 9871-9883, 1993.

Burridge R., and L. Knopoff, Model and theoretical seismicity, *Bull. Seis. Soc. Am., 57,* 341-371, 1967.

Carlson, J.M., and J.S. Langer, Mechanical model of an earthquake fault, *Phys. Rev., A40*, 6,470-6,484, 1989.

Carlson, J.M., J.S. Langer, and B.E. Shaw, Dynamics of earthquake faults, *Rev. Mod. Phys., 66*, 657-670, 1994.

Dobrovolsky, I.R., S.I. Zubkov, and V.I. Miachkin, Estimation of the size of earthquake preparation zones, *Pure Ap. Geophys.*, 117, 1025-1044, 1979.

Drossel, B., and F. Schwabl, Self-organized criticality in a forest-fire model, *Phys. Rev. Lett.*, 69, 1629-1632, 1992.

Frohlich, C., and S.D. Davis, Teleseismic b values; or, much ado about 1.0, *J. Geophys. Res., 98,* 631-644, 1993.

Geller, R.J., D.D. Jackson, Y.Y. Kagan, and F. Mulargia, Earthquakes cannot be predicted, *Science*, *275*, 1,616-1,617, 1997.

Gutenberg, B., and C.F. Richter, *Seismicity of the Earth and Associated Phenomenon,* 2nd ed., Princeton University Press, Princeton, 1954.

Harris, R.A., Forecasts of the 1989 Loma Prieta, California, earthquake, *Bull. Seis. Soc. Am., 88,* 898-916, 1998.

Huang, J., G. Narkounskaia, and D.L. Turcotte, A cellular automaton, slider-block model for earthquakes. 2. Demonstration of self-organized criticality for a two-dimensional system, *Geophys. J. Int., 111*, 259-269, 1992.

Huang, J., and D.L. Turcotte, Are earthquakes an example of deterministic chaos? *Geophys. Res. Let., 17*, 223-226, 1990.

Johansen, A., D. Sornette, H. Wakita, U. Tsnnogai, W.I. Newman, and H. Saleur, Discrete scaling in earthquake precursory phenomena: Evidence in the Kobe earthquake, Japan, *J. Phys. I France 6,* 1391-1402, 1996.

Kadanoff, L.P., S.R. Nagel, L. Wu, and S.M. Zhou, Scaling and universality in avalanches, *Phys. Rev., A39*, 6,524-6,533, 1989.

Keilis-Borok, V.I., The lithosphere of the earth as a nonlinear system with implications for earthquake prediction, *Rev. Geophys., 28*, 19-34, 1990.

Keilis-Borok, V.I., Intermediate-term earthquake prediction, *Proc. Natl. Acad. Sci., 93*, 3,748-3,755, 1996.

Keilis-Borok, V.I., and V.G. Kossobokov, Premonitory activation of earthquake flow, Algorith, M8, *Phys. Earth Planet. Int., 61*, 73-83, 1990.

Keilis-Borok, V.I., and I.M. Rotwain, Diagnosis of time of increased probability of strong earthquakes in different regions of the world: Algorithm CN., *Phys. Earth Planet. Int., 61*, 52-72, 1990.

Knopoff, L., T. Levshina, V.I. Keilis-Borok, and C. Mattone, Increased long-range intermediate-magnitude earthquake activity prior to strong earthquakes in California, *J. Geophys. Res., 101,* 5779-5796.

Kossobokov, V.G., L.L. Romashkova, V.I. Keilis-Borok, and J.H. Healy, Testing earthquake prediction algorithms: statistically significant advance prediction of the largest earthquakes in the Circum-Pacific, 1992-1997, *Phys. Earth Planet. Int., 111*, 187-196.

Kossobokov, V.G., and D.L. Turcotte, A systematic global assessment of the seismic hazard (abstract), *EOS, 77,* F480, 1996.

Lorenz, E.N., Deterministic nonperiodic flow, *J. Atmos. Sci., 20,* 130-141, 1963.

Malamud, B.D., G. Morein, and D.L. Turcotte, Forest fires: An example of self-organized critical behavior, *Science, 281*, 1840-1842, 1998.

Mandelbrot, B.B., How long is the coast of Britain? Statistical self-similarity and fractional dimension, *Science, 156*, 636-638, 1967.

Mandelbrot, B.B., *The Fractal Geometry of Nature,* Freeman, San Francisco, 1992.

May, R.M., Simple mathematical models with very complicated dynamics, *Nature, 261,* 459-467, 1976.

Molchan, G.M., O.E. Dmitieva, I.M. Rotwain, and J. Dewey, Statistical analysis of the results of earthquake prediction, based on bursts of aftershocks, *Phys. Earth Planet. Int., 61*, 128-139, 1990.

Nakanishi, H., Statistical properties of the cellular automata model for earthquakes, *Phys. Rev., A43*, 6,613-6,621, 1991.

Otsuka, M., A simulation of earthquake occurrence, *Phys. Earth Planet. Int., 6,* 311-315, 1972.

Rotwain, I., and O. Novikova, Performance of the earthquake prediction algorithm CN in 22 regions of the world, *Phys. Earth Planet. Int., 111*, 207-213.

Rundle, J.B., and D.D. Jackson, Numerical simulation of earthquake sequences, *Seis. Soc. Am. Bull, 67,* 1363-1377, 1977.

Schreider, S., Formal definition of premonitory seismic quiescence, *Phys. Earth Planet. Int., 61*, 113-127, 1990.

Strahler, A.N., Quantitative analysis of watershed geomorphology, *Am. Geophys. Un. Trans., 38*, 913-920, 1957.

Sykes, L.R., and S.C. Jaume, Seismic activity on neighboring faults as a long-term precursor to large earthquakes in the San Francisco Bay area, *Nature, 348*, 595-599, 1990.

Turcotte, D.L., *Fractals and Chaos in Geology and Geophysics*, 2nd ed., Cambridge University Press, Cambridge, 1997.

Turcotte, D.L., B.D. Malamud, G. Morein, and W.I. Newman, An inverse-cascade model for self-organized critical behavior, *Physica, A268,* 629-643, 1999.

Turcotte, D.L., and D.A. Spence, An analysis of strain accumulation on a strike slip fault, *J. Geophys. Res., 79,* 4407-4412, 1974.

Varnes, D.J., Predicting earthquakes by analyzing accelerating precursory seismic activity, *Pure Ap. Geophys., 130*, 661-686, 1989.

Varnes, D.J., and C.G. Bufe, The cyclic and fractal seismic series prededing an m_b 4.8 earthquake on 1980 February 14 near the Virgin Islands, *Geophys. J. Int., 124*, 149-158, 1996.

D.L. Turcotte, Snee Hall, Cornell University, Ithaca, NY 14853

W.I. Newman, Department of Earth and Space Sciences, UCLA, Box 851567, Los Angeles, CA 90024

A. Grabrielov, Department of Earth and Atmospheric Sciences, Purdue University, West Lafayette, IN 47907

Earthquakes: Friction or a Plastic Instability?

D.C. Roberts and D.L. Turcotte

Department of Geological Sciences, Cornell University
Ithaca, New York 14853

The mechanical friction hypothesis is widely accepted as the universal explanation for earthquakes in the earth's crust. However, there are a number of observations that are inconsistent with this hypothesis. In addition, it is generally accepted that the mechanical friction hypothesis is not applicable to deep earthquakes in the subducted lithosphere. However, earthquakes occur at all depths with no evidence for a change in mechanism between deep and shallow events. In this paper the applicability of the plastic instability responsible for shear bands in metals to earthquakes is examined. This instability combines work hardening with thermal softening. We numerically solve the applicable equations for this rupture instability and by adjusting two free parameters, we obtain results that are consistent with observations of both crustal and mantle earthquakes.

1. INTRODUCTION

It is generally accepted that earthquakes occur on preexisting faults and that faults must exhibit stick-slip behavior in order to generate earthquakes. The widely accepted mechanism for this behavior in the earth's crust is velocity weakening mechanical friction. Laboratory studies of friction have provided an understanding of the dependence of friction on slip velocity and surface characteristics. There is general agreement on frictional behavior under well-constrained laboratory conditions. A variety of rate and state friction laws have been generated empirically to satisfy the laboratory observations [*Dieterich*, 1978, 1979, *Ruina* 1983, *Blanpied and Tullis* 1986, *Weeks* 1993]. These empirical formulations clearly represent the available laboratory data.

However, there are serious questions regarding the applicability of the mechanical friction laws to actual faults [*Scholz* 1990, pp. 91-96]. It is generally accepted that mechanical friction is not applicable to deep earthquakes because of the very high confining pressures. It is popular to attribute deep earthquakes to phase transitions in the mantle [*Liu* 1997, *Marone and Liu* 1997, *Green and Houston* 1995]. However, if this were the case, then the occurrence of deep earthquakes would be expected to correlate with the depths at which the major phase changes occur, 410 and 660 km. The dependence of the number of earthquakes on depth has been given by *Frolich* [1989]. The number of earthquakes decreases exponentially with depth to a depth of about 300 km. From this depth to a depth of 660 km the number of earthquakes remains about constant with an increase between 500 and 600 km, particularly associated with the Tonga-Kemadec subduction zone. Any deep earthquake mechanism must be able to produce earthquakes over this entire depth range.

One mechanism that can produce earthquakes over a wide range of depths is the plastic instability studied by *Hobbs and Ord* [1988]. Adiabatic plastic shear instabilities have been observed in many materials. They result in localized catastrophic slip events which strongly

GeoComplexity and the Physics of Earthquakes
Geophysical Monograph 120

resemble frictional instabilities. *Hobbs and Ord* [1988] found that below a critical pressure dependent temperature that mantle rock can be strain rate (velocity) softening and may undergo catastrophic plastic shear. These authors also give arguments why plastic shear band instabilities can be expected in mantle and crustal rocks. These include low thermal conductivity, a stress sensitivity to temperature, and low strain hardening.

Hobbs et al. [1986] associate plastic faulting with pseudotachylytes. They studied a crustal shear zone in central Australia and found cyclic generation and deformation of pseudotachylytes. They proposed that pseudotachylytes and associated ultramylonites develop in the crust as ductile (plastic) instabilities. Based on their observations they concluded that plastic shear instabilities are an important mechanism for generating earthquakes at intermediate crustal depths.

White [1996] carried out microstructural and microcompositional analyses of host mylonites and primary and deformed pseudotachylytes from the Outer Hebrides thrust fault using light and electron microscopy. This author concluded that the deformational microstructures and estimated pressure-temperature conditions were consistent with a catastrophic plastic instability as proposed by *Hobbs and Ord* [1988]. We examine the applicability of the plastic mechanism to essentially all earthquakes.

Below we list some of the problems associated with the application of the mechanical friction hypothesis to earthquake rupture that can be explained by the plastic hypothesis:

1. Stress drop verses mean stress. Seismic and geodetic studies measure stress drop. They clearly indicate a stress drop of less than 10 MPa, even in large earthquakes (with a few exceptions). However, there is also observational evidence that the absolute stress is low. For many years this was known as the heat flow paradox, the measured heat flow adjacent to the San Andreas fault is nearly an order of magnitude less than that predicted by the direct application of laboratory friction experiments [*Lachenbruch and Sass* 1992]. A major justification for the Cajon Pass drill hole was to test the high vs low stress hypotheses, the results conclusively favored the low-stress hypothesis [*Zoback and Healy* 1992]. These observations provide conclusive evidence that the stress drop in major earthquakes is at least 50% of the initial stress and is probably closer to 80 – 90% of the initial stress. In the laboratory, the stress drop in mechanical friction experiments is less than 5% [*Dietrich* 1972, 1978, 1979]. A large stress drop is a natural consequence of the plastic rupture hypothesis. Because the yield stress has exponential sensitivity to temperature, it would be expected to drop a large percentage during rupture before healing.

2. Depth of nucleation. Strike-slip earthquakes, such as those in California, often nucleate at depths of 10 km or more where the lithostatic pressure in greater than 250 MPa, a typical failure stress on such a fault is 10 MPa, giving a coefficient of static friction f = 0.04, while typical laboratory values of the mechanical coefficient of friction are close to 0.6 [*Byerlee*, 1977]. Recognizing this major discrepancy a number of authors have proposed that the low stress is due to a high fluid pressure [*Byerlee*, 1990; *Blanpied* et al., 1992; Sleep and *Blanpied*, 1992, 1994; *Sleep*, 1997]. This requires an impermeable fault zone and several mechanisms have been proposed, but at best, this proposal must be considered ad hoc. With plastic rupture, failure is associated with the yield stress which is sensitive to both temperature and pressure whereas the mechanical static friction is only a function of pressure.

3. Earthquakes often nucleate at the deepest point on a earthquake rupture zone. This point has the highest ambient (lithostatic) pressure and would be expected to be the strongest in terms of the mechanical friction law. Nucleation at the deepest point is a natural consequence of the plastic mechanism due to the dependence of the yield stress on both temperature and pressure.

4. Between earthquakes on a fault, there is little evidence for seismic or aseismic displacement on the fault. While there are exceptions, observations of such fault offsets either from earthquakes or geodetically, are quite rare. The locked northern and southern sections of the San Andreas fault are not slipping. Since the lithostatic normal force increases linearly with depth, it is also expected that the frictional resistance to slip also increases with depth. As the stress on a fault increases during an earthquake cycle, it would be expected that slip would occur on the upper portion of the fault while it remains locked at depth [*Lorenzetti and Tullis*, 1989; *Rice*, 1993]. Once again observations are not consistent with the predictions of the laboratory friction laws. With a plastic failure mechanism the weakest point on the fault would have the highest temperature below the critical temperature T_c, thus the deepest point on the rupture zone.

5. *Heaton* [1990] suggested that self-healing pulses of slip occur in earthquake rupture. This mode of rupture was strongly confirmed by observations of slip distributions during the 1992 Landers earthquake [*Wald and Heaton*, 1994]. As pointed out by Heaton, this type of self-healing pulse is inconsistent with the friction laws described above. The plastic mechanism provides healing naturally wheras mechanical friction does not.

6. With a large stress drop an underdamped (ringing) oscillation on the fault would be expected. This is never seen in earthquakes. The healing associated with the plastic mechanism prevents ringing.

7. Very few earthquakes are recorded in the upper five kilometers of the crust. Also there is little evidence of aseismic creep in this region. Strain meters on many

sections of the San Andreas fault indicate that the upper portion of the fault is locked. Friction laws give a failure stress that increases linearly with depth. Thus they are inconsistent with these observations. With the plastic faulting hypothesis the strength decreases with depth due to thermal effects consistent with these observations

Adiabatic shear is the name given to localization phenomena that are important in many problems involving high rate deformations of solids. Adiabatic shear bands form in a deforming material when thermal softening due to plastic heating is stronger than work hardening and rate hardening combined so that a regime with net strain softening occurs [*Clifton et al.* 1984, *Wright and Batra* 1985]. The quantitative treatment of adiabatic shear bands has been based on the general theory of themoplasticity given by *Green and Naghdi* [1965]. Although studies of adiabatic shear bands have been primarily applied to studies of metals they are certainly applicable to other materials. We use the general theory for the localization of adiabatic shear bands as given by *Wright and Ockendon* [1992, 1996] in our quantification of the plastic earthquake instability.

2. EQUATIONS FOR A PLASTIC INSTABILITY

In this paper we will consider the equations that have been used to understand adiabatic shear bands in metals [*Wright and Ockendon*, 1996] The applicable equation of motion is

$$\rho \frac{\partial u}{\partial t} = \frac{\partial \tau}{\partial y}, \tag{1}$$

where τ is the shear stress, u is the velocity in the shearing layer, and ρ is the density. The velocity is in the x-direction and the variables are functions of y and t in this two-dimensional formulation. The left side of (1) is inertia and the right side is the body force due to the variation of the shear stress.

The heat conduction equation is

$$\rho c \frac{\partial T}{\partial t} = k \frac{\partial^2 T}{\partial y^2} + \tau \frac{\partial u}{\partial y} \tag{2}$$

where c is specific heat, T temperature, and k thermal conductivity. The first term is heat storage, the second term heat conduction, and the third term is frictional heating. The plastic constitutive equation is given by

$$\tau = \tau_y(T, p) \left[1 + b \frac{\partial u}{\partial y} \right]^m \tag{3}$$

where p is the pressure, b is the normalization factor for strain rate, m is the normalization factor for strain rate sensitivity and τ_y is the yield stress for plastic flow. If $\tau < \tau_y$ there is no deformation. The temperature and pressure dependent yield stress is given by

$$\tau_y = A \exp\left(\frac{E^* + pV^*}{RT} \right), \tag{4}$$

where E^* is the activation energy, V^* is the activation volume and R is the universal gas constant. When (3) and (4) are combined we have strain hardening and thermal softening.

The instability mechanism associated with these equations is easily illustrated. If the shear stress exceeds the yield stress in (3) we obtain a strain. This strain in (2) heats the slip zone. The increase in temperature in (4) reduces the yield stress and this results in an increase in strain from (3).

3. APPROXIMATIONS

In this section, we simplify the partial differential equations introduced in the last section into coupled algebraic equations that we can solve numerically. The configuration is symmetrical about the center line. The half-width of the plastic slip zone β is specified. All plastic deformation takes place in this zone. Elastic energy is stored over a half width L, $L >> \beta$. The initial stress in the elastic zone is τ_{yo} and the initial displacement across this zone is $\tau_{yo}L/G$ where G is the shear modulus. During an earthquake plastic deformation takes place in the plastic slip zone. This displacement reduces the displacement across the elastic zone and reduces the shear stress τ. The displacement at the boundary of the slip zone is w. The reduction of the shear stress is Gw/L and the equation of motion (1) is replaced by the force equation

$$\tau = \tau_{yo} - \frac{Gw}{L}, \tag{5}$$

In writing this equation we have neglected inertia and radiated energy. During each time step δt there is an incremental displacement δw across the plastic slip zone. By symmetry the velocity $u = 0$ at the center of the plastic slip zone. The velocity gradient $\partial u/\partial y$ is not a function of y in the slip zone. Thus the velocity at the boundary of the slip zone is $\beta\ (\partial u/\partial y)$ and the displacement on the boundary is given by

$$w = \sum_{n=1}^{\max t} \beta \left(\frac{\partial u}{\partial y} \right)_n \delta t, \tag{6}$$

where the summation is over the time steps associated with the slip event.

We next consider the energy equation (2). Because of the relative short time scale of the slip event we will neglect heat conduction. We will justify this assumption when we consider a specific example. The temperature increase in the plastic zone during each time interval δt is given by

$$\delta T = \frac{\tau}{\rho c}\frac{\partial u}{\partial y}\delta t. \quad (7)$$

Finally, to close the problem, we solve the constitutive equation (3) for the velocity gradient

$$\frac{\partial u}{\partial y} = \frac{1}{b}\left[\left(\frac{\tau}{\tau_y}\right)^{\frac{1}{m}} - 1\right]. \quad (8)$$

At the shallow depths associated with crustal earthquakes it is appropriate to assume that E^* is much greater than pV^* so the temperature dependent yield stress simplifies to

$$\tau_y = A\exp\left[\frac{E^*}{RT}\right]. \quad (9)$$

With (5-9) we have five equations for the five variables w, $\partial u/\partial y$, T, τ, and τ_y.

4. INITIAL CONDITIONS

Before solving for the structure of a slip event we require initial conditions to start the simulation. Let us perturb the stress, temperature, and velocity around their equilibrium values τ_{yo}, T_o, and $u = 0$ using $\tau = \tau_{y0} + \tau'$, $T = T_0 + T'$, *and* $u = u'$, where the primed quantities are small perturbations. Substituting these relations into (3) we obtain

$$\tau_{y0} + \tau' = A\exp\left[\frac{E^*}{R(T_0 + T')}\right]\left[1 + b\frac{\partial u'}{\partial y}\right]^m \quad (10)$$

This equation becomes

$$\tau_{y0} + \tau' = A\exp\left(\frac{E^*}{RT}\right)\exp\left(\frac{-E^*T'}{RT_0^2}\right) \quad (11)$$
$$\left[1 + b\frac{\partial u'}{\partial y}\right]^m,$$

by using the power series expansion:

$$\frac{E^*}{R(T_0 + T')} = \frac{E^*}{RT_0\left(1 + \frac{T'}{T_0}\right)} \approx \frac{E^*}{RT_0}\left(1 - \frac{T'}{T_0}\right). \quad (12)$$

Linearizing (11) and using $\tau_{y0} = A\ exp\ (E^*/RT_0)$, we obtain

$$\tau_{yo} + \tau' = \tau_{y0}\left(1 - \frac{E^*T'}{RT_0^2}\right)\left(1 + mb\frac{\partial u'}{\partial y}\right). \quad (13)$$

We can solve for the perturbation stress τ' by keeping only first order terms with the result

$$\tau' = \tau_{yo}\left(mb\frac{\partial u'}{\partial y} - \frac{E^*T'}{RT_0^2}\right). \quad (14)$$

In order to initiate the slip event, we prescribe a small increment of applied stress τ' above the yield stress τ_{yo}. We assume (14) can be used to determine the initial $(\partial u'/\partial y)_0$ before there is any temperature increment, we obtain

$$\left[\frac{\partial u'}{\partial y}\right]_o = \frac{\tau'}{\tau_{yo}mb} \quad (15)$$

Using this result and (6), we obtain the initial displacement,

$$w_o = \beta\left[\frac{\partial u'}{\partial y}\right]_o \delta t, \quad (16)$$

which, combined with (5), can be used to find the initial stress,

$$\tau_o = \tau_{yo} + \tau' - \frac{Gw_0}{L}. \quad (17)$$

The initial temperature can then be expressed as

$$T_1 = T_0 + \frac{\tau_o}{\rho c}\left[\frac{\partial u'}{\partial y}\right]_o \delta t \quad (18)$$

by using (7) and the above initial conditions.

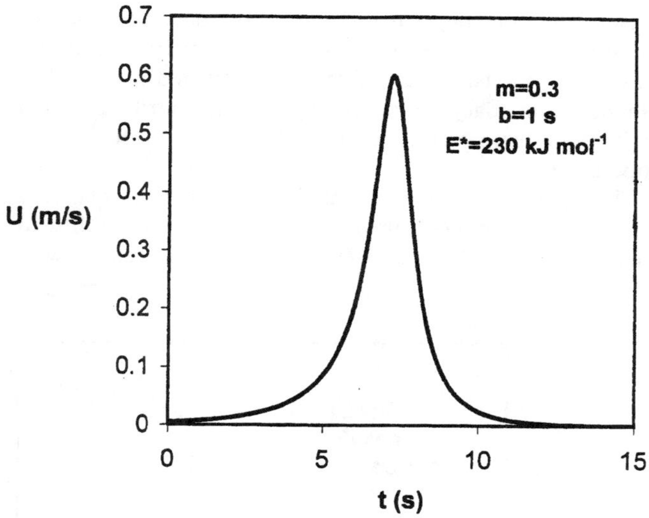

Figure 1. Velocity of slip u as a function of time t using $m = 0.3$, $b = 1$ s, and $E^* = 230$ kJ mol^{-1} (wet quartzite).

5. RESULTS

Before solutions can be obtained, values for parameters must be specified. For a typical deep crustal earthquake we take $\rho = 2700$ kg m^{-3}, c = 1 kJ kg^{-1}, $\tau_{yo} = 10$ MPa, L = 5 km, $\beta = 100$ mm, G = 3 x 10^{10} Pa and T = 700 K. We first take an activation energy appropriate for the crust (wet quartzite) with $E^* = 230$ kJ mol^{-1}. Finally we must specify the strain hardening parameters m and b. Very little information is available concerning these parameters for rock. As an example we take $m = 0.3$ and $b = 1$ s. A variety of other values of these parameters would give similar results.

Results for these parameters are given in Figures 1-4. From Figure 1 we see that the total displacement on the model fault is $w = 1.32$ m and the rupture time is approximately 5 seconds. From Figure 2 we see that the maximum slip velocity is $u = 0.61$ ms^{-1}. This is a typical rupture velocity on a fault. The dependence of the shear stress on time is given in Figure 3. The stress drops from $\tau = 10$ MPa to $\tau = 2.2$ MPa when healing has taken place. The temperature on the fault is given in Figure 4. Frictional heating increases the temperature from $T = 700$ K to $T = 730$ K. These values are all reasonable. The characteristic time for heat conduction across the slip zone is $\beta^2\kappa^{-1}$. With $\kappa = 1$ mm^2 s^{-1} we have $\beta^2\kappa^{-1} = 10^4 s$. Since this is large compared with the rupture time our neglect of heat conduction is appropriate.

We also give results for a mantle (olivine) rheology and take $E^* = 523$ kJ mol^{-1}. For this application we take $m = 0.3$ and $b = 100$ s. Results are given in Figures 5-8. Again, reasonable values are obtained for all variables.

6. DISCUSSION

We have obtained solutions for plastic faulting using a plastic rheology. The rheology includes work hardening and thermal softening. The values of slip velocity, stress drop, and heating are reasonable for both crustal and mantle earthquakes.

Our results must be considered preliminary in that we use rheological parameters that are chosen arbitrarily. Laboratory based parameters are not available so that we argue that the parameters chosen may represent the

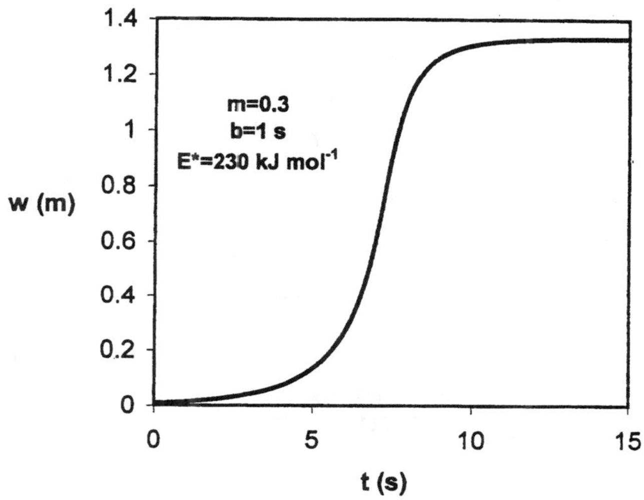

Figure 2. Displacement w as function of time t using $m = 0.3$, $b = 1$ s, and $E^* = 230$ kJ mol^{-1} (wet quartize).

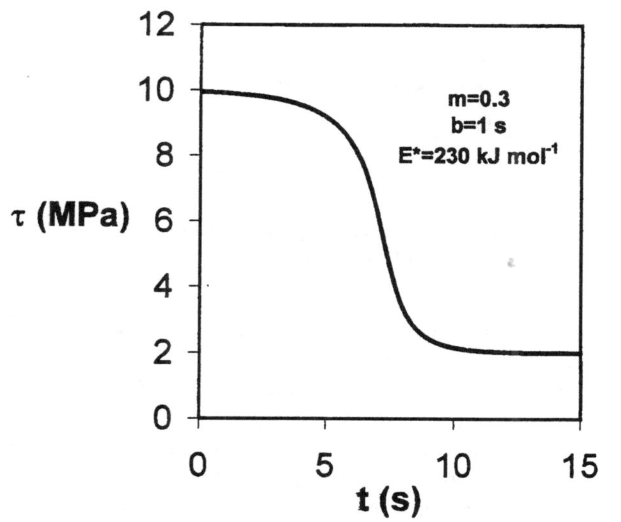

Figure 3. Stress τ as a function of time using $m = 0.3$, $b = 1$ s, and $E^* = 230$ kj mol^{-1} (wet quartzite).

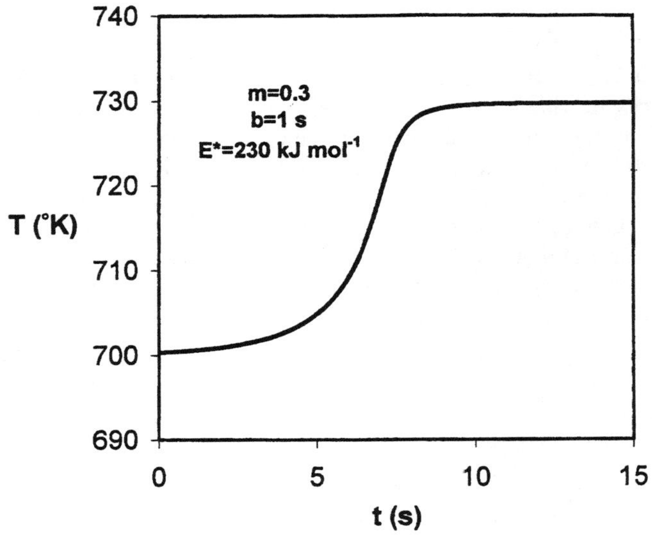

Figure 4. Temperature T as a function of time t using m = 0.3, b = 1 s, and E^* = 230 kJ mol^{-1} (wet quartzite).

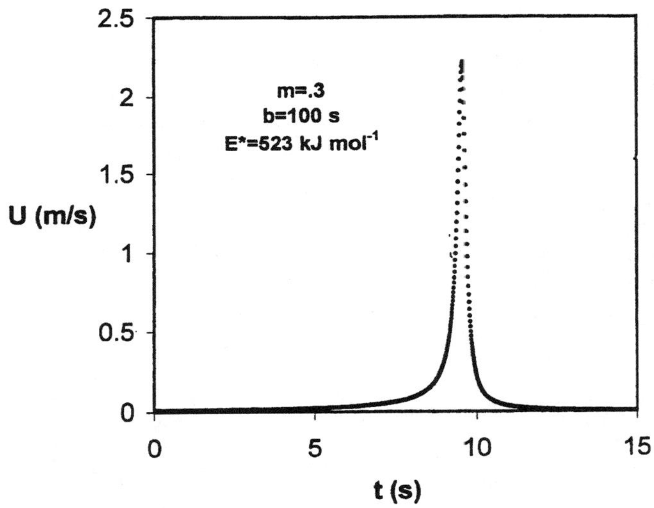

Figure 5. Velocity of slip u as function of time t using m = 0.3, b = 100 s, and E^* = 523 kJ mol^{-1} (dry olivine).

complex damage rheology involving dislocations and granular recrystalization appropriate to faulting at depth. It is clear that plastic rupture can be obtained for a very wide range of parameter values.

We argue that a plastic failure mechanism, such as that considered here, is in much better agreement with observations of seismicity than mechanical friction. Under typical (say on the San Andreas fault in California) there are very few earthquakes shallower than 5 km and essentially none deper than 15 km. The deep boundary of seismicity can be attributed to a transition from a plastic instability to steady-state thermally activated creep. The shallow boundary of seismicity can be attributed to a minimum temperature for the plastic instabilities to be operative.

The plastic faulting mechanism provides a relatively large stress drop (≈ 80%) while retaining the ability to heal. Because the yield (failure) stress decreases with

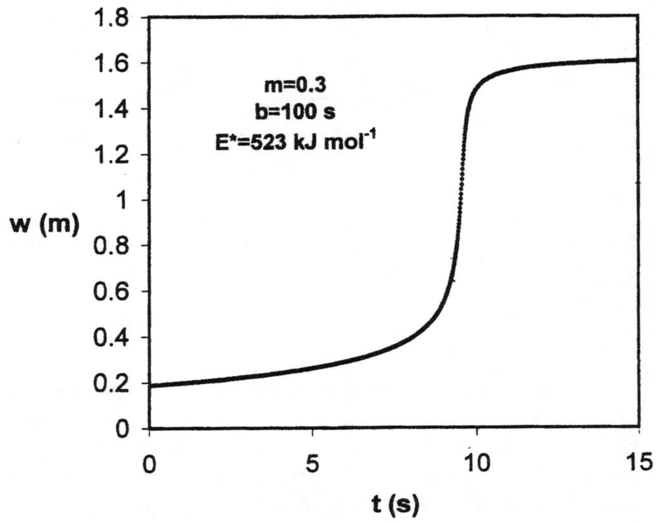

Figure 6. Displacement w as a function of time t using m = 0.3, b = 100 s, and E^* = 523 kJ mol^{-1} (dry olivine)..

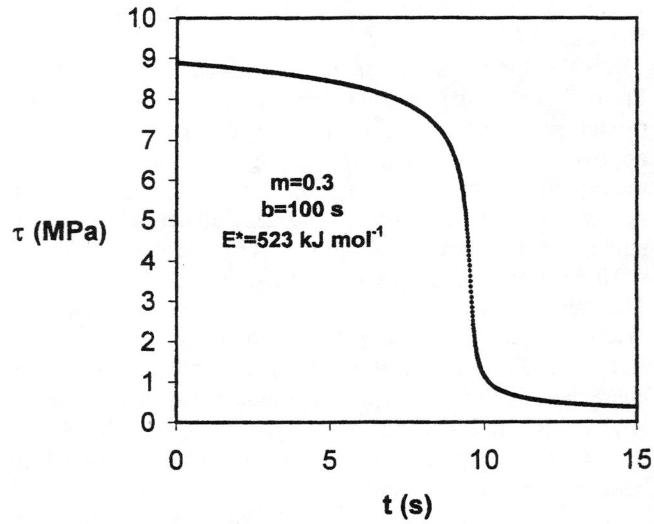

Figure 7. Stress τ as a function of time t using m = 0.3, b = 100 s, and E^* = 523 kJ mol^{-1} (dry olivine).

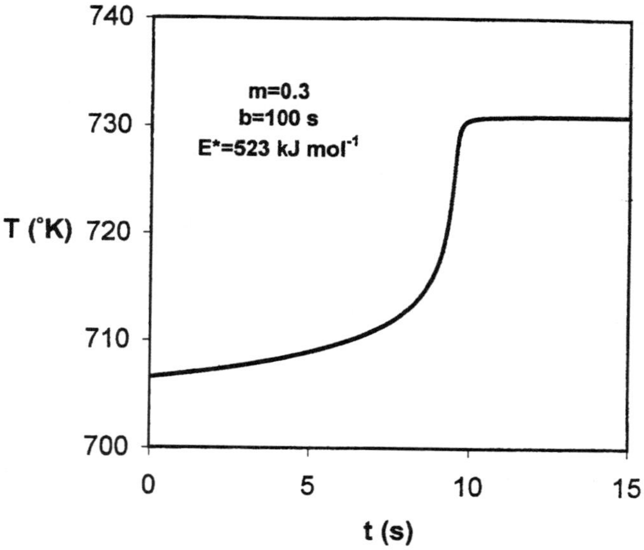

Figure 8. Temperature T as a function of time t using $m = 0.3$, $b = 100$ s, and $E^* = 523$ kJ mol^{-1} (dry olivine).

increasing temperature, earthquake nucleation at the base of the seismogenic zone follows naturally.

REFERENCES

Blanpied, M.L., D.A. Lockner, and J.D. Byerlee, An earthquake mechanism based on rapid sealing of faults, *Nature, 358*, 574-576, 1992.

Blanpied, M.L. and T.E. Tullis, The stability of a frictional system with a two state variable constitutive law, *Pageoph., 124*, 413-333, 1986.

Byerlee, J.D., Friction of rocks, in *Experimental Studies of Rock Friction with Applications to Earthquake Prediction,* J.F. Evernden, ed., pp. 55-77, U.S. Geological Survey, Menlo Park, CA, 1977.

Byerlee, J.D., Friction, overpressure and fault normal compression, *Geophys. Res. Let., 17*, 2109-2112, 1990.

Clifton, R.J., R.J. Duffy, K.A. Hartley, and T.G. Shawki, On critical conditions for shear band formation at high strain rates, *Scripta Metallurgica, 18*, 448-465, 1984.

Dieterich, J.H., Time-dependent friction in rocks, *J. Geophys. Res.*, 77, 3690-3697, 1972.

Dieterich, J.H., Time-dependent friction and the mechanics of stick-slip, *Pageoph., 116*, 790-806, 1978.

Dieterich, J.H., Modeling of rock friction, 1. Experimental results and constitutive equations, *J. Geophys. Res., 84*, 2161-2168, 1979.

Frolich, C., The nature of deep focus earthquakes, *An. Rev. Earth Planet. Sci., 17*, 227-254, 1989.

Green, A.E. and P.M. Naghdi, A general theory of a elastic-plastic continum, *Arch. Rat. Mech. Anal., 18*, 252-270, 1965.

Green, H.W. and H. Houston, The mechanics of deep earthquakes, *An. Rev. Earth Planet. Sci., 23*, 169-213, 1995.

Heaton, T.H., Evidence for and implications of self-healing pulses of slip in earthquake rupture, *Phys. Earth Planet. Int., 64*, 1-20, 1990.

Hobbs, B.E. and A. Ord, Plastic instabilities: Implications for the origin of intermediate and deep focus earthquakes, *J. Geophys. Res., 93*, 10,521-10,540, 1988.

Hobbs, B.E., A. Ord and C. Teyssies, Earthquakes in the ductile regime? *Pure Ap. Geophys., 124*, 309-336, 1986.

Lachenbruch, A.H. and J.H. Sass, Heat flow from Cajon Pass, fault strength, and tectonic implications, *J. Geophys. Res., 97*, 4995-5015, 1992.

Liu, M., A constitutive model for olivine-spinel aggregates and its application to deep earthquake nucleation, *J. Geophys. Res., 102*, 5295-5312, 1997.

Lorenzetti, D. and T.E. Tullis, Geodetic predictions of a strike-slip fault model: Implications for intermediate and short-term earthquake prediction, *J. Geophys. Res., 94*. 12,343-12,361, 1989.

Marone, C. and M. Liu, Transformation shear instability and the seismogenic zone for deep earthquakes, *Geophys. Res. Let., 24*, 1887-1890, 1997.

Rice, J.R., Slip instability and state variable friction laws, *J. Geophys. Res., 98*, 9885-9907, 1993.

Ruina, A., Slip instability and state variable friction laws, *J. Geophys. Res., 88,* 10,359-10,370, 1983.

Scholz, C.H., *The Mechanics of Earthquakes and Faulting,* Cambridge University Press, Cambridge, 1990.

Sleep, N.H., Application of a unified and state friction theory to the mechanics of fault zones with strain localization, *J. Geophys. Res., 102*, 2875-2895, 1997.

Sleep, N.H. and M.L. Blanpied, Creep, compaction and the weak rheology of major faults, *Nature, 359*, 687-692, 1992.

Sleep, N.H. and M.L. Blanpied, Ductile creep and compaction: A mechanism for transiently increasing fluid pressure in mostly sealed fault zones, *Pure Ap. Geophys., 143*, 9-40, 1994.

Wald, D.J. and T.H. Heaton, Spatial and temporal distributions of slip for the 1992 Landers, California, earthquake, *Seis. Soc. Am. Bull., 84*, 668-691, 1994.

Weeks, J.D., Constitutive laws for high-velocity frictional sliding and their influence on stress drop during unstable slip, *J. Geophys. Res., 98*, 17,637-17,648, 1993.

White, J.C., Transient discontinuities revisited: pseudotachylyte, plastic instability and the influence of low pore fluid pressure on deformation processes in the mid-crust, *J. Struc. Geol., 18*, 1471-1486, 1996.

Wright, T.W. and C. Batra, The initiation and growth of adiabatic shear bands, *Int. J. Plasticity, 1*, 205-212, 1985.

Wright, T.W. and H. Ockendon, A model for fully formed shear bands, *J. Mech. Phys. Solids, 40*, 1217-1226, 1992.

Wright, T.W. and H. Ockendon, A scaling law for the effect of inertia on the formation of adiabatic shear bands, *Int. J. Plasticity, 12*, 927-934, 1996.

Zoback, M.D. and J.H. Healy, In situ stress measurements to 3.5 km depth in the Cajon Pass scientific research borehole: Implications for the mechanics of crustal faulting, *J. Geophys. Res., 97*, 5039-5057, 1992.

D.C. Roberts and D.L. Turcotte, Snee Hall, Cornell University, Ithaca, Ny 14853

Lattice Solid Simulation of the Physics of Fault Zones and Earthquakes: the Model, Results and Directions

P. Mora, D. Place, S. Abe and S. Jaumé

QUAKES, Department of Earth Sciences, The University of Queensland, Brisbane, Australia

The lattice solid model is a particle-based numerical model that is being developed with the aim of eventually allowing all relevant physical phenomena underlying earthquake dynamics to be realistically simulated. In its present form, the model is capable of simulating the evolution and dynamics of complex fault zones consisting of a brittle solid and/or granular gouge region. Elasticity is modeled in solid regions and friction is simulated between solid regions in contact. Solid regions can fracture and break down allowing the evolution of fault zones to be studied. Some thermal effects are simulated including thermo-mechanical and thermo-porous feedback which respectively allow studies of the effects of thermal expansion of the solid and increased pore fluid pressure when frictional heating occurs. We review the lattice solid model and discuss major results obtained so far. The model has been applied to studies of the dynamics and evolution of fault gouge zones. For inter-particle friction ranging from 0.4 to 1.0 and pressure changing by a factor of 10, the effective fault friction is found to remain approximately constant with a value of $\bar{\mu}_f \sim 0.6$. In some cases, the active slip band became highly localized after a large displacement and $\bar{\mu}_f$ dropped dramatically to around 0.3. These results suggest that a self-regulation process controlling the balance between slip, rolling and fracture is responsible for the almost constant friction of 0.6 measured in the laboratory for different rock types. They also suggest it is possible for a gouge zone to self-organize to a structure in which slip is highly localized and the fault zone is anomalously weak, thus providing a comprehensive potential explanation for the heat flow paradox. Simulations show that thermal effects play a role on rupture dynamics with a relatively minor influence originating from thermal expansion and a larger effect due to the increase in pore pressure. Long simulations involving a fractured zone show accelerating Benioff strain sequences and evolving event-size statistics in the periods prior to large simulated earthquakes which are consistent with the critical point hypothesis for earthquakes. The simulation results using the present lattice solid model demonstrate the potential

GeoComplexity and the Physics of Earthquakes
Geophysical Monograph 120

of the approach to address fundamental questions regarding the physics of earthquakes from nucleation studies to studies of the dynamics of interacting fault systems. The model results suggest that earthquake forecasting or intermediate term earthquake prediction is a realistic goal and that the statistical physics analogies of the earthquake problem have relevance.

1. INTRODUCTION

Earthquakes involve physical processes occurring over a wide range of space and time scales (Figure 1). The processes include fracture of heterogeneous solids, granular dynamics, friction between rough brittle rock surfaces, solid-fluid dynamics, lubrication, phase transformations such as mineralogical or solid-liquid, stress accumulation, finite-strain elastic or plastic deformation, elastic stress transfer via seismic wave propagation, seismic wave radiation from the crustal zone of interest, heat conduction and thermo-mechanical feedback. Numerical models capable of simulating such phenomena would provide a virtual laboratory to probe complex earthquake behavior and offers a new means to advance earthquake science. The wide range of space and time scales coupled with the complexity of the physical phenomena involved renders formidable the construction of such a simulation model and has motivated establishment a major international collaborative effort involving Australia, China, Japan and USA (The APEC Cooperation for Earthquake Simulation, http://quakes.earthsciences.uq.edu.au/ACES).

Inter-fault regions approximate elastic, viscous or attenuative solids rendering simulation of such regions tractable to classical continuum-mechanics-based approaches such as finite-differences, finite-elements, pseudo-spectral and spectral-elements. However, to model the complete earthquake process, the fault zone and inter-fault region must also be well simulated. Furthermore, macroscopic simulations of earthquake phenomena require an effective fault constitutive relation for fault zones, but these are only partially accessible by direct observations (see Figure 1). The need for a fault constitutive relation necessitates extrapolations based on laboratory observations or indirect inferences based on compatibility between scaling relations derived from macroscopic simulation studies and field data. A numerical model capable of simulating processes at the micro-scale and meso-scale covering the nucleation processes zone of Figure 1 would help bridge this crucial gap in knowledge and has motivated the development of the lattice solid model (Mora, 1992 [16]; Mora and Place, 1993 [17]; Mora and Place, 1994 [18]).

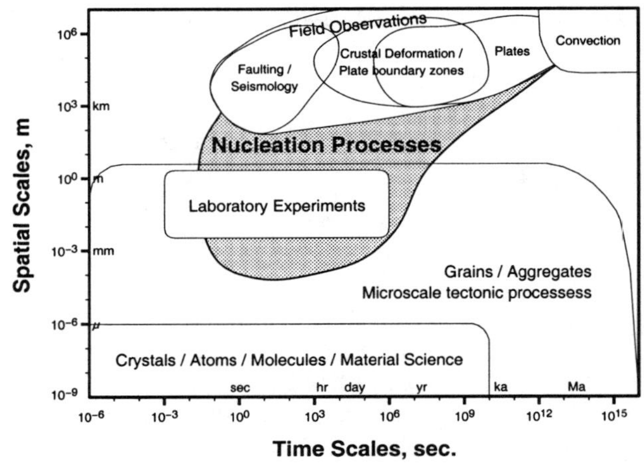

Figure 1. Space and time scales relevant to the earthquake cycle. Physical processes occurring at the microscopic scale (Grains/Aggregates Micro-tectonic processes) control earthquake generation. This domain is only partially accessible by direct observations. The meso-scale where earthquake nucleation processes occur is difficult to access with direct observations and lies between the microscopic and macroscopic realms (note: boundaries on the plot should be considered as fuzzy rather than sharp). From the APEC Cooperation for Earthquake Simulation (ACES) proposal, Mora, Matsu'ura, Minster and Yin, 1997.

In its present form, the lattice solid model can simulate many elements of solid behavior including elastic stress transfer, fracture, friction, heat generation and flow, thermo-mechanical coupling and thermo-porous feedback (Mora and Place, 1998 [19]; Place and Mora, 1999a [24]; Abe et al, 1999 [1]). It can simulate evolving fractures and discontinuities as well as the dynamics of a gouge layer.

The model has been used to simulate the frictional instability and dynamic rupture processes (Mora and Place, 1994 [18]), localization processes in a fault zone (Place and Mora, 1999b [25]) and the dynamics of a fracture zone with dynamics approaching that of an interacting fault system (Mora and Place, 1999b [21]; Mora, Jaumé and Place, 1999 [22]). Simulations of fault gouge layers demonstrated the possibility that gouge may self-organize into an anomalously weak state, thus offering a potential explanation for the heat flow paradox (Mora and Place, 1998 [19]; Mora and Place, 1999a [20]). Long simulations of multiple ruptures in a complex fracture zone have demonstrated that long term earthquake phenomena in the model evolve in a manner consistent with the critical point hypothesis for earthquakes (Mora, Jaumé and Place, 1999 [22]) and observations of accelerating Benioff strain release (Jaumé and Sykes, 1999 [12]). In the following, we review key results and discuss their implications.

2. THE LATTICE SOLID MODEL

Motivated by molecular dynamics, the Lattice Solid Model (LSM) was initially proposed in 1992 (Mora, 1992 [16]). It consists of a system of particles whose interactions are specified such that they represent the basic units of the system being simulated.

For example, to model rocks at the meso-scale, interactions should be specified corresponding to those between cemented or touching rock grains. The particle discretization of matter makes it tractable to simulate the essential complex phenomena underlying fault dynamics such as fracture of heterogeneous solids and friction between rough brittle surfaces. In the initial LSM, solid material was discretized in 2D as particles bonded by elastic brittle bonds. Even with such simple interactions, a variety of phenomena relevant to earthquakes were observed including mode II fracture (Mora and Place, 1993 [17]), slick-slip behavior and occurrence of rupture as a slip-pulse (Mora and Place, 1994 [18]).

In the most recent version of the LSM (Mora and Place, 1998 [19]; Place and Mora, 1999 [24]), the smallest pieces of material are discretized as particles arranged into a regular triangular lattice in 2D or a close-packed lattice in 3D (Abe, Mora and Place, 1999 [1]). Theoretical studies of the elastic properties of the lattice solid demonstrate that use of a triangular lattice in 2D or face-centered hypercube (FCHC) lattice in 3D yield isotropic elastic behavior for each piece of model material. These pieces of material are used to represent rock grains or blocks. Bonded particles undergo linear elastic attractive and repulsive forces depending on the inter-particle separation. Grains are bonded together by breakable bonds to enable fracture processes to be simulated. Different pieces of material interact with one another if they come into contact through the frictional and repulsive forces that occur between their surface particles. The frictional work done on surface particles generates heat and the heat-flow equation is solved within solid regions. Thermal expansion is included and Darcian fluid flow and the effect on friction of increased pore fluid pressure due to heating is modeled.

2.1. Elastic Behavior

Bonds between lattice solid particles can be described using a linear or nonlinear relationship. So far, we have preferred to use a simple linear relationship so a bond acts like a linear spring and exerts a force on the bonded particles given by

$$\mathbf{F}^\alpha = k(r_0 - r^\alpha)\mathbf{e}^\alpha \quad , \tag{1}$$

where k is the spring constant, r_0 is the equilibrium separation, r^α is the distance to the α-th bonded particle and $\mathbf{e}^\alpha$ is the unit vector pointing to the α-th bonded particle. By expressing particle displacement in terms of a continuous differentiable infinitesimal displacement field $u_i \to 0$, the i-th component of the total force on the particle can be written as

$$\begin{aligned} F_i &= ma_i = \sum_\alpha F_i^\alpha \\ &= \sum_\alpha \sum_j \sum_k \sum_\ell \frac{kr_0^2}{2!} e_i^\alpha e_j^\alpha e_k^\alpha e_\ell^\alpha \frac{\partial^2 u_j}{\partial x_k \partial x_\ell} \quad , \end{aligned} \tag{2}$$

where m is used to denote the particle mass and higher order terms in u_i have been neglected. By comparing the above expression with the equation for the acceleration a_i of a point in an elastic solid

$$a_i = \frac{c_{ijk\ell}}{\rho} \frac{\partial^2 u_j}{\partial x_k x_\ell} \quad , \tag{3}$$

we observe that in macroscopic limit ($r_0 \to 0$) when the expansion of particle displacement in terms of a continuous field u_i is valid, the lattice solid behaves as an elastic solid with Hooke tensor given by

$$c_{ijk\ell} = \frac{1}{\widehat{\rho}} \sum_\alpha k \frac{r_0^2}{2!} e_i^\alpha e_j^\alpha e_k^\alpha e_\ell^\alpha \tag{4}$$

where $\widehat{\rho}$ is the density of the lattice which depends on the particle mass and lattice geometry (e.g. see Mora and Place, 1994 [18]).

It is numerically convenient to arrange particles within solid regions into a regular lattice structure such as a triangular lattice in 2D or a close packed lattice in 3D which both have shear elasticity due to their lattice geometry. However, while the 2D triangular lattice yields an isotropic Hooke tensor with elastic moduli

Figure 2. Velocity surfaces for the 3D FCC lattice without 2nd order interactions (the part above the x-y plane is cut off).

$\lambda = \mu$ (i.e. similar to rocks), the close packed lattice in 3D is anisotropic.

In order to achieve isotropic behavior in the 3D case using the simple central force model described by Equation (1), it is necessary to implement a lattice which contains second order interactions in a face centered cubic (FCC) structure which is equivalent to use of the face centered hypercube (FCHC) structure in the lattice gas methods (Frisch et al. 1987 [10]). First order or nearest neighbor interactions involve particles separated by a distance d^α smaller than a given maximum distance $d^\alpha \leq r^1_{max}$ whereas second order interactions involve particles separated by a distance d^α within a range $r^2_{min} \leq d^\alpha \leq r^2_{max}$ where $r^2_{min} > r^1_{max}$ (Donzé 1994 [9]). For the 3D FCC lattice without second order interactions, the wave velocities V_p, V_{s1} and V_{s2} can been computed numerically by inserting the $\mathbf{e}^\alpha$ for the lattice into Equation (4), computing the acoustic tensor Q_{ik} for different directions $\vec{n}$, and solving for the eigenvalues (Abe, Mora and Place, 1999 [1]). The resulting velocity surfaces shown in Figures 2 and 3 demonstrate the strong anisotropy of the shear wave velocity of the FCC structure and the isotropy using the second order interactions (c.f. FCHC structure). Isotropy of shear wave propagation also results for the triangular lattice in 2D.

The isotropic lattice solid structures has Lamé elastic moduli given by

$$\lambda = \mu = \frac{\sqrt{3}}{4}k \quad , \tag{5}$$

which is equivalent to that of a 3D solid with Poisson's ratio of 0.25 which is similar to most crustal rocks. The density of the 2D triangular lattice is given by

$$\rho = \frac{2}{\sqrt{3}r_0^2}m \quad , \tag{6}$$

which is obtained by dividing the particle mass by the volume of space filling elements (hexagons for the triangular lattice). A similar expression for density can be obtained for the 3D FCC lattice.

To achieve arbitrary elastic properties (specific anisotropy or arbitrary λ and μ), more complex elastic interactions between particles are required. However, this added complexity adds computational cost.

Furthermore, other aspects such as friction, fracture, gouge dynamics, heat and porous effects are likely to have a stronger influence on the nonlinear dynamical behavior of a fault zone than the exact elastic coefficients. For this reason, we have so far preferred to retain the simplicity of the radial only or central forces model. This allows individual particles to be treated as point masses and results in a very simple and efficient computational model and numerical integration of the equations of motion for particles. Shear elasticity is present as a consequence of the lattice structure and is similar to that of rocks (i.e. $V_s = V_p/\sqrt{3}$).

A single particle carries no angular momentum and hence, particle rotation is meaningless using this simple approach. However, groupings of bonded particles can

Figure 3. Velocity surfaces for the isotropic 3D FCHC lattice with 2nd order interactions (the part above the x-y plane is cut off). Only two surfaces are visible because the 2 slower velocities are identical ($V_{s1} = V_{s2}$).

rotate and conserve angular momentum. To ensure that rotation and shear elasticity is simulated at the lowest scale in the model, which may have a fundamental effect on system dynamics, particles are grouped into small unbreakable "grains" at the lowest scale. Each grain consists of several particles arranged into the lattice structure and therefore, represents an isotropic elastic piece of material. Using this approach, grains can be considered as the atoms of the model (smallest indivisible unit of model matter) and the particles can be considered as the quarks (a simple unit from which atoms can be made but which can never exist in isolation). Grains are bonded together to make up an aggregate material representing the lattice solid.

2.2. Artificial Viscosity

An artificial viscous force proportional to particle velocity is applied to avoid the buildup of kinetic energy in the closed system that may modify system dynamics (Mora and Place, 1994 [18]). This mimics the effect of seismic wave radiation to outside a given crustal fault region. The viscous forces on each particle are given by

$$\mathbf{F}_\nu = -\nu \dot{\mathbf{x}} \quad , \tag{7}$$

where ν is the viscosity coefficient and $\dot{\mathbf{x}}$ the velocity of the particle. The viscosity is frequency independent and does not fundamentally alter the dynamics of the system if carefully chosen (Mora and Place, 1994 [18]). The energy lost to the artificial viscosity is generally considered as kinetic energy lost from the system but recent work (Mora and Place, 1999b [21]) has shown that this assumption leads to an overestimate of the seismic wave radiation during rupture. This is because some of the energy lost to artificial viscosity is converted to other energy forms such as heat and strain energy in the absence of viscous damping.

2.3. Fracture and Healing

Brittle fracture behavior occurs under most conditions relevant to the earthquake process so bonds between different grains are allowed to break irreversibly according to a given criterion. Similarly, it is possible to reform bonds to simulate healing or chemical processes. In the limit, molecular interactions can be used such as a radial Lennard-Jones potential function (e.g. Mora and Place, 1993) or a more complex function. With the exception of initial work using the Lennard-Jones potential and some unpublished tests on the influence of healing due to chemical interactions, most simulations have been conducted with fracturing only (i.e. no healing mechanism). In these, we have chosen a simple fracture criterion that depends only on the separation between bonded particles although more complex criteria (e.g. temperature and pressure dependent) would be easy to implement if required. Namely, bonds are be broken irreversibly when the inter-particle separation exceeds a given threshold breaking distance r_b. The elastic force between unbonded particles coming into contact (e.g. surface particles of unbonded grains) is purely repulsive

$$\mathbf{F}^\alpha = \begin{cases} k(r_0 - r)\mathbf{e}^\alpha & r \leq r_0 \\ 0 & r > r_0 \end{cases} . \tag{8}$$

A limitation of modeling fracturing behavior in a discrete lattice is imposed by the desire to use the tractable and elastically isotropic regular lattices (e.g. triangular in 2D and FCHC in 3D). Such regular lattice can be compared to a crystal lattice and exhibit anisotropic fracturing behavior (e.g., Mora and Place, 1993 [17]) that is not associated with most crustal rocks. However, this unrealistic behavior is suppressed in the lattice solid model by the introduction of irregularity through grouping of particles into unbreakable "grains" with random shapes. The size of particles within grains can also be varied to further add to the irregularity of the lattice solid aggregate material. This approach of grouping particles into irregular grains has been shown to yield realistic fracture behavior in simulations using the DEM particle based computational approach (e.g. Sakaguchi and Mühlhaus, 1999 [28]).

2.4. Friction

Unbonded particles that come into contact (i.e. $r < r_0$) undergo linear elastic repulsion of the same form as Equation (8) and a simple frictional force. These are termed touching particles. The intrinsic frictional force opposes the relative tangential velocity between the touching particles and is applied at the center of particles rather than at the surface. This simplification avoids the need to model rotation at the particle scale and makes the computational problem more tractable. It is not a fundamental drawback considering groupings of particles can rotate, and torque on such groupings due to forces on surface particles will be modeled. Any desired frictional force can in principle be specified (e.g. depending on physical conditions such pressure and temperature or dynamic variables such as slip velocity etc.). Modeling an arbitrary friction requires no slip when the tangential force between a surface particle pair is insufficient to overcome the static friction, and application of the specified dynamical frictional force to resist tangential slip otherwise. In most experiments so far we have retained the simple Coulomb-like friction or modified this slightly (e.g. to model the effect of increased pore pressure).

In the case of the Coulomb-like frictional interactions, the force on a particle due to dynamic frictional contact with particle β is given by

$$\mathbf{F}_D^\beta = -\mu\sigma_n^* r_0 \mathbf{e}_T^\beta \quad , \tag{9}$$

where μ is the coefficient of friction between particles, σ_n^* is the effective normal stress between particle pair computed as $\sigma_n^* = k(r - r_0)/r_0$, and $\mathbf{e}_T^\beta = \dot{\mathbf{x}}_T^\beta / |\dot{\mathbf{x}}_T^\beta|$ is the unit vector pointing in the direction of the relative tangential velocity $\dot{\mathbf{x}}_T^\beta$ between the particle and touching particle β given by

$$\dot{\mathbf{x}}_T^\beta = (\dot{\mathbf{x}} - \dot{\mathbf{x}}^\beta) - \left((\dot{\mathbf{x}} - \dot{\mathbf{x}}^\beta)\cdot \mathbf{e}^\beta\right)\mathbf{e}^\beta \quad . \tag{10}$$

The dynamic frictional force given by Equation (9) represents an upper bound for the effective frictional force on a given particle due to the frictional interaction with another particle β. For example, if the two particles are in static frictional contact (zero relative tangential velocity), they will remain static until the external forces exceed the dynamic frictional force given by Equation (9). In other words, to maintain a static frictional contact, a frictional force which is lower in magnitude than the so called dynamic frictional force is effectively being applied. We denote this effective frictional force as $\mathbf{F}_f^\beta$. The computation of the effective frictional force is non-trivial and involves resolution of a nonlinear system as described in Place and Mora (1999a) [24]. The approach involving resolution of the nonlinear system allows all dynamical and static frictional forces to be precisely calculated at each time step of the numerical integration algorithm. The LSM is similar to the particle based Discrete Element Method (DEM) of Cundall and Strack (1979 [8]) but the independent roots have led to a vastly different computational approach. Comparisons between the two methods (Place and Mora, 1999a [24]) demonstrate that under the given LSM assumptions of stopping slip exactly between static surface particles, the LSM is more precise and efficient than the DEM method. The frictional work done is converted to heat in the model.

2.5. Numerical Integration

Once the forces on each particle are computed from the positions and velocities of all particles as described above, the acceleration of each particle can be calculated by summing the different forces (e.g. elastic, friction and viscous)

$$\ddot{\mathbf{x}} = \frac{1}{m}\left(\sum_\alpha \mathbf{F}^\alpha + \sum_\beta \mathbf{F}_f^\beta + \mathbf{F}_\nu\right) \quad . \tag{11}$$

In a manner similar to that used in molecular dynamics, the positions and velocities of particles can then be extrapolated in time using a finite-difference scheme. The scheme is based on the velocity-Verlat approach (e.g., Allen and Tildesley, 1987 [2])

$$\mathbf{x}(t+\Delta t) = \mathbf{x}(t) + \Delta t\,\dot{\mathbf{x}}(t) + \frac{\Delta t}{2!}\,\ddot{\mathbf{x}}(t) \quad , \tag{12}$$

$$\dot{\mathbf{x}}(t+\Delta t) = \dot{\mathbf{x}}(t) + \Delta t\left(\frac{\ddot{\mathbf{x}}(t) + \ddot{\mathbf{x}}(t+\Delta t)}{2}\right) \quad . \tag{13}$$

However, one observes that the simple velocity-Verlat scheme requires acceleration at the next time step in order to extrapolate the velocity to the next time step. This depends on the elastic, frictional and viscous forces at the next time step. The elastic force at the next time step can be computed as this depends only on the position of particles which is calculated in the first step of the velocity-Verlat algorithm. However, the artificial viscous force specified by Equation (7) depends on velocity at the next time step which is not yet calculated.

For this reason, an iterative approach was developed (Mora and Place, 1994 [18]) in the case of no intrinsic friction. When intrinsic friction is present as in this study, the finite-difference scheme for numerical integration becomes more complex in order to accurately treat particles in static frictional contact and particles that pass between dynamic and static frictional contact during a time step (Place and Mora, 1999a [24]). As mentioned previously, the effective frictional force on a particle due to touching particle β that must be applied during the time step is generally lower than the dynamic frictional force given by Equation (9). This effective frictional force depends on all other forces on the particle, including the yet to be calculated effective frictional forces $\mathbf{F}_f^\beta$ due to other particles β that touch the particle. Consequently, a coupled system involving all touching particles must be resolved to determine the $\mathbf{F}_f^\beta$. Because the elastic forces acting on particles change after bond-breaking, the system must be resolved twice at each time step denoted t in order to determine the effective frictional forces before and after the bonds are broken. As an added complexity, a special case must be treated when the relative tangential velocity, and hence frictional force, reverses direction during a time step due to elastic forces acting on the particles. This adds non-linearity to the coupled system that must be resolved. To achieve a precise and efficient numerical integration, an adaptive time step that varies according the level of dynamical activity in the model was utilized. A half time-step scheme is used to take into account the discontinuity in frictional forces that may occur at a time step t due to bond breaking.

2.6. Energy Conservation

Total energy - computed as the sum of heat, potential energy, kinetic energy, fracture energy and applied work done has been verified to remain constant to a high precision - typically to within 1% after 10^6 time steps - and provides a useful implementation test.

2.7. Heat Flow

The behavior of temperature in a solid with heat sources can be described by the heat flow equation

$$\frac{\partial T}{\partial t} = \kappa_{ij}\frac{\partial^2 T}{\partial x_i \partial x_j} + \frac{1}{\rho c_p}H(\vec{x},t) \quad , \tag{14}$$

where ρ is the density of the solid, c_p is the heat capacity and $H(\vec{x},t)$ the local heat production or source term. In the case of an isotropic medium, the thermal diffusivity tensor κ_{ij} is diagonal so a scalar κ can be used and Equation (14) simplifies to

$$\frac{\partial T}{\partial t} = \kappa\frac{\partial^2 T}{\partial \vec{x}^2} + \frac{1}{\rho c_p}H(\vec{x},t) \quad . \tag{15}$$

In the Lattice Solid Model, the source term represents the frictional heat being generated by particles rubbing together. The solution of Equation (15) is split into 2 steps. First the heat produced by interparticle friction is added to the particles and then the heat is extrapolated in time by an explicit finite-difference method, namely

$$T(t) \quad = \quad T(t) + \frac{1}{\rho c_p}H(t) \tag{16}$$

$$T(t+\Delta t) \quad = \quad T(t) + \xi\kappa\Delta t\sum_\alpha \frac{T^\alpha(t) - T(t)}{(r^\alpha)^2} \quad ,\tag{17}$$

where $T(t)$ is the temperature of the particle at time t, $T(t+\Delta t)$ the temperature at the time $t+\Delta t$, ξ is a constant imposed by the lattice geometry and r^α is the distance between the particle and touching particle α. This method, which is equivalent to the well known FTCS Finite Difference scheme (forward time, centered space), is numerically stable under the condition $\frac{\kappa\Delta t}{(r^\alpha)^2} < \frac{1}{2}$. The constraint $\Delta t \le \epsilon\frac{r_0}{V_p}$ (e.g., Mora and Place 1994 [18], Place and Mora 1999a [24]) is imposed on the timestep by the computation of the elastic interactions in the model. Typically $\epsilon < 0.2$ must be chosen to ensure numerical stability and sufficient accuracy of the dynamic computations. Considering the thermal diffusivity of typical rocks is approximately $\kappa = 10^{-6}m^2s^{-1}$, and assuming a P-wave velocity $V_p = 5000ms^{-1}$ this guarantees that the stability condition is satisfied for $r_0 > 4\cdot 10^{-11}m$ which is always the case (Abe, Mora and Place 1998 [1]).

The only heat source implemented in the model is frictional heating, i.e. there is no heat generated by the dissipation of elastic wave energy in the present version of the model. The constant ξ for the 2D triangular lattice and 3D close packed lattice respectively equals 1.5 and 2 (see Abe, Mora and Place, 1999 [1]).

2.8. Thermal Expansion

If a fault system is heated so that the temperature distribution becomes heterogeneous, geometry will be changed by thermal expansion of the material. This has an influence on the distribution of stress and thus potentially on the dynamics of fault slip and rupture. This could be importance, particularly if the system is close to a critical state where even the small thermally induced geometric changes, which would not exceed 1%, can have an influence.

The simplest approach for implementation of thermal expansion would be to use a relation between the temperature and the radius of a particle. The high temperature increases of several hundred degrees which can occur during large slip events make it necessary to use a nonlinear relationship such as

$$r_0(T) = R_0(1 + c_0 + c_1(T - T_0) + c_2(T - T_0)^2 + \ldots) \ , \tag{18}$$

(see Abe, Mora and Place (1999) [1]), because at those

temperatures the contribution of the higher order terms becomes significant.

A cubic relation with the parameters for polycrystalline quartz (e.g, Touloukian et al.,1977 [30])

$$\begin{aligned} c_0 &= -0.00236 \\ c_1 &= 6.912 \cdot 10^{-6} \\ c_2 &= 5.559 \cdot 10^{-11} \\ c_3 &= 1.312 \cdot 10^{-11} \quad , \end{aligned}$$

has used been used in Abe, Mora and Place (1999) [1]. Using these coefficients, at 800K - a representative temperature which can be reached during slip events (e.g., McKenzie and Brune, 1972 [15]) - the contribution of the higher order terms is approximately 70% of the expansion.

2.9. *Temperature Dependent Pore Fluid Pressure*

If pore fluid is present in a fault zone, the heat generated during a slip event can cause thermal expansion of the fluid, thus leading to a local increase in pore fluid pressure along the fault. This can reduce the effective normal stress and the resulting frictional force. If the reduction in frictional force is significant, then this process will have an influence on the dynamics of the fault slip.

In the presence of a pore fluid, the heat flow equation (15) becomes

$$\frac{\partial T}{\partial t} = \kappa \frac{\partial^2 T}{\partial \vec{x}^2} + \frac{1}{\rho c_p} H(\vec{x}, t) + \frac{1}{\rho c_p} \frac{\partial}{\partial \vec{x}} (\rho_{fl} c_{fl} \vec{w} T) \quad , \tag{19}$$

where ρ and c_p are the density and heat capacity of the aggregate (i.e. rock and fluid together), ρ_{fl} and c_{fl} are the parameters for the fluid only and $\vec{w}$ is the convective velocity of the pore fluid. As shown by Lachenbruch (i.e. Lachenbruch 1980 [14]), the convective term can be ignored safely so Equation (19) simplifies to

$$\frac{\partial T}{\partial t} = \kappa \frac{\partial^2 T}{\partial \vec{x}^2} + \frac{1}{\rho c_p} H(\vec{x}, t) \quad , \tag{20}$$

which is identical to Equation (15). Assuming further that the porosity of the material is small enough that the contribution of the fluid to the heat capacity of the solid can be ignored, the same parameters as in the case without a pore fluid can be used in the numerical solution of Equation (20).

The fluid pressure can be computed from

$$\frac{\partial P}{\partial t} = \frac{1}{\beta} \left[\gamma \frac{\partial T}{\partial t} - \dot{D} \right] + \alpha_P \frac{\partial^2 P}{\partial \vec{x}^2} \quad , \tag{21}$$

where β represents the compressibility, γ the thermal expansion of the fluid and α_P the Darcian diffusivity (Lachenbruch 1980 [14]). If the dilatational strain rate of the pore space $\dot{D}$ is assumed to be sufficiently small and therefore ignored, Equation (21) simplifies to

$$\frac{\partial P}{\partial t} = \frac{\gamma}{\beta} \frac{\partial T}{\partial t} + \alpha_P \frac{\partial^2 P}{\partial \vec{x}^2} \quad . \tag{22}$$

The first term on the right hand side represents the source term, i.e. the pressure rise caused by the increase in temperature and the second one is the transport term. Thus Equation (22) has the same structure as the heat flow equation (14) and is solved numerically using the same method. The same considerations about numerical stability also apply, i.e. for realistic values of α_P, the constraints placed on the time step by the computations of the dynamics ensure that the numerical solution of the pressure equation is stable.

Assuming that the pore fluid is water, the value of the pressure increase for the conditions at seismogenic depths in the crust, i.e. 150 – 300MPa and 600 – 900K, is $\frac{\gamma}{\beta} = 1.5 - 2MPaK^{-1}$. Thus an immediate temperature increase of a few hundred degrees would generate a pore fluid pressure which exceeds the lithostatic pressure.

The relation between frictional stress τ, normal stress σ_n, coefficient of dynamic friction μ and the pore fluid pressure P can be expressed as (Nur and Byerlee, 1971 [23])

$$\tau = \mu \left(\sigma_n - \left(1 - \frac{\beta_g}{\beta_r} \right) P \right) \quad , \tag{23}$$

where β_g is the compressibility of the solid grains and β_r is the compressibility of the whole material including the fluid. Assuming that the compressibility of the solid grains is far smaller that the compressibility of the whole aggregate, i.e. the compression takes place at the expense of pore volume (Lachenbruch 1980 [14]), we can assume

$$\frac{\beta_g}{\beta_r} \longrightarrow 0 \quad , \tag{24}$$

in which case Equation (23) simplifies to

$$\tau = \mu(\sigma_n - P) \quad . \tag{25}$$

Assuming friction vanishes if the fluid pressure exceeds the normal stress, and ignoring effects occurring at high fluid pressures such as hydraulic fracturing, the effective inter-particle friction coefficient used for the computation of frictional force between touching particles can be calculated as

$$\mu_{eff} = \begin{cases} \frac{\sigma_n - P}{\sigma_n} & \textit{if } P < \sigma_n \\ 0 & \textit{if } P \geq \sigma_n \end{cases} . \tag{26}$$

3. NUMERICAL SIMULATION OF FAULT ZONES

To study the behavior and evolution of fault gouge zones, numerical experiments have been conducted involving a weak heterogeneous region placed between two elastic blocks with rough surfaces (Mora and Place, 1999a [20]; Place and Mora, 1999b [25]). These simulations were performed prior to the introduction of temperature related effects which implies that the model is scaleless, or in other words, phenomena occurring in the model occur independently of the scale. For this reason, we do not use absolute units but numerically convenient model units. The results can be compared to laboratory experiments by considering any appropriate relative scales (e.g. gouge layer width relative to minimum grain size, speed of driving blocks past one another relative to rupture speed, etc.).

The gouge zone was initialized as unbreakable grains of different sizes and shapes. Grains were bonded to one another by breakable bonds with a given breaking criterion r_b and single particles were removed from the gouge zone (Plate 1). Therefore, the gouge region is weak and breaks down during simulations. Large grains, composed of groupings of smaller grains which are bonded together, are formed naturally early in the simulation.

The model consisted of 256 × 64 particles and the gouge layer thickness was 50 particle diameters = $50 \times r_0$ which is about 12 times the longest grain length. Four different grain shapes were used: elongated hexagon (10 particles), hexagon (7 particles), diamond (4 particles) and triangle (3 particles). A distribution of grains inversely proportional to grain size (Size is measured in terms of number of particles.) was used to randomly fill the gouge region. Diamond and triangle grain shapes are then used to fill as much of the remaining porosity as possible. Roughness of the model fault surfaces were initialized using a power law spatial spectrum of surface heights with exponent 0.5 (Mora and Place, 1994 [18]) using a maximum height variation of 5 particle diameters.

Each block was pushed from the outer driving plates at a constant velocity of $V_{plate} = 0.00025 V_p$ where V_p represents the P-wave velocity (approximately equal to 1.0 for a spring constant $k = 1$ and a particle mass $m = 1$, see Mora and Place, 1994 [18]) while being subjected to a constant normal stress. Simulations were performed over displacements of up to approximately 1400 particle diameters (more than 21 full rotations of the circular model along the x-axis) where displacement is defined as the distance moved by the upper block relative to the lower block.

In the following, we applied two normal stresses, one higher and one lower, and use two different breaking separations. Scaling these for a solid with $V_p = 3\sqrt{3}$km/s and density $\rho = 3000$kg.m^{-3}, the higher normal stress would correspond to 1500 MPa and the lower normal stress to 150 MPa (see Place and Mora, 1999 [24] for an explanation of the scaling between model and MKS units). For comparison, using $V_p = 3$km/s and $\rho = 2200$kg.m^{-3}, the two normal stresses would correspond to 367 MPa and 36.7 MPa. In view of the somewhat arbitrary value of the normal stress in physical units depending on the assumed elastic moduli and density, we will not label these different runs with a numerical value of normal stress but only as high and low normal stress. For the runs with the higher normal stress, we have used a higher breaking separation of $r_b = 1.075 r_0$ and in the lower normal stress runs, we have used two different values being $r_b = 1.0075$ and $r_b = 1.015 r_0$. The results will be compared and discussed.

3.1. *Effective Fault Friction*

The shear stress on the driving plates divided by the constant normal stress is termed the instantaneous fault friction μ_f. Figure 4 shows the instantaneous fault friction as a function of the displacement for numerical experiments using coefficients of friction ranging from $\mu = 0.4$ through $\mu = 1.0$ using the higher normal stress and $r_b = 1.075 r_0$. Characteristic sawtooth shapes of stick-slip cycles can be clearly seen in the magnification box. During the stick phase, the stress builds up in the solid. When enough energy is stored in the deformation of blocks to overcome the static friction or fracture criterion of bonds along a given rupture surface within the gouge layer, a dynamic rupture event occurs. The stress is suddenly released during this slip phase. To obtain the effective friction of the gouge layer (termed the fault friction), the saw-tooth instantaneous friction

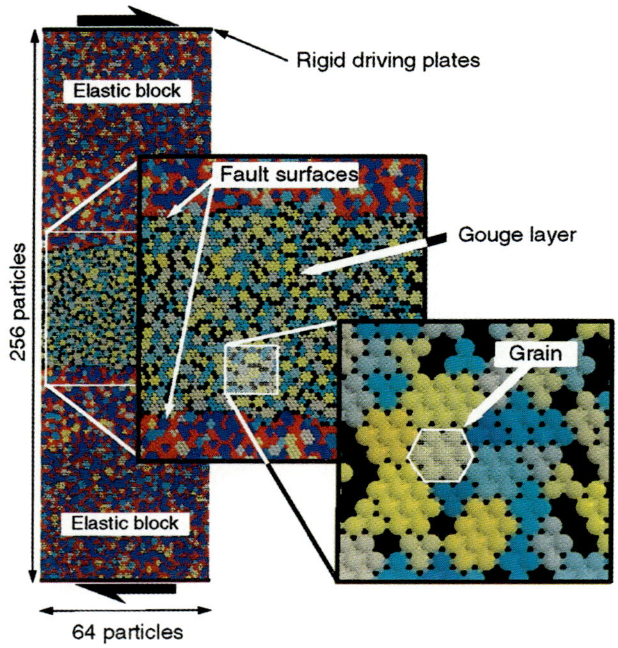

Plate 1. Experimental setup: the simulated fault gouge is initialized by specifying a weak region composed of unbreakable grains of rock which are weakly bonded together. Boundary conditions are circular in the x-direction.

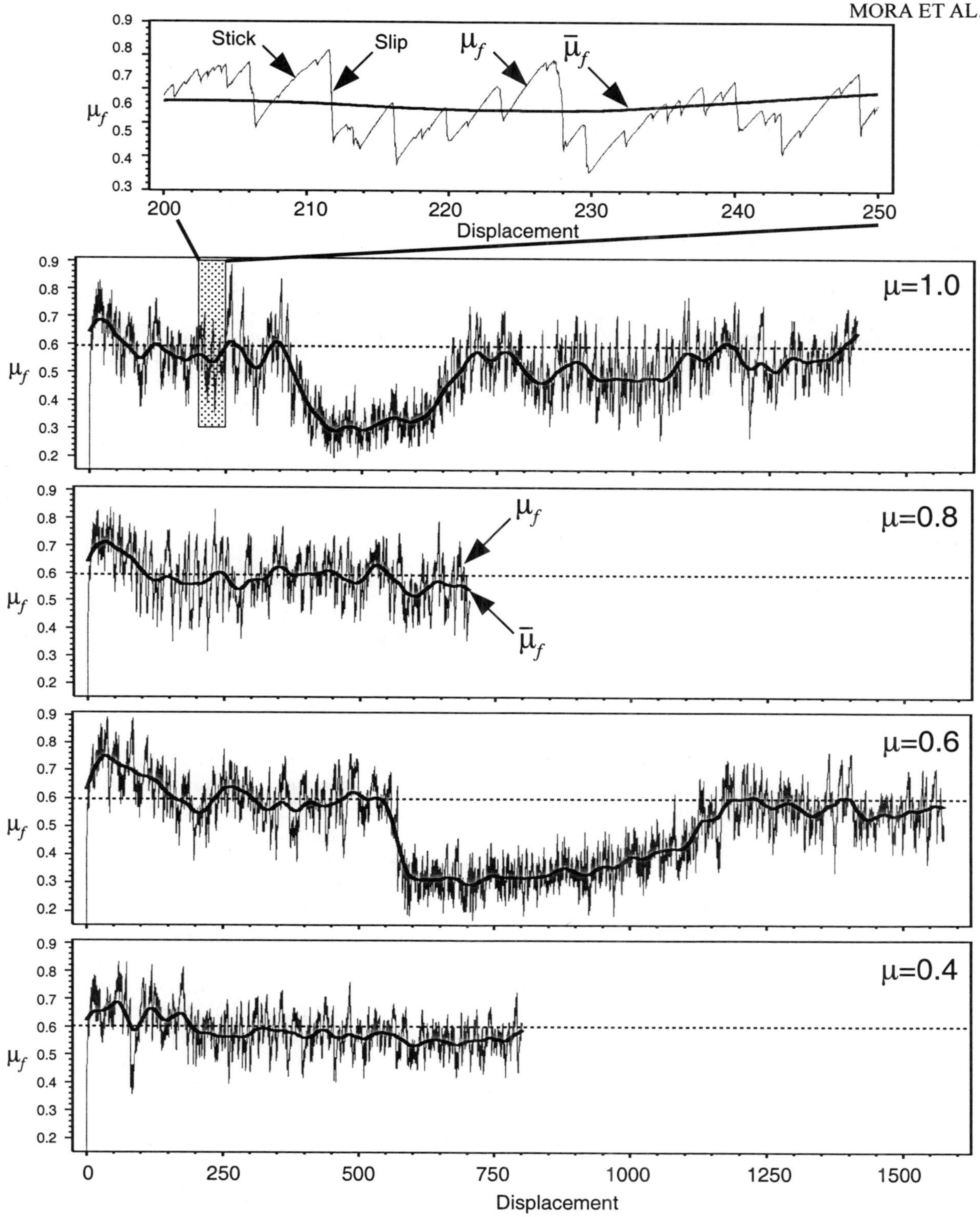

Figure 4. Plot of the instantaneous fault friction μ_f (thin light line) and fault friction $\bar{\mu}_f$ (solid bold line) for different values of inter-particle friction μ. The magnification box allows the characteristic saw-tooth shapes of the stick-slip cycles to be seen. The displacement is the distance moved by the upper block relative to the lower block and is measured in terms of particle diameters (r_0).

curve is smoothed using a running average using triangular weights. The fault friction is denoted as $\bar{\mu}_f$.

The fault friction initially rises as the system is stressed until the stress is sufficiently high for the gouge layer to start to break down and slip.

At this time, the fault friction reaches a maximum value, usually about 0.7, which is above the subsequent long term average which is around 0.6. After this initial period, the fault friction decreases as fracturing occurs in the active shear zone. The rate of breakdown of large grains into smaller grains which can better accommodate slip is highest during this initial period when $\bar{\mu}_f$ is decreasing. Subsequently, the fault friction reaches a minimum and then has gradual oscillations about the long term average. The behavior so far is very similar to that of gouge layers in long displacement laboratory experiments (e.g. Beeler et al., 1996 [4]).

3.1.1. Self-regulation.. It is remarkable that despite the inter-particle friction μ ranging by a factor of 2.5, the long term average friction oscillates around a constant value of 0.6 except when a self-organization process (described below) occurs. Hence, the balance between grain surface sliding and grain rolling-type movements self-regulates such that the fault friction is always a constant of about 0.6. This offers an explanation as to why all rock types seem to have roughly similar static friction despite very different chemical compositions and surface properties. It suggests that the effective friction of faults is not due to rock surface friction. Rather, it is due to the complex dynamics of a granular zone in which the different modes of slip (sliding, fracture and rolling) self-regulate such that the effective friction remains almost constant.

A similar behavior has also been observed in earlier studies involving a thin gouge layer consisting of unbreakable grains (Mora and Place, 1998 [19]) where plots were obtained of the long term average fault friction as a function of inter-particle friction. These studies demonstrated that average fault friction was relatively insensitive to inter-particle friction μ for a wide range of μ (see Figure 5). For μ ranging from 0.2 through to 2.0, the average fault friction remained around 0.5 in one case studied, *Gouge 2*, which contained a distribution of grain shapes. In the second case studied (*Gouge 1*), the average fault friction varied by a greater amount over the range of μ considered but these variations do not directly correlate with μ and remain small. Combined, the thick breakable gouge layer results and the thin gouge layer results suggest the balance between rolling and sliding is able to account for the self-regulation process. The balance between these processes and fracturing may also be important and is expected to have a modifying influence on the overall behavior. The same model fault but with the gouge grains removed yielded a high friction as expected due to the combined effect of model fault roughness and the intrinsic friction between grains.

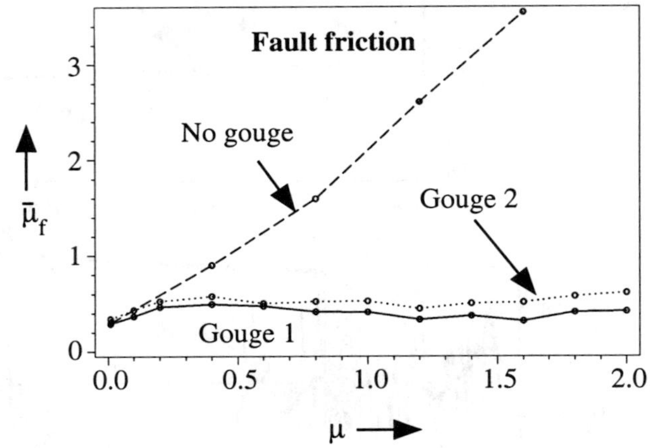

Figure 5. Plots of average fault friction $\bar{\mu}_f$ as a function of inter-particle friction for suites of experiments involving thin gouge layers containing unbreakable grains and unbreakable upper and lower blocks (from Mora and Place, 1998). Curves for two different grain distributions are plotted denoted *Gouge 1* and *Gouge 2*. The case when no gouge is present is denoted *No gouge*.

3.1.2. Self-organization. In two of the four cases shown in Figure 4, the fault friction dramatically drops to $\bar{\mu}_f \sim 0.3$ about half way through the run and remains stable for a considerable time. Movie sequences and snapshots demonstrate that the dramatic weakening effect is due to slip becoming more localized along a very thin basal shear band where slip occurs through a highly efficient rolling-type mechanism similar to that observed in thin gouge layer experiments (Mora and Place, 1998 [19]). The long time for self-organization into this weak state is a possible reason why weak gouge has not been observed in the laboratory, and offers a comprehensive potential explanation for the heat flow paradox (Mora and Place, 1998; Mora and Place, 1999a [20]). We also note that in some simulations, we have not seen the self-organization process before the end of the runs (e.g. the cases with $\mu = 0.4$ and $\mu = 0.8$). When we have seen the self-organization process, the runs were recontinued to study whether the weak state was a stable configuration, hence the longer simulation times for the cases $\mu = 0.6$ and $\mu = 1.0$.

The snapshots shown in Figure 6 illustrate the localization and self-organization processes that are occur-

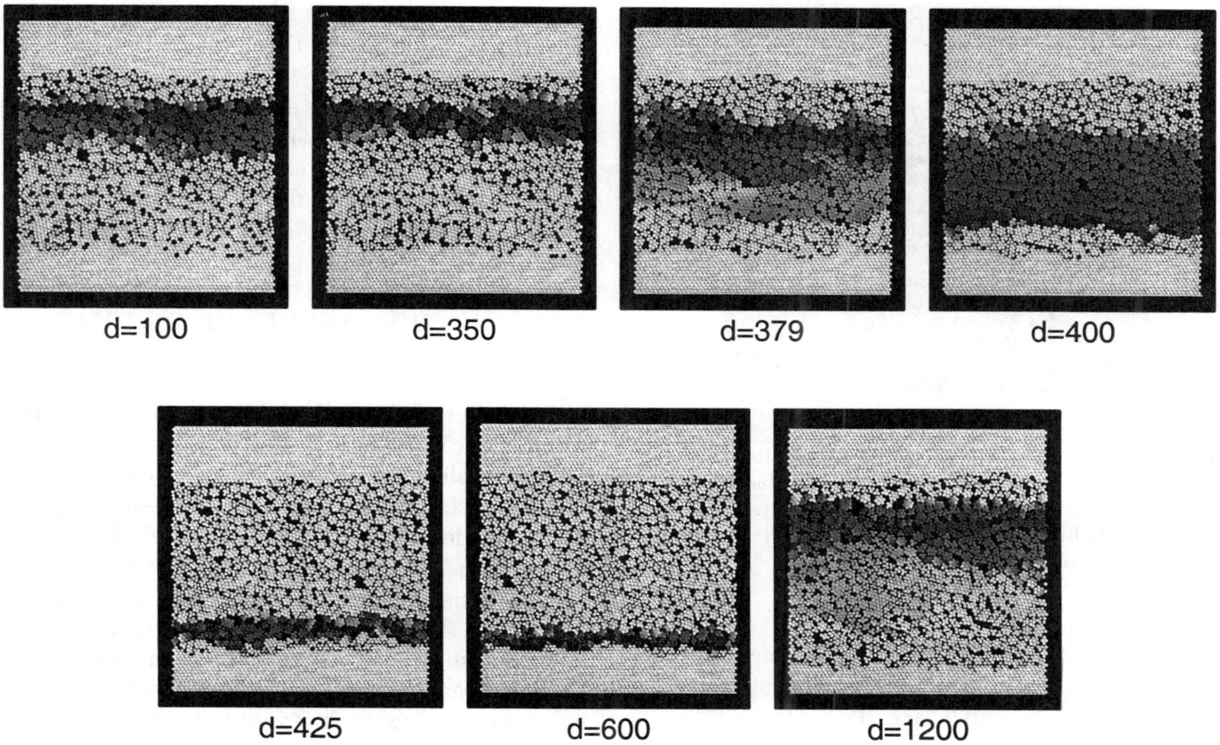

Figure 6. Snapshots of the localization experiment using $\mu = 1.0$. The shades of grey represent the time averaged horizontal particle speed. Dark grey represents small velocities and highlights the active shear zone. The lightest grey corresponds to the speed of the upper and lower blocks (which are being driven in opposite directions past one another) as well as any gouge material that is moving in unison with the blocks (i.e. locked zones of the gouge layer that are not shearing).

ring in the numerical experiment. During the initial breakdown period, slip is distributed throughout the weak zone. Subsequently, slip becomes localized in a relatively wide shear band ($d = 100$ to $d = 379$) that is about 25% of the gouge layer width. This broad active shear band gradually meanders as the simulation progresses. Slip events are occurring on rupture surfaces within this active shear zone such as illustrated in Figure 7. After a large displacement, the active shear band rapidly migrates to the base of the gouge zone and re-localizes into a narrower shear zone (snapshot $d = 379$ to $d = 425$ in Figure 6). When this occurs, fault friction $\bar{\mu}_f$ drops dramatically from $\bar{\mu}_f \sim 0.6$ to $\bar{\mu}_f \sim 0.3$. The suddenness of this migration and re-localization in movie sequences makes it appear visually like a phase transformation phenomenon. During the period when slip is highly localized at the base of the gouge layer, fault rupture occurs along this narrow horizontal band rather than along the shears occurring earlier in the simulation which are typically oriented at about 10–20 degrees from the horizontal (similar to the theoretical R_1 Reidel shear angle, see Place and Mora, 1999b [25]).

3.1.3. Heat generation. Thin gouge layer simulation results (Mora and Place, 1998 [19]) have demonstrated that rolling and jostling movements of grains to accommodate slip can lead to low heat production. However, it was unclear from this earlier work whether or not a thin weak shear band is able to form on natural faults or in simulations involving a thicker gouge layer. Recent work (Place and Mora, 1999 [25]) has proven that similar phenomena also occur in simulations during the period when slip is highly localized. Figure 8 compares the amount of heat that would theoretically be produced if the fault friction was equal to the inter-particle friction, to the heat actually generated in the run with $\mu = 0.6$. The large drop in heat observed during period p_2 when slip is highly localized coupled with the low fault friction during this periods suggests the numerical results may offer a comprehensive explanation of the heat flow paradox. The

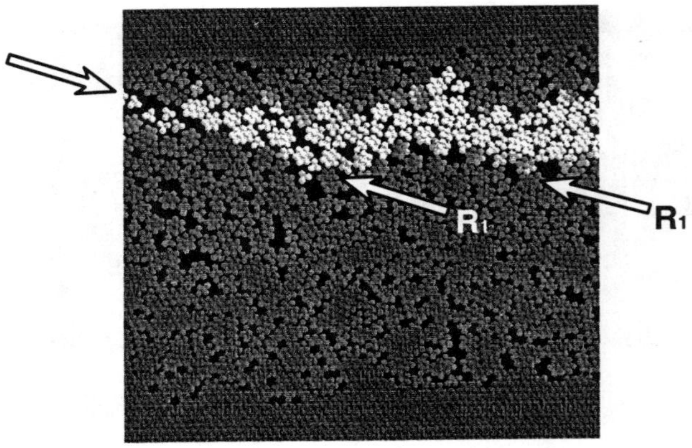

Figure 7. Snapshot of a simulation showing Reidel shears. The shade of grey represents the particle displacement during 10,000 time steps. Hence, lines separating areas with contrasted colors highlight shear zones. Two R_1 shears zones can be distinguished in the snapshot.

numerical model is capable of matching all constraints surrounding the heat flow paradox including low stress drops during simulated earthquakes, low heat and low fault strength (Mora and Place, 1998 [19]; Mora and Place, 1999a [20]). Recent work has shown that the seismic efficiencies during a typical rupture event is a few percent, a value which is substantially lower than the earlier estimates (Place and Mora, 1999b [25]).

3.1.4. Effect of normal stress and fracturing. Figure 9 illustrates the effect of decreasing the normal stress. When the normal stress is lower, slip events occur more easily and hence more frequently. Thus, it was necessary to lower the driving rate of the plates by a large factor in order for slip events to be well separated in time (If slip events overlap in time, the experiment would be more comparable to a high velocity experiment to study dynamic friction rather than low strain rate laboratory studies of rock friction.). The results presented in Figure 9 were obtained using $V_{plate} = 0.00005V_p$, five times lower than the runs described in the previous sections (c.f. the normal stress was ten times lower). The two runs were calculated with a different breaking criterion $r_b = 1.0075$ in the upper plot and $r_b = 1.015$ in the lower plot. In each case, an inter-particle friction of $\mu = 0.6$ was used.

The results again show that the fault friction is relatively constant although the long term average is slightly higher than in the previous runs using higher normal stress. The effect of the breaking criterion r_b can also be seen. For the lower value of r_b, bonds can break more easily and the initial bump seen on earlier plots prior to a breakdown period (see Figure 4) is suppressed. With the higher value of r_b, the form of the fault friction curve $\bar{\mu}_f$ is very similar to the previous numerical experiments. These results suggest the bond breaking criterion has an effect but this is mainly manifested early in the run when the breakdown is occurring. Subsequently, the effective fault friction is regulated by the balance between slip and rolling-type movements of the gouge grains.

4. THE INFLUENCE OF HEAT

4.1. Thermo-mechanical Coupling

A number of simulations have been performed to determine the influence of thermal expansion on the dynamics of slip events (Abe, Mora and Place, 1999 [1]). For these simulations, we used a 2D-model consisting of 64×64 identical particles with a well developed fault gouge layer initialized similar to that used in the numerical experiments described in the previous section. To restrict the influence of thermal expansion to a single major slip event, a long simulation was run without thermal expansion. The run was subsequently restarted just prior to a major slip event but with the thermal expansion effect turned on. Thus, the initial state of the model was identical in both cases allowing for a comparison between an event with and without the influence of thermal expansion.

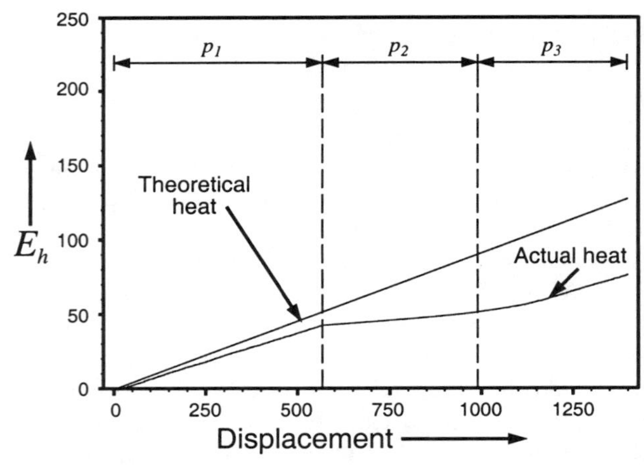

Figure 8. Plot of a theoretical calculation of heat generated assuming the fault friction is equal to the inter-particle friction, and the heat actually generated during the simulation with inter-particle friction $\mu = 0.6$. The p_i indicate the periods before (p_1), during (p_2), and after p_3), the time when slip was highly localized.

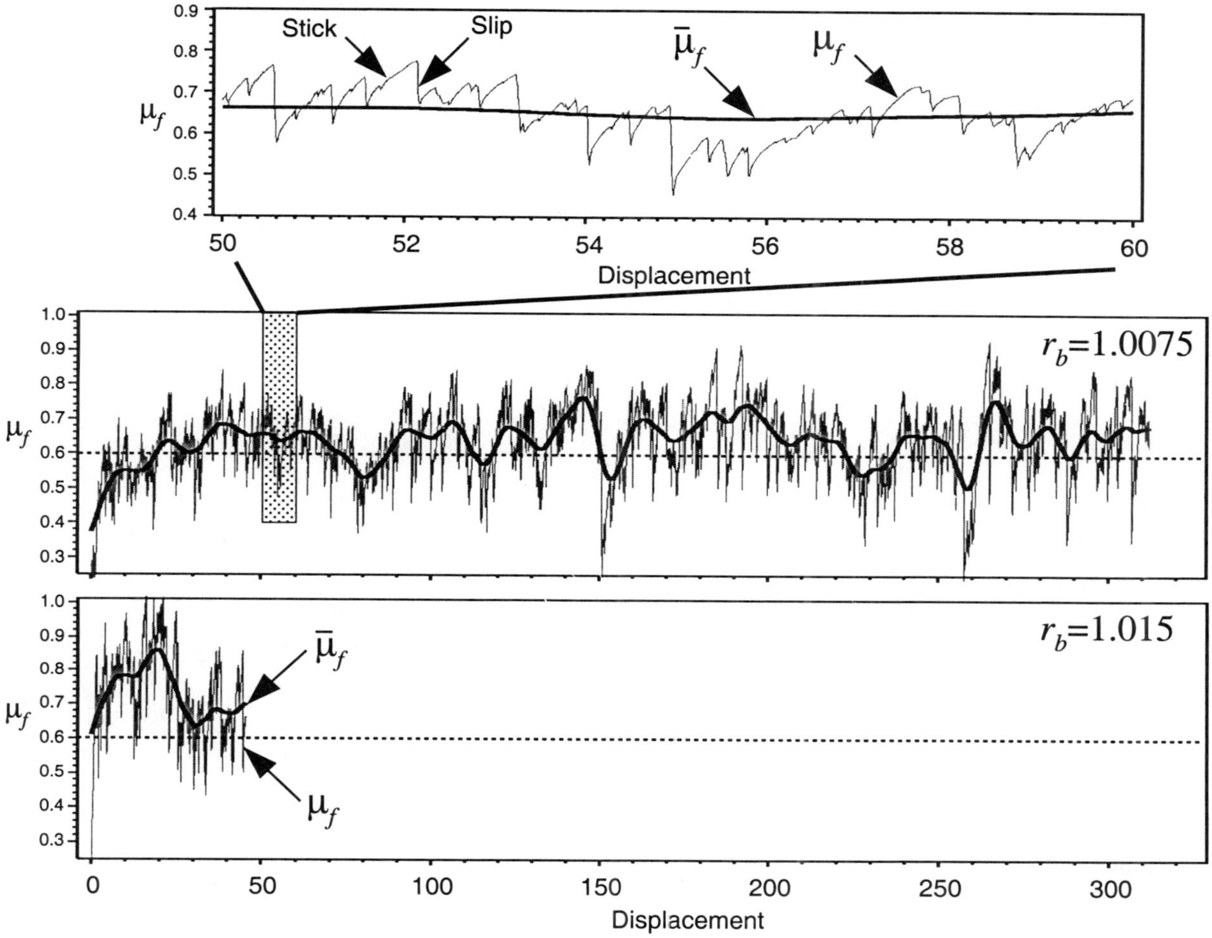

Figure 9. Plot of fault friction for simulations using the lower value of normal stress (i.e. σ_n here is 10× lower than for Figure 4).

Figure 10 shows that there is no significant difference in the size of the main events when the thermal expansion effect is modeled. However, event timing is influenced by the thermal expansion. One of the simulations with thermal expansion also showed a small precursory event which did not occur when the simulation was started from the same initial conditions but without thermal expansion.

4.2. Thermo-porous Coupling

To investigate the influence of the pore fluid pressure on the dynamics of a fault, two sets of simulations have been performed. First, single slip events have been investigated by restarting the model immediately before a slip event with and without the pore fluid pressure effect on friction being modeled. The comparison of these slip events starting from identical initial conditions shows that the inclusion of the pore fluid pressure has a significant influence on the dynamics of large events, whereas small events are not significantly affected. The reduction of friction during major events generally leads to an increase in size for those events (Figure 11). This can be seen in the event size distributions obtained in long runs generated with and without the pore pressure effect turned on (Figure 12).

Figure 12 shows that the magnitude-frequency relation remains unchanged for small events, but that there is an increase in the number of large events and maximum event size. The tendency towards an increase in the size of large events can be explained by the reduction in friction caused be the frictional heating and the

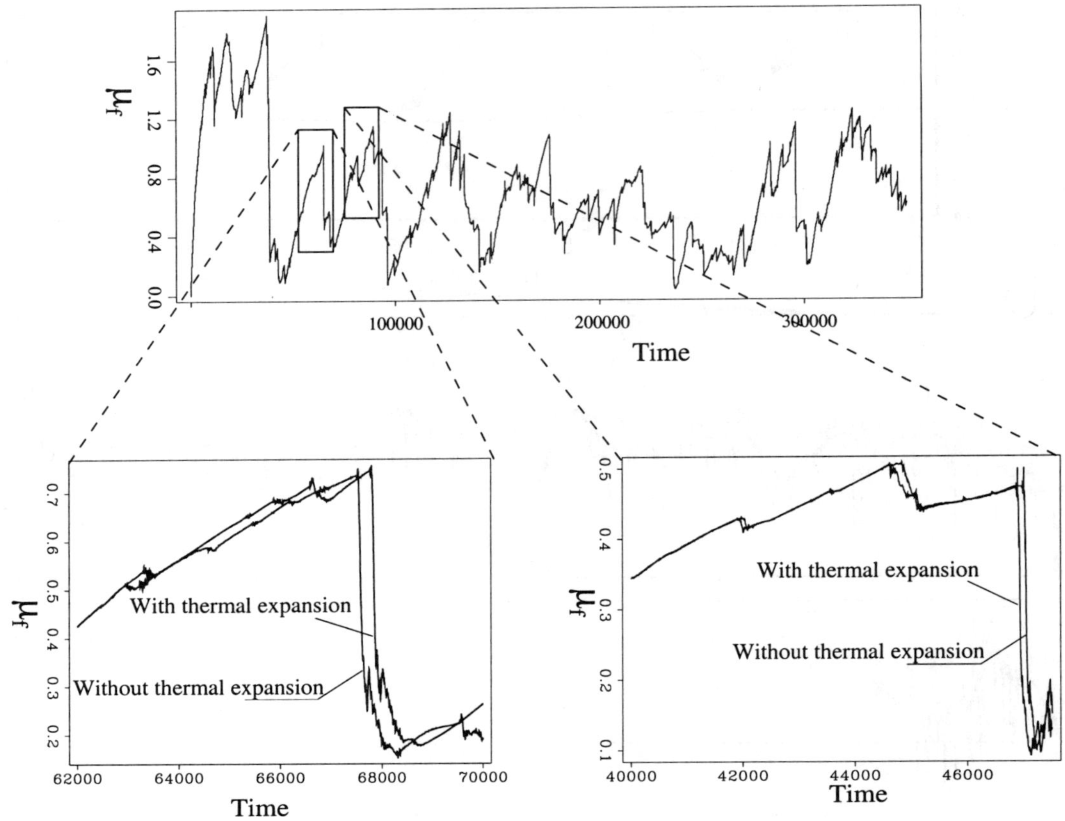

Figure 10. Instantaneous fault friction $\mu_f = \tau/\sigma_n$ during slip events with and without thermal expansion.

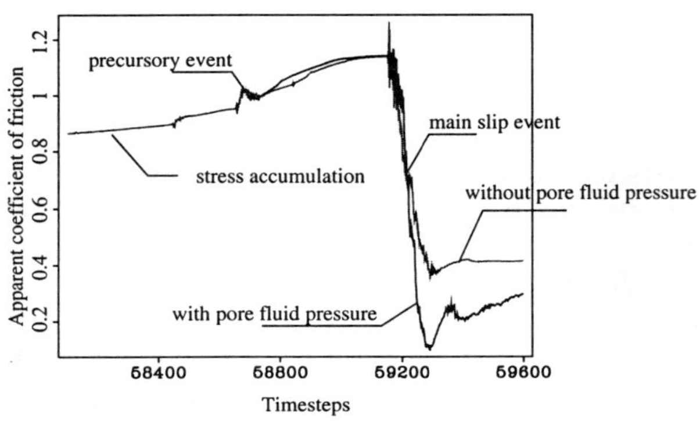

Figure 11. Difference of the development of slip events with and without reduced friction due to an increased pore fluid pressure.

resultant increased pore fluid pressure. Small events do not generate enough heat to cause a significant increase in pore fluid pressure and are therefore not significantly affected.

5. SIMULATIONS OF MODEL FAULT SYSTEMS

5.1. The Critical Point Hypothesis for Earthquakes

The observed power law frequency-size distributions for earthquakes are widely used as a basis for probabilistic hazard analysis (i.e. the assumption of a linear fit of the log of the number of events versus magnitude is used to make a prediction of the rate of large earthquakes). The power law observation has led some researchers to propose that the earth's crust is in a critical state when a large earthquake can occur at any time (Bak and Tang, 1989 [3]). If so, then the Poisson process assumption underlying traditional hazard analysis would be appropriate (i.e. that the probability of an earthquake in time does not change). The Poisson process or random hypothesis contradicts expectations based on the simple physics principles that stress would need to build up on a fault following a large earthquake before another event can occur.

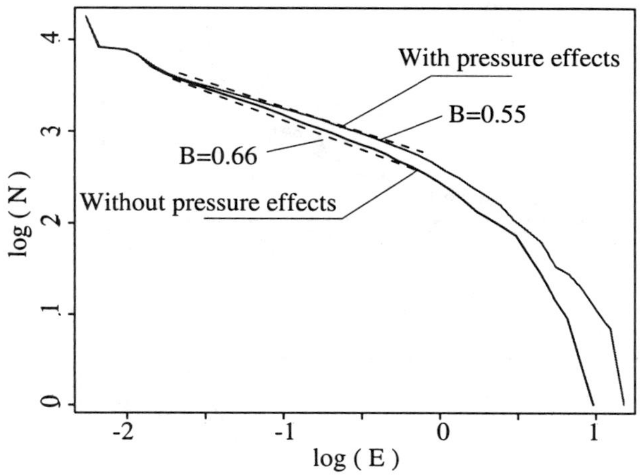

Figure 12. Change in the event size distribution caused by the friction reduction due to pore fluid pressure increases.

An increasing number of observations worldwide suggests a more evolutionary behavior. A power law time-to-failure function has been found to fit well the cumulative Benioff strain release in the lead-up to some large events, and during such periods, the statistics of earthquakes evolves (Bufe and Varnes, 1993 [6]; Bufe et al., 1994 [7]; Jaumé and Sykes, 1999 [12]; Jaumé, 1999 [13]). This has led various researchers to propose at least some parts of the crust may behave in a similar way to a critical point (Sornette and Sammis, 1995 [29]; Bowman et al., 1998 [5]; Huang et al., 1998 [11]; Saleur et al., 1998 [26]; Sammis et al., 1996 [27]).

Under this hypothesis, long-range correlations in stress (or closeness to failure) in the crust gradually build up as earthquakes occur within a given region. Once a large earthquake occurs, these long range correlations are destroyed. This hypothesis explains the observed power law time-to-failure fits and provides a possible physical basis for earthquake forecasting (i.e. intermediate term earthquake prediction).

Renormalization group theory yields complex exponents which implies the power law time-to-failure function should have log-periodic fluctuations (Sornette and Sammis, 1995 [29]).

5.2. Simulation Results

In the numerical experiments involving a thick gouge layer described in previous sections, ruptures were seen to occur anywhere within an active shear band. These typically occurred at orientations similar to Reidel R_1 shears (Place and Mora, 1999b [25]) such as shown in Figure 7. Hence, the model is simulating ruptures in a 2D system which can be considered as a simplied low resolution model of an interacting fault system. For this reason, we have studied the cumulative Benioff strain in a long sequence. The sequence analyzed is the case with $r_b = 1.0075$ shown in Figure 9.

We fit a power law time-to-failure function with log-period fluctuations using least-squares to cumulative Benioff strain (i.e. sum of the square root of kinetic energy release in synthetic earthquake events) preceding large simulated earthquakes. The power law time-to-failure function (Sornette, 1995 [29]) is given by

$$\epsilon(t) = A + B(t_f - t)^c \left[1 + C\cos\left(2\pi\frac{\log(t_f - t)}{\log\lambda} + \psi\right)\right] , \tag{27}$$

where ϵ is the cumulative Benioff strain release, t_f is the predicted time of the critical point, c is the power law exponent, and A, B, C, λ and ψ are parameters. All parameters of the power law function including t_f were obtained by least squares fitting of the cumulative Benioff strain release calculated from the synthetic earthquake catalog. The model data was found to contain numerous sequences that match well with the power law time-to-failure function (i.e. to have a low C-value where C is defined as the RMS error of the power law fit divided by the RMS error of a linear fit). Figure 13shows several such sequences. It is interesting that small sub-cycles with almost identical power law exponents are often superimposed on longer cycles. The predicted critical point time is slightly prior to the large earthquakes and aftershocks occurring at the end of the sequences. This result is consistent with the critical point hypothesis that the largest earthquakes can only occur after the system has reached a critical point (i.e. after the largest scale stress correlations have been built up within the system). To confirm whether the good fits can be ascribed to chance, we studied a random catalog with the same event-size statistics. We found that the random data could not reproduce the good power law fits observed using the lattice solid generated data (Mora, Jaumé and Place, 1999 [22]).

The ratio of the RMS error of the power law time-to-failure fit divided by the RMS error of a linear fit provides a measure of goodness of fit of the time-to-failure fits (Bowman et al., 1998 [5] used the same measure). Table 1 shows the power law exponents and the goodness of fit measures obtained by the least squares fitting algorithm. The power law exponents show remarkable consistency. In all cases, the log-periodic fluctuations significantly improved the goodness of fit measure (i.e $C_{lp} < C$).

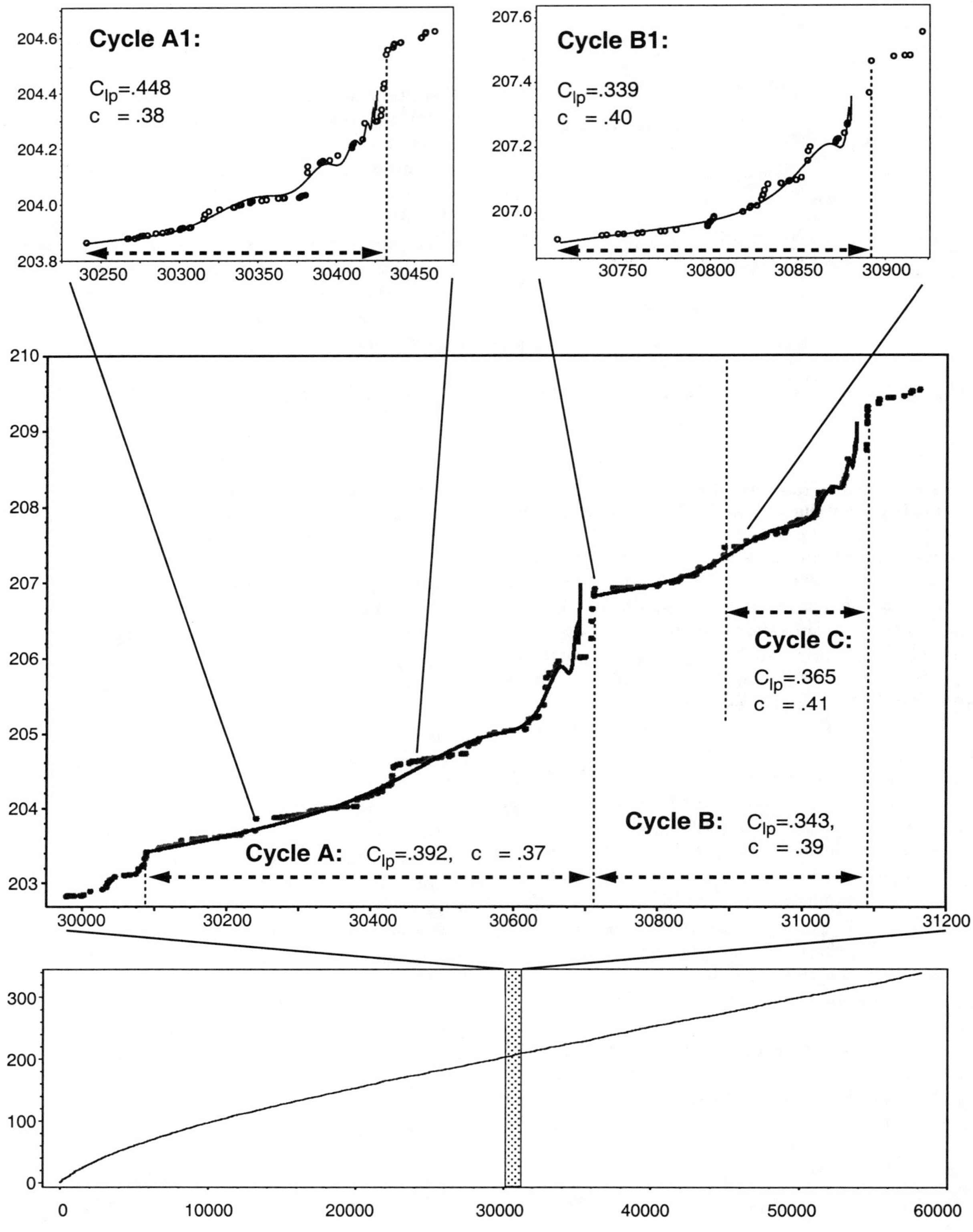

Figure 13. **Bottom:** cumulative Benioff strain release for the entire sequence with $\mu = 0.6, V_{plate} = 0.00005$ that is shown in Figure 9. **Middle:** A typical part of the cumulative Benioff strain sequence (open circles) and power law time-to-failure fits (solid line) for the two adjacent accelerating sequences or cycles (labeled A and B) which are contained in this section of the data. **Top:** Two accelerating sub-sequences labeled A1 and B1 which are respectively contained within cycles A and B.

Table 1.

Cycle	C	C_{lp}	c
A1	0.545	0.448	0.38
A	0.511	0.392	0.37
B1	0.568	0.339	0.40
B	0.486	0.343	0.39
C	0.535	0.365	0.41

Best fit power law fit exponents (c) and the goodness of fit measures obtained by the least squares fitting algorithm. C and C_{lp} respectively denote the measures without and with the log-periodic fluctuations.

The statistics of earthquakes in synthetic earthquake sequences were observed to evolve (Figure 14). In the sequences studied so far, the evolution typically manifested itself as evolution from a higher b-value of around 0.9 early in the sequence to a lower b-value of around 0.65 in the latter part of the sequence (e.g. see cycles A and A1). In some cases (e.g. cycles B and B1), we also observed an overall increase in seismicity in the latter part of the sequence with the b-value remaining almost unchanged.

When searching for an optimal power law fit preceding the last event of cycle B, we observed a local

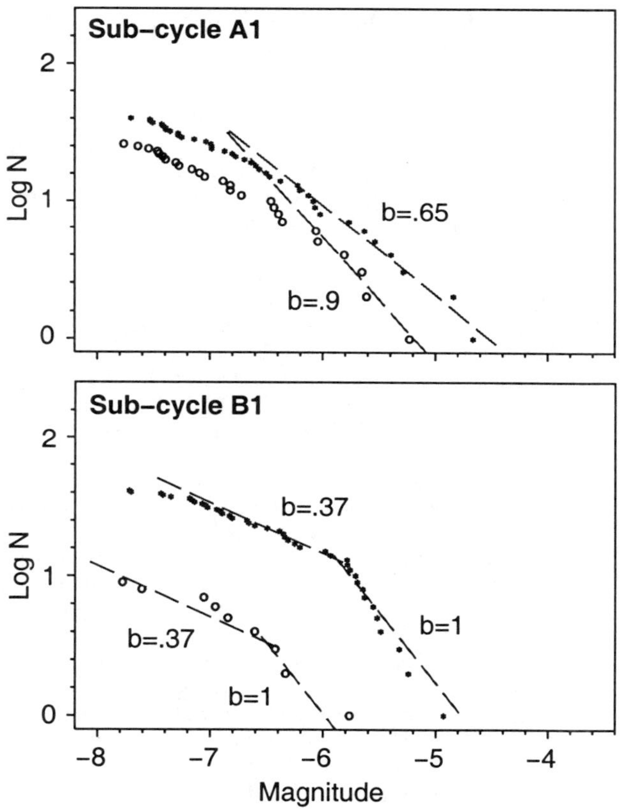

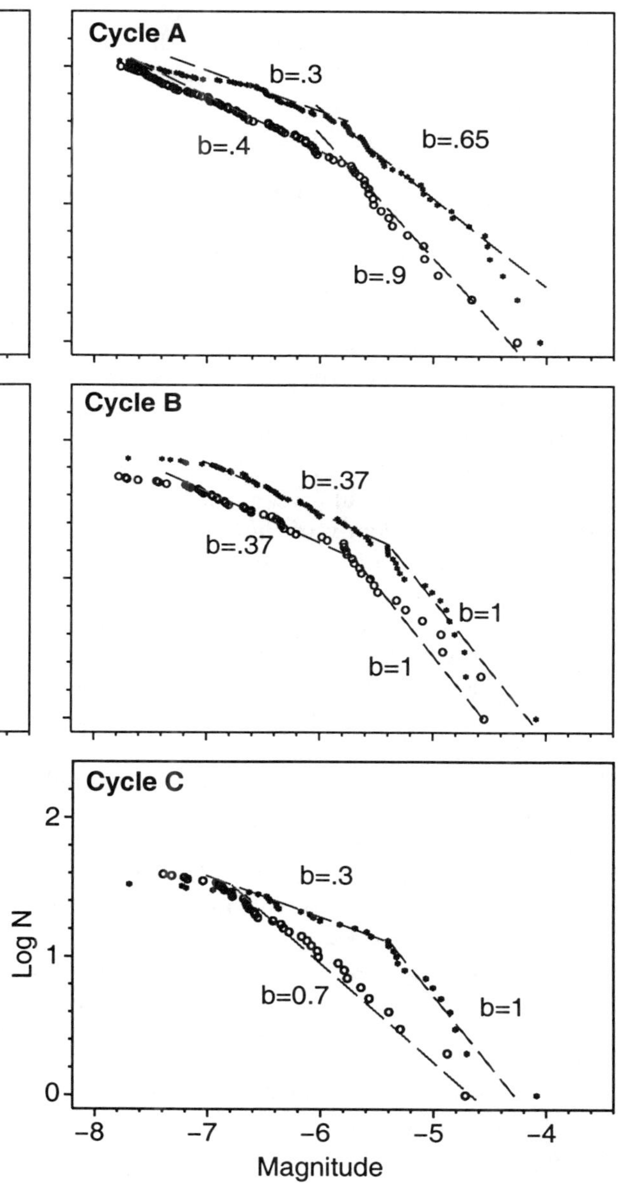

Figure 14. Cumulative frequency versus magnitude plots for simulation data generated using the lattice solid model shown in Figure 13. Open circles show the size-frequency statistics in the first half of the sequence (period of the power-law fit) and stars show the statistics in the last half of the sequence. As the time-to-failure predicted using the power-law fitting always occurred prior the mainshock, the statistics exclude the mainshock and aftershocks. For reference, straight line segments with their corresponding negative slope or b-value are plotted as broken lines.

minimum in least squares error for a shorter sequence (cycle C) that did not span sub-sequence B1. To test whether the overall increase in seismicity also occurs in the sequence identified as cycle C, we also plotted the statistics for this case. The results show an evolution effect in cycle C for the smaller to medium magnitudes ($M < -5.5$) that is more similar to that of cycle A. Namely, the smallest magnitudes have the same rate in the first and last half of the cycle whereas the rate of medium sized events increases. However, in the medium to large event-size range, there appears to be an overall increase in rate.

The above results are consistent with recent observational studies (Jaumé, 1999). Namely, in most cases when accelerating moment release is observed prior to real earthquakes, there is an increase in the rate of moderate to large events in the latter part of these sequences, but there are also some cases when the seismicity rate may increase at all magnitudes.

6. CONCLUSIONS

Particle based methods such as the lattice solid model offer a powerful means to study the nonlinear physics of earthquakes. Simulations show that the friction of gouge layers depends primarily on the dynamics of the system rather than the properties of rock surfaces within a gouge layer. A self-regulation process seems to balance the different micro-mechanical slip mechanisms such as grain rolling versus slip between grain surfaces leading to a universal friction coefficient of approximately 0.6 independently of the normal stress or breaking criterion. After a long time, the gouge microstructure can self-organize such that slip occurs on a narrow band mainly through rolling-type movements resulting in an anomalously weak fault. Such structures can be stable over long durations. If such a self-organization process occurs in nature, it offers a comprehensive solution to the heat flow paradox.

Simulations involving a model with multiple internal rupture surfaces exhibit accelerating rates of Benioff strain release and an evolution in event-size statistics in the lead-up to large events. The consistency of the lattice solid results with the critical point hypothesis for earthquakes and observations prior to real earthquakes suggests that the statistical physics analogy of earthquake phenomena is relevant. The evolution effects seen in the physically based lattice solid numerical model suggest that intermediate-term earthquake prediction is a realistic goal of earthquake science, and that numerical simulation can play an important role in unraveling the puzzle.

Acknowledgments. This research was supported by the Australian Research Council, The University of Queensland, and the sponsors of QUAKES. Computations were made on the SGI Origin 2000 at QUAKES.

REFERENCES

Abe, S., Mora, P., and Place, D. (1999) Extension of the Lattice Solid model to incorporate temperature related effects, *Pure Appl. Geophys.*, submitted.

Allen, M.P., and Tildesley, D.J. (1987) *Computer simulation of liquids*, (Oxford Univ. Press, New York).

Bak, P., and Tang, C. (1989) Earthquakes as a self-organised critical phenomena,*J. Geophys. Res., 94*, 15,635–15,637.

Beeler, N. M., Tullis, T., Blanpied and Weeks (1996) Frictional behavior of large displacement experimental faults, *J. Geophys. Res. 101*, 8697–8715.

Bowman, D. D., Ouillon, G., Sammis, C. G., Sornette, D., and Sornette, A. (1998) An observational test of the critical earthquake concept,*J. Geophys. Res. 103*, 24,359–24,372.

Bufe, C. G., and Varnes, D. J. (1993) Predictive modeling of the seismic cycle in the greater San Francisco Bay region,*J. Geophys. Res. 98*, 9871–9983.

Bufe, C. G., Nishenko, S. P., and Varnes, D. J. (1994) Seismicity trends and potential for large earthquakes in the Alaska-Aleutian region, *Pure Appl. Geophys. 142*, 83–99.

Cundall, P.A. & Strack, O.D.L. (1979) A discrete numerical model for granular assemblies, *Géotechnique, 29*, 47–65.

Donzé, F.V. (1994) *A discrete numerical model for brittle rock deformation.* PhD thesis

Frisch, U., d'Humiére, D., Hasslacher, B., Lallemand, P., Pomeau, Y. and Rivet, J. (1986) Lattice Gas Hydrodynamics in Two and Three Dimensions,*Complex Systems, 1*, 649–707.

Huang, Y., Saleur, H., Sammis, C., and Sornette, D. (1998) Precursors, aftershocks, criticality and self-organized criticality,*Europhys. Lett. 41*, 43–48.

Jaumé, S.C. and L.R. Sykes (1999) Evolving towards a critical point: a review of accelerating seismic moment/energy release prior to large and great earthquakes, *Pure Appl. Geophys., 155*, 279–305.

Jaumé, S.C. (1999) Changes in earthquake size-frequency distributions underlying accelerating seismic moment/energy release, submitted to this volume (AGU Physics of Earthquakes book).

Lachenbruch, A.H. (1980) Frictional Heating, Fluid Pressure, and Resistance to Fault Motion, *J. Geophys. Res. 85*, 6097–6112.

McKenzie, D. and Brune, J.N. (1972) Melting on Fault Planes during large earthquakes, *Geophys.J.R. Astr. Soc., 29*, 65–78.

Mora P. (1992) A lattice solid model for the nonlinear dynamics of earthquakes, *The Seismic Simulation Project Tech. Rep. # 4*, 3-28 (Institut de Physique du Globe, Paris).

Mora P. & Place, D. (1993) A lattice solid model for the nonlinear dynamics of earthquakes, *Int. J. Mod. Phys. C, 4*, 1059–1074.

Mora, P. & Place, D. (1994) Simulation of the frictional stick–slip instability, *Pure Appl. Geophys., 143*, 61–87.

Mora, P., & D. Place (1998) Numerical simulation of earthquake faults with gouge: towards a comprehensive explanation for the heat flow paradox, *J. Geophys. Res., 103*, 21,067–21,089.

Mora, P., & D. Place (1999a) The weakness of earthquake faults, *Geophys. Res. Lett., 26*, 123–126.

Mora, P., & Place, D. (1999b) Accelerating energy release prior to large events in simulated earthquake cycles: implications for earthquake forecasting, *in: 1-st ACES Workshop Proceedings*, pp 513–519 (APEC Cooperation for Earthquake Simulation, Brisbane).

Mora, P., Jaumé & Place, D. (1999) Accelerating energy release and evolution of event statistics in the lattice solid model, *Pure Appl. Geophys.*, submitted.

Nur, A. and Byerlee J.D. (1971) An exact effective stress law for elastic deformation of rock with fluids, *J. Geophys. Res., 76*, 6414–6419.

Place, D., & P. Mora (1999a) A lattice solid model to simulate the physics of rocks & earthquakes: incorporation of friction, *J. Comp. Phys., 150*, 1–41.

Place, D. and Mora, P. (1999b) Numerical simulation of localisation phenomena in a fault zone, *Pure Appl. Geophys.*, submitted.

Saleur, H., Sammis, C. G., and Sornette, D. (1996) Discrete scale invariance, complex fractal dimension, and log-periodic fluctuations in seismicity, *J. Geophys. Res. 101*, 17661–17677.

Sammis, C. G., Sornette, D., and Saleur, H. (1996) Complexity and earthquake forecasting,*In Reduction and Predictability of Natural Disasters, SFI Studies in the Sciences of Complexity*, (eds. Rundle. J. B., Klein. W., and Turcotte. D. L.) (Addison-Wesley, Raeding, MA 1996) pp. 143–156.

Sakaguchi, H. and Mühlhaus, H. (1999) Hybrid modelling of coupled pore fluid-solid deformation problems, *Pure Appl. Geophys.*, submitted.

Sornette, D. and Sammis, C.G. (1995) Complex critical exponents from renormalization group theory of earthquakes: implications for earthquake predictions, *J. Phys. I France, 5*, 607–619.

Touloukian, Y.S., Kirby,R.K., Taylor,R.E. and Lee, T.Y.R. (1977) *Thermophysical properties of Matter*, 13, (IFI/Plenum 1977).

S. Abe, S. Jaumé, P. Mora and D. Place, QUAKES, Department of Earth Sciences, The University of Queensland, Brisbane 4072, Qld, Australia.
(e-mail: mora@quakes.earthsciences.uq.edu.au; place@quakes.earthsciences.uq.edu.au; steffen@quakes.earthsciences.uq.edu.au; jaume@quakes.earthsciences.uq.edu.au)

Dynamics of Seismicity Patterns in Systems of Earthquake Faults

John B. Rundle

Department of Physics
and
Colorado Center for Chaos & Complexity
and
Cooperative Institute for Research in Environmental Sciences, University of Colorado, Boulder, CO

W. Klein

Department of Physics
and
Center for Computational Science, Boston University, Boston, MA

Kristy Tiampo and Susanna Gross

Colorado Center for Chaos & Complexity, Cooperative Institute for Research in Environmental Sciences, University of Colorado, Boulder, CO

Systems of earthquake faults frequently show space-time behavior that is difficult to interpret. We describe a new technique based upon a Karhunen-Loeve expansion that allows the observed space-time seismicity patterns to be understood as eigenstates of a suitably constructed correlation operator. The evolution of seismicity can then be viewed in terms of a "Phase Dynamical System" giving rise to a "Pattern Dynamics" that can be obtained directly from observed data. Since the method does not use information on stress, displacements, or other fault-related variables, it necessarily represents an incomplete description of the dynamics. In principle, these methods can be used to construct forecast algorithms. We then apply our methods to seismicity from both numerical simulations and data obtained from southern California. Finally, we perform statistical tests to measure the quality of the forecasts computed using simulation data.

1. INTRODUCTION

Earthquake faults can be considered to be driven nonlinear threshold systems comprised of interacting spatial networks of sliding surfaces that are subjected to persistent driving stresses. To first order, a fault segment slips when the shear

GeoComplexity and the Physics of Earthquakes
Geophysical Monograph 120

stress $\sigma(\mathbf{x},t)$on a cell at position $\mathbf{x}$ and time t rises to an effective threshold level at which frictional resistance is overcome. The result is an increase in the slip $s(\mathbf{x},t)$ of the segment, as well as a decrease in the stress supported by the segment to a lower, residual value. Failure thresholds, residual stresses, and slip magnitudes may be modified by the presence of spatial heterogeneity in physical properties. Elastic interactions between segments may be excitatory (positive) in the sense that failure of interacting neighbors brings a segment closer to sliding, or inhibiting (negative) in the opposite case. Observations, together with numerical simulations of these systems [Rundle and Klein, 1995; Fisher et al., 1997; Scholz, 1990], reveal spatial and temporal patterns $\Psi(\mathbf{x},t)$ of seismic activity that are richly complex, and difficult to understand from any deterministic point of view [*Nijhout*, 1997]. Understanding such patterns will increase our understanding of the physics of the fault systems themselves, since these space-time patterns are clearly emergent processes that reflect the structures, dynamics, and properties of the underlying high dimensional nonlinear system.

In earthquake fault systems, there exists the further difficulty that the deterministic dynamics, together with values of the stress $\sigma(\mathbf{x},t)$ and slip $s(\mathbf{x},t)$ state variables, are often unknown or effectively unobservable. Only very limited information is available at the earth's surface from drilling, GPS, InSAR, or other technologies. However, the associated pattern variables $\Psi(\mathbf{x},t)$, the seismic events themselves, *are* observable. Quantitatively, the seismicity $\Psi(\mathbf{x},t)$ can be represented as a set of time series at all positions $\mathbf{x}$, with $\Psi(\mathbf{x},t)$ defined by $\Psi(\mathbf{x},t) = 1$ if an earthquake occurs between $\{t,t+\Delta t\}$; $\Psi(\mathbf{x},t) = 0$ otherwise. The basic problem is illustrated in figure 1. The slip $s(\mathbf{x},t)$ at position $\mathbf{x}$ and time t evolves to a value $s(\mathbf{x},t + \Delta t)$ under the action of the deterministic dynamics D_t:

$$D_t \{ s(\mathbf{x},t) \} \rightarrow s(\mathbf{x},t + \Delta t) \qquad (1)$$

The stress $\sigma(\mathbf{x},t)$ also evolves to $\sigma(\mathbf{x},t+\Delta t)$. The values of the slip and stress state variables are effectively hidden from view, below the horizontal dashed line in figure 1.

However, the time, location, and magnitude of seismic events on fault systems are readily observable, and are tabulated routinely [*Hill et al.*, 1990; *Sieh et al.*, 1989; *Bakun and McEvilly*, 1984]. The seismicity gives rise to observable space-time patterns $\Psi(\mathbf{x},t)$, which should not be confused with the variables in which the underlying dynamics are formulated, $s(\mathbf{x},t)$ and $\sigma(\mathbf{x},t)$. These latter variables also display patterns and correlations, but these are not the patterns of interest to us here since they are effectively unobservable. Since both $\Psi(\mathbf{x},t)$ and $\Psi(\mathbf{x},t + \Delta t)$ are observable, it might be possible in principle to learn about D_t, $s(\mathbf{x},t)$ and $\sigma(\mathbf{x},t)$ by studying the evolution of seismicity patterns. Unfortunately, there exists no evidence at present that $\Psi(\mathbf{x},t)$ is uniquely related to $s(\mathbf{x},t)$ and $\sigma(\mathbf{x},t)$. In other words, there may be a large number of slip states $\sigma(\mathbf{x},t)$ that produce a similar seismicity pattern $\Psi(\mathbf{x},t)$, or a large number of patterns $\Psi(\mathbf{x},t)$ associated with similar states $s(\mathbf{x},t)$. This weak association between $s(\mathbf{x},t)$ and $\Psi(\mathbf{x},t)$ is represented schematically in figure 1 by a vertical dotted line.

In attempting to forecast the future evolution of slip on fault systems, a common approach is to attempt to 1) use observations of seismicity $\Psi(\mathbf{x},t)$ to infer $s(\mathbf{x},t)$; then 2) use $s(\mathbf{x},t)$ together with an assumed model for the dynamics D_t to calculate $s(\mathbf{x},t + \Delta t)$; and finally 3) use $s(\mathbf{x},t + \Delta t)$ to infer $\Psi(\mathbf{x},t + \Delta t)$, which represents the observable seismic pattern of interest. However, this method involves a long chain of inference and assumption, particular links of which may not be well-justified. As we will show, earthquakes can be considered to be an example of a phase dynamical system [*Mori and Kuramoto*, 1998], in which evolution of the observables can be associated with rotation of a state or pattern vector within a high dimensional correlation space. In this new approach, we propose the construction of an approximate dynamics based on Karhunen-Loeve (KL) methods that associates a future seisimicity pattern state $\Psi(\mathbf{x},t + \Delta t)$ with an earlier pattern state $\Psi(\mathbf{x},t)$ by means of an assumed "Pattern Dynamics" operator PD_t:

$$PD_t \{ \Psi(\mathbf{x},t) \} \Rightarrow \Psi (\mathbf{x},t + \Delta t) \qquad (2)$$

Moreover, we propose to construct PD_t without any knowledge of the true deterministic dynamics D_t, using only knowledge of the seismicity pattern states preceding $\Psi(\mathbf{x},t)$.

The pattern dynamics operator that we construct operates on a space of functions $\Psi_R(\mathbf{x},t)$ that we call "reconstructed pattern states". The $\Psi_R(\mathbf{x},t)$ are continuous-valued in time, and represent approximations, in the sense described below, to the pattern states $\Psi(\mathbf{x},t)$. In fact, the $\Psi_R(\mathbf{x},t)$ will be seen to represent probability amplitudes from which probability density functions for seismicity may be calculated. This identification also follows from the fact that $\Psi_R(\mathbf{x},t)$ evidently arises from phase dynamics [*Mori and Kuramoto*, 1998], which are known to have deep similarities to quantum systems. In contrast to the underlying deterministic dynamics D_t, which are most probably strongly nonlinear, the operator PD_t is assumed to be linear over small time intervals Δt. A convenient analogy for the relationship between PD_t and D_t is the correspondence between an equilibrium dynamical system, such as an Ising model that is governed by a nonlinear Langevin equation,

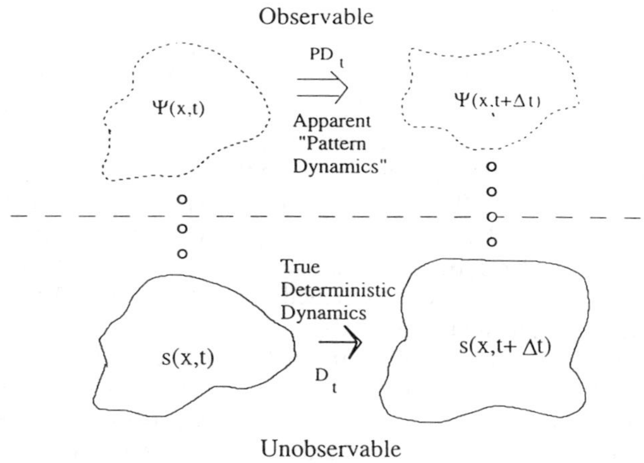

Figure 1. Schematic diagram of threshold systems, including unobservable state variables s(**x**,t) and observable pattern variables $\Psi\mu$(**x**,t).

and the corresponding functional Fokker-Planck equation [*Haken*, 1983; *Gardiner*, 1985]. Evolution of dynamical variables such as the spin density are governed by a strongly nonlinear Langevin equation, but the probability density functions evolve according to a linear functional Fokker-Planck equation.

The assumption of a linear dynamics governing the $\mathbf{\Psi}_R$(**x**,t) follows from the statistical time-stationary properties of the underlying dynamics, which in turn justifies the use of Karhunen-Loeve methods. We have shown in previous work [*Rundle et al.*, 1995; *Klein et al.*, 1997; *Ferguson et al.*, 1999] that mean field threshold systems such as earthquake faults, characterized by long range interactions between sites **x** and **x'**, can be treated as equilibrium systems and have dynamics that are statistically stationary over long time intervals. These long intervals are eventually punctuated by rare, large events that serve to reorder the entire system. As the range of interaction increases, these equilibrium-like time intervals increase. In the limit of infinite range interactions where mean field conditions hold and the system is ergodic, Poincare's theorem [*Huang*, 1987] implies that dynamical variables can be written as sums and integrals of complex exponentials. This result has recently been independently demonstrated by Egolf (2000). In this paper, we construct a pattern dynamics for these mean field earthquake fault systems.

Our approach is similar in some respects to previous methods that use correlation operators [*Nijhout et al.*, 1997; *Fukunaga*, 1990], but also has important differences. In most of these methods, a formalism is constructed that in effect uses space-time patterns to extrapolate future system behavior. However, our approach differs in that we retrieve a complete set of space-time patterns represented by the eigenvectors $\Phi_n(\mathbf{x}_i)$ and eigenfrequencies ω_n of an equal time, rate correlation function. The $\Phi_n(\mathbf{x}_i)$ provide information about spatial correlations of patterns, whereas the ω_n provide information regarding the frequency with which each spatial correlation (eigenvector) appears in the seismicity data. If the process is statistically stationary, as we have established for the mean field threshold systems we consider, these $\Phi_n(\mathbf{x}_i)$ and ω_n *will themselves be independent of time*. Hence $\Phi_n(\mathbf{x}_i)$ and ω_n can justifiably be determined from an equal time correlation operator. From these eigenvectors and eigenfrequencies, as well as initial-value data on the most recent event activity at each spatial location, we obtain the reconstructed pattern state $\mathbf{\Psi}_R$(**x**,t), which in turn is used to compute a probability density function P(**x**,t).

Forecasting future evolution of other nonlinear systems, such as climate, weather, or El Nino-Southern Oscillations, often involve the computation of an unequal time correlation function (UTC) over short time intervals [*Preisendorfer*, 1988; *Penland*, 1989; *Garcia and Penland*, 1991; *Penland and Sardeshmukh*, 1995; *Penland and Magorian*, 1993; *Broomhead and King*, 1986; *Vautard and Ghil*, 1989]. However, in these systems, which are not "threshold systems" in the sense we consider here, the state variables s(**x**,t) and σ(**x**,t) that define the true deterministic dynamics *can* be directly observed. These include wind speeds, temperatures, and pressures, and so forth. The dynamics, as represented by the Navier-Stokes equation, are essentially known. It would not be useful to construct a UTC operator for the state variables s(**x**,t) and σ(**x**,t), since these are unobservable. In the case of earthquake mainshocks, computing a UTC operator in the pattern variables $\mathbf{\Psi}$(**x**,t) is difficult because of the apparently random nature of the long intervals of quiescence that separate mainshocks. Long periods of activity are punctuated by short bursts of activity, making the definition of one or more time steps for the UTC problematic.

Our first attempts at forecasting the patterns seen in the earthquake simulations described below used just such a UTC technique, which are based on a linear operator acting on a state at time t to predict a state or probability at time t + Δt. As a result, the UTC forecast method would be inherently unable to distinguish between a precursory quiescence that precedes a large earthquake main shock and the interseismic quiescent interval that follows it. Yet the probability of the main shock would clearly be greater before the event than following. Moreover, there is often a wide spectrum of time scales among the pattern states, a property which is extremely difficult for a method using a one-time-step UTC state transition matrix to adequately

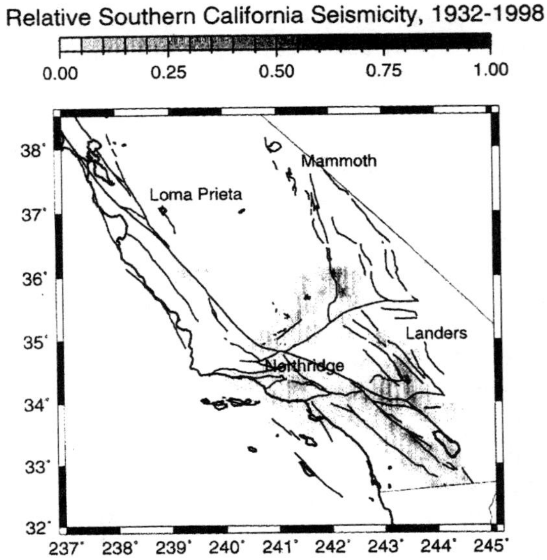

Figure 2. Relative seismicity in southern California from 1932-1998.

capture. For these reasons, our attempts at forecasting simulated earthquakes using an unequal time correlation operator were not satisfactory, and we were thus motivated to develop the new method described in the following.

2. EXAMPLES: DYNAMICS AND "PATTERN DYNAMICS"

As illustrations to which our methods will be applied below, we show two examples of pattern activity in earthquake fault systems. In the first example, natural seisimicity from southern California were obtained from the Caltech catalogue. Seismicity data between longitudes -115° and -121°, and latitudes 36° and 39° were considered. Only A and B quality events were accepted and all blast events were removed. Epicentral locations for seismic events were then binned into squares of 0.1^{o} latitude and 0.1^{o} longitude on a side. Figure 2 is a plot of the relative activity at all sites in southern California at which at least one earthquake occurred during the time interval 1932-1998. These are the data that define $\mathbf{\Psi}(\mathbf{x},t)$. Obviously, no information is available about slips or stresses, thresholds, residual stresses, or friction constitutive laws on the various faults in the region.

Figure 3 illustrates space-time patterns of activity produced by a realistic earthquake fault model [*Rundle*, 1988]. In this model, 80 fault segments are endowed with static and kinetic coefficients of friction, and are driven by a persistently increasing shear stress that, over long time intervals, produces slip events whose time-averaged rate of slip is equal to that observed in nature. In the model, there are 6 major faults: the San Andreas (SA), Imperial (I), Cerro Prieto (CP), San Jacinto (SJ), Elsinore (E), and Garlock (G). The top diagram is a plot with positions on all 6 faults concatenated together along the horizontal axis, and with event times (lines) on the vertical axis for a simulation run. The middle diagram is a plot of friction against position, and the bottom is a map view of the faults. For this simulation, the coarse-grained time interval $\Delta t = 1$ year. Details are given in appendix A.

Referring to the top diagram in figure 3, each horizontal line represents an earthquake that occurred at the time indicated on the vertical axis. These data correspond to the observable patterns $\mathbf{\Psi}(\mathbf{x},t)$ described above. At each position $\mathbf{x}$ along the earthquake faults, an activity time series can be constructed for which $\mathbf{\Psi}(\mathbf{x},t) = 1$ if an earthquake occurs at time t, and $\mathbf{\Psi}(\mathbf{x},t) = 0$ otherwise. Thus we have 80 time series, one for each fault segment, and each with 2000 time entries (2000 years in the simulation). These data are meant to be a somewhat realistic representation of the actual earthquake fault system. Like the real faults, the function $\mathbf{\Psi}(\mathbf{x},t)$ represents the observable data, and encodes all space-time patterns that may exist. In contrast, there are many unobservable parameters in the simulations. As described in the appendix, each fault segment has coefficients of friction, a slip state variable, two stress variables (shear and normal stress), and so forth. In the real earth, there is no realistic possibility of measuring values for these variables.

3. BASIC METHOD

Viewed over the time spans of hundreds to thousands of years characteristic of the earthquake cycle, it has long been observed that earthquake mainshocks are recurrent at quasiperiodic intervals [*Scholz*, 1990]. For example, in the Nankai Trough near southwest Japan, the average interval between mainshocks is 180 ± 67 years. For the San Andreas fault at Pallet Creek in the Big Bend region of southern California, the interval is 131 ± 10 years. Elsewhere along the San Andreas, deviations from the mean are more significant Scholz [1990]. At the famous Parkfield, California site, as well as along the Alaska-Aleutian trench, and along the central and South American subduction zones, average recurrence intervals are also well defined. However, temporal and spatial clustering is also evident in these data, as has been discussed elsewhere [*Scholz*, 1990; *Kagan and Jackson*, 1991; *Gross and Kisslinger*, 1994; *Rundle et al.*, 1996; *Sornette and Knopoff*, 1997] . Analysis in refs. Gross and Kisslinger [1994] and Rundle et. al. [1996] suggests that neither the recurrent

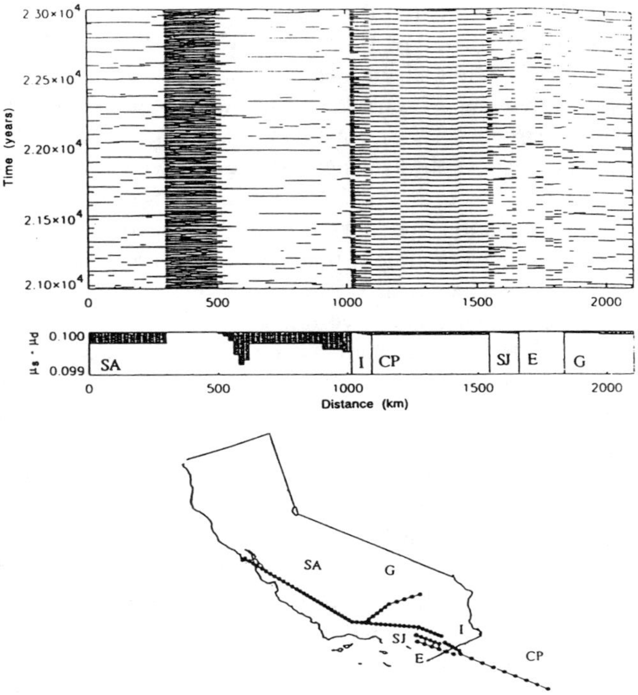

Figure 3. (Top). Time-distance plot of simulated seismic activity. N=80 fault segments (abbreviations in text) are plotted end to end, with north to left in all cases. Horizontal line over a given segment represents slip of that segment at time indicated. (Middle). Plot of difference in friction coefficients μ_s - μ_d as a function of distance along faults. (Bottom). Map view of fault segments shown superposed on a map of California. A segment is represented by the interval between dots.

nature of the mainshocks, nor the observed spatial-temporal clustering of activity, is compatible with a pure Poisson probability function.

Nevertheless, if we were to attempt the construction of a probability density function $P(\mathbf{x}_i,t)$ to forecast mainshocks on the ith fault segment $\mathbf{x}_i$, a logical approach would be to measure the average frequency of events on $\mathbf{x}_i$ by calculating:

$$\omega_i = \pi \, \nu_i = \pi \left\{ \frac{\text{Number of Events on } i^{th} \text{ Segment}}{\text{Time Interval}} \right\} \quad (3)$$

and then write an expression such as:

$$P(\mathbf{x}_i, t) = Z_i^{-1} \{\text{Cos}\, \omega_i \, (t - t_o)\}^2 \quad (4)$$

The constant t_o would be determined as the time of the latest slip event, and Z_i is a normalizing factor. This and other approaches have been examined in the literature [*Nijhout et al.,* 1997; *Gray et al.,* 1997; *Shatz et al.,* 1997; *Constantine-Paton et al.,* 1990; *Abeles,* 1991; *Gross and Rundle,* 1998; *Crutchfield and Hanson,* 1993; *Eneva and Ben-Zion,* 1997a,b; *Pepke and Carlson,* 1997; *Pepke et al.,* 1994; *Kossobokov and Carlson,* 1995; *Minster and Williams,* 1992; *Molchan and Kagan,* 1992; *Sammis et al.,* 1996], but success has so far been mixed.

We have therefore been motivated to search for an alternative method in which $P(\mathbf{x}_i,t)$ is constructed by taking account of the variability in the seismicity data, as well as space-time correlations that may exist with activity at other segments $\mathbf{x}_j$. We have developed a method based on a discrete Karhunen-Loeve expansion [Holmes et al., 1996; Funkunaga, 1990; Preisendorfer, 1988] that allows computation of the "normal modes" of the space-time seismicity pattern data from an equal-time correlation operator, yielding a set of eigenvectors $\phi_n(\mathbf{x})$ and a corresponding set of eigenfrequencies ω_n, n = 1,...,N. If the process is statistically stationary, as has been established for the mean field systems we consider, these $\phi_n(\mathbf{x})$ and ω_n *will themselves be independent of time*. Hence $\phi_n(\mathbf{x})$ and ω_n can justifiably be determined from an *equal time* correlation operator. It is important to stress here that, due to the long range nature of the stress transfer Green's function describing real faults, earthquake fault systems can be considered to be mean field systems.

In the case of figure 3, we have N = 80 such eigenvectors and eigenfrequencies. From these, we compute the complex-valued, reconstructed seismicity pattern state $\Psi_R(\mathbf{x}_i,t)$:

$$\Psi_R(\mathbf{x}_i,t) = \Sigma_{n=1,N} \; \alpha_n \, e^{-i\,\omega_n t} \, \phi_n(\mathbf{x}_i) \quad (5)$$

The constants α_n are determined by fitting $\Psi_R(\mathbf{x}_i,t)$ to the time t_{oi} of the most recent slip event on segment $\mathbf{x}_i$, $\Psi_R(\mathbf{x}_i,t_{oi}) = 1$ (see equation 21 below). Expression (5) can in principle describe phenomena in a system characterized by a large number of greatly diverse time scales $\tau_i = 2\pi/\omega_i$. Furthermore, equation (5) implies that time evolution of the pattern state $\Psi_R(\mathbf{x}_i,t)$ occurs via rotation in a complex, N-dimensional space spanned by the eigenfunctions $\phi_n(\mathbf{x}_i)$. In this example, seismicity is thus an example of a phase dynamical system [*Mori and Kuramoto,* 1998].

From $\Psi_R(\mathbf{x}_i,t)$, we take the real, observable part:

$$\Psi_{obs}(\mathbf{x}_i,t) = \frac{1}{\sqrt{2}} \{\Psi_R(\mathbf{x}_i,t) + \Psi_R^{*}(\mathbf{x}_i,t)\} \quad (6)$$

$\Psi_R(\mathbf{x}_i,t)$ is written in terms of a sum of eigenvectors $\phi_n(\mathbf{x})$, which arise from an equal-time correlation operator. A correlation function is related to a probability density function. In this sense, $\Psi_R(\mathbf{x}_i,t)$ thus represents the "square root" of a probability, therefore a probability amplitude:

$$P(\mathbf{x}_i,t) = | \Psi_{obs}(\mathbf{x}_i,t) |^2 = \{ \Psi_{obs}(\mathbf{x}_i,t) \}^2 \qquad (7)$$

Thus the *probability amplitudes* associated with the "Pattern Dynamics" evolve linearly over the time interval $\{t,t+\Delta t\}$, even though the *underlying dynamics* do not.

As an example, for a model with N independent, non-interacting segments, we have

$$\phi_n(\mathbf{x}_i) = \delta_{ni} \qquad (8)$$

where δ_{ni} is the Kronecker symbol. Using (5) - (8) for this case, we find an expression for $P(\mathbf{x}_i,t)$ that is identical to (4). The more interesting question arises when all of the segments in the fault system interact. In that case, it is not clear from the above discussion how the eigenvectors and eigenfrequencies can be computed. We address this topic in the following section.

4. CORRELATION OPERATORS

When segments of an earthquake fault system or other nonlinear system interact, it is likely that space-time correlations and patterns will develop [see, e.g., *Nijhout et al.*, 1997; *Ball*, 1999]. Correlations that exist in the slip $s(\mathbf{x},t)$ and stress $\sigma(\mathbf{x},t)$ state variables lead to correlations in the seismicity pattern variables $\Psi(\mathbf{x},t)$. To describe the correlations in seismicity, one can define any of a large number of equal time correlation operators. We consider two, a *static* correlation operator, and a *rate* correlation operator. The *static* correlation operator $C(\mathbf{x}_i,\mathbf{x}_j)$ between discrete segments centered at $\mathbf{x}_i$ and $\mathbf{x}_j$ is:

$$C(\mathbf{x}_i,\mathbf{x}_j) \equiv \frac{1}{T}\int_0^T dt \quad z(\mathbf{x}_i,t)\, z(\mathbf{x}_j,t) \qquad (9)$$

where $z(\mathbf{x}_i,t)$ is the mean-zero, univariant time series obtained from $\Psi(\mathbf{x}_i,t)$. The *rate* correlation operator $K(\mathbf{x}_i,\mathbf{x}_j)$ is:

$$K(\mathbf{x}_i,\mathbf{x}_j) \equiv -\frac{1}{T}\int_0^T \frac{\partial(\mathbf{x}_i,t)}{\partial t}\,\frac{\partial(\mathbf{x}_j,t)}{\partial t} \qquad (10)$$

Both of the operators (9) and (10) are symmetric, rank-N operators that can be diagonalized using standard methods. The eigenvalues of $C(\mathbf{x}_i,\mathbf{x}_j)$ represent relative weights (probabilities) p_n [*Holmes et al.*, 1996; *Fukunaga*, 1990; *Preisendorfer,* 1988], with corresponding eigenvectors $\chi_n(\mathbf{x}_i)$. The eigenvalues of $K(\mathbf{x}_i,\mathbf{x}_j)$ are the squares of inverse times $(1/\tau_n)^2$, therefore related to frequencies $\omega_n = 2\pi/\tau_n$, with corresponding eigenvectors $\phi_n(\mathbf{x}_i)$. The latter are just the eigenfrequencies and eigenvectors that we used to define the reconstructed pattern state $\Psi_R(\mathbf{x}_i,t)$ in (5). In Appendix B, we discuss the physical significance of the eigenvalues of $K(\mathbf{x}_i,\mathbf{x}_j)$.

Other correlation operators can be defined as well. For example, one can define a time series $y(\mathbf{x}_i,t)$ whose entries are the slips on segment $\mathbf{x}_i$ in the time interval $\{t , t + \Delta t \}$, and whose eigenvalues have the units of slip squared. Or one can define a time series $w(\mathbf{x}_i,t)$ whose entries are the number of events. The eigenvalues obtained from the correlation operator formed from $w(\mathbf{x}_i,t)$ are the squares of relative frequencies. Or one can define a covariance matrix if $z(\mathbf{x}_i,t)$ is not univariant, and so forth. Any of these possible operators can be examined using the methods we describe, some of which we are in the process of studying. Each have different physical interpretations and yield different information about correlations in the system. The eigenvectors of each contain information related to the spatial correlations between segments.

Since $\Psi(\mathbf{x},t)$ and $z(\mathbf{x}_i,t)$ represent time series whose elements are impulse functions, it is operationally difficult to compute the time derivatives in (10). Computing $K(\mathbf{x}_i,\mathbf{x}_j)$ is therefore difficult, whereas computing $C(\mathbf{x}_i,\mathbf{x}_j)$ is easy. As a result, we propose an alternative approach based on Fourier transforms.

$$z(\mathbf{x}_i,t) = \frac{1}{\sqrt{2\pi}}\int_{-\infty}^{\infty} d\varpi_i\; e^{i\,\varpi_i\,t}\;\hat{z}(\mathbf{x}_i,\varpi_i) \qquad (11)$$

Then:

$$C(\mathbf{x}_i,\mathbf{x}_j) = \int_{-\infty}^{\infty}\int_{-\infty}^{\infty} d\varpi_i\, d\varpi_j\; P_*(\mathbf{x}_i, \mathbf{x}_j, \varpi_i , \varpi_j) \qquad (12)$$

$$K(\mathbf{x}_i,\mathbf{x}_j) = \int_{-\infty}^{\infty}\int_{-\infty}^{\infty} d\varpi_i\, d\varpi_j\; \varpi_i\, \varpi_j\, P_*(\mathbf{x}_i, \mathbf{x}_j, \varpi_i , \varpi_j) \qquad (13)$$

$P_*(\mathbf{x}_i, \mathbf{x}_j, \varpi_i , \varpi_j)$ is the joint probability density of the two variables ϖ_i, ϖ_j:

$$P_*(\mathbf{x}_i, \mathbf{x}_j, \varpi_i , \varpi_j) = \hat{z}(\mathbf{x}_i,\varpi_i)\, \hat{z}(\mathbf{x}_j,\varpi_j) \left\{\frac{1}{2\pi T}\int_0^T dt\; e^{i\,(\varpi_i + \varpi_j)\,t}\right\} \qquad (14)$$

The covariance $< \varpi_i \; \varpi_j >$ is then:

$$<\varpi_i \varpi_j> = \frac{\int_{-\infty}^{\infty}\int_{-\infty}^{\infty} d\varpi_i \, d\varpi_j \, \varpi_i \, \varpi_j \; P_*(\mathbf{x}_i, \mathbf{x}_j, \varpi_i, \varpi_j)}{\int_{-\infty}^{\infty}\int_{-\infty}^{\infty} d\varpi_i \, d\varpi_j \; P_*(\mathbf{x}_i, \mathbf{x}_j, \varpi_i, \varpi_j)} \qquad (15)$$

If we assume that the joint spectra of the time series are simultaneously sharply peaked about the average frequencies $< \varpi_i >$ and $< \varpi_j >$:

$$\left| \frac{< (\varpi_i - < \varpi_i >)(\varpi_j - < \varpi_j >) >}{< \varpi_i >< \varpi_j >} \right| << 1 \qquad (16)$$

we obtain the result:

$$K(\mathbf{x}_i,\mathbf{x}_j) \approx C(\mathbf{x}_i,\mathbf{x}_j) < \varpi_i > < \varpi_j > \qquad (17)$$

The frequencies $< \varpi_i >$ are observables, they are simply:

$$< \varpi_i > = \pi \upsilon_i = \omega_i \qquad (18)$$

from (3).

Once $K(\mathbf{x}_i,\mathbf{x}_j)$ is found from (17), it can be diagonalized to obtain the normal modes of the patterns, the eigenfrequencies ω_n and the eigenvectors $\phi_n(\mathbf{x}_i)$. These eigenvectors provide information about specific patterns of spatial correlations between sites, and the eigenfrequencies provide information about how often these recur.

5. FORECAST METHODOLOGY

Our forecast methodology makes use of equations (5)-(7). Given an observed pattern $\Psi(\mathbf{x},t)$, we wish to compute $P(\mathbf{x}_i,t+\Delta t)$, the probability density for activity at the next time step. From (6) we have:

$$\Psi_{obs}(\mathbf{x}_i, t) = \frac{1}{\sqrt{2}} \{ \Psi_R(\mathbf{x}_i, t) + \Psi_R^*(\mathbf{x}_i, t) \}$$
$$= \Sigma_{n=1,N} \; \beta_n \; \mathrm{Cos} \{ \omega_n t \} \; \phi_n(\mathbf{x}_i) \qquad (19)$$

Our approach is to use the observed pattern data $\Psi(\mathbf{x},t)$ prior to $t+\Delta t$ to find $\phi_n(\mathbf{x}_i)$ and ω_n from the operator $K(\mathbf{x}_i,\mathbf{x}_j)$. Then if the β_n are known, (6) can be used to compute $P(\mathbf{x}_i,t+\Delta t)$:

$$P(\mathbf{x}_i,t+\Delta t) = | \Psi_{obs}(\mathbf{x}_i,t+\Delta t)) |^2 = \{ \Psi_{obs}(\mathbf{x}_i,t+\Delta t)) \}^2 \qquad (20)$$

To fix the N values β_n in (19), we find the most recent time t_{mi} prior to time t at which the segment or cell at $\mathbf{x}_i$ "fires". We then require that:

$$\Psi_{obs}(\mathbf{x}_i, t_{oi}) = (-1)^m \qquad (21)$$

where the first earthquake at site $\mathbf{x}_i$ in the sequence has m=0, the next m=1, and so forth (both values ± 1 correspond to an event). Condition (21) then leads to a set of N x N independent equations that can be solved by standard methods. This process is then repeated at time $t + \Delta t$ to calculate $P(\mathbf{x}_i,t+2\Delta t)$, the forecast for the next time $t + 2\Delta t$. One should always take care to use the most recent data to calculate the β_n. As discussed above, we assume that the forecast probability amplitudes evolve linearly over the time from t to $t + \Delta t$.

6. SPATIAL CORRELATION PATTERNS IN SOUTHERN CALIFORNIA SEISMICITY

As a first application of these methods, we demonstrate that the observed data shown in figure 2 from southern California may be decomposed into pattern basis states $\chi_n(\mathbf{x}_i)$ having relative probabilities p_n, by computing the static correlation operator $C(\mathbf{x}_i,\mathbf{x}_j)$. For this example, the coarse-grained time interval was .001 years, or approximately 8-hour units, yielding a total of 66,500 time steps over the interval 1932-present (see figure 2). Epicentral locations for seismic events were binned into squares of 0.1^o latitude and 0.1^o longitude on a side, yielding N = 1,329 boxes (sites) $\mathbf{x}_i$ in which at least one event occurred. At each site $\mathbf{x}_i$ and coarse grained time step t_q, $\Psi(\mathbf{x}_i, t_q) = 1$ if at least one event occurred, $\Psi(\mathbf{x}_i, t_q) = 0$ otherwise. The correlation operator $C(\mathbf{x}_i,\mathbf{x}_j)$ was then computed according to (12), then diagonalized to obtain (real) eigenvectors $\chi_n(\mathbf{x}_i)$ and eigenvalues (probability weights) p_n. The results for the first two eigenvectors $\chi_1(\mathbf{x}_i)$ and $\chi_2(\mathbf{x}_i)$ are shown in figure 4. Values of $\chi_n(\mathbf{x}_i)$ are scaled so that $1 \geq \chi_n(\mathbf{x}_i) \geq -1$. $\chi_n(\mathbf{x}_i) > 0$ is represented by dark gray, $\chi_n(\mathbf{x}_i) < 0$ by light gray.

The physical interpretations of p_n and $\chi_n(\mathbf{x}_i)$ are motivated by the fact that the eigenvalues p_n of a real, symmetric matrix like $C(\mathbf{x}_i,\mathbf{x}_j)$ are all positive. Therefore, all of the information about correlation between sites must be carried by the eigenvectors $\chi_n(\mathbf{x}_i)$. Since correlation operators are also probabilities, the p_n must represent the probability, or

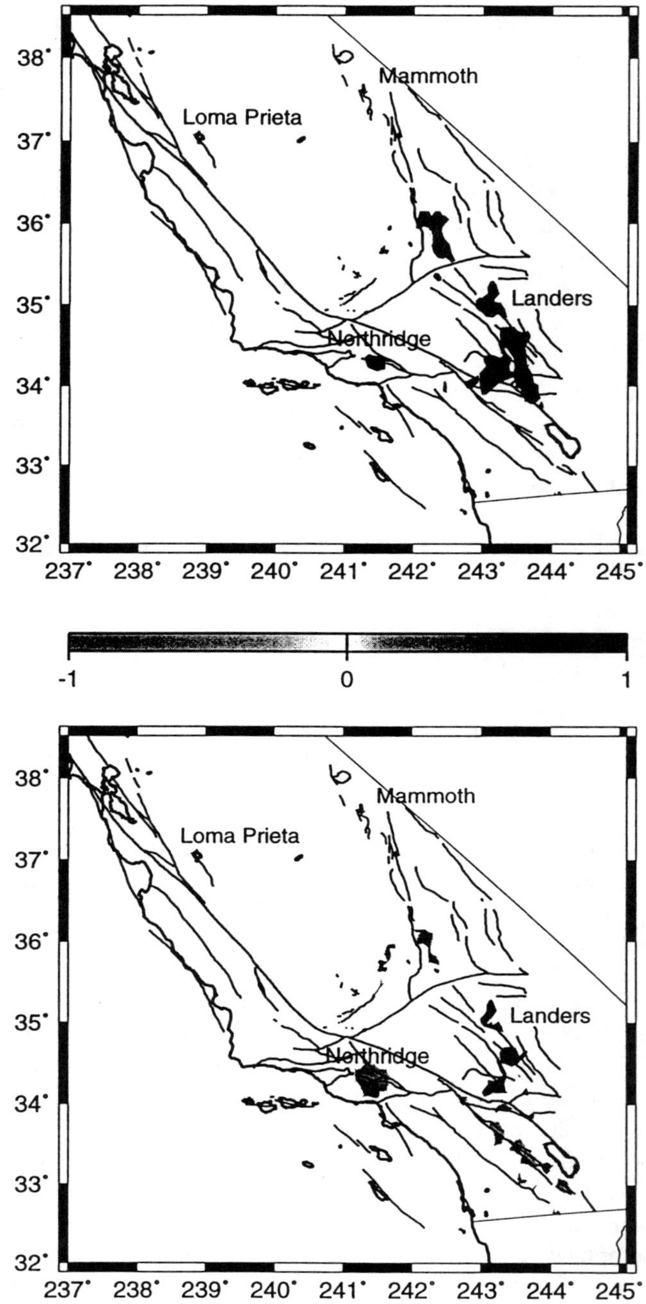

Figure 4. Eigenvectors for southern California seismicity. (Top) Eigenvector $\chi_1(\mathbf{x}_i)$. (Bottom) Eigenvector χ_2 $(\mathbf{x}_i)$.

fraction of the variance, that is associated with $\chi_n(\mathbf{x}_i)$. Also, two sites that have the same value, say $\chi_n(\mathbf{x}_i) = \chi_n(\mathbf{x}_j) = 1$, are either always active together, or always quiescent together. The same is true if $\chi_n(\mathbf{x}_i) = \chi_n(\mathbf{x}_j) = -1$ since $\chi_n(\mathbf{x}_i)$ may be multiplied by an overall constant without changing the physical content. On the other hand, if $\chi_n(\mathbf{x}_i) = -\chi_n(\mathbf{x}_j) = 1$, the physical interpretation is that site $\mathbf{x}_i$ tends to be active when $\mathbf{x}_j$ is not, and vice versa. Finally, if $\chi_n(\mathbf{x}_i) \neq 0$ and $\chi_n(\mathbf{x}_j) = 0$, activity at the two sites is uncorrelated. Thus one can use the information encoded in each $\chi_n(\mathbf{x}_i)$ to compare two or more sites for their tendencies to be simultaneously active in a given correlation mode.

Turning to figure 4 (a), we see that eigenvector $\chi_1(\mathbf{x}_i)$ reflects the relative intensity of seismicity in the region. On the other hand, eigenvector $\chi_2(\mathbf{x}_i)$ in figure 4(b) is more interesting: the sites near the 1992 Landers-Big Bear sequence are all dark gray, while the vast majority of sites in the remainder of southern California are light gray. This eigenvector can be interpreted in either of two ways. The first is that the Landers-Big Bear area is active only when the remainder of southern California is inactive. The second is the "Mogi donut" interpretation -- that the quiescent Landers area is surrounded by active regions. We note that the values of p_n corresponding to figure 4 are $p_1 = .11$, $p_2 = .045$, implying that $\chi_1(\mathbf{x}_i)$ is present in 11% of the southern California seismicity, while $\chi_2(\mathbf{x}_i)$ is present in 4.5% of the seismicity. Also, it should be pointed out that the value of this approach is that all possible spatial patterns are represented in the eigenvectors, which therefore serve as "pattern basis states" for the seismicity data. Many further results will be discussed in Tiampo et al. [2000].

Finally, we note that it is not possible to straightforwardly use this particular seismicity data set to make forecasts using the methods outlined in the previous section. In order to adequately define ϖ_n and $\phi_n(\mathbf{x}_i)$ from $K(\mathbf{x}_i,\mathbf{x}_j)$, at least several cycles of seismic activity are needed at each site, thus several thousand years of data would be necessary to define the $\Psi(\mathbf{x}_i, t_q)$ for the spatial sites shown. Instead, we illustrate below the forecast technique using the relatively realistic southern California simulation described in section 3, for which arbitrarily long synthetic seismicity catalogs can be obtained. This suggests that in order to eventually implement a forecast technique for a seismically active region such as southern California, one should use the combination of all available instrumental and historic seismicity data, together with a the good simulation data to define as well as possible the ϖ_n and $\phi_n(\mathbf{x}_i)$.

7. FORECASTING SYNTHETIC SEISMICITY IN SOUTHERN CALIFORNIA

As an illustration of the forecast method discussed in section 5, we use the relatively realistic model of an earthquake fault system described in the appendix. Selected results from tens of thousands of model years of simulations are shown in figure 3, which is a time-distance plot of all the

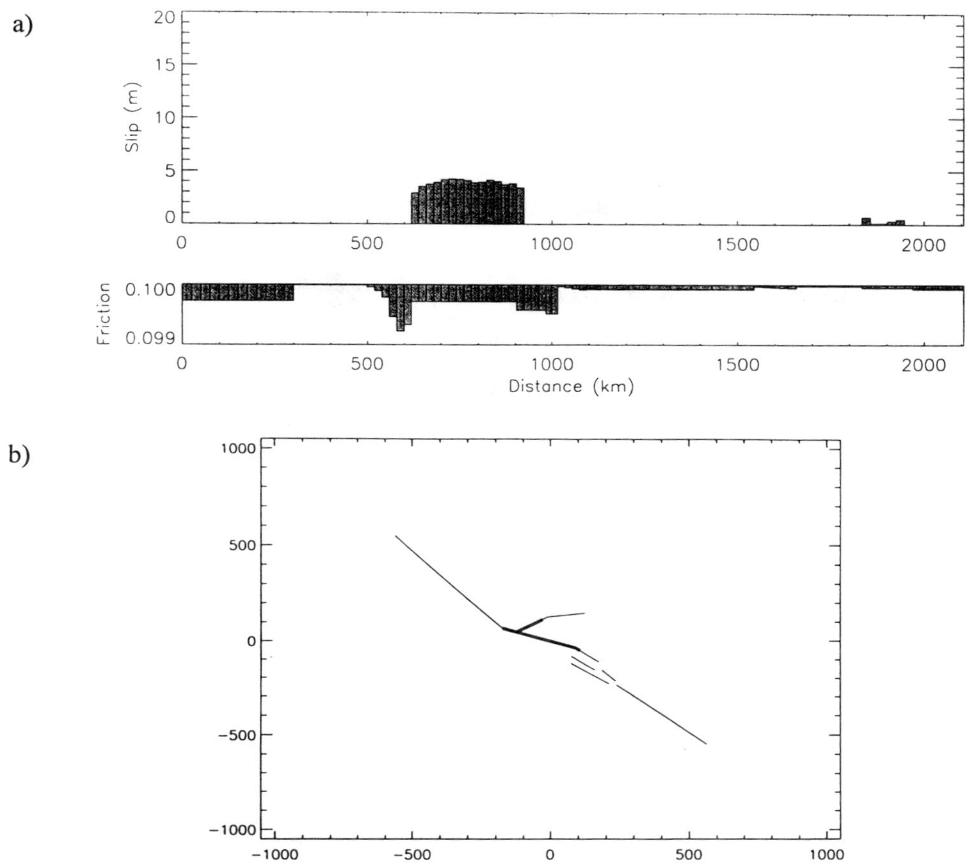

Figure 5. Events at year (time step) 21664. (a). Slip plotted against distance. (b) Map view of events.

events occurring in 2000 years of simulation data, with a reference map view of the model fault system. Figure 5 shows the events at year 21664, and in figure 6, the events at year 22554. The data shown in figure 3 (Top) were used to construct $C(\mathbf{x}_i,\mathbf{x}_j)$ and $K(\mathbf{x}_i,\mathbf{x}_j)$, from which were obtained the eigenvectors $\phi_n(\mathbf{x})$ and eigenfrequencies ω_n. The first three spatial eigenvectors for the 6000 year time period preceding the data shown in figure 3 are shown in figures 7-9. It is interesting that between distances ~ 500 km to ~1000 km, the eigenvector in figure 7 has 0 spatial nodes, the eigenvector in Figure 8 has 1 spatial node, and the eigenvector in figure 9 has 2 spatial nodes.

The probability $P(\mathbf{x}_i,t)$ for fault segment $\mathbf{x}_i$ to slip at time t was computed using the eigenstates and eigenperiods obtained from a simulation data set over the 6000 years prior to the data to be predicted ("training data set"). The results are shown in figure 10, which shows a 300 year subset of simulation data taken from that shown in figure 3. Figure 10 also shows contours of probability $P(\mathbf{x}_i,t)$ superposed on the previous simulation data. It can be seen that in many cases, there is reasonable agreement between probabilities and the time and location of events, although the "false alarm" rate is high.

8. "RANDOMIZED MODEL" AND UNCERTAINTY PRINCIPLE

Success in forecasting future events using the pattern dynamics method depends on the capability to accurately retrieve both the eigenvectors $\phi_n(\mathbf{x}_i)$, and the eigenfrequencies ϖ_n. For purposes of comparison, it is useful to have an alternative model, that we call the "Randomized Model", that illuminates trade-offs in different approaches. This new model defines the eigenfrequencies at each site much more precisely, but at the expense of losing important information about the spatial correlation patterns.

The impulsive time series defining the correlation operators consist of signals that are sharply localized in time. However, time series having signals localized in time

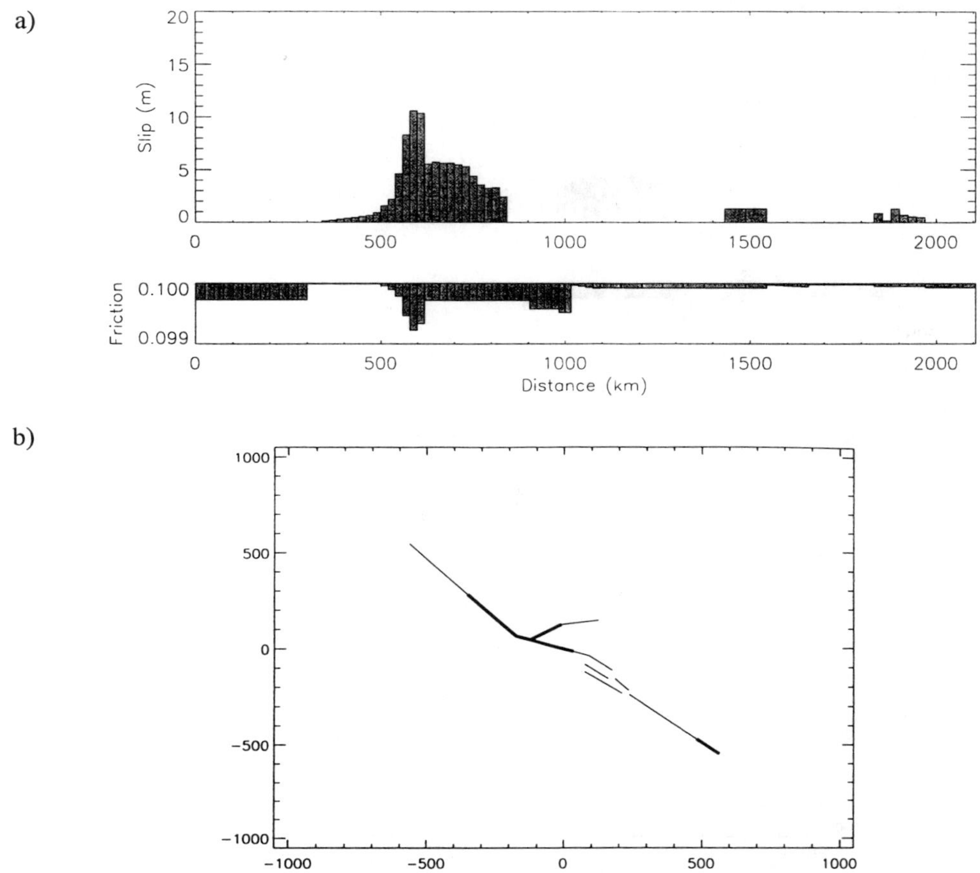

Figure 6. Events at year (time step) 22554. (a). Slip plotted against distance. (b) Map view of events.

have a relatively flat spectrum over a broad range of frequencies. The uncertainty principle for Fourier transforms [28] defined using equation (23) indicates that:

$$\{ < \Delta\varpi > \; < \Delta t > \} \; \geq \pi \qquad (22)$$

This result suggests a simple, but illustrative strategy to improve the frequency resolution of the method described above: that we add a random time increment, drawn from a uniform distribution of times about the original event time, to each event time of the $y_i(t_m)$. In this way, we degrade the temporal resolution without changing the average frequency υ_i or the mean event time. We should then expect that resolution of normal mode frequencies will improve. However, as remarked above, important information will be lost about the spatial correlations of the various sites.

To test this idea, we defined a set of new random time series $y'_i(t_m)$, in which the time of each slip event was altered by the addition of a random number uniformly distributed on the interval $[-1/2v_i, \; 1/2v_i]$ Because the randomized times are uniformly distributed around the original event times, a memory of the original event times is retained. We then repeated the entire procedure and constructed new forecast probabilities $P(\mathbf{x}_i,t)$. The eigenvectors corresponding to the three largest eigenvalues are shown in figures 11, 12, and 13. It can be seen from these figures that better frequency resolution is obtained at the cost of less knowledge about spatial correlations between sites $\mathbf{x}_i$ and $\mathbf{x}_j$.

Results for the new forecast probabilities are shown in figure 14 for the same simulation data as was used in figure 10. In several locations, for example at distances ranging from 0 to 300 km, and from 500 to 1000 km, forecasts in figure 14 are superior to those in figure 10. However, the Likelihood ratio test described below indicates that the original method with non-randomized times is more accurate overall.

a)

b)

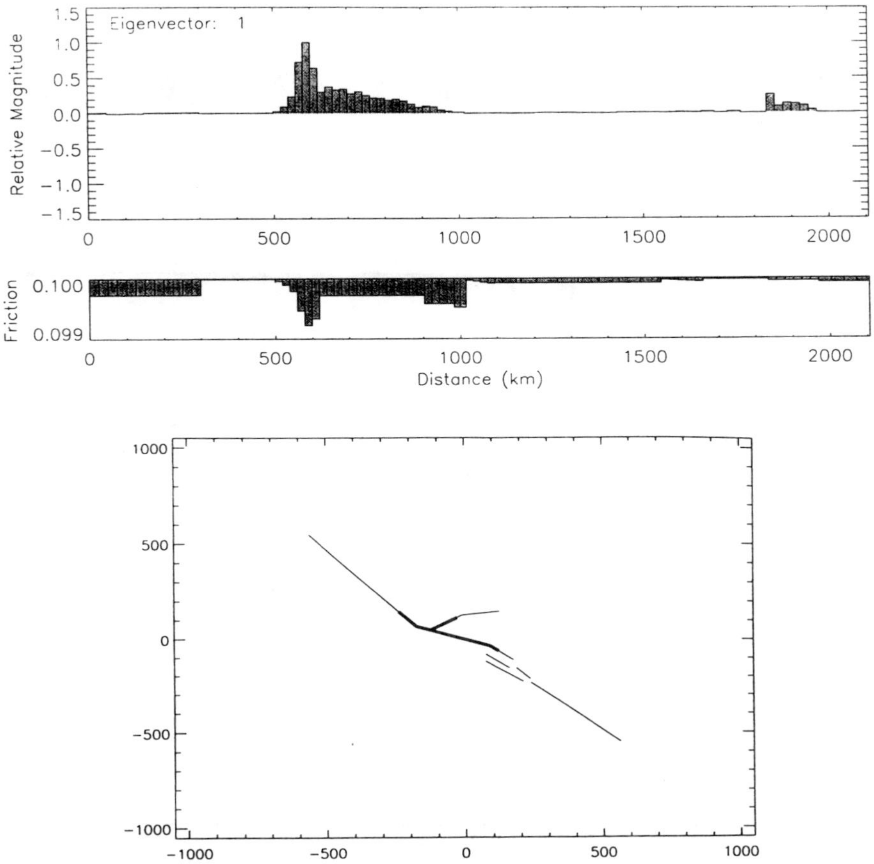

Figure 7. Eigenvector ϕ_1. (a) Eigenvector plotted against distance. For display purposes, ϕ_1 are scaled so that $\text{Max}\{|\phi_1|\} = 1$. (b) Map view. Values at all segments $\mathbf{x}_i$ at which $\phi_1(\mathbf{x}_i) > .05$ are shown with a bold line. Values at all segments $\mathbf{x}_i$ at which $\phi_1(\mathbf{x}_i) < -.05$ are shown with a dotted line. All other segments shown with a thin line. Period = 804 years.

9. STATISTICAL TEST OF FORECAST PROBABLILITIES

The method we describe can be tested by statistical means to assess the quality of the forecast. The predictive skill of a candidate method can be compared, for example, to a standard Poisson model by computing a Likelihood ratio. The same method has been used to evaluate other proposed forecast methods, including time-to-failure models, for real earthquakes [Sammis et al., 1996; Gross and Rundle, 1998]. The Likelihood test is based on the idea that a cost is incurred if a forecast method produces a false alarm, i.e. an event during a time interval when no event occurs. Using this test, we find that the Pattern Dynamics (PD) method associates significantly more probability measure with the actual simulation events than does a competing, Poisson probability model.

The Likelihood ratio method is described in detail elsewhere [*Gross and Rundle*, 1998], and only a brief summary is given here. Consider a given normalized probability density $P(\mathbf{x}, t)$. If the ith of N fault segments slips Q_i times, so that there are N_T total events on all segments, the Likelihood function is:

$$L = \prod_{i=1}^{N_T} \prod_{q=1}^{Q_i} P(\mathbf{x}_i, \mathbf{t}_q) \tag{23}$$

We wish to compare the quality of forecasts given by two probability distributions, for example the PD method $P_{PD}(\mathbf{x}_i t)$ and a reference time-independent Poisson model $P_P(\mathbf{x})$. The corresponding Likelihoods are designated as L_{PD} and L_P. To compare the two probability models, we

a)

b)

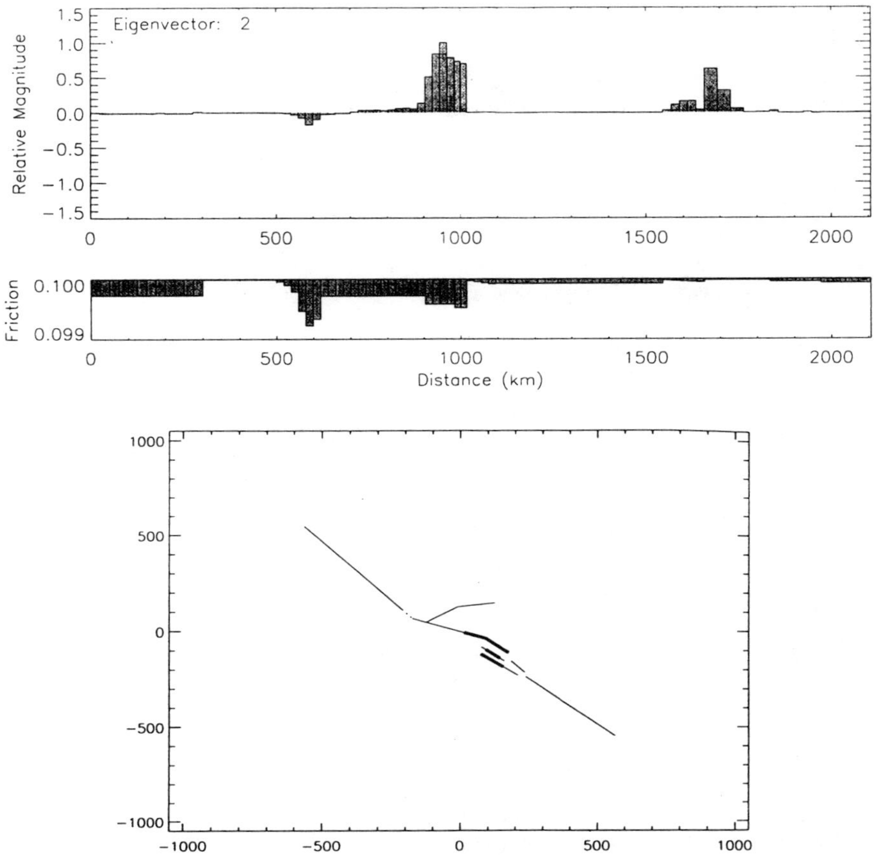

Figure 8. Eigenvector ϕ_2. As in figure 5 a,b. Period = 680 years.

define a composite Likelihood ratio that involves a mixing ratio r:

$$L(r) = \prod_{i=1}^{N_T} \prod_{q=1}^{Q_i} P(\mathbf{x}_i, t_q, r) \qquad (24)$$

where:

$$P(\mathbf{x}_i,t_q,r) = r\, P_{PD}(\mathbf{x}_i,t_q) + (1 - r)\, P_P(\mathbf{x}_i) \quad (25)$$

Equations (23)-(25) assumes that the earthquakes are all statistically independent, a standard assumption that is clearly not correct in the case of our simulations. However, statistical independence is an assumption routinely used in earthquake forecasting, and the Poisson model represents one of the current most widely used method for calculating earthquake hazard [*Scholz*, 1990; *Gross and Rundle*, 1998].

Significance of Results: In figure 15, we plot the log-likelihood ratio $\log_{10}\{L_R(r)\}$, which is defined by:

$$\log_{10}\{ L_R(r) \} = \log_{10} \left\{ \frac{L(r)}{L_P} \right\}$$
$$= \log_{10}\{ L(r) \} - \log_{10}\{ L_P \} \qquad (26)$$

In figure 15, we show a plot of $\log_{10}\{ L_R(r) \}$ against the mixing ratio r, corresponding to the forecasts and simulation data in figures 10 and 14. In both cases:

$$L(r) >> L_P \quad ; \quad r > 0 \qquad (27)$$

implying that $L_{PD} >> L_P$. These results show that the PD model is substantially better (~ 20 in figure 8, ~ 5 in figure 15 for large r) than the Poisson model at predicting the occurrence times and locations of the synthetic events. The

a)

b)

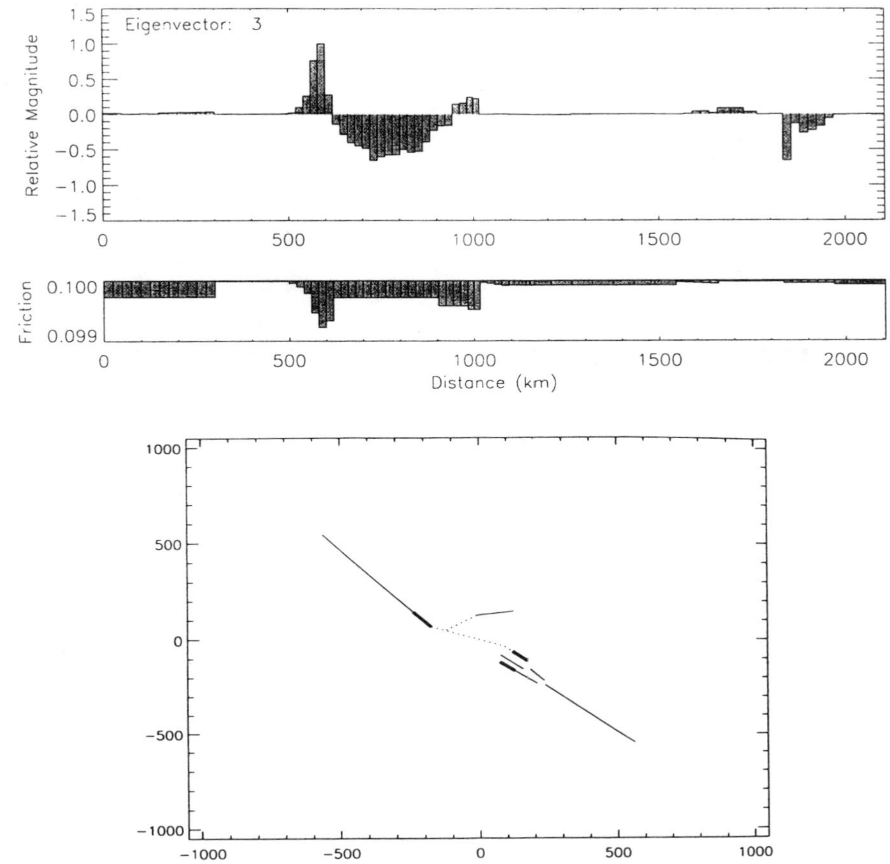

Figure 9. Eigenvector ϕ_3 . As in figure 5 a,b. Period = 597 years.

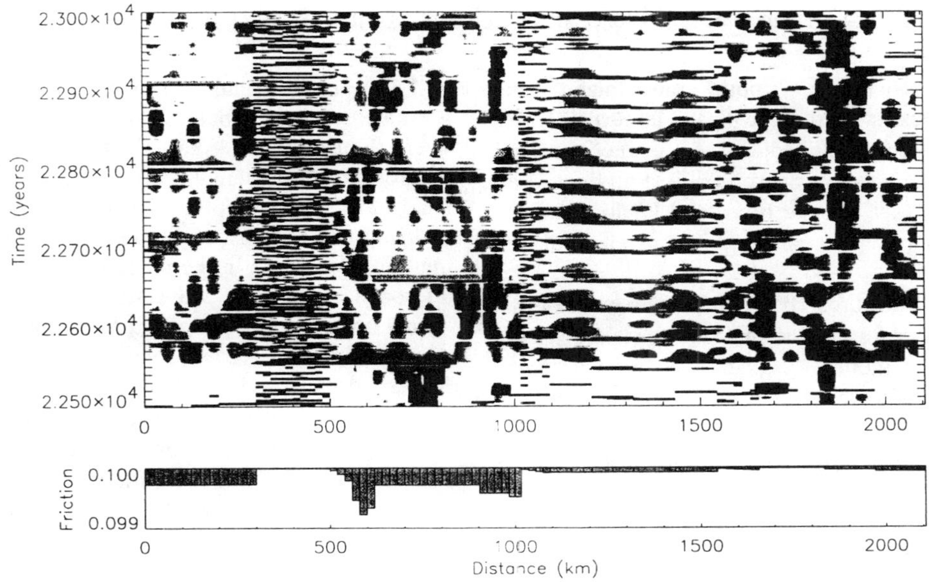

Figure 10. Superposition of events (solid horizontal lines) from year 22,500 to year 23,000 top with probabilities $P(\mathbf{x}_i, t)$ calculated according to (20). Light shaded regions are for $P(\mathbf{x}_i, t) > 1/3$, dark shaded regions are for $P(\mathbf{x}_i, t) > 1/2$. Method used was basic method described in section 5.

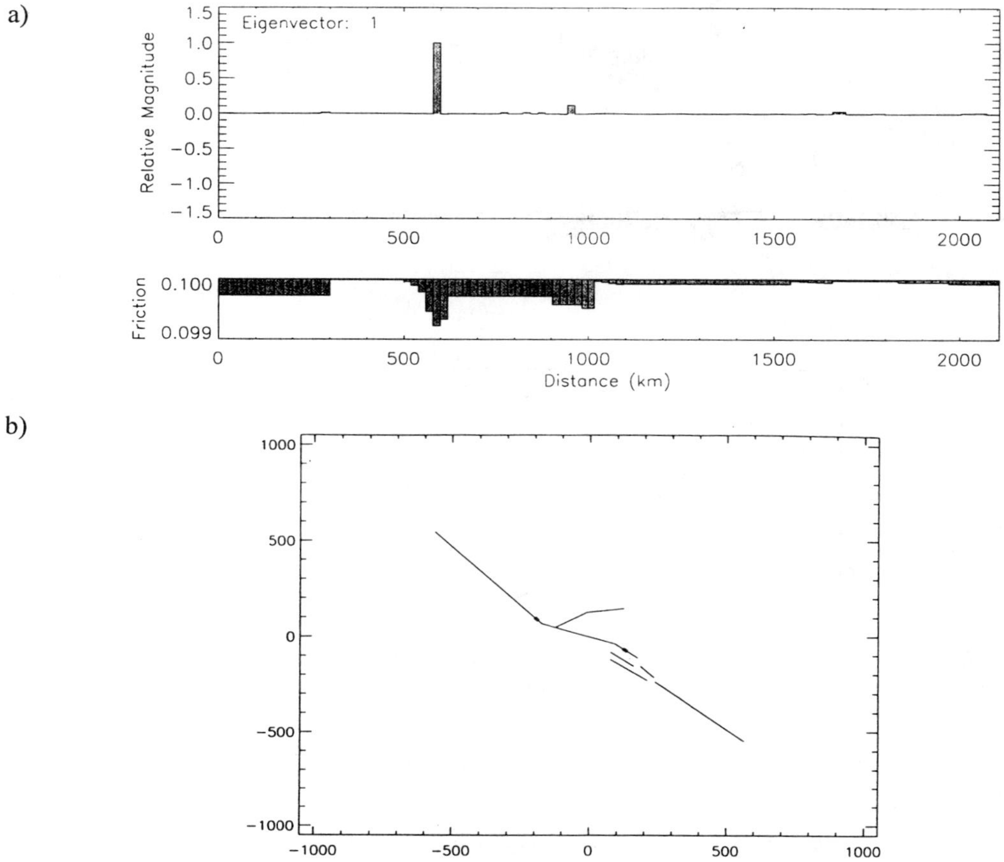

Figure 11. Eigenvector ϕ_1 for randomized model. As in figure 5 a,b. Period = 580 years.

best probability model for both randomized and nonrandomized forecasts are those with the largest r. Cases with r = 1 cannot be evaluated: if an event were not predicted, $P_{PD}(\mathbf{x}_i,t_q) = 0$ there. Since L_{PD} is a product of all the probabilities $P_{PD}(\mathbf{x}_i,t_q)$ at the times and locations of the events, L_{PD} would be zero, and $\log_{10}\{ L_{PD} \}$ would be undefined.

From examination of figures 10 and 14, it can be seen that there exist a number of false positive space-time forecasts in using the probability model $P_{PD}(\mathbf{x}_i,t_m)$, as well as some "failures to predict". However, the result (27) as shown in figure 15 means that $P_{PD}(\mathbf{x},t)$ still puts far more of its probability measure at the times of the slip events than does $P_P(\mathbf{x})$.

10. EXTENSIONS TO OBSERVED SEISMICITY IN SOUTHERN CALIFORNIA

Application of the forecast methods described above to observed seismicity in a region such as southern California will be more difficult, since one does not have available accurate seismicity data spanning many earthquake cycles over thousands of years. For example, in southern California, instrumental records began in 1932 with only a handful of seismographs. This network has been augmented and improved almost continually ever since. However, the methods described in the preceding sections suggest similar approaches that may prove fruitful. A variety of methods based upon using "pattern recognition" to forecast future events have been tried, recent examples being those of Keilis-Borok et al. [1988] and Press and Allen [1995].

We can explicitly make use of the idea that earthquakes can be considered as an example of phase dynamics, as in equation (5). Thus evolution of pattern states is via rotations in N-dimensional complex correlation space. This implies that the time average of such pattern states is zero. If we assume that the dynamics is represented by an incoherent superposition of such pattern states over a sufficiently large region, the spatial mean of pattern states will vanish also. Defining $R(\mathbf{x}_i,t)$ as the instantanteous rate of seisimicity (number of events of any size) at time t and

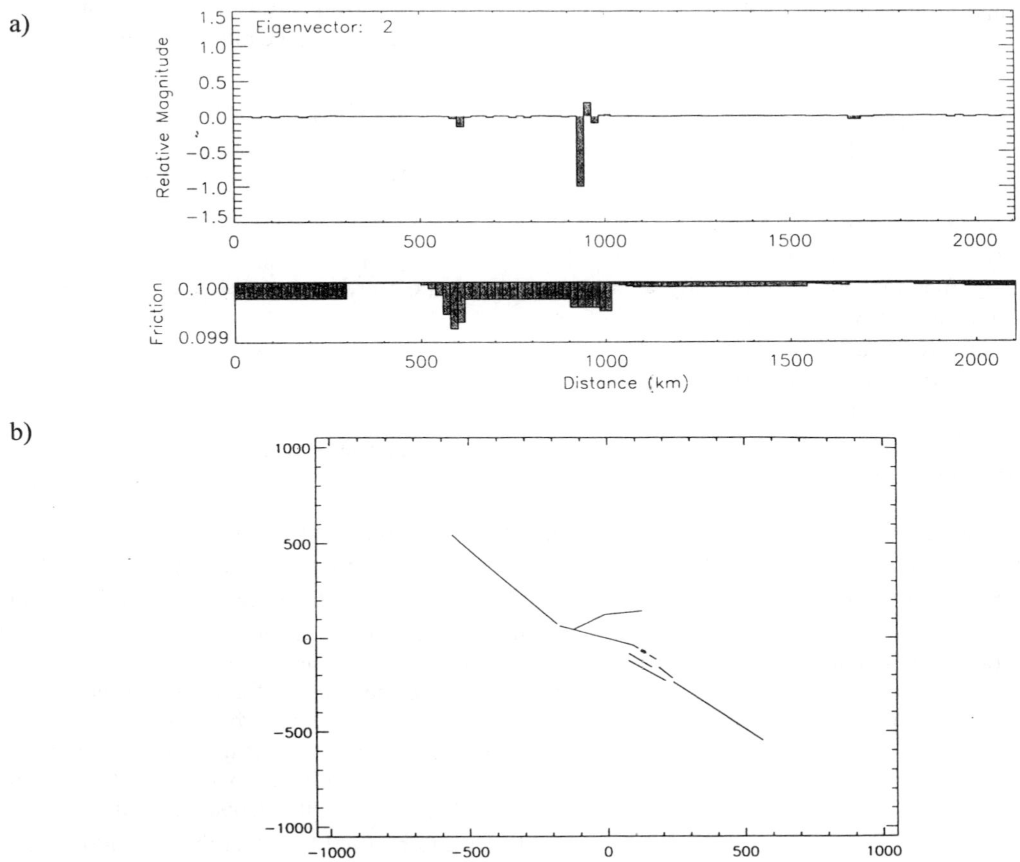

Figure 12. Eigenvector ϕ_2 for randomized model. As in figure 5 a,b. Period = 476 years.

location $\mathbf{x}_i$, and $< R(\mathbf{t}) >$ as the spatial average of $R(\mathbf{x}_i,t)$:

$$\Psi_{obs}(\mathbf{x}_i, t) \approx R(\mathbf{x}_i,t) - < R(t) > \qquad (28)$$

We also need to account for the fact that the seismicity $R(\mathbf{x}_i,t)$ in a real region such as southern California is an extremely noisy function [e.g., *Scholz*, 1990), and little information has been obtained about systematic behavior of the amplitude of this function in previous studies. However, over longer time intervals we can expect $S(\mathbf{x}_i,t_0,t)$ to be more stable (figure 2). In fact, the integral of $S(\mathbf{x}_i,t_0,t)$ over space corresponds to the "a" value in the Gutenberg-Richter relation, which approaches a constant as the area considered is increased. Thus it is usually practical to deal only with time-averaged quantities. We therefore define a time-averaged seismicity function $S(\mathbf{x}_i,t_0,t)$:

$$S(\mathbf{x}_i,t_0,t) = \frac{1}{t-t_0} \int_{t_0}^{t} dt' \, \Psi_{obs}(\mathbf{x}_i, t') \qquad (29)$$

$S(\mathbf{x}_i,t_0,t)$ thus represents the seismic activity throughout the southern California region over the time interval (t_0,t). $S(\mathbf{x}_i,t_0,t)$ is also presumably the real part of a complex-valued seismicity phase function $\Sigma(\mathbf{x}_i,t_0,t)$, the imaginary part of which is not defined from the data at hand. The function $S(\mathbf{x}_i,1932,1998)$ has the same spatial dependence as the temporal average of seismicity given in figure 2, but has zero spatial mean.

Now $S(\mathbf{x}_i,t_0,t)$ can be considered to be an N-dimensional vector $\mathbf{S}(t_0,t)$ that has an inner product and a well-defined L^2 norm, $\| \mathbf{S}(t_0,t) \|$ [e.g., *Bachman and Narici*, 1966]:

$$\| \mathbf{S}(t_0,t) \| = \{ \int d\mathbf{x} \, | \mathbf{S}(t_0,t) |^2 \}^{1/2} = \{ \int d\mathbf{x} \, \{ S(\mathbf{x}_i, t_0,t) \}^2 \}^{1/2} \qquad (30)$$

From these considerations, we can then define $\hat{\mathbf{S}}(t_0,t)$, which can be understood to be the unit vector in N-dimensional correlation space pointing in the direction of $S(\mathbf{x}_i,t_0,t)$, with tail fixed at the origin and head lying on the N dimensional unit sphere:

$$\hat{S}(t_0,t) = \frac{S(t_0,t)}{\|S(t_0,t)\|} \quad (31)$$

On the surface, equation (31) may seem to be only a minor alteration of (29), but this is not at all the case. As expressed by the activity function $\mathbf{S}(t_0,t)$, seismicity at $\mathbf{x}_i$ over the interval (t_0,t) depends only upon local conditions near $\mathbf{x}_i$. As expressed by $\hat{S}(t_0,t)$ however, seismicity now depends upon conditions everywhere in the system, because the norm $\| \mathbf{S}(t_0,t) \|$ is a non-local function. For a system in which the correlation length is small, the difference between (29) and (31) is unimportant. However, for a strongly correlated system, the difference may be profound.

In a phase dynamical system, the phase functions $\hat{S}(t_0,t)$ represent probability amplitudes, so we define the probability associated with the state $\hat{S}(t_0,t)$ to be:

$$P(\mathbf{x},t_0,t) = | S(\mathbf{x},t_0,t) |^2 = | \hat{S}(t_0,t) |^2 \quad (32)$$

In view of the definition (31) of $\hat{S}(t_0,t)$, we have the condition that the total probability mass at any time over all sites $\mathbf{x}$ is constant:

$$\int d\mathbf{x} \;\; P(\mathbf{x},t_0,t) = 1 \quad (33)$$

The physically meaningful changes in $\hat{S}(t_0,t)$ occur only via rotations about the origin. Over the time interval $(t_1, t_2 = t_1 + \Delta t)$, such a rotation $\Delta\hat{S}(t_1, t_2)$ can be denoted by:

$$\Delta\hat{S}(t_1, t_2) = \hat{S}(t_0,t_2) - \hat{S}(t_0,t_1) \quad (34)$$

Furthermore, it can be seen that, as $t_2 \rightarrow t_1$:

$$| \Delta\hat{S}(t_1,t_2)|^2 \propto (\Delta t)^2 \;\; | \boldsymbol{\Psi}_{obs}(\mathbf{x}_i, t_1) |^2 \quad (35)$$

In view of (7), we see that $|\frac{1}{2}\Delta\hat{S}|^2$ defines a change in probability $\Delta P(\mathbf{x},t_1,t_2)$ over the time interval (t_1,t_2). Moreover:

$$|\frac{1}{2} \Delta\hat{S}(t,t) |^2 = 0 \quad (36)$$

which implies that the change in total probability mass over all sites $\mathbf{x}$ is zero over any time interval $\Delta t = 0$. Because probability is conserved, we must have the condition:

$$\int d\mathbf{x} \;\; \Delta P(\mathbf{x_i}, t_1, t_2) = 0 \quad (37)$$

for all times (t_1,t_2). These conditions can all be satisfied if we take:

$$\Delta P(\mathbf{x_i}, t_1, t_2) = \left| \frac{1}{2}\Delta\hat{S}(t_1, t_2) \right|^2 - \mu_P(t_1, t_2) \quad (38)$$

where:

$$\mu_P(t_1, t_2) = \frac{1}{A}\int dx \; |\frac{1}{2}\Delta\hat{S}(t_1, t_2) |^2 \quad (39)$$

and A is the total area of all sites $\mathbf{x}$.

The physical meaning of these equations is that an increase in $\Delta P(\mathbf{x_i}, t_1, t_2)$ above the background level measured by μ_P should be interpreted as a tendency to form a spatially correlated region of seismic activity, and that such regions evidently must be present for larger earthquakes to occur. In work to be reported elsewhere [*Tiampo et al.*, 2000] we examine these methods using data from the southern California seismic network over the time interval 1932-present. We find that the probability function $\Delta P(\mathbf{x_i}, t_1, t_2)$ evidently can be used to detect the formation and existence of regions correlated in space and time that are subsequently the sites of major earthquakes.

11. FINAL COMMENTS

A variety of methods have been proposed for understanding the space-time patterns in driven nonlinear systems including Principal Component *Analysis [Penland and Magorian*, 1993], Principal Oscillation Pattern [*Preisendorfer,* 1988; *Penland,* 1989; *Garcia and Penland,* 1991] analysis, and Singular Spectrum Analysis [*Fukunaga*, 1990; *Broomhead and King*, 1986; *Vautard and Ghil*, 1989] However, in these systems, the variables that define the underlying dynamics are directly amenable to observation, in contrast to most driven threshold systems such as neural systems, earthquakes, disordered magnetic systems and so forth. In the latter, only the associated "derivative variables" such as the correlated-event activity (e.g., earthquakes, action potentials, depinning avalanches), can be directly observed. Thus the standard methods cannot be applied to threshold systems without serious qualifications. Other methods that are being examined elsewhere are

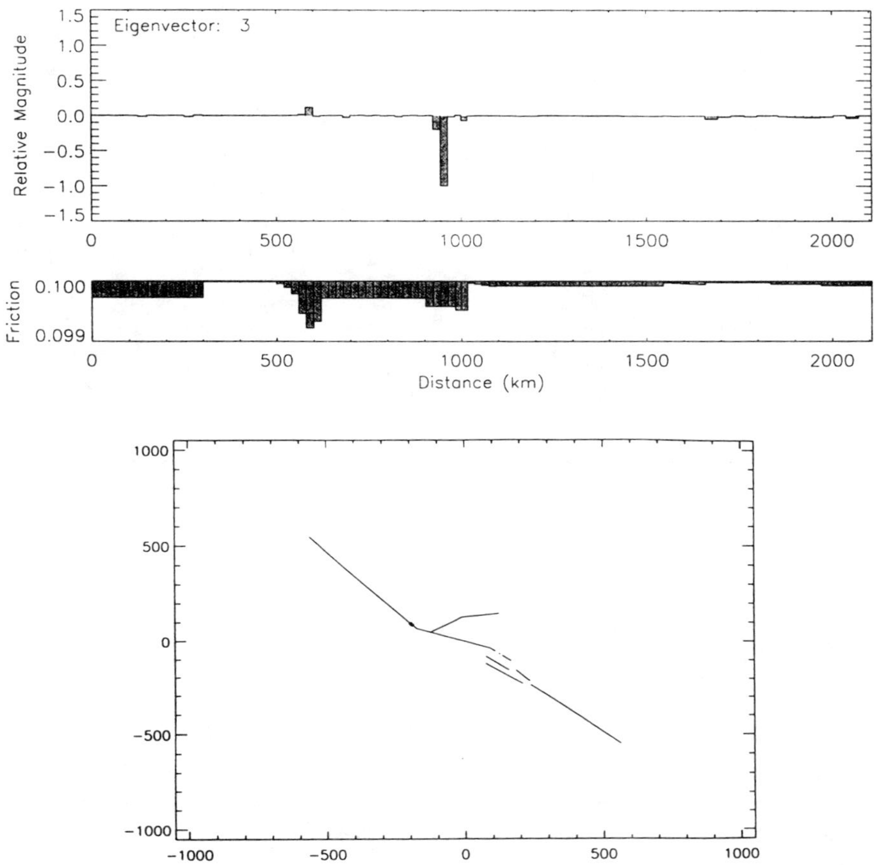

Figure 13. Eigenvector ϕ_3 for randomized model. As in figure 5 a,b. Period = 473 years.

described in references [*Eneva and Ben-Zion*, 1997a,b; *Pepke and Carlson*, 1994; *Pepke et al.*, 1994; *Kossobokov and Carlson*, 1995; *Minster and Williams*, 1992; *Molchan and Kagan*, 1992; *Sammis et al.*, 1996; *Ouchi,* 1993], but these have so far not yielded adequate results for threshold systems.

Understanding the patterns of activity in earthquake fault systems must proceed via the construction of representative simulations, and the development of analysis methods for the patterns that result. Once the analysis methods are in place, they should be applied to forecasting and understanding data from natural systems. In particular, it is often the case that the largest events are the most infrequent, but the most important to understand, since they usually act to slave the entire system. For earthquake models, these correspond to the largest and most destructive events. Detailed pattern analysis may allow the patterns of smaller, more frequent events leading up to the largest events to be detected and recognized. If this can be demonstrated, reliable forecasting of the largest events may be possible.

12. APPENDIX A: EARTHQUAKE FAULT SYSTEM MODEL

The dynamical model of the earthquake fault system was developed in Rundle [1988]. In contrast to many current models in the literature describing a single planar fault, the model in Rundle [1988] includes all of the major faults in southern California, albeit at a relatively crude scale (figure 3, bottom). In this Cellular Automaton-type model, each of N = 80 fault segments is driven at its own particular long term rate V of a few centimeters/year.

Segment i slips when the shear stress $\sigma_{sh,i}(t) = \mu_s \sigma_{norm,i}(t)$, where μ_s is a preassigned coefficient of "static" friction, and $\sigma_{norm,i}(t)$ is the normal stress on the segment. At failure, slip occurs that is sufficient to reduce $\sigma_{sh,i}(t)$ to the value $\sigma_{sh,i}(t) = \mu_d \sigma_{norm,i}(t)$, where μ_d is a coefficient of "dynamic" friction. The difference in frictional coefficients μ_s - μ_d is tuned so that realistic event sequences on the middle ("big bend") and southern parts of the model

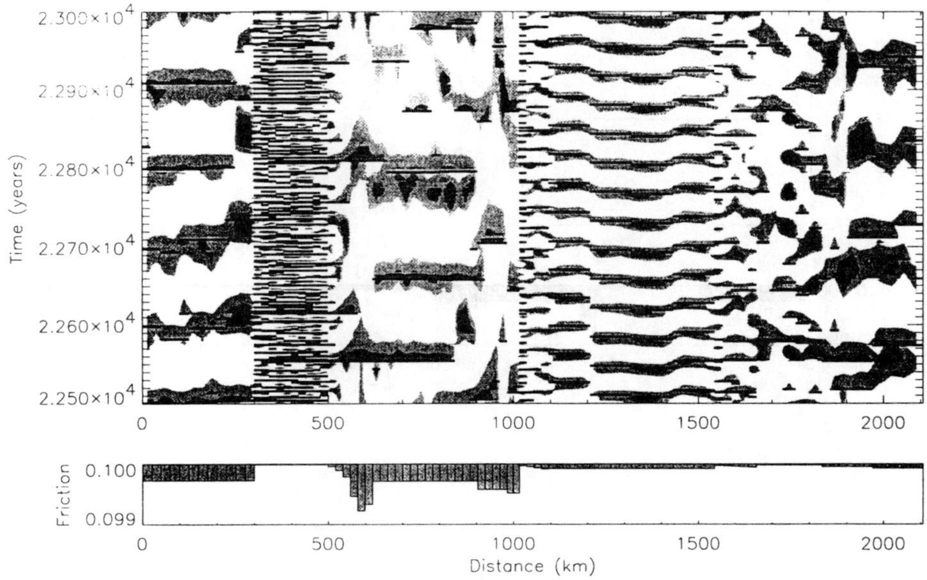

Figure 14. Superposition of events (solid horizontal lines) from year 22,500 to year 23,000 top with probabilities $P(\mathbf{x}_i, t)$ calculated according to (20) using "improvements" described in section 7. Light shaded regions are for $P(\mathbf{x}_i, t) > 1/3$, dark shaded regions are for $P(\mathbf{x}_i, t) > 1/2$.

San Andreas (SA) fault occur at realistic intervals of ~ 150 years. Some tuning was also carried out to produce "realistic" sequences on the Imperial (I), the San Jacinto (SJ), the Elsinore (E) and Garlock (G) faults, in contrast to frictional properties on the Cerro Prieto (CP) and Northern San Andreas fault segments, which were not tuned. Time steps of 1 year are used. Interactions between segments allow stress transfer so that one slipping segment can induce others to slip as well in the same event. Realistic directionally-dependent elastic and viscoelastic interactions are included by means of the stress Green's functions. In figure 3 (Bottom) we show a map view of the 80 segments of the model, each of which is a vertical rectangular slip surface. These segments extend to a depth of 18 km in an elastic plate of thickness 30 km. The plate overlies, and is in welded contact with, a Maxwell viscoelastic half space in which shear stress can relax via viscoelastic flow. The reader is referred to Rundle [1988] for other details.

13. APPENDIX B: EIGENVALUES OF $K(\mathbf{x}_i,\mathbf{x}_j)$

In this appendix we give an argument as to why the eigenvalues of the rate correlation matrix $K(\mathbf{x}_i,\mathbf{x}_j)$ should equal the square of the frequencies of the eigenmodes. The main assumptions we make for this illustration is that each fault segment that can fail has only one frequency, and that the time averaging period can be approximated by infinity.

Consider first the simple case where each fault segment has a different frequency ϖ_i in the Fourier domain. Referring to equation (11), we can represent each time series as:

$$z(\mathbf{x}_i,t) = \mathrm{Re}\ \{\ z(\mathbf{x}_i,0)\ e^{i\varpi_j t}\ \} \tag{B1}$$

In this case, with normalization, the static correlation matrix is the identity matrix. The rate correlation matrix is also diagonal with elements of ϖ_i^2. Consequently in this simple case the eigenvectors of the static correlation matrix and the rate correlation matrix are the same, and the eigenvalues of the rate matrix are simply the frequencies.

In the slightly more complicated case where the frequencies can be the same the matrices are no longer diagonal. However, it is simple to see that any two rows have the following property; either they are identical or they are disjoint. By disjoint we mean that the non-zero elements in one row, say row one, are zero in any other row that is not identical to row one. Since the numbering of the fault segments is arbitrary we can renumber so that all of the rows that are identical are grouped together. This will put the matrices in Jordan-normal form. That is the matrices, both static and rate correlation are now made up of blocks of non-zero elements arranged along the diagonal. Each block has the same non-zero element which is simply the frequency squared of some element in the rate correlation

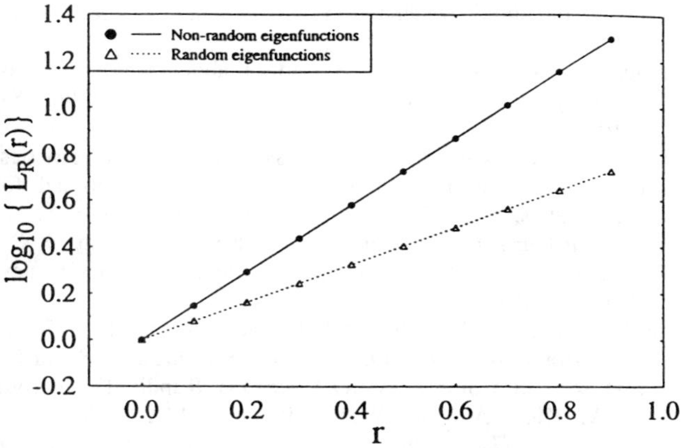

Figure 15. Plot of log-likelihood ratio $\log_{10}\{ L_R(r) \}$ against mixing ratio r corresponding to event forecasts in figures 8 (solid, circles) and 12 (dashed, triangles).

matrix and one in the static matrix. Clearly, the eigenvalues are the rate matrix are the square of the frequencies times the eigenvalues of the static matrix.

In the case of real data, it is certainly true that the time averaging period is not infinite and the number of frequencies for an individual fault element is not one. However, there is evidence that we discussed in the body of the paper that supports the point of view that the assumptions we made are good approximations. It is clear that there will be modifications that will need to be made to incorporate additional modes and finite averaging times if we wish to forecast over long times and that the less periodic the individual fault elements are, the more important the corrections will be.

Acknowledgements. The authors would like to acknowledge helpful conversations with C. Penland and J. Hopfield. Research by JBR has been supported by NASA grant NAG5-5168 (simulations), and by USDOE grant DE-FG03-95ER14499 (theory), to Univ. of Colorado. Research by WK has been supported by USDOE grant DE-FG02-95ER14498 to Boston University. Research by KT has been supported by NGT5-30025 (NASA).

REFERENCES

Abeles, M. , *Corticonics, Neural Circuts of the Cerebral Cortex*, Cambridge University Press, Cambridge, UK, 1991.

Bachman, G. and L. Narici, *Functional Analysis*, Academic Press, New York, 1966.

Bakun,W. and T.V. McEvilly, Recurrence models and Parkfield, California, earthquakes, *J. Geophys. Res.*, *89*, 3051-3058, 1984.

Ball, P., *The Self-Made Tapestry, Pattern Formation in Nature*, Oxford University Press, Oxford, UK, 1999.

Broomhead, D.S. and G.P. King, Extracting qualitative dynamics from experimental data, *Physica D*, *20*, 217-236, 1986.

Constantine-Paton, M., H.T. Cline and E. Debski, Patterned activity, synaptic convergence, and the NMDA receptor in developing visual pathways, *Ann. Rev. Neurosci.*, *13*, 129-154, 1990.

Crutchfield, J.P. and J.E. Hanson, Turbulent pattern bases for cellular automata, *Physica D.*, *69*, 279-301, 1993.

Egolf, D.A., Equilibrium regained: From nonequilibrium chaos to statistical mechanics, *Science*, *287*, 101-104, 2000

Eneva, M. and Y. Ben-Zion, Application of pattern recognition techniques to earthquake catalogs generated by model of segmented fault systems in three-dimensional elastic solids, *J. Geophys. Res.*, *102*, 24513-24528, 1997a.

Eneva, M. and Y. Ben-Zion, Techniques and parameters to analyze seismicity patterns associated with large earthquakes, *J. Geophys. Res.*, *102*, 17785-17795, 1997b.

Ferguson, C.D., W. Klein and J.B. Rundle, Spinodals, scaling and erogodicity in a model of an earthquake fault with long-range stress transfer, *Phys. Rev E.*, *60*, 1359-1373, 1999.

Fisher, D.S., K. Dahmen, S. Ramanathan and Y. Ben-Zion, Statistics of earthquakes in simple models of heterogeneous faults, *Phys. Rev. Lett.*, *78*, 4885-4888, 1997.

Fukunaga, K., *Introduction to Statistical Pattern Recognition*, Academic Press, NY, 1990. (See especially chapter 9)

Garcia, A. and C. Penland, Fluctuating hydrodynamics and principal oscillation pattern analysis, *J. Stat. Phys.*, *64*, 1121-1132, 1991.

Gardiner, C.W. , *Handbook of Stochastic Methods*, Springer-Verlag, Berlin, 1985.

Gray, C.M., p. 93, in H.F. Nijhout, L. Nadel and D.L. Stein, eds., *Pattern Formation in the Physical and Biological Sciences*, Lecture Notes V, Santa Fe Inst., Addison Wesley, Reading, MA, 1997.

Gross, S.J. and C. Kisslinger, Tests of models of aftershock rate decay, *Bull. Seism. Soc. Am.*, *84*, 1571-1579, 1994.

Gross, S. and J.B. Rundle, A systematic test of time-to-failure analysis, *Geophys. J. Int.*, *133*, 57-64, 1998.

Haken, H., *Synergetics, An Introduction*, Springer-Verlag, Berlin, 1983.

Hill, D., J.P. Eaton and L.M. Jones, in R.E. Wallace, ed., *The San Andreas Fault System, California*, US Geol. Surv. Prof. Paper 1515, 1990.

Holmes, P.J., J.L. Lumley and G. Berkooz, *Turbulence, Coherent Structures, Dynamical Systems and Symmetry*, Cambridge University Press, Cambridge, UK, 1996.

Huang, K., *Statistical Mechanics*, John Wiley, New York, 1987.

Kagan, Y.Y. and D.D. Jackson, Long term earthquake clustering, *Geophys. J. Int.*, *104*, 117-133, 1991.

Keilis-Borok, V.I., L. Knopoff, I.M. Rotwain and C.R. Allen, Inter-mediate-term prediction of occurrence times of strong earthquakes, *Nature*, *371*, 1988.

Klein, W., J.B. Rundle and C. Ferguson, Scaling and nucleation in models of earthquake faults, *Phys. Rev. Lett.*, *78*, 3793-3796, 1997.

Kossobokov, V. and J.M. Carlson, Active zone size versus activity: a study of different seismicity patterns in the context of the pre- diction algorithm M8, *J. Geophys. Res.*, *100*, 6431-6441, 1995.

Minster, J.-B. and N. Williams, M8 intermediate-term earthquake prediction algorithm; performance update for M>7.5, 1985-1996, *Trans. Am. Geophys. Un. EOS*, *73(43)*, Fall Mtg. Suppl., 456-456, 1996.

Molchan, G.M. and Y.Y. Kagan, Earthquake prediction and its optimization, *J. Geophys. Res.*, *97*, 4823-4838, 1992.

Mori, H. and Y. Kuramoto, *Dissipative Structures and Chaos*, Springer-Verlag, Berlin, 1998.

Nijhout, H.F., L. Nadel and D.L. Stein, *Pattern Formation in the Physical and Biological Sciences*, Lecture Notes V, Santa Fe Inst., Addison Wesley, Reading, MA, 1997.

Ouchi, T., Population dynamics of earthquakes and mathematical modeling, *Pure Appl. Geoph.*, *140*, 15-28, 1993.

Penland, C., Random Forcing and Forecasting Using Principal Oscillation Pattern Analysis, *Monthly Weather Rev.*, *117*, 2165-2185, 1989.

Penland, C. and P.D. Sardeshmukh, The Optimal Growth of Tropical Sea Surface Temperature Anomalies, *J. Climate*, *8*, 1999-2024, 1995.

Penland, C. and T. Magorian, Prediction of Nino 3 Sea Surface Temperatures Using Linear Inverse Modeling, *J. Climate*, *6*, 1067-1075, 1993.

Pepke, S. L., J.M. Carlson and B.E. Shaw, Prediction of large events on a dynamical model of a fault, *J. Geophys. Res.*, *99*, 6769-6788, 1994.

Pepke, S.L. and J.M. Carlson, Predictability of self-organizing systems, *Phys. Rev. E.* , *50*, 236-242, 1994.

Preisendorfer, R.W., Principal Component Analysis in Meteorology and Oceanography, ed. C.D. Mobley, *Develop. Atm. Sci. 17*, Elsevier, Amsterdam, 1988.

Press, F and C.R. Allen, Patterns of seismic release in the southern California region, *J. Geophys. Res.*, *100*, 6421-6430, 1995.

Rundle, J.B., A physical model for earthquakes; 2, Application to Southern California, *Jour. Geophys. Res.*, *93*, 6255-6274, 1988.

Rundle, J.B., W. Klein, S. Gross, and D.L. Turcotte, Boltzmann fluctuations in numerical simulations of nonequilibrium lattice threshold systems, *Phys. Rev. Lett.*, *75*, 1658-1661, 1995.

Rundle, J.B. and W. Klein, New ideas about the physics of earthquakes, *Rev. Geophys. Space Phys. Suppl.*, July, *283*, 1995.

Rundle, J.B., W. Klein and S.J. Gross, Rupture characteristics, recurrence, and predictability in a slider-block model for earthquakes, pp. 167-203 in *Reduction and Predictability of Natural Hazards*, Santa Fe. Inst. ser. Sci. Complexity, XXV, ed. J.B. Rundle, D. Turcotte and W. Klein, Addision-Wesley, Reading, MA., 1996.

Sammis, C.G., D. Sornette and H. Saleur, pp. 143-156 in *Reduction and Predictability of Natural Hazards*, Santa Fe. Inst. ser. Sci. Complexity, XXV, ed. J.B. Rundle, D. Turcotte and W. Klein, Addision-Wesley, Reading, MA., 1996.

Scholz, C. H., *The Mechanics of Earthquakes and Faulting*, Cambridge, UK, 1990.

Shatz, C.J., p. 299, in H.F. Nijhout, L. Nadel and D.L. Stein, eds., *Pattern Formation in the Physical and Biological Sciences*, Lecture Notes V, Santa Fe Inst., Addison Wesley, Reading, MA, 1997.

Sieh, K., M. Stuiver and D. Brillinger, A more precise chronology of earthquakes produced by the San Andreas Fault in Southern California, *J. Geophys. Res.*, *94*, 603-623, 1989.

Sornette, D. and L. Knopoff, The paradox of the expected time until the next earthquake, *Bull. Seism. Soc. Am.*, *87*, 789-798, 1997.

Tiampo, K., J.B. Rundle, S. McGinnis, W. Klein, and S.J. Gross, Observation of systematic variations in non-local space-time seismicity patterns in southern California prior to major earthquakes, this volume, 2000.

Vautard, R. and M. Ghil, Singular spectrum analysis in nonlinear dynamics, with applications to paleoclimatic time series, *Physica D*, *35*, 395-424, 1989.

J.B. Rundle, K. Tiampo and S. Gross, Colorado Center for Chaos and Complexity, Cooperative Institute for Research in Environmental Sciences, University of Colorado at Boulder, Campus box 216, Boulder, Co 80309-0216.

W. Klein, Department of Physics and Computational Science, Boston University, Boston, MA 80309

Microscopic and Macroscopic Physics of Earthquakes

Hiroo Kanamori and Thomas H. Heaton

Seismological Laboratory, California Institute of Technology, Pasadena California 91125

Frictional melting and fluid pressurization can play a key role in rupture dynamics of large earthquakes. For faulting under frictional stress σ_f, the temperature increases with σ_f and the earthquake magnitude, M_W. If the thickness of the heated zone, w, is of the order of a few mm, then, even for a modest σ_f, the temperature rise, ΔT, would exceed 1000° for earthquakes with M_W=5 to 6, and melting is likely to occur, and reduce friction during faulting. If fluid exists in a fault zone, a modest ΔT of 100 to 200° would likely increase the pore pressure enough to significantly reduce friction for earthquakes with M_W=3 to 4. The microscopic state of stress can be tied to macroscopic seismic parameters such as the seismic moment, M_0, and the radiated energy, E_R, by averaging the stresses in the microscopic states. Since the thermal process is important only for large earthquakes, the dynamics of small and large earthquakes can be very different. This difference is reflected in the observed relation between the scaled energy $\tilde{e}=E_R/M_0$ and M_W. The observed $\tilde{e}$ for large earthquakes is 10 to 100 times larger than for small earthquakes. Mature fault zones such as the San Andreas are at relatively moderate stress levels, but the stress in the plate interior can be high. Once slip exceeds a threshold, runaway rupture could occur, and could explain the anomalous magnitude-frequency relationship observed for some mature faults. The thermally controlled slip mechanism would produce a non-linear behavior, and under certain circumstances, the slip behavior at the same location may vary from event to event. Also, slip velocity during a large earthquake could be faster than what one would extrapolate from smaller earthquakes.

INTRODUCTION

Modern broad-band seismic data have allowed seismologists to determine important seismic source parameters such as seismic moment, M_0, radiated energy, E_R, rupture parameters, and stress drops of earthquakes over a large magnitude range. However, at short length scales, resolution of seismic methods is limited because of the complex propagation and wave attenuation effects near the Earth's surface, and it is difficult to determine the details of rupture process below some length scale. The complex wave forms at high frequency must be controlled by microscopic processes on a fault plane. Such microscopic processes include frictional melting [*Jeffreys*, 1942; *McKenzie and Brune*, 1972; *Richards*, 1977; *Sibson*, 1977; *Cardwell et al.*, 1978], fluid pressurization [*Sibson*, 1973; *Lachenbruch*, 1980; *Mase and Smith*, 1985, 1987], acoustic fluidization [*Melosh*, 1979, 1996], dynamic unloading effects [*Schallamach*, 1971; *Brune et al.*, 1993; *Weertman*, 1980; *Ben-Zion and Andrews,* 1998; *Mora and Place*, 1998, 1999] and geometrical effects [*Scott*, 1996].

GeoComplexity and the Physics of Earthquakes
Geophysical Monograph 120

The importance of thermal processes in earthquake mechanics has long been recognized. *Sibson* [1977] discussed the implication of frictional heating for fault dynamics. He suggested that melt formation and transient increases in fluid pressure caused by frictional heating may decrease the friction to near-zero values once slip is initiated. Here, we extend the model discussed by Sibson in light of recent seismological data. A recent study of the deep Bolivian earthquake (M=8.3, depth=637 km) [*Kanamori et al.*, 1998] presented an interesting observational case which suggests a dominant role of thermal processes during faulting. For this earthquake, the released potential energy, 1.4x10^{18} J, is at least 30 times larger than the radiated energy, with a large amount of non-radiated energy (comparable to the total thermal energy released during the 1980 Mount St. Helens eruption) deposited in a relatively small fault zone over a time scale of less than a minute.

The thermal process during faulting would cause a complex sequence of events including local melting, freezing, fluid pressurization, micro-fracturing and injection of fluids. Although these microscopic processes are important for understanding rupture dynamics, it is difficult to determine how these processes work in detail during faulting because of the limited resolution of seismic methods.

In this paper, we investigate the effects of frictional melting and fluid pressurization and relate them to macroscopic seismic source parameters such as M_0 and E_R. This approach is somewhat similar to that of statistical mechanics in which the physics applied to small-scale processes is used to determine the average macroscopic parameters such as pressure and temperature.

THERMAL BUDGET DURING FAULTING

The possibility of frictional melting during faulting has been suggested by several investigators. In particular, *McKenzie and Brune* [1972] quantitatively investigated this problem as a one-dimensional heat conduction problem. They assumed that the fault surface is simultaneously heated during slippage (*i.e.* infinite rupture speed) over a finite time, and concluded that if both the frictional and driving stresses are of the order of 1 kbar, melting can occur for fault slips as small as one millimeter. *Richards* [1977] solved elasto-dynamic equations for a propagating elliptical crack, estimated frictional heating rate behind the rupture front, and showed that if the driving stress is 100 bars and the fault particle velocity is 10 cm/sec at nucleation, a temperature rise of about 1000° can occur within a few seconds. These studies indicate that frictional melting is likely to occur during seismic faulting, at least locally.

Here we consider a gross thermal budget during faulting under a frictional stress σ_f. Let S and D be the fault area

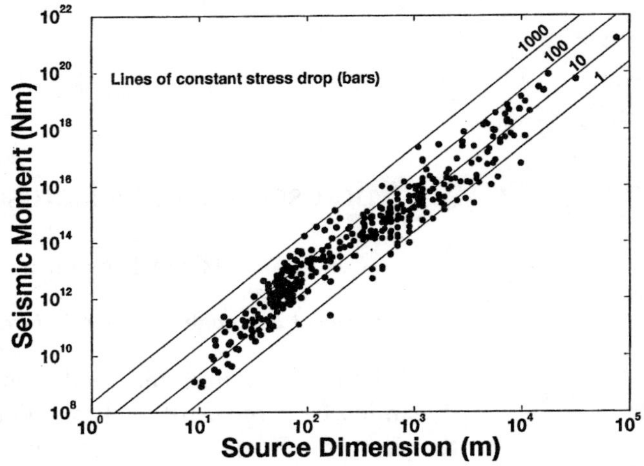

Figure 1. Static stress drop of earthquakes. Modified from [*Abercrombie and Leary*, 1993].

and the displacement offset respectively. Then the total heat generated during faulting is $Q=\sigma_f DS$. If we assume that the heat is distributed during seismic faulting within a layer of thickness w around the rupture plane, the average temperature rise ΔT is given by

$$\Delta T= Q/C\rho Sw= \sigma_f D/C\rho w \qquad (1)$$

where C is the specific heat, and ρ is the density. In general D increases with the earthquake magnitude, M_W. Here we use a simple circular model in which the static stress drop is $\Delta\sigma_s$ [*Eshelby*, 1957]. Then,

$$D = (16/7)^{2/3}(1/\pi)M_0^{1/3}\Delta\sigma_s^{2/3}/\mu \qquad (2)$$

where M_0 is the seismic moment and μ is the rigidity. From (1) and (2), we obtain

$$\Delta T = (16/7)^{2/3}(1/\pi)\sigma_f\Delta\sigma_s^{2/3}M_0^{1/3}/\mu C\rho w \qquad (3)$$

The seismic moment M_0 is related to M_W by

$$\log M_0=1.5M_W+9.1 \quad (M_0 \text{ in Nm}) \qquad (4)$$

The static stress drop, $\Delta\sigma_s$, for most earthquakes is in the range of 10 to 100 bars, as shown in Figure 1 [*Kanamori and Anderson*, 1975, *Hanks*, 1977, *Abercrombie and Leary*, 1993]. However, higher stress drops have been reported for some earthquakes for which the source dimension was determined well [*e.g. Kanamori et al.*, 1990; *Wald*, 1992]. Also, there is evidence that the stress drop can be locally very high (up to 25 kbar) around small asperities [*Nadeau and Johnson*, 1998]. Since the

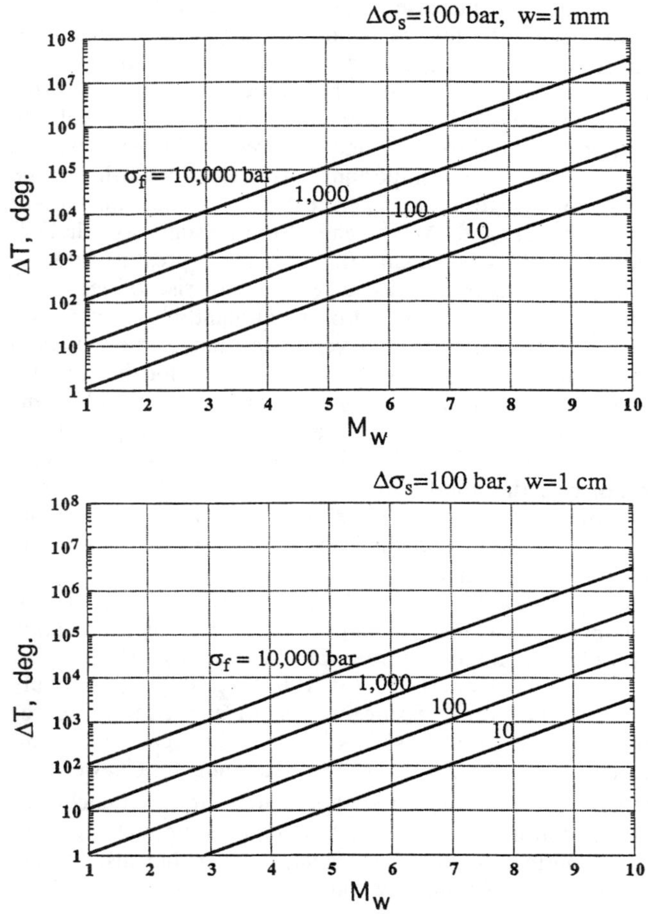

Figure 2. Temperature rise, ΔT, in a fault zone as a function of magnitude, M_W, with the frictional stress, σ_f, as a parameter. The static stress drop $\Delta\sigma_s$ is assumed to be 100 bars. The upper and lower figures correspond to the cases of w (thickness of the heated zone)=1 mm and 1 cm, respectively.

thermal process considered here is most important at high-stress spots, we use $\Delta\sigma_s$=100 bars and μ=0.3 Mbar for estimation of ΔT.

The thickness w cannot be determined with seismological methods. Here, w refers to the thickness of the coseismic slip zone, not the width of the shear zone along a fault. Fault surface breaks in bed rocks are often extremely sharp suggesting that the coseismic slip must have occurred in a very narrow zone. An example of a thin slip surface within a fault zone is described in *Hubbert and Rubey* [1959]. Laboratory studies by *Goldsby and Tullis* [1998, 1999] found that when the displacement is large, slip tends to be localized in a thin layer. Also, some pseudotachylytes layers are very thin, less than a millimeter [*Otsuki*, 1998], suggesting that the slip zone is thin at least locally.

It is true that wide shear zones are often found along a fault. *Otsuki* [1978] and *Hull* [1988] have shown that the width of the shear zone is about 1 % of the total offset of the fault. Also, recent seismological studies suggest a 40 to 200 m thick zone with a low seismic velocity along several major faults [*Li et al.* 1994, 1997]. However, these wide fault-zone structures are probably formed by long-term processes involving deformation, chemical alteration, and aftershocks, and are not directly involved in coseismic slip.

If the slip zone is thin and heat is carried by thermal diffusion, then w would be of the order of $\sqrt{k\tau_0}$, the thermal diffusion distance, where k is the thermal diffusivity, and τ_0 is the time scale of faulting. Since τ_0 is less than 10 sec for most seismic events, w would be at most a few mm. A simple scaling shows that D is proportional to the time scale of faulting, τ_0, while w is proportional to $\tau_0^{1/2}$; then, we obtain from (1), $\Delta T \propto D^{1/2}$. This means that the thermal process would become increasingly important for large earthquakes.

Figure 2 shows ΔT calculated from (3) as a function of magnitude M_W for two representative values of w, 1 mm and 1 cm. We used C=1 J/g°C, and ρ=2.6 g/cm^3. If w=1 mm, ΔT exceeds 1000 °C at M_W=5 even for a modest value of friction, σ_f =100 bars. Even for w=1 cm, ΔT exceeds 1000 °C at M_W=7 for the same value of friction. If σ_f>100 bars, ΔT exceeds 1000 °C at a lower M_W. Thus, thermal process becomes important for large earthquakes.

Depending on whether fluid exists or not in a fault zone, two distinct thermal processes can happen. If there is no fluid in a fault zone, the temperature can rise to cause frictional melting. Figure 2 shows that if no fluid exists, frictional melting is likely to occur for earthquakes with M_W =5 to 7. This general conclusion appears unavoidable even if the values of $\Delta\sigma_s$, σ_f, and w used in (3) are varied over fairly large, but plausible, ranges.

Many investigators have found pseudotachylytes in cataclasites and presented them as evidence for frictional melting. Although pseudotachylytes are not commonly found [*Sibson*, 1975], the following are well-document examples. *Lin* [1994a, 1994b] reported on glassy pseudotachylytes from the Fuyun fault zone, China, which he believed to have been formed during seismic faulting at a minimum temperature of 1,450 °C. *Obata and Karato* [1995] examined ultramafic pseudotachylytes from the Ivrea-Verbano zone, Italy, which are about 1 cm thick, and exhibit evidence for melting and cooling on a time scale of about 100 sec or less under a differential stress of 3 kbar. *Otsuki* [1998] examined cataclasites from the Nojima fault, Japan, on which the 1995 Kobe earthquake occurred. According to *Otsuki* [1998], these cataclasites were not formed during the recent Kobe earthquake, but they were formed from granites at a depth of about 3 km, and the original texture is exceptionally well preserved. They have a sandwich structure with alternating layers of pseudotachylytes (amorphous phase) and fine-grained rock. The pseudotachylyte layer, about 0.2 to 1 mm thick, exhibits evidence for melting at temperatures above 1,100 °C followed by rapid cooling.

Some pseudotachylytes are believed to be formed by crushing during faulting rather than melting [*Lin et al.*, 1994; *Lin,* 1996]. It is also possible that pseudotachylytes were formed during faulting but they have been altered to clay minerals or crystallized into some kind of mylonites. We do not necessarily think that seismic faulting always causes melting. If a fault zone is highly crushed, the thermal energy can be distributed over a large volume and no melting occurs. Also, if fluid exists in a fault zone or some dynamic process such as acoustic fluidization or dynamic unloading occurs, the friction may drop before melting occurs. However, the existence of pseudotachylytes indicates that melting is an important process, at least locally, during faulting.

Melting does not necessarily mean reduction of friction. Once a thin melt layer is formed, high viscous friction may prevail depending on the thickness of the layer and the viscosity of the melt [*Scholz*, 1980]. In fact, *Tsutsumi and Shimamoto* [1997] performed high-velocity friction experiments and found a sharp increase in friction at the initiation of visible frictional melting. However, as shown by *Spray* [1993], the viscosity of molten pseudotachylytes is low and drops rapidly with the temperature so that friction is likely to drop eventually, though the details can be complex. Recent laboratory studies by *Beeler et al.* [1996] and *Goldsby and Tullis* [1998, 1999] demonstrated that friction dropped significantly when displacement was large. *Goldsby and Tullis* [1998, 1999] (details described in *Tullis and Goldsby* [1998]) found that, at a normal pressure of 1.12 kbar, the coefficient of friction dropped to 0.14 when a large displacement, 1.6 m, occurred at a relatively high slip velocity of 3.2 mm/sec. These experiments were performed under confining pressure and the condition is close to that of natural earthquakes. From the observations of the sliding surface, they suggested that melting may have occurred during sliding.

Whether fluid exists in fault zones or not is still a materr of debate, but it is generally believed that some fault zones contain fluids and many mechanisms have been proposed to maintain high fluid pressure in fault zones [*e.g. Irwine and Barnes*, 1975; *Byerlee*, 1990; *Rice*, 1992]. If fluid exists in a fault zone, fluid pressurization could occur. This concept was introduced to seismology by *Sibson* [1973], and analyzed in great detail by *Lachenbruch* [1980], and *Mase and Smith* [1985, 1987]. Under the pressure-temperature conditions at the seismogenic depths, the thermal expansivity of water is of the order of 10^{-3} °C, and significant increase in pore pressure with temperature could occur. If fluid does not escape (small permeability) and the surrounding rock is not compressive, the pressure increase would be of the order of 10 bars/deg [*Lachenbruch*, 1980]. In actual fault zones, permeability and compressibility vary and the pressure increase may be less. The most important parameter controlling the pressure change is the permeability. The analysis of Lachenbruch and Mase and Smith suggests that if permeability is less than 10^{-18} m^2, fluid pressurization is most likely to occur with a temperature rise of less than 200 °C, and friction will drop significantly. Permeability in the crust varies over a very wide range, more than a factor of 10^{10}. Figure 3 shows the results for the samples taken from the Cajon drilling site in California [*Morrow and Byerlee*, 1992], and the Nojima fault, Japan [*Ito et al.*, 1998]. *Ito et al.* [1998] show that permeability is very small near the middle of the shear zone, where the grain size of rocks is small. Ito et al.'s results are at a pressure of 500 bar (corresponding to a depth of 1.5 km), and suggest even smaller values in the deeper seismogenic zone. Although the distribution of permeability can be complex, these results suggest that pressure fluidization can play an important role, at least locally, in reducing friction. A modest ΔT of 100 to 200° would likely increase the pore pressure enough to significantly reduce friction. Figure 2 shows that this can occur for earthquakes with M_W=3 to 5. According to *Chester and Chester* [1998], the internal structure of the Punchbowl fault, California, implies that earthquake ruptures were not only confined to the ultracataclasite layer, but also largely localized to a thin prominent fracture surface. They suggest that mechanisms that are consistent with extreme localization of slip, such as thermal pressurization of pore fluids, are most compatible with their observations.

Since a fault zone is probably complex and heterogeneous in stress, fluid content, permeability, porosity, and compressibility, no single process is likely to dominate. In other words, we do not necessarily expect a single continuous layer of melting and pressurization; we envision, instead, a fault zone that consists of many microfaults (subfaults) where different mechanisms are responsible for slip at different stress levels, producing complex rupture patterns as observed.

EARTHQUAKE ENERGY BUDGET

We consider the energy budget for each subfault. The energy budget of earthquakes has been extensively studied by many investigators [*e.g. Knopoff*, 1958; *Dahlen*, 1977; *Kostrov*, 1974; *Savage and Walsh*; 1978]. Following these studies, and referring to *Orowan* [1960] and *Savage and Wood* [1971], here we consider a simple stress-release model. The simplest case is shown in Figure 4a which

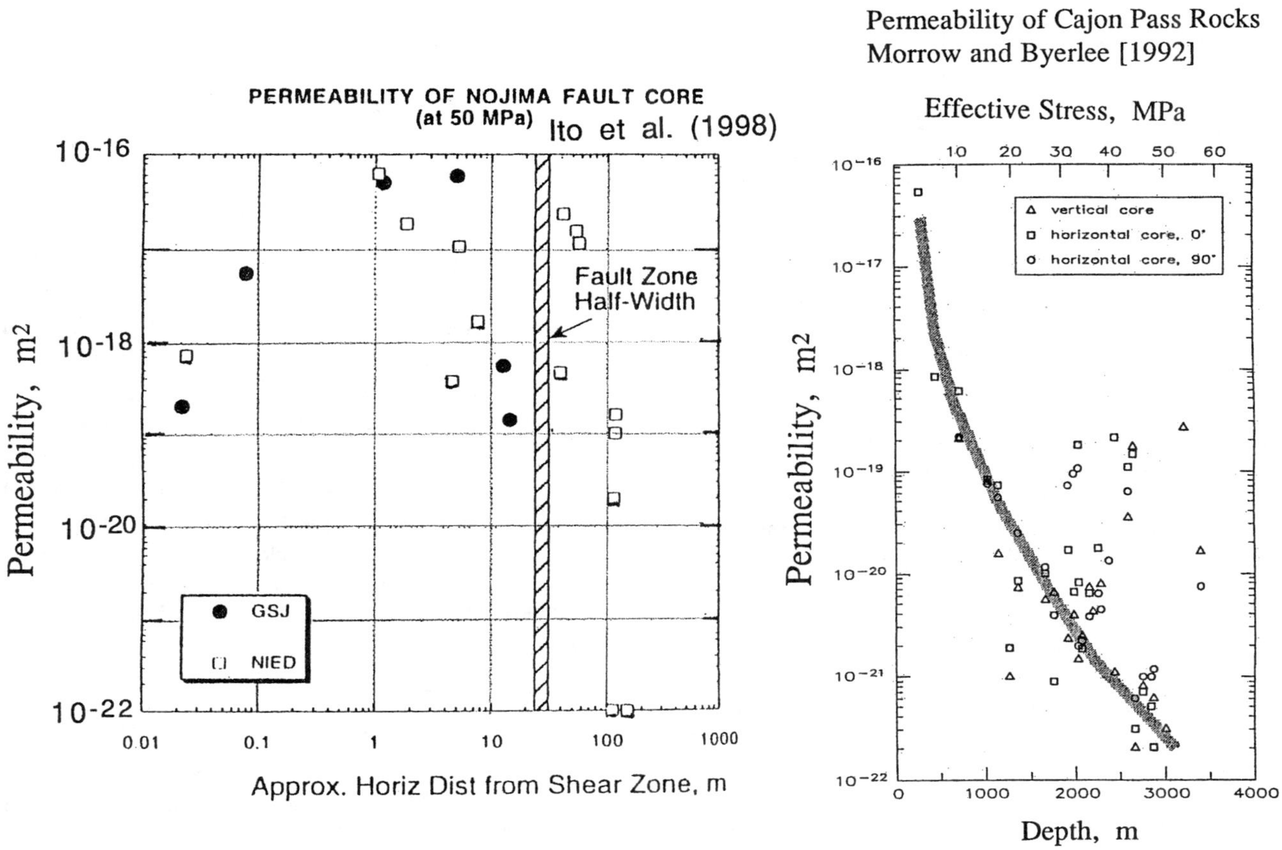

Figure 3. Permeability near a fault zone. a) Nojima fault, Japan [*Ito et al.* 1998]. b) Cajon Pass, California [*Morrow and Byerlee*, 1992].

shows the stress on the fault plane as a function of slip. An earthquake is viewed as a stress release process on a surface S where, at the initiation of an earthquake, the initial (before an earthquake) shear stress on the fault plane σ_0 drops to a constant dynamic friction σ_f. If the condition for instability is satisfied [*Brace and Byerlee*, 1966; *Scholz*, 1990], rapid fault slip motion begins and eventually stops. At the end, the stress on the fault plane is σ_1 (final stress) and the average slip (offset) is D. For the example shown in Figure 4a, $\sigma_f=\sigma_1$. The difference $\Delta\sigma_s=\sigma_0-\sigma_1$ is the static stress drop, and the difference $\Delta\sigma_d=\sigma_0-\sigma_f$ is the driving stress of fault motion and is usually called the dynamic stress drop or effective tectonic stress [*Brune*, 1970]. During this process, the potential energy (strain energy plus gravitational energy) of the system, W, drops to $W-\Delta W$ where ΔW is the strain energy drop, and seismic wave is radiated carrying energy E_R. Then the energy budget can be written as

$$\Delta W = E_R + E_F + E_G \tag{5}$$

where E_F is the frictional energy loss given by $E_F=\sigma_f DS$, and E_G is the fracture energy. *Knopoff* [1958], *Dahlen* [1977] and *Kostrov* [1974] showed that $\Delta W = \bar{\sigma} DS$ where $\bar{\sigma}=(\sigma_0+\sigma_1)/2$ is the average stress during faulting. From (5), we obtain

$$E_R=(\sigma_0+\sigma_1)DS/2-\sigma_f DS-E_G=(1/2)(2\Delta\sigma_d-\Delta\sigma_s)DS-E_G$$

$$=M_0(2\Delta\sigma_d-\Delta\sigma_s)/2\mu-E_G \tag{6}$$

where $M_0=\mu DS$ is the seismic moment, and μ is the rigidity. This is a simple but fundamental relationship which does not involve major assumptions. As we will show later, the fracture energy E_G can be ignored for large shallow earthquakes, and (6) can be written as

$$E_R=M_0(2\Delta\sigma_d-\Delta\sigma_s)/2\mu \tag{6'}$$

This relation can be derived with a simple analogous spring system, and can be shown to be consistent with that

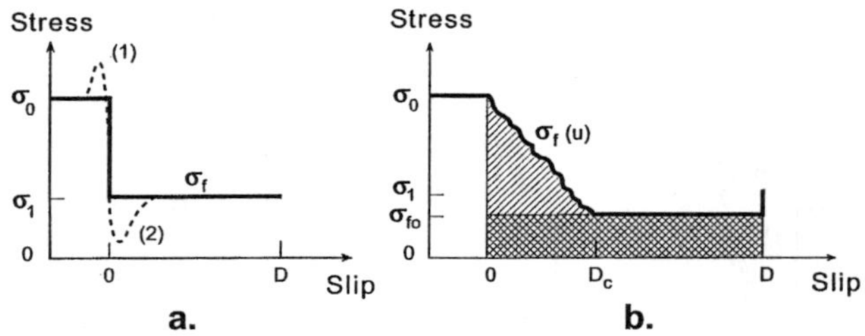

Figure 4. Illustration of simple stress release patterns during faulting. a) Simple case. b) Slip-weakening model. Hatched and cross-hatched areas indicate the fracture energy and frictional energy loss, respectively.

derived from more rigorous relations for continuum. A similar relation has been used in seismology [*e.g. Savage and Wood*, 1971], but this particular form introduced here is useful because E_R is expressed in terms of the specific physical parameters $\Delta\sigma_s$ and $\Delta\sigma_d$ which directly characterize the stress release process on the fault plane.

The variation of stress during faulting can be more complex than shown in Figure 4a. For example, the stress may increase in the beginning of the slip motion (curve (1) in Figure 4a) because of loading caused by advancing rupture, or of a specific friction law such as the state-rate dependent friction law [*Dieterich*, 1979a, 1979b]. In fact, seismological inversion studies have shown this increase [*Quin*, 1990; *Miyatake*, 1992; *Mikumo and Miyatake*, 1993; *Beroza and Mikumo*, 1996; *Ide*, 1997; *Bouchon*, 1997]. However, this increase is of short duration and the amount of slip during this stage is small so that little energy is radiated. Thus, we will not include it in our energy budget.

Also, the friction may not be constant during faulting. It may drop drastically in the beginning and later resume a somewhat larger value (curve (2) in Figure 4a), or it may decrease gradually to a constant level (Figure 4b). The latter is called a slip-weakening process. These models have been considered in *Brune* [1970], *Heaton* [1990], *Kikuchi and Fukao* [1988], *Kikuchi* [1992], *Kanamori* [1994], *Winslow and Ruff* [1999], and *Thio* [1996].

If the friction is not constant, the rupture dynamics is complicated, but for the energy budget considered here, we formulate this problem referring to a simple case shown in Figure 4b. The friction σ_f gradually drops to a constant value σ_{f0} until the slip becomes D_c. In general the final stress σ_1 can be different from σ_{f0}. Then, we define the average friction $\bar{\sigma}_f$ by

$$\bar{\sigma}_f = \frac{1}{D}\int_0^D \sigma_f(u)du \qquad (7)$$

where u is the slip (offset) on the fault plane. Then, equation (6') can be written as

$$E_R = M_0\left(2\Delta\bar{\sigma}_d - \Delta\sigma_s\right)/2\mu \qquad (8)$$

where

$$\Delta\bar{\sigma}_d = \sigma_0 - \bar{\sigma}_f \qquad (9)$$

Here, $\Delta\bar{\sigma}_d$ defined by (9) can be called the average dynamic stress drop. If friction drops rapidly, $\Delta\bar{\sigma}_d$ is the same as $\Delta\sigma_d$, but if friction drops very gradually to σ_1, then fault motion becomes quasi-static with no energy radiation, and $\bar{\sigma}_f$ defined above would be close to the average stress $(\sigma_0+\sigma_1)/2$. Then $\Delta\bar{\sigma}_d = (1/2)\Delta\sigma_s$, and $E_R \approx 0$ from (8). We will use $\Delta\bar{\sigma}_d$ in this paper, but the following alternative interpretation is also useful.

We can interpret the slip weakening process in terms of the breakdown process at the advancing front of an earthquake rupture. Then the total energy loss, $\bar{\sigma}_f DS$ can be divided into two parts,

$$\bar{\sigma}_f DS = S\left[\sigma_{f0}D + \int_0^{D_c}\left(\sigma_f(u) - \sigma_{f0}\right)du\right] \qquad (10)$$

The first term can be interpreted as frictional energy (cross-hatched area in Figure 4b), and the second term, the fracture energy (hatched area in Figure 4b). Then, equation (6') can be written as

$$E_R = M_0\left(2\Delta\sigma_d - \Delta\sigma_s\right)/2\mu - S\int_0^{D_c}\left(\sigma_f(u) - \sigma_{f0}\right)du \qquad (11)$$

where

$$\Delta\sigma_d = \sigma_0 - \sigma_{f0} \qquad (12)$$

Here, the definition of the dynamic stress drop is the same as the traditional one, but the fracture energy $E_G = S\int_0^{D_c}\left(\sigma_f(u) - \sigma_{f0}\right)du$ needs to be subtracted from the right-hand side of (6') to obtain E_R.

FRACTURE ENERGY

The estimates of fracture energy for earthquakes vary over a wide range. The specific fracture energy G^* (fracture energy per unit area) ranges from 1 to 10^8 J/m^2 [*Kostrov* *and Das*, 1988]. The largest values are derived from seismic data on the assumption that rupture is arrested by a barrier [*Aki*, 1979] and may not be representative of the average fracture energy of earthquakes. *Husseini* [1977] estimated G^* to be on the order of 10^5 J/m^2. *Scholz* [1990] quotes a range 10^6 to 10^7 J/m^2.

The fracture energy can be related to the rupture velocity. For simplicity, we use a Mode III (longitudinal shear) crack model in the following, but we can qualitatively develop a similar argument for other crack geometries.

We take a Cartesian coordinate system (x, y, z), and consider an infinitely long crack extending in z direction. The crack growth is in x direction. Let $2c$ be the width of the crack in x direction. The crack is under uniform stress σ_0 and friction σ_f, both in z direction. In actual faulting, σ_f is likely to vary during faulting, but here we assume it to be constant. Then,

$$\Delta W=(\sigma_0+\sigma_f)DS/2=(\sigma_0-\sigma_f)DS/2+\sigma_f DS$$
$$=\Delta W_1+\sigma_f DS \qquad (13)$$

where

$$\Delta W_1=(\sigma_0-\sigma_f)DS/2=\pi c^2(\sigma_0-\sigma_f)^2/2\mu \qquad (14)$$

In the above, the relations S=2c and $(\sigma_0-\sigma_f)=2\mu D/\pi c$ [*Knopoff*, 1958] are used.

The static energy release rate (specific fracture energy) G^* is given by

$$G^* = K^2/2\mu = \pi c(\sigma_0-\sigma_f)^2/2\mu \qquad (15)$$

where $K=(\sigma_0-\sigma_f)(\pi c)^{1/2}$ is the stress intensity factor [*Dmowska and Rice*, 1986; *Lawn*, 1993; *Freund*, 1998]. From (14) and (15),

$$\mathrm{d}(\Delta W_1)=2G^*\mathrm{d}c \qquad (16)$$

Following *Kostrov* [1966], *Eshelby* [1969], and *Freund* [1972], the energy release rate, G, for a crack growing at a rupture speed V is given approximately by

$$G=G^*g(V) \qquad (17)$$

where $g(V)$ is a universal function of V. For a Mode III crack, it is given by

$$g(V)=[(\beta-V)/(\beta+V)]^{1/2} \qquad (18)$$

where β is S-wave velocity. Then, the fracture energy is

$$E_G=\int G dS = 2\int_0^c G^* g(V)dc$$
$$= g(V)\int d(\Delta W_1) = g(V)\Delta W_1 \qquad (19)$$

which becomes small compared with the strain energy involved as V increases to the limiting velocity β, because $g(V)$ approaches 0 in this limit. For most large shallow earthquakes, it is generally established that the rupture velocity is about 75 to 85 % of β, [*Heaton*, 1990] and we can neglect E_G.

LINK BETWEEN MICROSCOPIC AND MACROSCOPIC PROCESSES

Our fault model consists of many faults (microfaults or subfaults) each one of which radiates seismic energy following the stress release process described above. We cannot distinguish every fault, but what we observe seismologically is the total energy radiated from all of them. Using equation (8), the total energy is given by

$$E_R=\sum E_{R_i}=\sum M_{0_i}\left(2\Delta\bar{\sigma}_{d_i}-\Delta\sigma_{s_i}\right)/2\mu$$
$$=M_0\left(2\Delta\bar{\sigma}_d-\Delta\sigma_s\right)/2\mu \qquad (20)$$

where the average dynamic stress drop, $\Delta\bar{\sigma}_d$, and the average static stress drop, $\Delta\sigma_s$, are the macroscopic parameters defined by

$$\Delta\bar{\sigma}_d=\sum M_{0_i}\Delta\bar{\sigma}_{d_i}/M_0 \qquad (21)$$

and

$$\Delta\sigma_s=\sum M_{0_i}\Delta\sigma_{s_i}/M_0 \qquad (22)$$

Here subscript i denotes the i-th subfault. Equations 21 and 22 show that the macroscopic stress drops $\Delta\bar{\sigma}_d$ and $\Delta\sigma_s$ are given as weighted averages of the stress drops for each subfault. The weight is the seismic moment of each subfault.

With this interpretation, we can tie the microscopic processes occurring on a fault plane to the macroscopic parameters, such as M_0 and E_R, measurable with seismological methods. This is similar to the treatise in the kinetic theory of gas, in which macroscopic thermodynamic parameters like temperature and pressure are tied to the kinetic energy of molecules.

Since the rupture pattern on a fault plane is so complex that we cannot use a simple stress pattern shown in Figure 4 to represent the entire faulting. However, we can use the static and dynamic stress drops defined by (21) and (22) to represent the overall state of stress during seismic rupture.

INTERPRETATION

We use the macroscopic seismic parameters, M_0 and E_R, which are now tied to microscopic processes through equations 20, 21 and 22 for interpreting seismic data. Specifically, we use the ratio $\tilde{e}=E_R/M_0$. This ratio, $\tilde{e}$, multiplied by μ was introduced in seismology in the 1960's as "apparent stress" [*Aki*, 1966; *Wyss and Brune*, 1968; *Wyss*, 1970a, 1970b]. It is usually expressed as a product of the efficiency η and the average stress $\overline{\sigma}=(\sigma_0+\sigma_1)/2$, neither of which can be directly determined seismologically. Nevertheless, the apparent stress, combined with static stress drop, provided useful information for the state of stress in different regions. *Wyss* [1970a] showed that the apparent stress of earthquakes on ridges do not differ much from those in trenches. The difficulty with the apparent stress was in difficulty in accurately computing the radiated energy. Although this difficulty still exists [*e.g. Singh and Ordaz*, 1994], the accuracy of energy estimates has improved [*Choy and Boatwright*, 1995], and we revive the use of $\tilde{e}$. In this paper, using equation 20, the relationship is cast in terms of the static and dynamic stress drops as follows.

$$\tilde{e}=E_R/M_0=(2\Delta\overline{\sigma}_d-\Delta\sigma_s)/2\mu \tag{23}$$

The quantity $\tilde{e}$ can be interpreted as a non-dimensional radiated energy scaled with M_0, the static size of the earthquake, and is called the scaled energy.

Qualitatively, if the friction drops rapidly, fault motion would be accelerated rapidly, and more energy will be radiated for a given M_0, and results in large $\tilde{e}$. In contrast, if the friction drops gradually, the fault motion is accelerated slowly thereby radiating less energy than the case for sudden drop in friction; this would result in small $\tilde{e}$. Thus, $\tilde{e}$ which can be determined with the conventional seismological method can be used to infer the rupture behavior.

We can state the above behavior more quantitatively as follows. As shown in Figure 4a, if the friction drops rapidly, $\Delta\overline{\sigma}_d$ is comparable, or larger than $\Delta\sigma_s$, and $\tilde{e}$ given by equation 23 is of the order of $\Delta\sigma_s/2\mu$. In contrast, if friction drops gradually, $\overline{\sigma}_f$ defined by (7) approaches the average stress $(\sigma_0+\sigma_1)/2$; then $\Delta\overline{\sigma}_d\approx(1/2)\Delta\sigma_s$, and $\tilde{e}\approx 0$.

Figure 5a shows the observed relation between E_R and M_0, and Figure 5b shows $\tilde{e}$ as a function of M_W. Although the determination of M_0 can be made accurately, the determination of E_R is still subject to large uncertainties. The values of E_R estimated for the same earthquake by different investigators often differ by more than a factor of 10 [*Singh and Ordaz*, 1994; *Mayeda and Walter*, 1996]. In particular, the values determined from

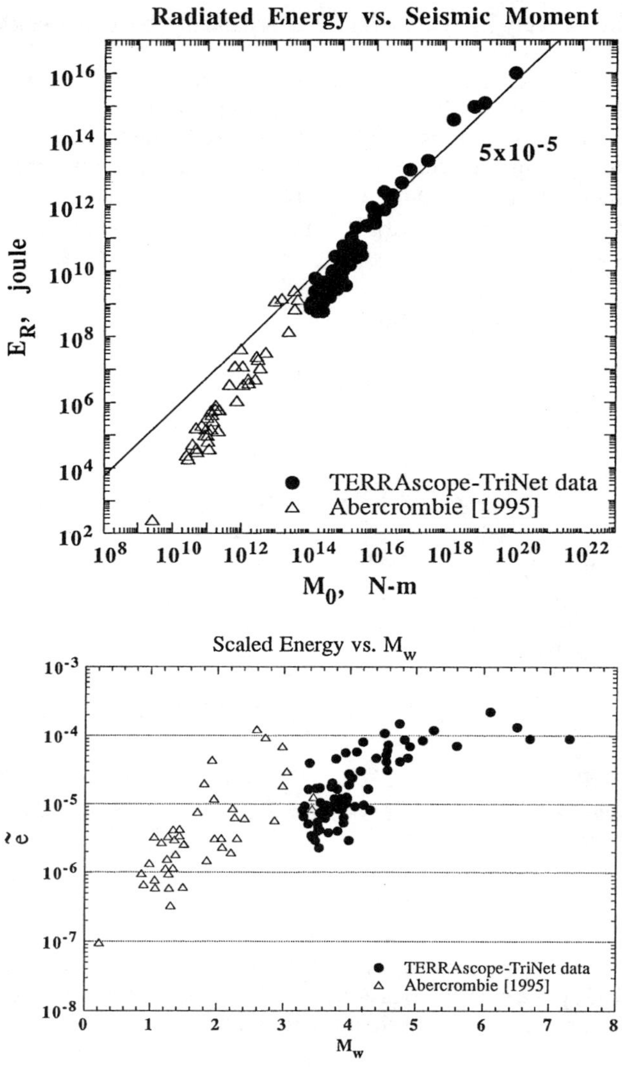

Figure 5. a). Relation between the radiated energy E_R and the seismic moment M_0. The data for large earthquakes (solid circle) are from southern California [updated from *Kanamori et al.*, 1993], and those for small earthquakes (open triangles) are taken from *Abercrombie* [1995]. b) The scaled energy, $\tilde{e}=E_R/M_0$, computed as a function of M_W. Note that the values of $\tilde{e}$ for small earthquakes are 10 to 100 times smaller than those for large earthquakes.

teleseismic data tend to be consistently smaller than those determined from regional data. This difficulty is mainly due to the complex propagation effects. Because of these uncertainties, the relation between E_R and M_0 has not been given close attention.

Figures 5a and 5b include two data sets. The data for large earthquakes ($M_W \geq 3.5$) are obtained in southern California using broad-band seismic data. The results obtained by *Kanamori et al.* [1993] have been slightly revised and updated using more recent data from TriNet, a broad-band seismic network in southern California [*Mori et al.*, 1998]. In these studies, broad-band data at relatively short distances were used, and the propagation and site effects were removed empirically. The results of a recent study by *Mayeda and Walter* [1996] who used coda waves to determine the radiated energy agree with those of *Kanamori et al.* [1993] within a factor of 2, with Mayeda and Walter's values being slightly larger. With the recent deployment of a large number of broad-band instruments in southern California (TriNet, *Mori et al.* [1998]), the propagation and site effects can be calibrated more accurately with many high-quality data at short distances. The new calibration data suggest that the results obtained earlier with TERRAscope are probably accurate within a factor of 3.

The data for smaller earthquakes in Figures 5a and 5b were obtained by *Abercrombie* [1995] using the down-hole (2.5 km deep) seismic data recorded in the Cajon drilling site in southern California [*Zoback and Lachenbruch*, 1992]. A distinct advantage of using down-hole data is that they are free from the complex free-surface effects and the large attenuation near the recording site. These are the main factors that cause the large uncertainties in the results obtained with surface instruments, especially for small earthquakes. Although only one station was available, the data set covers a fairly large azimuthal range (approximately 150°) so that the effects of radiation pattern and directivity were averaged out. Most events are within relatively short distances, 25 km, and the wave forms exhibit clean impulsive characters. Thus, these observations are considered among the most reliable for small earthquakes.

Large ($M_W \geq 4.5$) Earthquakes

Figure 5b shows that the values of $\tilde{e}$ is about 5×10^{-5} to 2×10^{-4} for large earthquakes. If the static stress drop $\Delta\sigma_s$ is 10 to 100 bars, this result indicates (equation 8) that the dynamic stress drop, $\Delta\overline{\sigma}_d$, is 20 to 110 bars for large earthquakes, comparable to, or slightly larger than, the static stress drop $\Delta\sigma_s$.

Our interpretation is that, for large earthquakes, melting and fluid pressurization reduce dynamic friction thereby causing rapid brittle failure resulting in a relatively large $\tilde{e}$. Since both $\Delta\sigma_s$ and $\Delta\overline{\sigma}_d$ are of the order of 100 bars, and the friction is low, the entire process must be occurring at a stress level comparable to the static and dynamic stress drops, about 100 bars (Figure 4a). This is consistent with the result of *Beroza and Zoback* [1993] and *Zoback and Beroza* [1993] who found from the diversity of aftershock mechanisms that the friction during the 1969 Loma Prieta, California, earthquake was very low. Also *Spudich* [1992] and *Spudich et al.* [1998] inferred from the rotation of slip vectors that the absolute stress during faulting of several earthquakes is comparable to stress drops. The assumption in these studies is that the slip direction is subparallel to the frictional stress on the fault plane.

Small (M_W<2) Earthquakes

A striking feature seen in Figures 5b is that the ratio, $\tilde{e}$, for small earthquakes is approximately 10 to 100 times smaller than that for large earthquakes, *i.e.* small earthquakes appear to be less efficient in wave radiation than large earthquakes. Even if we allow for the potentially large uncertainties in energy estimation, this difference appears to be too large to be attributed to experimental errors, and probably reflects the real difference in the rupture dynamics between small and large earthquakes. The transition occurs between M_W=2.5 and 5. Although Figures 5a and 5b show the results only from the two specific data sets, many other studies show a similar transition over this magnitude range [*e.g. Thatcher and Hanks*, 1973; *Fletcher and Boatwright*, 1991; *Boatwright et al.*, 1991; *Mayeda and Walter*, 1996; *Thio*, 1996; *Zhu*, 1998]. It is interesting to note that figure 7 of *Thatcher and Hanks* [1973] showing the relation between M_0 and M_L, if combined with their figure 10 (E_R *vs.* M_L), could be interpreted as showing this transition.

Referring to equations 7 and 9, we interpret this result in terms of a gradual drop in friction. Because the thermal energy involved is not large enough to reduce friction, the stress change can be gradual as shown in Figure 4b. This means that fracture energy, E_G, given in equation 6 or that defined by the second term of equation 10 is large for small earthquakes. As mentioned earlier, the fracture energy for large earthquakes is considered to be small but, for small earthquakes, there is no direct evidence for small fracture energy, *i.e.* small earthquakes could be significantly less brittle than large earthquakes.

McGarr [1999] suggests, on the basis of the data for the apparent stress, that the upper bound of efficiency of earthquakes is about 0.06. Our conclusion on small earthquakes is qualitatively consistent with *McGarr's* [1999]. However, our conclusion suggests that the efficiency for large earthquakes (*e.g.* M_W>6) could be considerably higher than that for small earthquakes.

Models for Small and Large Earthquakes

Using the results obtained above, we present here a schematic model contrasting small and large earthquakes.

First we assume that σ_0 is the same everywhere along a fault zone. Then, Figure 6 illustrates representative stress variations for small and large earthquakes. We assume that the stress on the fault plane drops linearly from σ_0 until the slip reaches a critical value, D_C, where the stress is equal to a constant frictional stress σ_{f0} . The stress eventually drops to almost 0 if slip exceeds D_T when melting or pressurization reduces friction. Figure 5b shows that this transition occurs at M_W=2.5 to 5, which suggests that D_T=2 to 30 cm (equation 2).

We let D_S and D_L be the total displacement for small and large earthquakes, respectively. For small earthquakes, $D_C<D_S<D_T$, and for large earthquakes, $D_L>>D_T$. Then for large earthquakes,

$$E_R = \Delta W - E_F - E_G$$
$$= D_L S\left[\frac{\sigma_0+\sigma_1}{2} - \sigma_{f0}\left(\frac{D_T}{D_L}\right) - \frac{\sigma_0-\sigma_{f0}}{2}\left(\frac{D_C}{D_L}\right)\right] \quad (24)$$

Since $D_L>>D_T>D_C$, and $\sigma_1 \approx 0$,

$$\tilde{e} = \frac{\Delta\sigma_s}{2\mu}, \quad \text{where} \quad \Delta\sigma_s = \sigma_0 \quad (25)$$

In contrast, for small earthquakes,

$$E_R = \Delta W - E_F - E_G$$
$$= D_S S\left[\frac{\sigma_0+\sigma_{f0}}{2} - \sigma_{f0} - \frac{\sigma_0-\sigma_{f0}}{2}\left(\frac{D_C}{D_S}\right)\right]$$
$$= D_S S\frac{(\sigma_0-\sigma_{f0})}{2}\left(1-\frac{D_C}{D_S}\right) \quad (26)$$

Thus,

$$\tilde{e} = \frac{\Delta\sigma_s}{2\mu}\left(1-\frac{D_C}{D_S}\right), \quad \text{where} \quad \Delta\sigma_s = \sigma_0 - \sigma_{f0} \quad (27)$$

In this case, $\Delta\sigma_s$(small earthquakes)<$\Delta\sigma_s$(large earthquakes), but the difference would be small, about a factor of 2 or so. On the other hand, the scaled energy, $\tilde{e}$, can be very different. If D_S is comparable to D_C for small earthquakes, then $\tilde{e}$ can be very small. This is the reason why we have the large difference in $\tilde{e}$ between large and small earthquakes, even if the static stress drop is about the same. Actually, σ_0 may vary considerably along a fault zone. If σ_0 is large for small earthquakes, then we can have $\Delta\sigma_s$(small)=$\Delta\sigma_s$(large), yet $\tilde{e}$ can be still very small for small earthquakes if $D_S \approx D_C$. Actually, a large σ_0 for small earthquakes may not be unreasonable considering the possibility of local stress concentration. In any case, the actual condition can be very heterogeneous, but as a whole, some mechanism as illustrated in Figure 6 is probably responsible for the difference between large and small earthquakes.

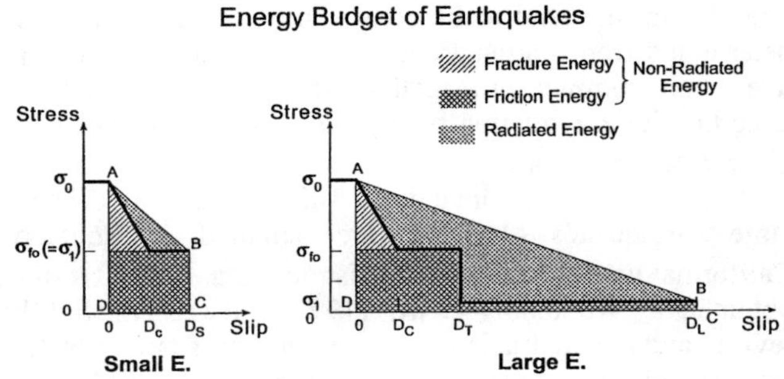

Figure 6. Schematic stress release patterns for small (left) and large earthquakes (right). Hatched, cross-hatched, and dotted areas represent the fracture energy, frictional energy loss and radiated energy, respectively.

IMPLICATIONS

State of Stress

The results obtained for large earthquakes suggest that the average stress level along mature faults where large earthquakes occur must be low because of the dominant thermal effects such as frictional melting and fluid pressurization. Because of melting or pressurization, a fault zone is self-organized into a low stress state. That is, even if the stress was high in the early stage of fault evolution, it would eventually settle in a low stress state after many large earthquakes. This state of stress is consistent with the generally held view that the absence of heat flow anomaly along the San Andreas fault suggests a shear strength of about 200 bars or less [*Brune et al.*, 1969; *Lachenbruch and Sass*, 1980].

The stress in the crust away from active mature faults can be high as has been shown by many in-situ measurements of stress [*McGarr*, 1980; *Brudy et al.*, 1997]. The stress difference is large, and a kbar type stress may be involved in small earthquakes, but the events are in general so small that it is hard to determine the stress parameters accurately. What is important, though, is that as long as the length of the fault is small, the state of

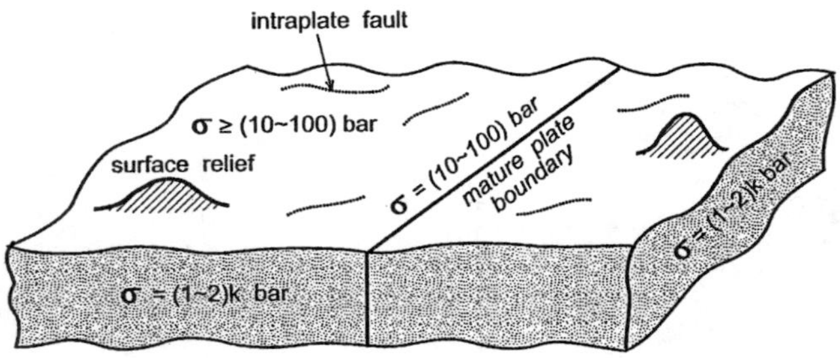

Figure 7. Schematic diagram showing the magnitude of the stresses in Earth's crust. Modified from [*Kanamori,* 1980].

stress in the fault zone would not affect the regional stress drastically. However, as the fault grows to some length (*e.g.* Japanese intra-plate earthquakes like Tango, Tottori, Nobi *etc.*), then some sort of self-organization occurs and the fault settles at a stress level somewhat higher than that on more active plate boundaries.

This type of stress distribution has been suggested from seismic data [Figure 7, *Kanamori*, 1980], and from stress orientations near major plate boundaries [*Mount and Suppe,* 1987; *Zoback et al.*, 1987].

Magnitude-frequency Relationship for Mature Faults

One probable consequence of sudden reduction in friction when slip exceeds a threshold value would be runaway rupture. In this context, an interesting observation is the magnitude-frequency relationship for some mature plate boundaries such as the San Andreas fault and some subduction zones. For example, the absence of events with magnitude between 6.5 and 7.5 on the San Andreas fault in southern California, despite the occurrence of magnitude 8 earthquake in 1957 (Fort Tejon earthquake) and the average repeat time of about a few hundred years [*Sieh,* 1984], has been thought somewhat odd. Figure 8a shows the magnitude-frequency relation taken from *Wesnousky* [1994]. Earthquakes with M from 6 to 7 appear to be fewer than expected for the conventional magnitude-frequency relationship. A similar observation has been made for the Nankai trough in Japan [*Masataka Ando*, 1999, personal communication] as shown in Figure 8b. In this region, many earthquakes with $M \geq 8$ are documented well (Figure 9), but almost no earthquakes with $7 < M < 8$ have occurred there since 1900. These observations can be interpreted in terms of the runaway process discussed above. As the magnitude exceeds a threshold value, about 6.5 for the San Andreas and 7 for the Nankai trough, the friction drops and fault slip cannot stop until it reaches some limit imposed by the regional seismogenic structure or loading geometry. This is a runaway situation caused by dynamic effects of faulting.

If the specific fracture energy, G^*, is constant, the Griffith type cracks are inherently unstable, *i.e.* if the crack length exceeds a threshold, the crack will runaway. So, in this sense all earthquakes, small and large, can get into runaway rupture. However, G^* is not constant in real fault zones, and the place where G^* is large acts as a barrier to stop rupture propagation [*Aki*, 1979]. Then the question is what is the probability of some barriers stopping the rupture. The easiest way to look at this situation is to use the stress intensity factor K which is given by $(\sigma_0 - \sigma_f)(\pi l)^{1/2}$ for a Mode III crack [*e.g. Dmowska and Rice,* 1986] where l is the crack length. As the fault grows, l and D increase. When D exceeds D_T, friction, σ_f, drops (see Figure 6). The combined effect of the decreasing σ_f and increasing l increases K. Since the crack extension force is proportional to K^2, the fault rupture becomes harder to stop and runaway rupture is more likely to occur.

The magnitude-frequency relationship is usually understood as a manifestation of heterogeneity of fault structure [*Scholz and Aviles,* 1986; *Okubo and Aki*, 1987; *Aviles et al.*, 1987]. In addition to this static feature, slip-controlled dynamic runaway process could be an important element that determines the earthquake statistics for mature faults.

Seismic Breakaway Phase

In a series of papers, *Ellsworth and Beroza* [1995, 1998] and *Beroza and Ellsworth* [1996] showed that the moment rate of many earthquakes is initially low but after some time it grows rapidly. They called this sudden increase in the moment rate a breakaway phase. The breakaway phase could be a manifestation of the slip-controlled runaway rupture. However, our model has a highly heterogeneous distribution of strength and would not explain the scaling relation proposed by *Ellsworth and Beroza* [1995, 1998] and *Beroza and Ellsworth* [1996]. Similar observations, on various time scales, have been made by *Umeda* [1990, 1992], and *Kikuchi* [1997].

Slip Behavior of a Plate Boundary

The thermally-controlled model discussed above is inherently non-linear in the sense that slip controls the slip behavior. In such a non-linear system, it is possible that an infinitesimally small perturbation in the initial condition may lead to a significantly different behavior. In this context, the historical sequence along the Nankai trough is interesting. Figure 9 shows the sequence determined by a series of studies of *Imamura* [1928], *Ando* [1975], and *Ishibashi* [1998]. An interesting event is the one in 1605. This event caused widespread tsunami along the Japanese coast, but no significant evidence for shaking has been documented [*Ishibashi*, 1981]. Although the evidence is qualitative, the historical data for this region are generally considered reliable. This evidence suggests

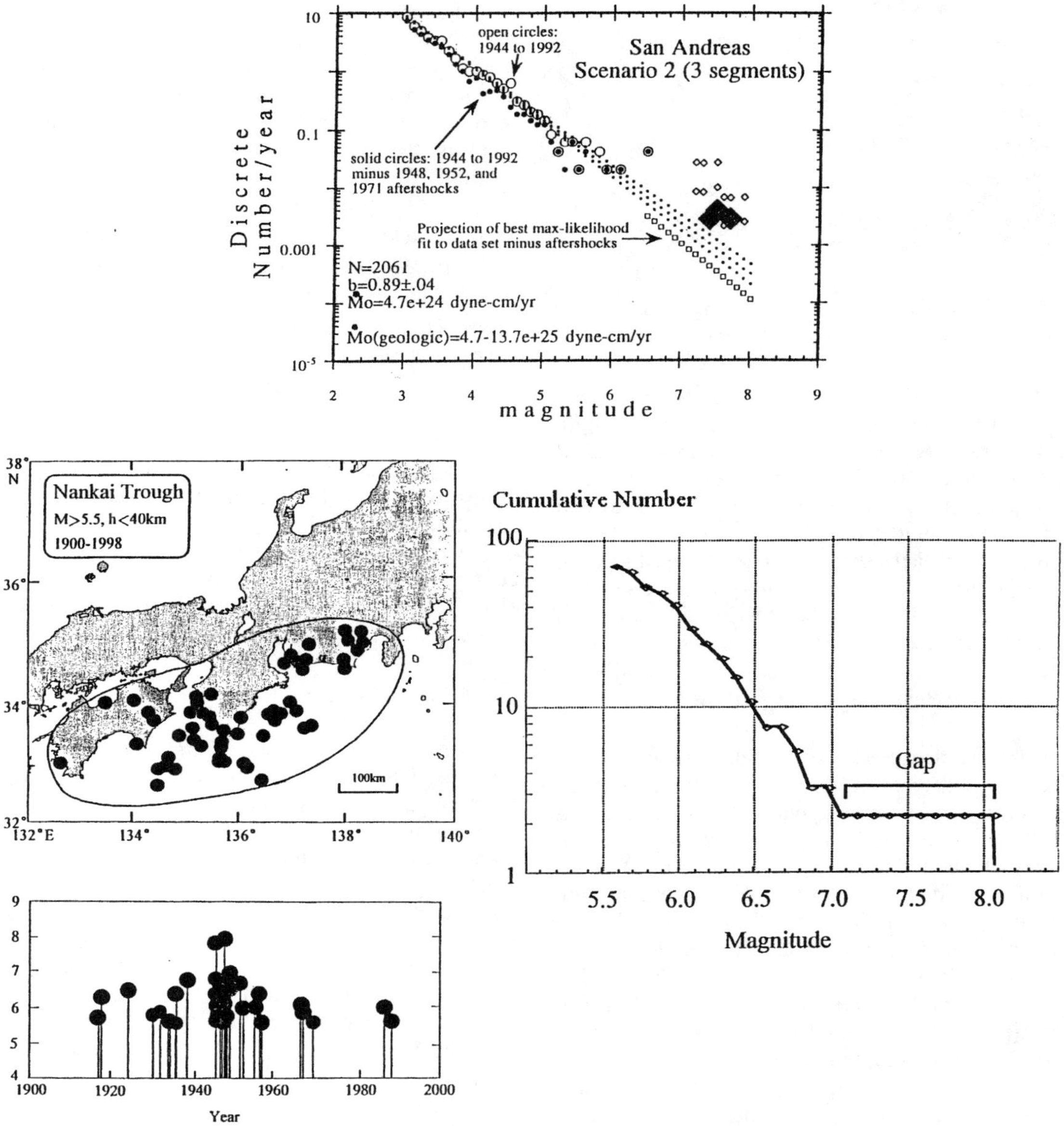

Figure 8 a). Magnitude-frequency relationship for the San Andreas fault, California [*Wesnousky,* 1994]. b). The same for the Nankai trough, Japan [*Masataka Ando*, written communication, 1999].

that the 1605 earthquake was a tsunami earthquake in the sense defined by *Kanamori* [1972]. We suggest that the general style of earthquakes along a plate boundary (*i.e.* brittle ordinary earthquake, slow tsunami earthquake, or creep) is determined by the properties of the boundary (*e.g.* age of the subducting plate, sediment structure, roughness of the subducting plate, convergence rate *etc*., [*Ruff and Kanamori*, 1983; *Uyeda and Kanamori*, 1979; *Scholz and Campos*, 1995]), but considerable perturbation from the average behavior could occur because of the non-linear nature of thermally-controlled mechanism. A slow tsunami earthquake could occasionally occur at a plate boundary where ordinary earthquakes regularly occur. This is a speculative interpretation, and the possibility that the 1605 earthquake was caused by some other mechanisms remains, but considering the significant thermal effects on fluid-filled subduction boundaries, this interpretation is plausible. *Yamashita* [1998] showed that fluid migration in a porous fault zone with spatially heterogeneous fracture strength can produce irregular event sequences. Although

Yamashita's model is for a quasi-static case, and is not directly applicable to dynamic rupture propagation, a similar model would produce variable dynamic rupture patterns.

Ground Motion from Large Earthquakes

The effect of a pulse-like near-field ground motion on large structures is becoming an important engineering problem [*Heaton*, 1990; *Heaton et al.*, 1995; *Hall et al.*, 1995]. However, very few recordings of near-field ground motion from large earthquakes exist. In modeling studies, the records from small earthquakes are used to estimate ground motions from hypothetical large earthquakes. This is a reasonable approach but the possibility exists that the slip velocity during very large earthquakes could be significantly larger than that for small earthquakes because of reduction of friction caused by large displacement.

CONCLUSION

The thermal budget during seismic slip suggests that frictional melting and fluid pressurization can play a key role in rupture dynamics of large earthquakes. In a simple model of faulting under frictional stress σ_f, the temperature increases with σ_f and the earthquake magnitude, M_W. If the slip zone is thin and heat transfer is mainly by conduction, the thickness of the heated zone, w, is of the order of a few mm for a seismic time scale of about 10 sec. Then, even for a modest σ_f, the temperature rise, ΔT, would exceed 1000° for earthquakes with M_W=5 to 7, and melting is likely to occur, and reduce friction during faulting. Another important process is fluid pressurization. If fluid exists in a fault zone, a modest ΔT of 100 to 200° would likely increase the pore pressure enough to significantly reduce friction for earthquakes with M_W=3 to 5. The microscopic state of stress caused by local melting and pressurization can be tied to macroscopic seismic parameters such as, M_0 and E_R, by averaging the stresses in the microscopic states. Since the thermal process is important only for large earthquakes, the dynamics of small and large earthquakes can be very different. This difference is reflected in the observed relation between the ratio $\tilde{e}=E_R/M_0$ and M_W. The available seismic data show that $\tilde{e}$ for large earthquakes is 10 to 100 times larger than for small earthquakes. According to this model, mature fault zones such as the San Andreas are at relatively moderate stress levels, but the stress in the plate interior can be high. The fault dynamics described here suggests that once slip exceeds a threshold, runaway rupture could occur. This could explain the anomalous magnitude-frequency relationship observed for some mature faults. Since the thermal state is controlled by the amount of slip, the slip behavior is controlled by the slip itself. This would produce a non-linear behavior, and under certain circumstances, the slip behavior at the same location may vary from event to event. Another important implication is that slip velocity during a large earthquake could be faster than what one would extrapolate from smaller earthquakes.

Acknowledgments. We thank Kenshiro Otsuki for sharing his insight into cataclasites and pseudotachylytes with us. Discussions with Rachel Abercrombie helped us assess the accuracy of energy measurements for small earthquakes. We benefited from the comments on the early version of the manuscript by Yoshio Fukao, Masayuki Kikuchi, Minoru Takeo, and Emily Brodsky. We also thank Toshihiko Shimamoto, Lee Silver, and Yuri Fialko for helpful discussions at various stages of this work. We thank Masataka Ando and Hisao Ito for allowing us to use some of their unpublished figures. This research was partially supported by the U. S. Geological Survey grant 99HQGR0035. Contribution #8635, Division of Geological and Planetary Sciences, California Institute of Technology.

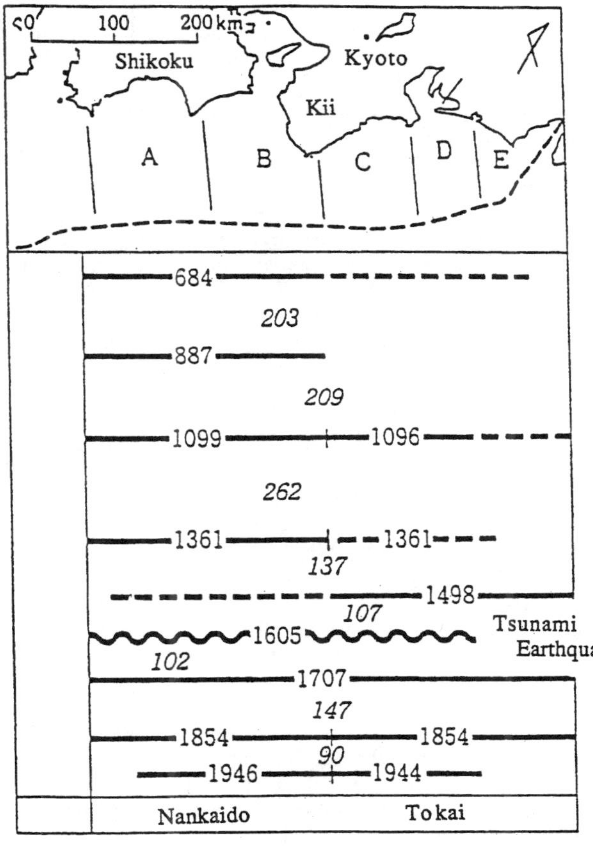

Figure 9. Large earthquakes along the Nankai trough, Japan [*Ishibashi and Satake*, 1998].

REFERENCES

Abercrombie, R., Earthquake source scaling relationships from -1 to 5 M_L using seismograms recorded at 2.5-km depth, *J. Geophys. Res.*, *100*, 24,015-24,036, 1995.

Abercrombie, R., and P. Leary, Source parameters of small earthquakes recorded at 2.5 km depth, Cajon Pass, Southern California: Implications for earthquake scaling, *Geophys. Res. Lett.*, *20*, 1511-1514, 1993.

Aki, K., Generation and propagation of G waves from the Niigata earthquake ofJune 16, 1964. Part 2. Estimation of earthquake moment, from the G wave spectrum, *Bull. Earthquake Res. Inst. Tokyo Univ.*, *44*, 73-88, 1966.

Aki, K., Characterization of barriers on an earthquake fault, *J. Geophy. Res.*, *84*, 6140-6148, 1979.

Ando, M., Source mechanisms and tectonic significance of historical earthquakes along the Nankai trough, Japan, *Tectonophysics*, *27*, 119-140, 1975.

Aviles, C. A., C. H. Scholz, and J. Boatwright, Fractal analysis applied to characteristic segments of the San Andreas fault, *J. Geophys. Res.*, *92*, 331-344, 1987.

Beeler, N. M., T. E. Tullis, M. L. Blanpied, and J. D. Weeks, Frictional behavior of large displacement experimental faults, *J. Geophys. Res.*, *101*, 8697-8715, 1996.

Ben-Zion, Y., and D. J. Andrews, Properties and implications of dynamic rupture along a material interface, *Bull. Seismol. Soc. Am.*, *88*, 1085-1094, 1998.

Beroza, G. C., and W. L. Ellsworth, Properties of the seismic nucleation phase, *Tectonophysics*, *261*, 209-227, 1996.

Beroza, G. C., and T. Mikumo, Short slip duration in dynamic rupture in the presence of heterogeneous fault properties, *J. Geophys. Res.*, *101*, 22,449-22,460, 1996.

Beroza, G. C., and M. D. Zoback, Mechanism diversity of the Loma Prieta aftershocks and the mechanics of mainshock-aftershock interaction, *Science*, *259*, 210-213, 1993.

Boatwright, J., J. B. Fletcher, and T. Fumal, A general inversion scheme for source, site and propagation characteristics using multiply recorded sets of moderate-sized earthquakes, *Bull. Seismol. Soc. Am.*, *81*, 1754-1782, 1991.

Bouchon, M., The state of stress on some faults of the San Andreas system as inferred from near-field strong motion data, *J. Geophys. Res.*, *102*, 11,731-11,744, 1997.

Brace, W. F., and J. D. Byerlee, Stick slip as a mechanism for earthquakes, *Science*, *153*, 990-992, 1966.

Brudy, M., M. D. Zoback, K. Fuchs, F. Rumell, and J. Baumgartner, Estimation of the complete stress tensor to 8 km depth in the KTB scientific drill holes: Implications for crustal strength, *J. Geophys. Res.*, *102*, 18,453-18,475, 1997.

Brune, J. N., Tectonic stress and spectra of seismic shear waves from earthquakes, *J. Geophys. Res.*, *75*, 4997-5009, 1970.

Brune, J. N., S. Brown, and P. A. Johnson, Rupture mechanism and interface separation in foam rubber models of earthquakes: A possible solution to the heat flow paradox and the paradox of large overthrusts, *Tectonophysics*, *218*, 59-67, 1993.

Brune, J. N., T. L. Henyey, and R. F. Roy, Heat flow, stress, and the rate of slip along the San Andreas fault,California, *J. Geophys. Res.*, *74*, 3821-3827, 1969.

Byerlee, J. D., Friction, overpressure and fault normal compression,, *Geophysical Research Letters*, *17*, 4,741-4,750, 1990.

Cardwell, R. K., D. S. Chinn, G. F. Moore, and D. L. Turcotte, Frictional heating on a fault zone with finite thickness, *Geophys. J. Roy. Astron. Soc.*, *52*, 525-530, 1978.

Chester, F. M., and J. S. Chester, Ultracataclasite structure and friction processes of the Punchbowl fault, San Andreas system, California, *Tectonophysic*, *295*, 199-221, 1998.

Choy, G. L., and J. L. Boatwright, Global patterns of radiated energy and apparent stress, *J. Geophys. Res.*, *100*, 18205-18228, 1995.

Dahlen, F. A., The balance of energy in earthquake faulting, *Geophys. J. R. astr. Soc.*, *48*, 239-261, 1977.

Dieterich, J. H., Modeling of rock friction 1. Experimental results and constitutive equations, *J. Geophys. Res.*, *84*, 2161-2168, 1979a.

Dieterich, J. H., Modeling of rock friction 2. Simulation of preseismic slip, *J. Geophys. Res.*, *84*, 2169-2175, 1979b.

Dmowska, R., and J. R. Rice, Fracture theory and Its seismological applications, in *Theories in Solid Earth Physics*, edited by R. Teisseyre, pp. 187-255, PWN-Polish Publishers, Warzawa, 1986.

Ellsworth, W. L., and G. C. Beroza, Seismic evidence for an earthquake nucleation phase, *Science*, *268*, 851-855, 1995.

Ellsworth, W. L., and G. C. Beroza, Observation of the seismic nucleation phase in the Ridgecrest, California, earthquake sequence, *Geophys. Res. Lett.*, *25*, 401-404, 1998.

Eshelby, J. D., The determination of the elastic field of an ellipsoidal inclusion and related problems, *Proceedings of the Royal Soc. London*, *241*, 376-396, 1957.

Eshelby, J. D., The elastic field of a crack extending non-uniformly under general anti-plane loading, *J. Mech. Phys. Solids*, *17*, 177-199, 1969.

Fletcher, J. B., and J. Boatwright, Source parameters of Loma Prieta aftershocks and wave propagation characteristics along the San Francisco peninsula from a joint inversion of digital seismograms, *Bull. Seismol. Soc. Am.*, *81*, 1783-1812, 1991.

Freund, L. B., *Dynamic Fracture Mechanics*, Cambridge Univ. Press, New York, 1-563, 1998.

Goldsby, D. L., and T. E. Tullis, Experimental observations of frictional weakening during large and rapid slip (abstract), *EOS, Transactions, American Geophysical Union*, *79*, F610, 1998.

Goldsby, D. L., and T. E. Tullis, Dramatic reduction in friction of quartz at rapid but subseismic slip (abstract), *Seismological Research Letters*, *70*, 247, 1999.

Guatteri, M., and P. Spudich, Coseismic temporal changes of slip direction: the effect of absolute stress on dynamic rupture, *Bull. Seismol. Soc. Am.*, *88*, 777-789, 1998.

Hall, J. F., T. Heaton, M. W. Halling, and D. J. Wald, Near-source ground motion and its effects on flexible buildings, *Earthquake Spectra*, *11*, 569-605, 1995.

Hanks, T. C., Earthquake stress drops, ambient tectonic stresses and stresses thatderive plate motions, *Pure Appl. Geophys.*, *115*, 441-458, 1977.

Heaton, T., Evidence for and implications of self-healing pulses of slip in earthquake rupture, *Physicsof the Earth and Planetary Interiors*, *64*, 1-20, 1990.

Heaton, T., J. F. Hall, D. J. Wald, and M. W. Halling,

Response of high-rise and base-isolated buildings to a hypothetical M_w7.0 blind thrust earthquake, *Science*, *267*, 206-211, 1995.

Hubbert, M. K., and W. W. Rubey, Role of fluid pressure in mechanics of overthrust faulting, *Bull. Geol. Soc. Am.*, *70*, 115-166, 1959.

Hull, J., Thickness-displacement relationships for deformation zones, *Journal of Structural Geology*, *10*, 431-435, 1988.

Husseini, M. I., Energy balance for formation along a fault, *Geophys. J. R. astr. Soc.*, *49*, 699-714, 1977.

Ide, S., and M. Takeo, Determination of constitutive relations of fault slip based on seismic wave analysis, *J. Geophys. Res.*, *102*, 27,379-27,391, 1997.

Imamura, A., On the seismic activity of central Japan, *Jap. J. Astron. Geophys.*, *6*, 119-137, 1928.

Irwin, W. P., and I. Barnes, Effectof geologic structure and metamorphic fluids on seismic behavior of the San Andreas fault system in central and northern California, *Geology*, *3*, 713-716, 1975.

Ishibashi, K., Specification of a soon-to-occur seismic faulting in the Tokai district, central Japan, based upon seismotectonics, in *Earthquake Prediction, An International Review*, edited by D. W. Simpson and P. G. Richards, pp. 297-332, American Geophysical Union, Washington, D. C., 1981.

Ishibashi, K., Inferred Tokai earthquake paired with the 1361 Shohei Nankai earthquake (abstract), *Programme and Abstracts, The Seismological Society of Japan, 1998, Fall Meeting, P125*, 1998.

Ito, H., H. Naka, D. Lockner, T. Kiguchi, H. Tanaka, R. Ikeda, T. Ohtani, K. Fujimoto, and Y. Kuwahara, Permeability of the Nojima fault: Comparison of borehole results with core measurements (abstract), *Programme and Abstracts, The Seismological Society of Japan, 1998, Fall Meeting, B20*, 1998.

Jeffreys, H., On the mechanics of faulting, *Geol. Mag.*, *79*, 291-295, 1942.

Kanamori, H., Mechanism of tsunami earthquakes, *Phys. Earth Planet. Inter.*, *6*, 346-359, 1972.

Kanamori, H., The state of stress in the Earth's lithosphere, in *Phys. Earth's Int., Course LXXVIII*, edited by A. M. Dziewonski and E. Boschi, pp. 531-554, North-Holland Pub. Co., Amsterdam, 1980.

Kanamori, H., Mechanics of Earthquakes, *Ann. Rev. Earth & Planetary Sciences*, *22*, 207-237, 1994.

Kanamori, H., and D. L. Anderson, Theoretical basis of some empirical relations in seismology, *Bull. Seis. Soc. Amer.*, *65*, 1073-1095, 1975.

Kanamori, H., T. H. Anderson, and T. H. Heaton, Frictional melting during the rupture of the 1994 Bolivian Earthquake, *Science*, *279*, 839-842, 1998.

Kanamori, H., E. Hauksson, L. K. Hutton, and L. M. Jones, Determination of Earthquake Energy Release and M_L Using TERRAscope, *Bull. Seismol. Soc. Am.*, *83*, 330-346, 1993.

Kanamori, H., J. Mori, and T. H. Heaton, The 3 December 1988, Pasadena earthquake (M_L=4.9) recorded with the very broadband system in Pasadena, *Bull. Seismol. Soc. Am.*, *80*, 483-487, 1990.

Kikuchi, M., Strain drop and apparent strain for large earthquakes, *Tectonophysics*, *211*, 107-113, 1992.

Kikuchi, M., Size dependence of moment-rate function (abstract), in *1997 IASPEI meeting*, Thesaloniki, Greece, 1997.

Kikuchi, M., and Y. Fukao, Seismic wave energy inferred from long-period body wave inversion, *Bull. Seismol. Soc. Am.*, *78*, 1707-1724, 1988.

Knopoff, L., Energy release in earthquakes, *Geophysic. Jour.*, *1*, 44-52, 1958.

Kostrov, B. V., Unsteady propagation of longitudinal shear cracks, *J. Appl. Math. Mech. (transl. P. M. M.)*, *30*, 1241-1248, 1966.

Kostrov, B. V., Seismic moment and energy of earthquakes, and seismic flow of rock (translated to English), *Izv. Earth Physics*, *1*, 23-40, 1974.

Kostrov, B. V., and S. Das, *Principles of Earthquake Source Dynamics*, Cambridge University Press, Cambridge, 1-286, 1988.

Lachenbruch, A. H., Frictional heating, fluid pressure, and the resistance to fault motion, *J. Geophys. Res.*, *85*, 6097-6112, 1980.

Lachenbruch, A. H., and J. H. Sass, Heat flow and energetics of the San Andreas fault zone, *J. Geophys. Res.*, *85*, 6185-6222, 1980.

Lawn, B., *Fracture of Brittle Solids - Second Edition*, Cambridge University Press, Cambridge, 1-378, 1993.

Li, Y. G., K. Aki, D. Adams, A. Hasemi, and W. H. K. Lee, Seismic guided waves trapped in the fault zone of the Landers, California, earthquake of 1992, *J. Geophys. Res.*, *99*, 11,705-11,722, 1994.

Li, Y. G., K. Aki, and F. L. Vernon, San Jacinto fault zone guided waves: A discrimination for recently active fault strands near Anza, California, *J. Geophys. Res.*, *102*, 11,689-11,701, 1997.

Lin, A., Glassy pseudotachylyte veins from the Fuyun fault zone, northwest China, *Journal of Structural Geology*, *16*, 71-83, 1994a.

Lin, A., Microlite morphology and chemistry in pseudotachylyte from the Fuyun fault zone, China, *Journal of Geology*, *102*, 317-329, 1994b.

Lin, A., Injection veins of crushing-originated pseudotachylyte and fault gouge formed during seismic faulting, *Engineering Geology*, *43*, 213-224, 1996.

Lin, A., T. Matsuda, and T. Shimamoto, Pseudotachylyte from the Iida-Matsukawa fault, Nagano prefecture: Pseudotachylyte of crush origin? (in Japanese), *Kozo Chishitsu (Structural Geology)*, *39*, 51-64, 1994.

Mase, C. W., and L. Smith, Pore-fluid pressures and frictional heating on a fault surface, *Pure and Applied Geophysics*, *122*, 583-607, 1985.

Mase, C. W., and L. Smith, Effects of frictional heating on the thermal, hydrologic, and mechanical response of a fault, *J. Geophys. Res.*, *92*, 6249-6272, 1987.

Mayeda, K., and W. R. Walter, Moment, energy, stress drop, and source spectra of western United States earthquakes from regional coda envelopes, *J. Geophys. Res.*, *101*, 11,195-11.208, 1996.

McGarr, A., Some constraints on levels of shear stress in the crust from observations and theory, *J. Geophys. Res.*, *85*, 6,231-6238, 1980.

McGarr, A., On relating apparent stress to the stress causing earthquake fault slip, *J. Geophys. Res.*, *104*, 3,003-3,011, 1999.

McKenzie, D. P., and J. N. Brune, Melting on fault planes during large earthquakes, *Geophys. J. R. astr. Soc.*, *29*, 65-78, 1972.

Melosh, J., Acoustic fluidization: a new geologic process?, *J. Geophys. Res.*, *84*, 7513-7520, 1979.

Melosh, H. J., Dynamical weakening of faults by acoustic fluidization, *Nature*, *379*, 601-606, 1996.

Mikumo, T., and T. Miyatake, Dynamic rupture processes on a dipping fault, and estimates of stress drop and strength excess from the results of waveform inversion, *Geophys. J. Int.*, *112*, 481-496, 1993.

Miyatake, T., Dynamic rupture process of inland earthquakes in Japan: Weak ad strong asperities, *Geophys. Res. Lett.*, *19*, 1041-1044, 1992a.

Miyatake, T., Reconstruction of dynamic rupture process of an earthquake with constraints of kinematic parameters, *Geophys. Res. Lett.*, *19*, 349-352, 1992b.

Mora, P., and D. Place, Numerical simulation of of earthquake faults with gauge: Towards a comprehensive explanation for the heat flow paradox, *J. Geophys. Res.*, *103*, 21,067-21,089, 1998.

Mora, P., and D. Place, The weakness of earthquake faults, *Geophys. Res. Lett.*, *26*, 123-126, 1999.

Mori, J., H. Kanamori, J. Davis, E. Hauksson, R. Clayton, T. Heaton, L. Jones, and A. Shakal, Major improvements in progress for southern California earthquake monitoring, *EOS Trans. American Geophysical Union*, *79*, 217-221, 1998.

Morrow, C. A., and J. D. Byerlee, Permeability of core samples from Cajon Pass scientific drill hole: Results from 2100 to 3500 m depth, *J. Geophys. Res.*, *97*, 5145-5151, 1992.

Mount, V., and J. Suppe, State of stress near the San Andreas fault: Implications for wrench tectonics, *Geology*, *15*, 1143-1146, 1987.

Nadeau, R. M., and L. R. Johnson, Seismological studies at Parkfield VI: Moment release rates and estimates of source parameters for small repeating earthquakes, *Bull. Seismol. Soc. Am.*, *88*, 790-814, 1998.

Obata, M., and S. Karato, Ultramafic pseudotachylite from the Balmuccia peridotite, Ivrea-Verbano zone, northern Italy, *Tectonophysics*, *242*, 313-328, 1995.

Okubo, P. G., and K. Aki, Fractal geometry in the San Andreas fault system, *J. Geophys. Res.*, *92*, 345-355, 1987.

Orowan, E., Mechanism of seismic faulting in rock deformation, *Geol. Soc. Am. Mem.*, *79*, 323-345, 1960.

Otsuki, K., On the relationship between the width of shear zone and the displacement along fault, *Jour. of Geolog. Soc. Japan*, *84*, 661-669, 1978.

Otsuki, K., Physical process of seismic frictional sliding observed in samples from fault zones (translated from Japanese), *Chikyu Monthly*, 213-224, 1998.

Quin, H., Dynamic stress drop and rupture dynamics of the October 15, 1979 Imperial Valley, California, earthquake, *Tectonophysics*, *175*, 93-117, 1990.

Richards, P. G., Dynamic motions near an earthquake fault: a three-dimensional solution, *Bull. Seismol. Soc. Am.*, *66*, 1-32, 1976.

Rice, J. R., Fault stress states, pore pressure distributions, and the weakness of the San Andreas fault, in *Fault Mechanics and Transport Properties of Rocks: A Festschrift in Honor of W. F. Brace*, edited by B. Evans and W. Teng-fong, pp. 475-503, Academic Press, New York, 1992.

Ruff, L., and H. Kanamori, Seismic coupling and uncoupling at subduction zones, *Tectonophysics*, *99*, 99-117, 1983.

Savage, J. C., and J. B. Walsh, Gravitational energy and faulting, *Bull. Seismol. Soc. Am.*, *68*, 1613-1622, 1978.

Savage, J. C., and M. D. Wood, The relation between apparent stress and stress drop, *Bull. Seismol. Soc. Am.*, *61*, 1381-1388, 1971.

Schallamach, A., How does rubber slide?, *Wear*, *17*, 301-312, 1971.

Scholz, C. H., Shear heating and the state of stress on faults, *J. Geophys. Res.*, *85*, 6174-6184, 1980.

Scholz, C. H., *The mechanics of earthquake faulting*, Cambridge University Press, New York, 1-438, 1990.

Scholz, C. H., and C. Aviles, The fractal geometry of faults and faulting, in *Earthquake Source Mechanics, AGU Geophys. Mono. 37*, edited by S. Das, J. Boatwright and C. Scholz, pp. 147-155, American Geophysical Union, Washington, D. C., 1986.

Scholz, C., and J. Campos, On the mechanism of seismic decoupling and back-arc spreading at subduction zones, *J . Geophys . Res.*, *100*, 22103-22115, 1995.

Scott, D. R., Seismicity and stress rotation in a granular model of the brittle crust, *Nature*, *381*, 592-595, 1996.

Sibson, R. H., Interactions between temperature and fluid pressure during earthquake faulting -- A mechanism for partial or total stress relief, *Nature*, *243*, 66-68, 1973.

Sibson, R. H., Generation of pseudotachylite by ancient seismic faulting, *Geophys. J. R. astr. Soc.*, *43*, 775-794, 1975.

Sibson, R. H., Kinetic shear resistance, fluid pressures and radiation efficiency during seismic faulting, *Pure and Applied Geophysics*, *115*, 387-400, 1977.

Sibson, R. H., Power dissipation and stress levels on faults in the upper crust, *J. Geophys. Res.*, *85*, 6239-6247, 1980.

Sieh, K., Lateral offsets and revised dates of large prehistoric earthquakes at Pallett Creek, southern California, *J. Geophys. Res.*, *89*, 7641-7670, 1984.

Singh, S. K., and M. Ordaz, Seismic energy release in Mexican subduction zone earthquakes, *Bull. Seismol. Soc. Am.*, *84*, 1533-1550, 1994.

Spray, J. G., Viscosity determinations of some frictionally generated silicate melts - Implications for fault zone rheology at high-strain rates, *J. Geophys. Res.*, *98*, 8053-8068, 1993.

Spudich, P. K. P., On the inference of absolute stress levels from seismic radiation, *Tectonophysics*, *211*, 99-106, 1992.

Spudich, P., M. Guatteri, K. Otsuki, and J. Minagawa, Use of fault striations and dislocation models to infer tectonic shear stress during the 1995 Hyogo-ken Nanbu (Kobe) earthquake, *Bull. Seismol. Soc. Am.*, *88*, 413-427, 1998.

Thatcher, W., and T. Hanks, Source parameters of southern California earthquakes, *J. Geophys. Res.*, *78*, 8547-8576, 1973.

Thio, H. K., 1 - Using Short-Period Surface Waves to Study

Seismic Source and Structure. 2 - Source Complexity of Large Strike-Slip Earthquakes, Ph.D thesis, pp., California Institute of Technology, 1996.

Tsutsumi, A., and T. Shimamoto, High-velocity frictional properties of gabbro, *Geophys. Res. Lett.*, *24*, 699-702, 1997.

Tullis, T. E., and D. L. Goldsby, Laboratory experiments on rock friction focussed on understanding earthquake mechanics, USGS Technical Report, Vol. 40, http://erp-web.er.usgs.gov/reports/VOL40/pt/g3034.htm, 1998.

Umeda, Y., High-amplitude seismic waves radiated from the bright spot of an earthquake, *Tectonophysics*, *141*, 335-343, 1990.

Umeda, Y., The bright spot of an earthquake, *Tectonophysics*, *211*, 13-22, 1992.

Uyeda, S., and H. Kanamori, Back-arc opening and the mode of subduction, *J. Geophys. Res.*, *84*, 1049-1061, 1979.

Wald, D. J., Strong motion and broadband teleseismic analysis of the 1991 Sierra Madre, California, earthquake, *J. Geophys. Res.*, *97*, 11,033-11,046, 1992.

Weertman, J., Unstable slippage across a fault that separates elastic media of different elastic constants, *J. Geophys. Res.*, *85*, 1455-1461, 1980.

Wesnousky, S. G., The Gutenberg-Richter or Characteristic Earthquake Distribution, Which is it?, *Bull. Seismol. Soc. Am.*, *84*, 1940-1959, 1994.

Winslow, N., and L. Ruff, A hybrid method for calculating the radiated wave energy of deep earthquakes, *Physics of the Earth and Planetary Interiors*, *115*, 181-190, 1999.

Wyss, M., Apparent stress of earthquakes on ridges compares to apparent stresses of earthquakes in trenches, *Geophys. J. R. astr. Soc.*, *19*, 479-484, 1970a.

Wyss, M., Stress estimates for south American shallow and deep earthquakes, *J. Geophys. Res.*, *75*, 1529-1544, 1970b.

Wyss, M., and J. N. Brune, Seismic moment, stress, and source dimensions for earthquakes in the California-Nevada region, *J. Geophys. Res.*, *73*, 4681-4694, 1968.

Yamashita, T., Simulation of seismicity due to fluid migration in a fault zone, *Geophys. J. Int.*, *132*, 674-686, 1998.

Zhu, L., Broadband Waveform Modeling and Its Application to the Lithospheric Structure of the Tibetan Plateau, Ph.D thesis, pp., California Institute of Technology, 1998.

Zoback, M. D., and G. C. Beroza, Evidence for near-frictionless faulting in the 1989 (M 6.9) Loma Prieta, California, earthquake and its aftershocks, *Geology*, *21*, 181-185, 1993.

Zoback, M., and A. H. Lachenbruch, Introduction to the special section on the Cajon Pass scientific drilling project, *J. Geophys. Res.*, *97*, 4991-4994, 1992.

Zoback, M. D., M. L. Zoback, V. S. Mount, J. Suppe, J. Eaton, D. Healy, P. Oppenheimer, L. Reasenberg, L. Jones, B. Raleigh, I. Wong, O. Scotti, and C. Wentworth, New evidence on the state of stress of the San Andreas fault system, *Science*, *238*, 1105-1111, 1987.

Hiroo Kanamori, Seismological Laboratory, California Institute of Technology, Pasadena, California 91125, (626) 395-6914, FAX 626-564-0715, hiroo@gps.caltech.edu.

Thomas H. Heaton, Seismological Laboratory, California Institute of Technology, Pasadena, CA 91125, (626) 6897, FAX 626-564-0715, heaton@gps.caltech.edu.

Depth Dependent Rupture Properties in Circum-Pacific Subduction Zones

Susan L. Bilek and Thorne Lay

Institute of Tectonics and Earth Sciences Dept., University of California, Santa Cruz

Depth dependence of the source rupture duration of interplate thrust earthquakes is examined for seven subduction zones around the Pacific to explore variations in faulting properties. Multi-station deconvolutions of teleseismic P waves for moderate size earthquakes yield estimates of the source time function and centroid depth for each event. Analysis of 17 to 75 earthquakes in each region reveals a consistent trend of decreasing source duration (inferred from the source time functions, after correction for differences in total energy release) with increasing depth. Rupture duration patterns vary somewhat between subduction zones as well as along strike within a given zone, and the data have large scatter, implying significant variation in rupture processes along the interplate megathrusts, but the depth dependence appears to be robust. The rupture duration variations prompt consideration of two end-member models: 1) depth-dependent rupture velocity is caused by variations of rigidity of materials in the fault zone, while static stress drop is constant, and 2) static stress drop varies with depth while material properties and rupture velocity are constant. For the first model, the volumetrically averaged rigidity of the fault zone must increase with depth in each region by a factor of 5 between depths of 5 to 20 km. If rupture velocity is constant, the stress drop must increase by an order of magnitude over the same depth range. This systematic variation in rupture behavior with depth may reflect spatial variations in the amount, compaction and porosity of sediment in the fault zone, topography on the subducting plate, phase transitions in the fault zone materials, thermal structure of the megathrust, and varying presence of fluids in the fault zone. Such physical variations appear to control the physics of rupture propagation, leading to intrinsic dependence of rupture velocity on materials and fluids within the fault zone.

INTRODUCTION

Approximately 90% of global seismic energy release occurs in subduction zone earthquakes, primarily involving thrusting motions on the interplate megathrust faults [*Pacheco et al.*, 1993]. Variations in earthquake rupture complexity and total energy release have been related to gross properties of subduction zone geometry and kinematics, with the underlying premise that earthquake faulting varies with the megathrust environment [*Ruff and Kanamori*, 1980; *Lay et al.*, 1982; *Kanamori*, 1986; *Ruff*, 1992]. While some intriguing general associations have become evident, we are far from a detailed understanding of what controls faulting

GeoComplexity and the Physics of Earthquakes
Geophysical Monograph 120

in subduction zones. It is becoming increasingly well documented that frictional properties of the megathrust are very complex, and the entire spectrum of convergent motions must be considered. For example, some regions experience large amounts of slip occurring over periods as long as one year after large thrust events. This has been observed in Japan, where *Heki et al.* [1997] used geodetic measurements to detect postseismic slip equal to the amount of coseismic slip in the year following the December 24, 1994 Sanriku-oki event. Similarly, significant postseismic slip has been observed for the 1992 Sanriku-oki event in Japan [*Kawasaki et al.*, 1995].

Several subduction zones have also experienced destructive tsunami earthquakes, which produce a larger tsunami than expected given their seismic magnitude [*Kanamori*, 1972]. These events commonly release seismic waves that are depleted in high frequency energy, apparently as a result of having unusually slow rupture propagation [*Pelayo and Wiens*, 1990; *Kanamori and Kikuchi*, 1993; *Satake*, 1994; *Tanioka and Satake*, 1996; *Johnson and Satake*, 1997; *Ihmlé et al.*, 1998]. The distinctive seismic wave spectra of these events has even served as a basis for near real-time detection of tsunami earthquakes involving calculation of spectral energy ratios [*Newman and Okal*, 1998; *Shapiro et al.*, 1998]. The observation that tsunami events rupture at very shallow depths where weak sediments are likely to be present in the fault zone has prompted speculation that rupture in low rigidity materials causes slow rupture velocities and results in a larger faulting displacement for a given seismic moment. These factors can explain the spectral characteristics and enhanced tsunami excitation of tsunami earthquakes [e.g., *Kanamori and Kikuchi*, 1993]. If this is correct, one might expect shallower thrust events in the interplate thrust zone to have longer rupture durations than deeper events, and relatively long duration ruptures should be observed in regions where significant sediment is subducting compared to regions where no sediment is present. It is not clear what to expect in terms of complexity of rupture, other than the possibility that the shallow environment is intrinsically more heterogeneous in material properties and stress than the deeper portion of the interplate thrust zone.

There have been several investigations of earthquake rupture duration variations with depth in subduction zones, primarily focused on intraplate events within the subducted slab. *Vidale and Houston* [1993] stacked short period P waveforms to estimate source duration of 160 intermediate to deep events (depth > 100 km). They found evidence for a decrease of source duration with increasing depth accompanied by enhanced scatter in duration for intermediate depth events. *Bos et al.* [1998] and *Houston et al.* [1998] stack broadband records for events from 100 to 650 km deep, finding a weak trend of decreasing source duration with increasing depth that can be completely accounted for by the expected increase in shear wave velocity with depth. *Houston et al.* [1998] find evidence of asymmetry and complexity in the stacked time functions for events from 350-550 km depth. *Campus and Das* [Comparison of the rupture and radiation characteristics of intermediate and deep earthquakes, submitted, *J. Geophys. Res.*] do not detect any unusual depth dependence for intermediate and deep focus events for events in Fiji and Japan. Overall, no dramatic depth dependence of rupture properties has been revealed within the intraplate environment of intermediate and deep focus earthquakes.

Many studies of earthquake rupture have been made for individual or suites of events on the megathrust zones, and some have focused on depth dependent systematics. *Tichelaar and Ruff* [1991, 1993] examined variations in maximum depth of interplate thrust events to assess controls on seismic coupling. *Zhang and Schwartz* [1992] analyzed the depth distribution of moment release in different zones to assess variations in thermal structure and stress. *Ekström and Engdahl* [1989] determined source parameters for many earthquakes in the central Aleutian Islands to examine variations in stress distribution, and *Tanioka et al.* [1996] considered lateral variations in earthquake rupture and slab morphology along the Japan trench. *Bilek and Lay* [1998, 1999a] focus on earthquake rupture duration variations for shallow events in the Japan and Middle America subduction zones, finding that the shallowest events have anomalously long source durations in both subduction zones. *Bilek and Lay* [1999b] show evidence for similar behavior in other regions, and propose several possible mechanisms to account for this observation, involving physical attributes of the subduction zone such as roughness of the subducting plate, amount and type of sediment being subducted, thermal structure, and fluid processes in the fault zone. Earthquake behavior in the interplate seismogenic zone can thus guide interpretations of the thermal and petrological structure of subduction zones [e.g., *Peacock*, 1993; *Hyndman et al.*, 1997; *Oleskevich et al.*, 1999]. This study extends the examination of depth dependent variations of earthquake rupture duration in seven subduction zones: Japan, Kuriles, Alaska-Aleutian Islands, Mexico, Middle America, Peru, and Chile.

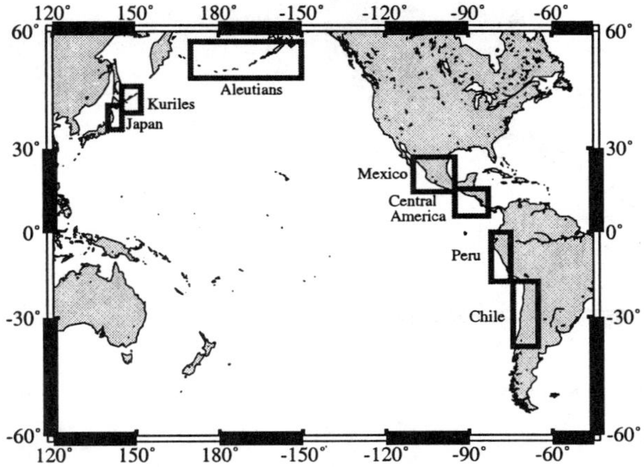

Figure 1. Map showing subduction zones studied. Boxes enclose the areas of event locations.

METHODOLOGY

The subduction zones around the Pacific that we analyze (Figure 1) are selected primarily on the basis of abundant shallow interplate seismicity. Our focus is on frictional and faulting processes along the megathrust zones, and several criteria are used to select events located on the main thrust fault. Earthquakes are initially selected based on: (1) close proximity to the main thrust zone of interest, (2) having a faulting mechanism (from the Harvard Centroid Moment Tensor (CMT) catalog) with strike, dip, and rake consistent with underthrusting of the subducting plate (typically, events have a strike within 20-30° of the local strike of the trench, a dip $<$ 30-35°, and a rake of 90°±30°), (3) having a moment magnitude (M_w) of 5.0-7.5, and (4) availability of at least 4 good quality broadband teleseismic P wave recordings that are well distributed azimuthally from the source (between 7 and 15 P wave recordings are used for most events). These criteria are similar to those used by other authors in earlier studies of depth dependence of subduction zone properties [e.g., *Zhang and Schwartz*, 1992; *Tichelaar and Ruff*, 1991, 1993].

For the selected events, we obtain all available teleseismic vertical component broadband recordings from the IRIS data center with time windows appropriate for the direct P phase and associated depth phases (pP, sP) needed for accurate source depth determination. Ground displacement traces are obtained by deconvolving the instrument response, and the P wave onsets are manually picked on each record. A multistation deconvolution method is then used to determine the source time function that represents the time history of seismic moment release from the source [*Kanamori and Ruff*, 1983; *Ruff*, 1989; *Tichelaar and Ruff*, 1991; *Ruff and Miller*, 1994]. The deconvolution method is based on computing synthetic P wave Green's functions for each event using the best double couple of the Harvard CMT solution for a model with a water layer over a uniform half space with a P wave velocity of 6.0 km/s. We deconvolve the P waves by Green's functions generated for a range of 15-25 point source depths, obtaining source functions at each trial depth. The depth at which the deconvolution minimizes the misfit between the data and synthetic seismograms is preferred. While simultaneous inversion for a revised focal mechanism may reduce some uncertainties, it is not viable for most of our events given the limitations of the available P wave data. In general, tradeoffs between source mechanism and either source depth or source time function are not too severe for shallow thrusting events, and the CMT solutions are probably fairly robust for most of our events. The general processing sequence is shown in Figure 2 for an event in the Aleutian Islands region. The stations used are well-distributed azimuthally from the source, which ensures a range of wave shapes and corresponding Green's functions that reduces the severe trade-offs between depth and source function. From the deconvolution procedure, performed for 23 depths in Figure 2, we determine an optimal depth of 31 km based on the minimum in the misfit curve. The corresponding source function has a simple trapezoidal shape with a duration of 7 s, followed by some low amplitude oscillations. The latter oscillations are highly variable, and represent instabilities caused by accumulating inaccuracies in the Green's functions with lapse time into the signal. This event and the associated P wave signals are well characterized by the strong trapezoidal pulse, although there is intrinsically some subjectivity as to when the true source radiation was finished.

Figures 3-9 show final source time functions for all events in the 7 subduction zones. The panel for each event shows the source time function for the optimal source depth and our preferred depth and source duration. In some cases, the depth determination is robust, with a single distinct minimum in the misfit curve. However, in other cases, there are double or multiple minima in the misfit curves, or a range in depths with relatively uniform misfit. For these cases, we examine the source time function produced for each depth, and choose the depth which yields the simplest time function, with most of the moment released early in the signal [*Christensen and Ruff*, 1985].

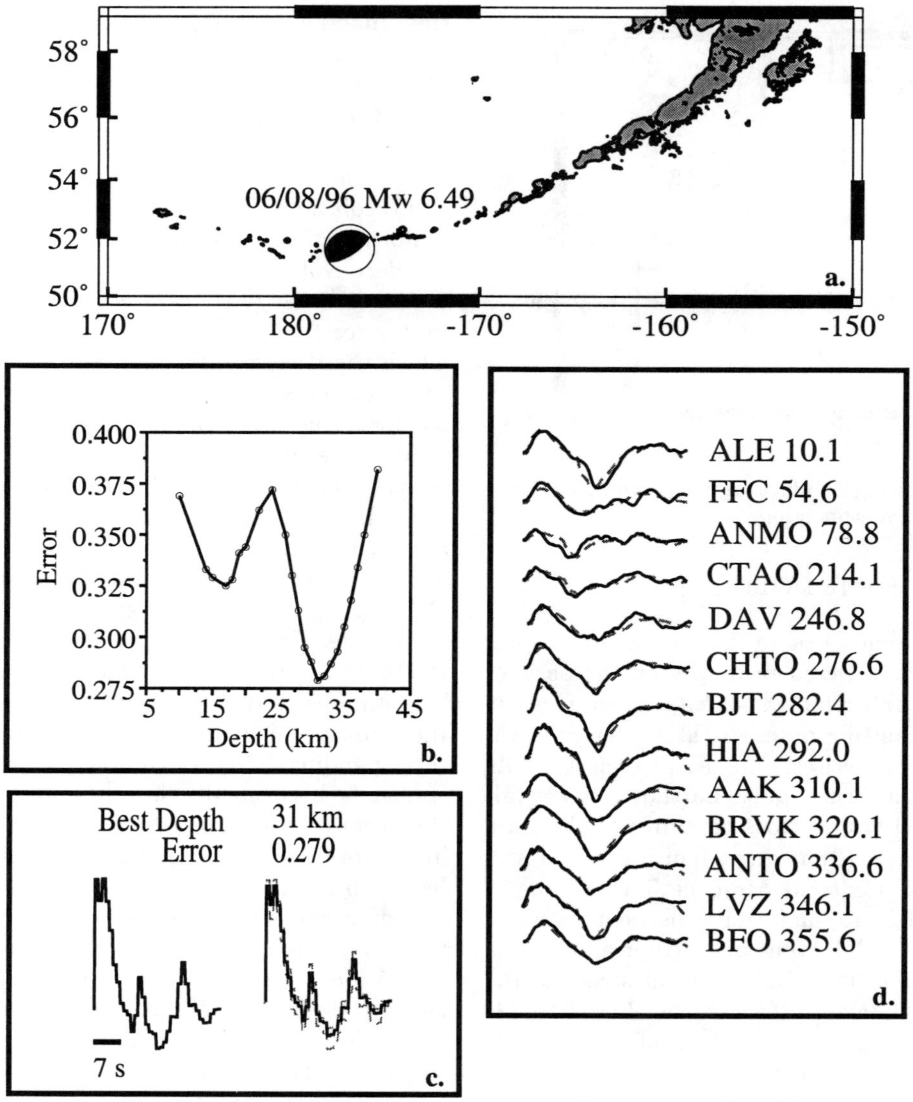

Figure 2. Example of data processing for M_w=6.49 event in the Aleutian Islands region. (a) Focal mechanism of the event taken from the Harvard CMT catalog. The event was chosen because of its underthrusting mechanism (strike 262°, dip 24°, rake 114°), close proximity to the trench, and moderate magnitude. (b) Error or misfit as a function of depth for the deconvolutions. We perform the deconvolution for 23 different depths to minimize the misfit between the data and synthetics. Minimum occurs at 31 km depth. (c) Source time function for deconvolution at 31 km depth. On left is source time function, on right is source time function with misfit bounds in gray dashed lines. Source duration is measured from the first large peak of the source time function; the black bar indicates the measured duration of 7 s. (d) Data (solid black lines) and synthetic seismograms (dashed) shown with the station code and azimuth from the event.

The point-source time functions in Figures 3 to 9 further illustrate the difficulties intrinsic to measuring the time interval of source energy release. As in Figure 2, the deconvolutions sometimes show significant amplitude oscillations after the main pulse of moment rate, or significant negative overshoot following the primary pulse. For the largest events, unparameterized spatial finiteness may contribute to these features, but their sensitivity to source depth suggests that inaccuracy of the later portions of the Green's functions is the primary

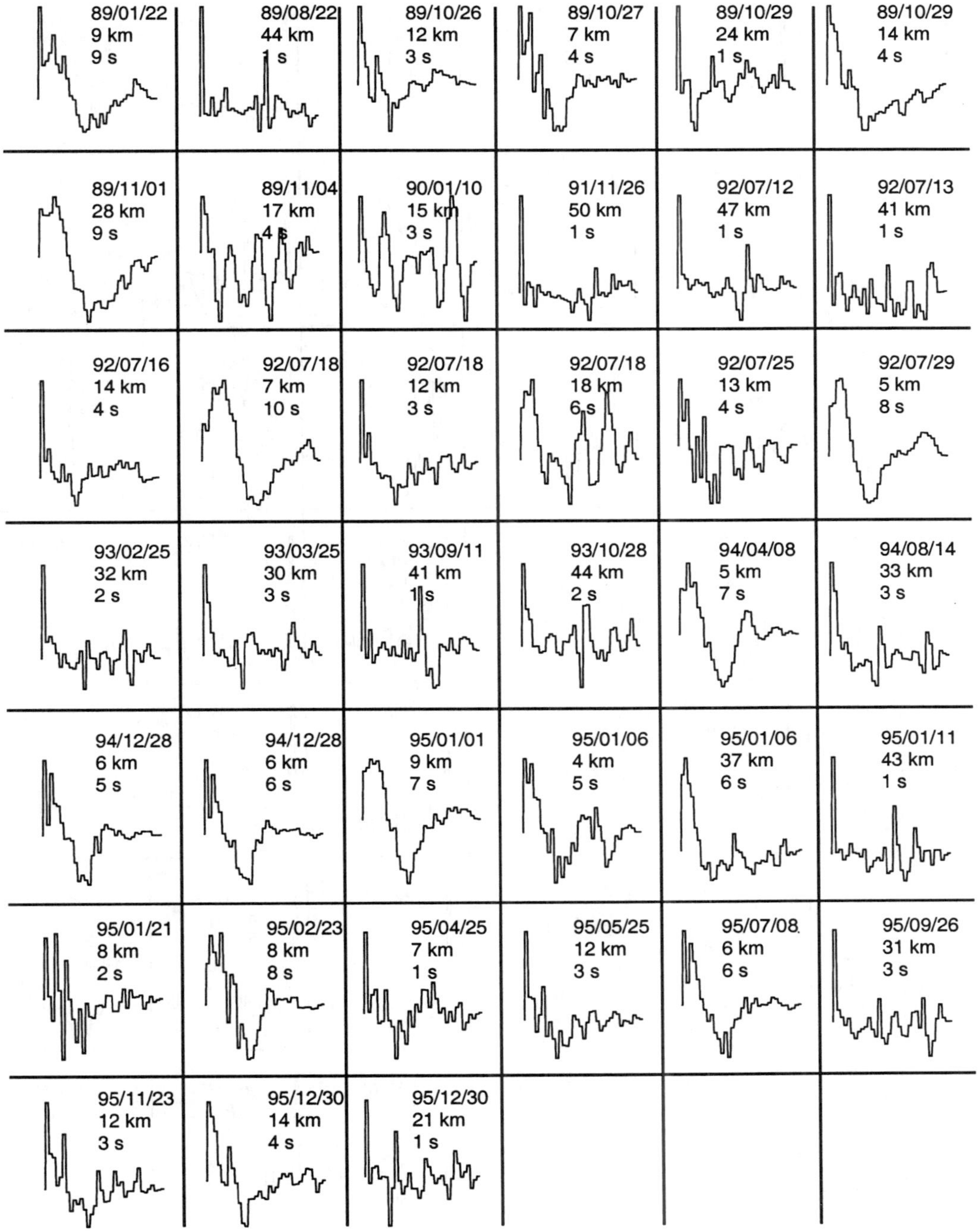

Figure 3. Source time functions for Japan events. Each event panel lists the event date, the best source time function, and the preferred depth and duration. Duration is measured from the source time functions, with each horizontal step of 1 s.

culprit. This is not unexpected given possible errors in the focal mechanism and the very simple velocity model assumed in computing the Green's functions. Until realistic near-surface and wedge velocity structure is known in each region and fully three-dimensional Green's functions can be computed for such structures, it will be hard to improve the accuracy of the time function estimates. The time function complexity makes it difficult

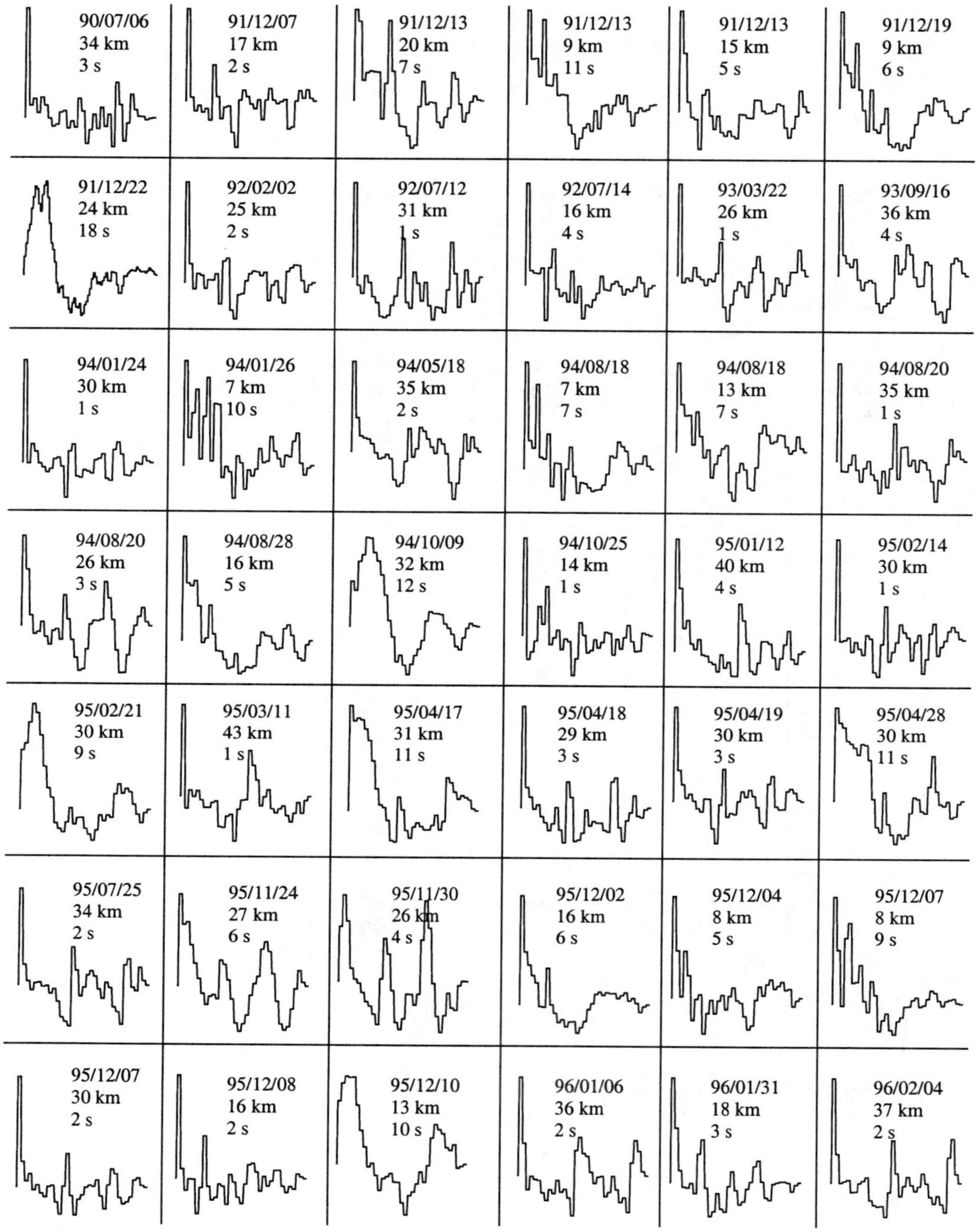

Figure 4. Source time functions for Kurile events, in same format as Figure 3.

to define the cessation of coherent energy release from the source, and in some cases there may be isolated secondary pulses of source radiation. However, every case shown does have a primary pulse of moment rate in the early part of the signal that accounts well for the predominant features of the teleseismic P waves. We proceed by measuring the duration of the primary positive pulse of the source time function, which usually suffices to characterize the energy release history. We attempt to be consistent in measuring the width of this

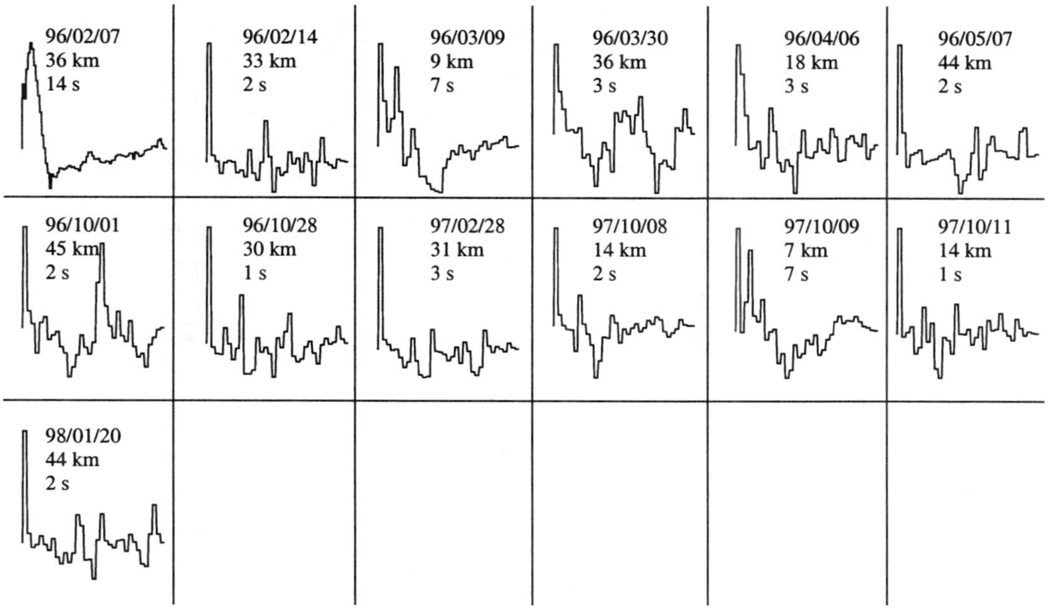

Figure 4. Continued

pulse, defining the end as when the pulse returns to a baseline level. We incorporate error bars on the duration estimates to subjectively reflect the difficulties in measuring the time function duration.

There are several trade-offs in determining the source duration and depth parameters. Source time function duration and depth have particularly strong trade-offs for individual stations [e.g., *Christensen and Ruff*, 1985]; however, by performing the deconvolution for many well distributed stations, the ability to separate these parameters is greatly improved. Additionally, there are trade-offs between velocity model and source depth. The choice of a half space velocity of 6.0 km/s undoubtedly biases our depth estimates. Tests using a different average velocity show that a 10% change in average velocity leads to a corresponding 8-12% change in source depth (approximately 3-4 km), similar to the results of *Tichelaar and Ruff* [1991]. Given the likelihood of low velocities in the sedimentary wedge above the thrust plane, we probably overestimate true depths by several kilometers, particularly for the shallowest events. Our depth estimates still tend to be shallower than those in the CMT catalog (Figure 10), partly due to the higher average crustal and upper mantle velocity of the Preliminary Reference Earth Model (PREM) [*Dziewonski and Anderson*, 1981] structure used in the CMT inversion. There are also differences caused by the practice of fixing CMT depths at either 15 km or 33 km for shallow earthquakes when the depth resolution is not good. Holding the focal mechanism constant is another source of possible error in the depth determinations, but tests of the effects of focal mechanism uncertainty indicate that the depth estimates are not very sensitive to small focal mechanism changes. The error bars that we assign to the depth estimates reflect these sources of error. The point-source assumption explicit in our deconvolutions clearly results in some averaging of spatial finiteness effects that may bias our source duration estimates. Estimation of source radiation duration from the source time functions involves an approximation that there is a simple rupture process that yields negligible directivity effect. As most of our events are of moderate size, and the deconvolution process emphasizes wave periods longer than 1 or 2 s, the directivity effects in teleseismic *P* waves are likely to be below the resolution of our measurements in almost all cases. The good fit of the point source synthetics to the observations gives direct support for this assertion. The residual waveform mismatch error varies somewhat, but not in a fashion simply linked to the shape of the source time functions or the event moment (see Figure 2). As a first-order approximation, we take the duration of the main pulse of the source time function as an estimate of the actual rupture duration. We correct the rupture duration estimates for the effect of varying seismic moment (M_o), as it has been empirically shown that the duration is proportional to the cube root of M_o [*Kanamori and Anderson*, 1975; *Houston et al.*, 1998; *Campus and*

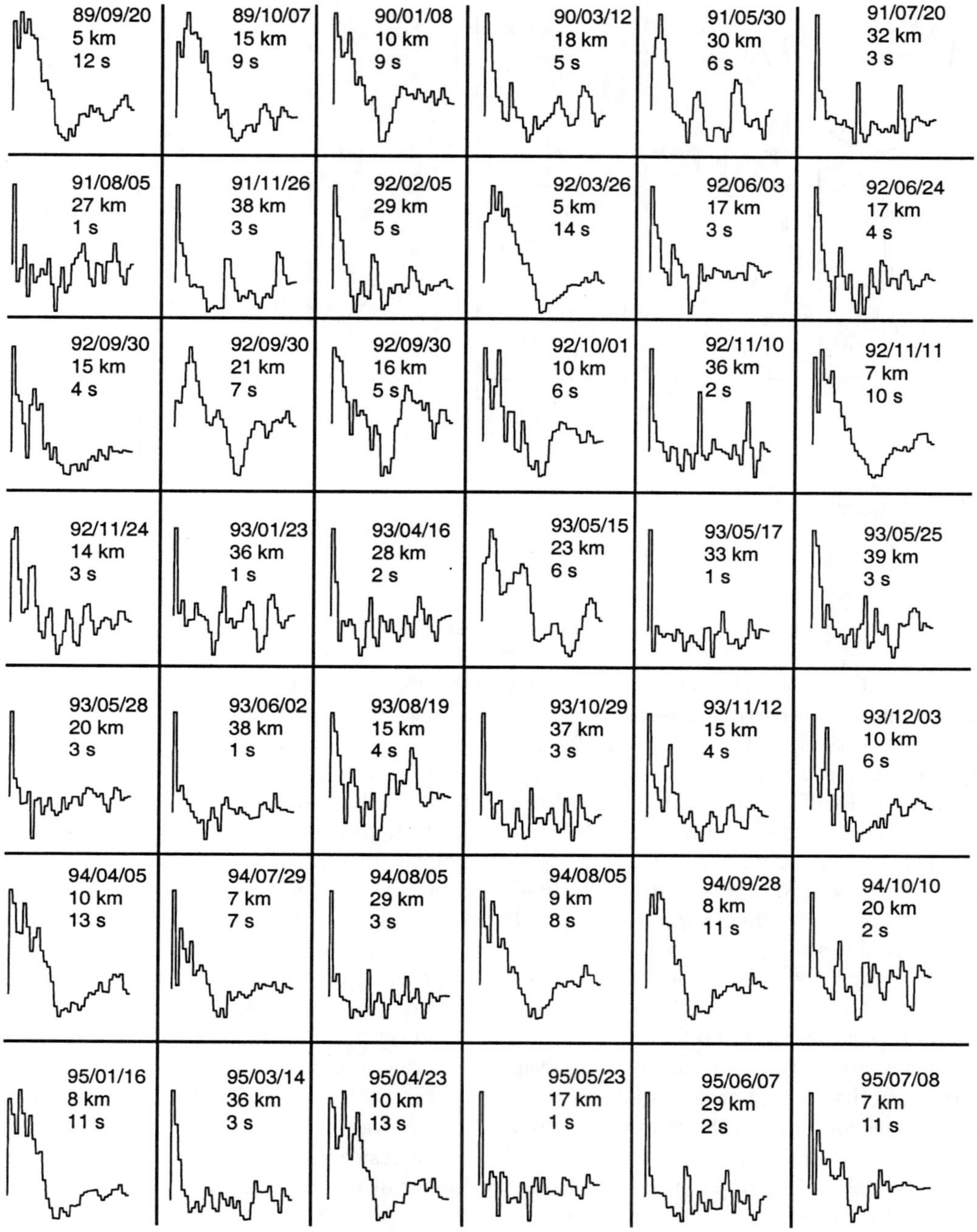

Figure 5. Source time functions for Alaska-Aleutian Islands events, in same format as Figure 3.

Das, 1999]. Each duration estimate is divided by the cube root of the associated M_o from the Harvard CMT catalog, normalized to a moment magnitude (M_w) 6.0 event. The CMT moments are used for scaling rather than the moments determined from the deconvolution process, as we feel the CMT moments are more stable estimates because they are derived from large numbers of long-period data. Our deconvolution moments tend to underpredict the values found by Harvard, in part due to the velocity model used, and possibly due to

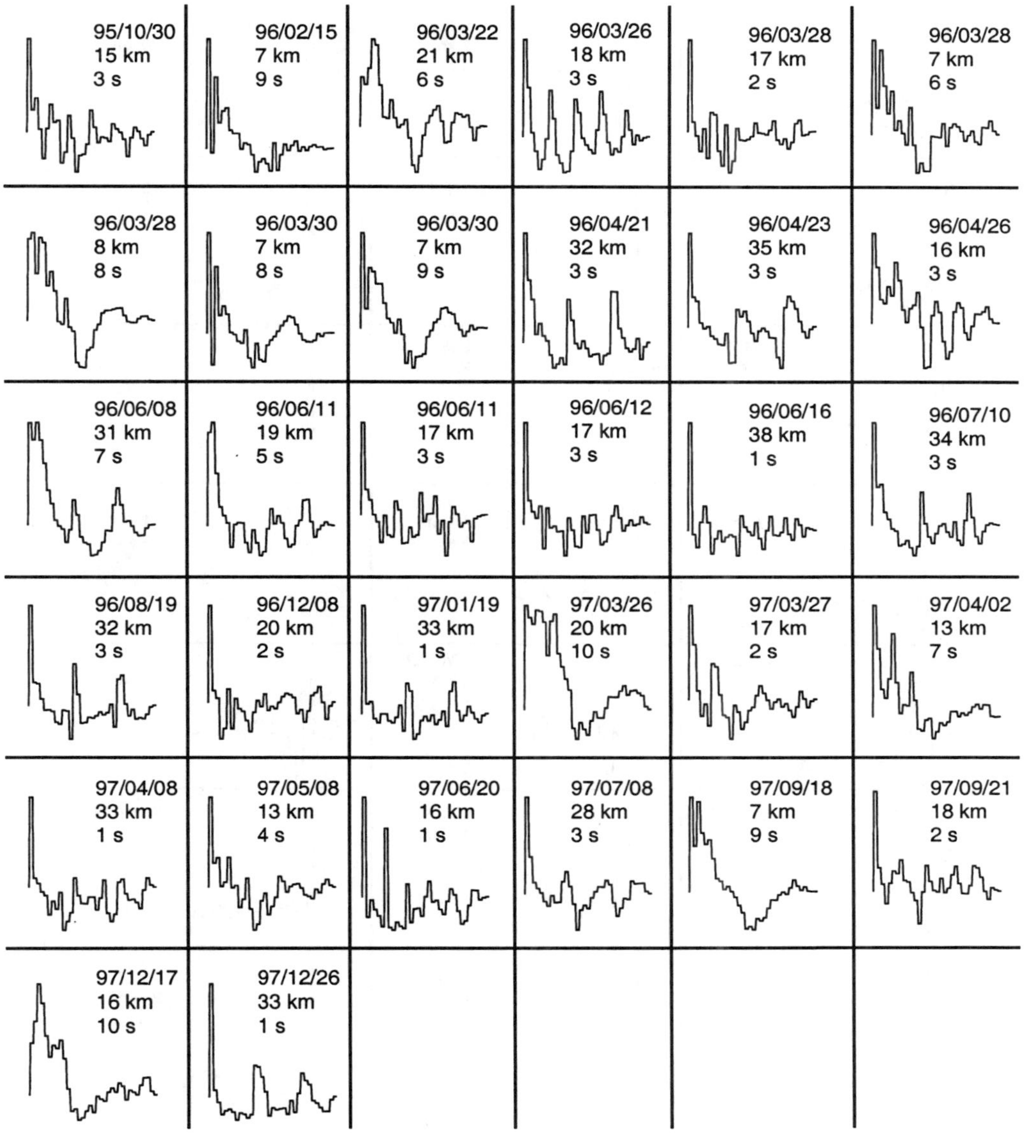

Figure 5. Continued.

omission of later low amplitude energy release. Scaling by the cube root of moment is widely practiced, and allows our results to be easily compared with other studies [*Ekström and Engdahl*, 1989; *Vidale and Houston*, 1993; *Bos et al.*, 1998; *Houston et al.*, 1998], after taking into account the differences in the reference moment used in the various studies. Figure 11a shows the observed relationship between our duration estimates and M_o. The unscaled durations show a clear increase with increasing M_o, while the scaled durations are free of any dependence on event size. Our data are also free of any systematic relationship between source depth and seismic moment (Figure 11b), which is important in assessing the results.

RESULTS

Figure 11c shows normalized duration estimates as a function of source depth for the entire dataset, representing 354 events. We include 68 events from a

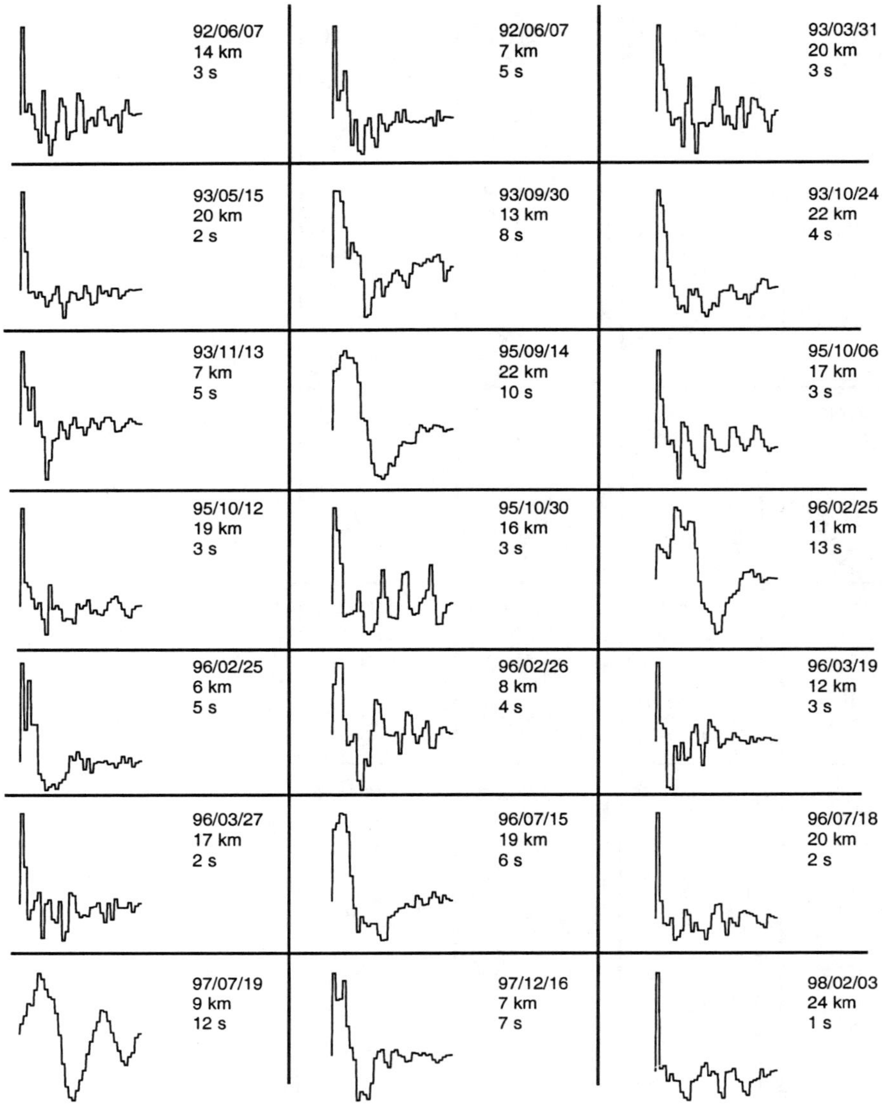

Figure 6. Source time functions for Mexico events, in same format as Figure 3.

study of maximum coupling depth in subduction zones [*Tichelaar and Ruff*, 1991; *Tichelaar and Ruff*, 1993], which provided depth and duration estimates by the same general procedure used here. Those duration estimates are also scaled using cube root of moment scaling. The trend of decreasing duration with increasing depth is very clear, even in this composite plot which combines data from subduction zones with different regional characteristics. A source duration of about 3 s is typical of an $M_w = 6.0$ event, but scaled durations as long as 18 s are found for events shallower than 15 km depth. With typical error bars being ± 1.5 s, the long duration estimates appear to be real anomalies.

Figure 12 shows the same data, subdivided by geographic region and with error bars on depth and duration values. The trend of decreasing source duration with increasing depth is apparent for each region, although some subduction zones have stronger patterns than others. For instance, data for the Alaska-Aleutian Islands region show a very dramatic decrease in duration down to a depth of 15-20 km, then a flatter distribution at depths deeper than 20 km. The Kuriles,

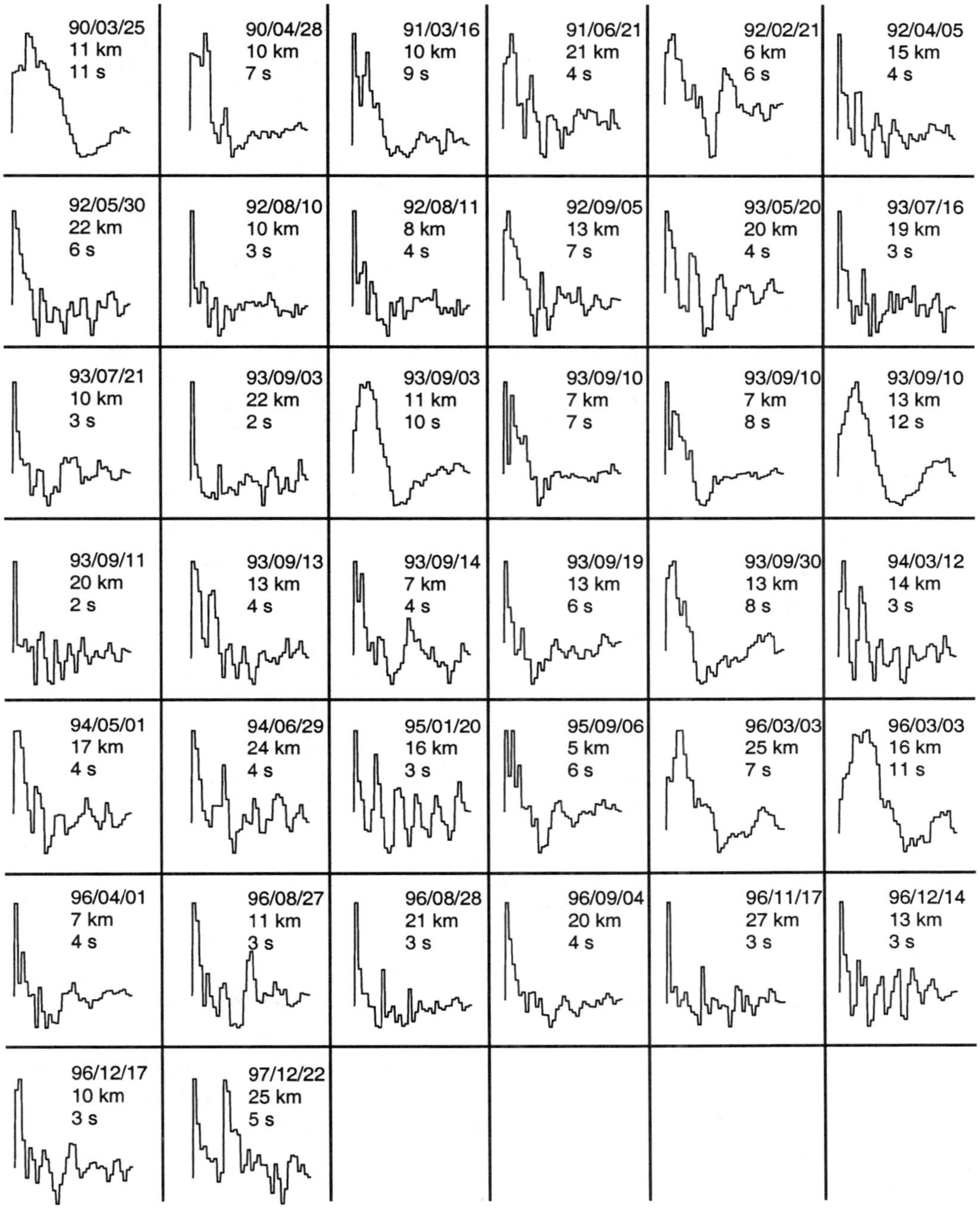

Figure 7. Source time functions for Central America events, in same format as Figure 3.

Mexico and Peru show similar abrupt changes at depths shallower than 25 km, but these three regions have fewer events than the Alaska-Aleutian Islands region. The South American zones have slightly shorter duration plateaus at depths greater than 20 km than do Japan and the Aleutians, which may reflect differences in crustal structure overlying the subduction zones. Central America displays the weakest variation with depth, with events less than 10 km deep showing some increase in duration. The scatter in duration is

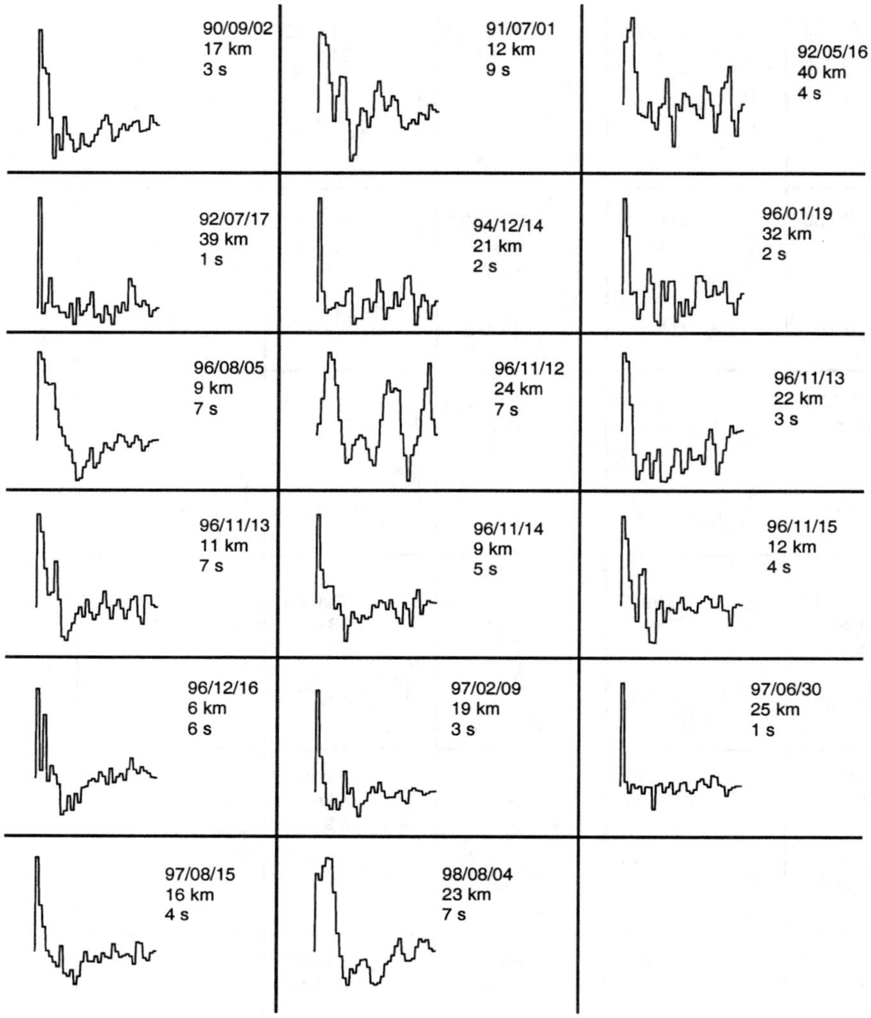

Figure 8. Source time functions for Peru events, in same format as Figure 3.

comparable between regions for depths greater than 10 km, and indicates the intrinsic variability in earthquake rupture processes and errors in duration estimation.

Given that our depth estimates differ from both the CMT values and earthquake bulletin values, we seek to confirm that the events are truly on the megathrust, and not intraplate ruptures. Plate 1 shows our depth and duration data plotted as a function of distance perpendicular to the regional trench axis. This distance was measured using bathymetry maps to track the trench axis. Events located within the dashed gray boxes are considered likely to lie on the main plate interface, which may itself involve multiple thrust planes, within a ±5-10 km region. Those events which lie outside the box may either be mislocated (laterally or in depth), are intraplate earthquakes with thrust mechanisms similar to underthrusting at the plate boundary, or are thrust events occurring in the accretionary wedge above the seismogenic zone. The events define dipping plate interfaces in each region, with some of the scatter attributable to changes in plate geometry. The distributions are much tighter than provided by either the CMT or earthquake bulletin locations, so our depth estimation process generally appears to have succeeded. In regions that exhibit a lot of scatter and where there is significant along-strike geometry change, we binned the data to ensure that we include only events that are confidently identified as interplate thrust events. The data still show a decrease in source duration with increasing depth along the plate interface for each region,

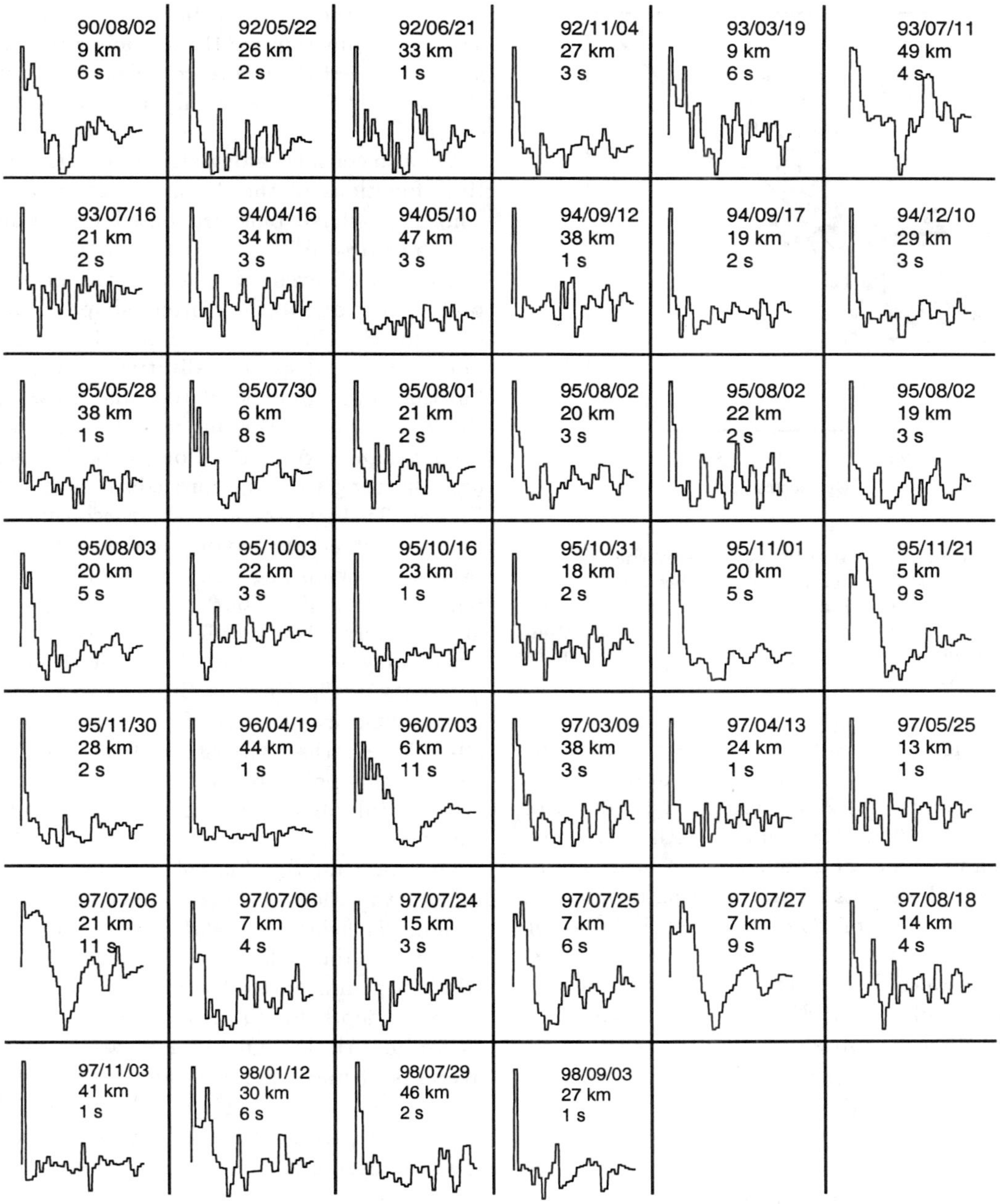

Figure 9. Source time functions for Chile events, in same format as Figure 3.

mainly because very few of the shallowest events have been excluded.

ALASKA-ALEUTIAN ISLANDS DATASET

The strong pattern and large number of data in Figure 12 for the Alaska-Aleutian Islands region prompted a more detailed look at the seismic parameters for events in this area. There have been many great earthquakes as well as tsunami earthquakes in this region, which emphasizes the importance of understanding the nature of the seismogenic zone here. Figure 13a shows the locations of the events analyzed for the Alaska-Aleutian Islands region; a total of 74 earthquakes from 1989 to 1997. We consider the patterns in the data in both depth and along-strike bins.

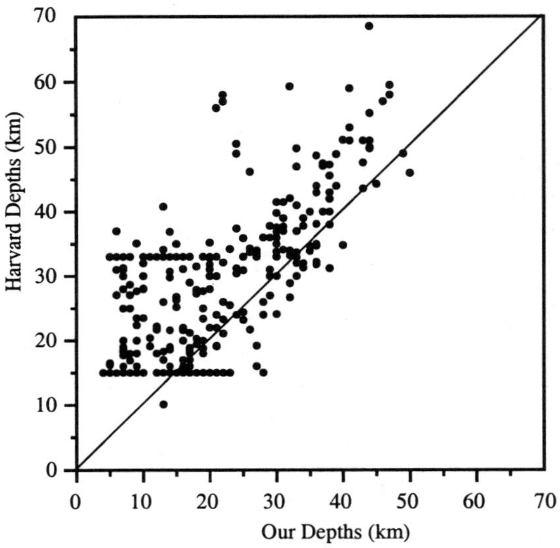

Figure 10. Comparison of depths as listed by the Harvard CMT catalog and depths determined in this study. Our depths tend to be more shallow than the Harvard CMT depths, and our depths do not show any bias towards certain depths, such as the large grouping of Harvard CMT depths of 15 km and 33 km.

Figure 13b shows the average source time functions along the entire Alaska-Aleutian arc for four depth bins of 0-10 km (20 events), 10-20 km (26 events), 20-30 km (10 events), and 30-40 km (18 events). Both the seismic moment rate amplitudes and the source durations were scaled to remove the effects of varying seismic moment, following the procedure of *Houston et al.* [1998], prior to binning the events and averaging the time functions. The time scales were divided by $\sqrt[3]{M_o}$, normalized by the reference M_w 6.0 event (M_o 1.16x10^{25} dyne-cm). The moment rate amplitudes were divided by the event $M_o^{2/3}$ in order to maintain the relationship between moment and area under the time function. Individual time functions were set to 0 prior to the beginning of the large pulse of energy in the source time function, and 0 after our pick of the termination point of the faulting radiation. The stacked source functions for each depth range have been scaled so that the area beneath each curve is equal to 1.16x10^{25} dyne-cm, to facilitate comparison. The most obvious feature is the dramatic difference in the shape of the source time function for the 0-10 km bin, as this visually demonstrates the general trend of decreasing source duration with increasing depth. The time function for the 0-10 km bin has a duration of 16.4 s while the time function for events greater than 30 km has the shortest duration of 7.0 s. The average source function for the shallow events suggests evidence of greater rupture complexity as well, although the subevent features are highly variable from event to event.

We also consider along-strike variations in the source time functions of the Alaska-Aleutian Islands events. The along-strike bins are based on distinct tectonic blocks as described by *Geist et al.* [1988]. Initially we considered all events in each spatial bin, but this was problematic due to the distinct shape of the time functions for the shallowest events. The shallowest events in each region display qualitatively similar broadening, but the total numbers are too small to average just the shallowest events. For this reason, we stack only events deeper than 10 km in the along-strike bins shown in Figure 13a using the same procedure as described above. Figure 13c shows the average source time functions for each subregion, plotted on same scale to facilitate comparison between regions. Four of the six bins look remarkably similar, but events in the Unimak Block and Shumagin Block differ somewhat from the others. There may still be some depth effect in these stacks, as events in the Unimak Block have mainly 10-20 km depths while events in the Shumagin Block have mainly 30-40 km depths. The cause of these variations is unclear at this point, but lateral variations along the arc are certainly less pronounced than the common depth variation.

Our data for the Alaska-Aleutian Islands events do suggest variations in maximum coupling depth for this region. *Tichelaar and Ruff* [1993] examined 11 events in the Aleutian Islands region and 5 events near the Alaskan peninsula to determine an average maximum coupling depth for the Aleutians of 35-41 km and 37-41 km for Alaska. Our larger data set supports these averages of maximum coupling depth, although it appears that there is significant variation in the coupling depth along strike of the trench, with some sections of the seismogenic zone having much deeper coupling than other sections. *Tichelaar and Ruff* [1993] mention that the easternmost portion of the 1957 earthquake rupture zone may have shallower coupling; this region corresponds approximately to the Unimak Block bin, where we see earthquakes only down to 20 km depth. We also observe the possibility of shallower coupling around longitude 180° within the Rat and Delarof Block bins, with deeper coupling found on either side. We are only considering events with M_w 5-7.7, so our results on coupling depth are appropriate for that magnitude range.

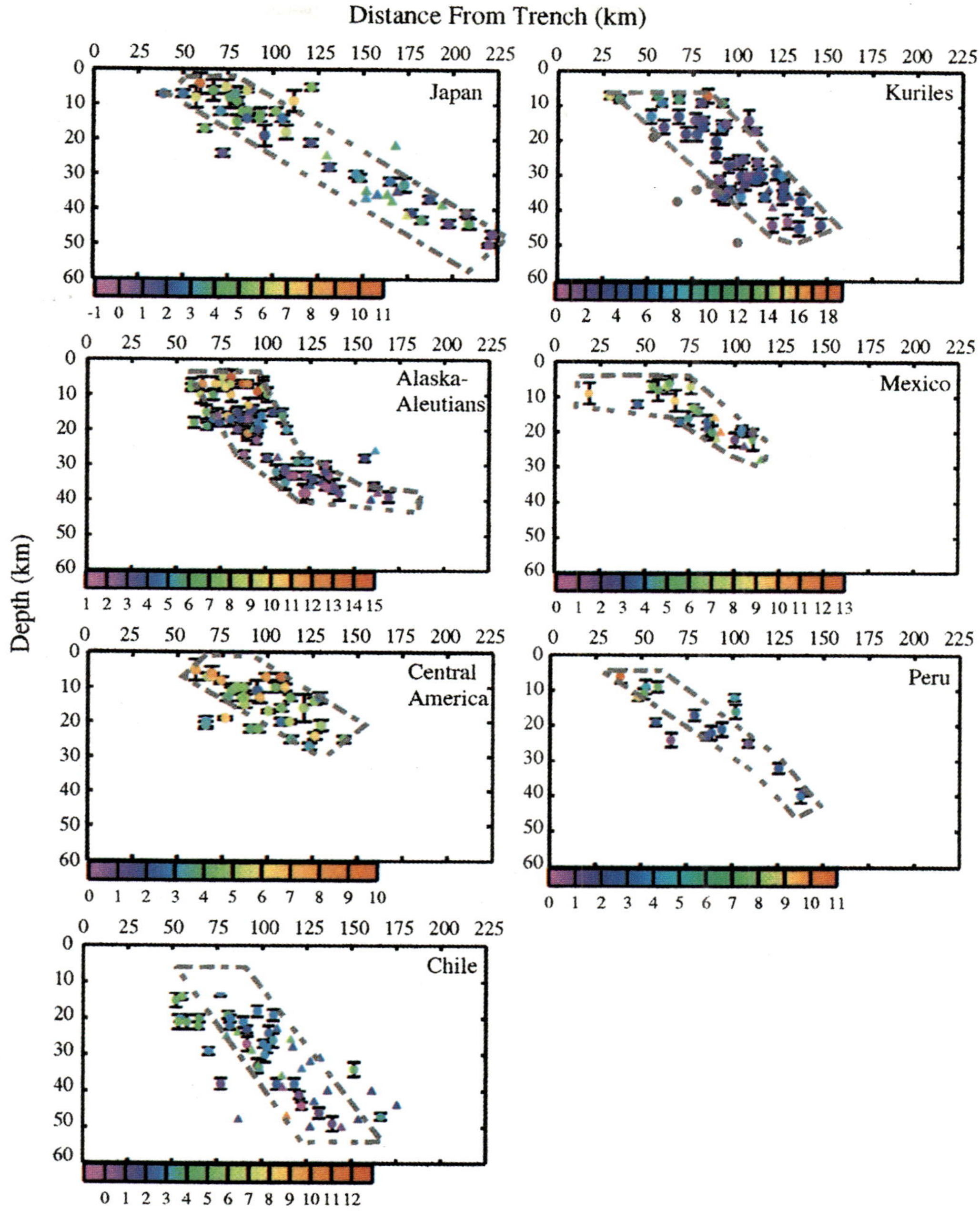

Plate 1. Normalized source duration and source depth as a function of distance from the respective trench. The color bar below indicates source duration for each symbol. Error bars show range in depth estimates from the range in misfit from the deconvolution processing. The dashed boxes enclose those events that define the plate interface, allowing for some scatter due to variation in the along-strike subduction zone geometry. Events which lie outside the box are likely either accretionary wedge or intraplate events (gray symbols in Figure 12). Triangles represent events determined by *Tichelaar and Ruff* [1991, 1993].

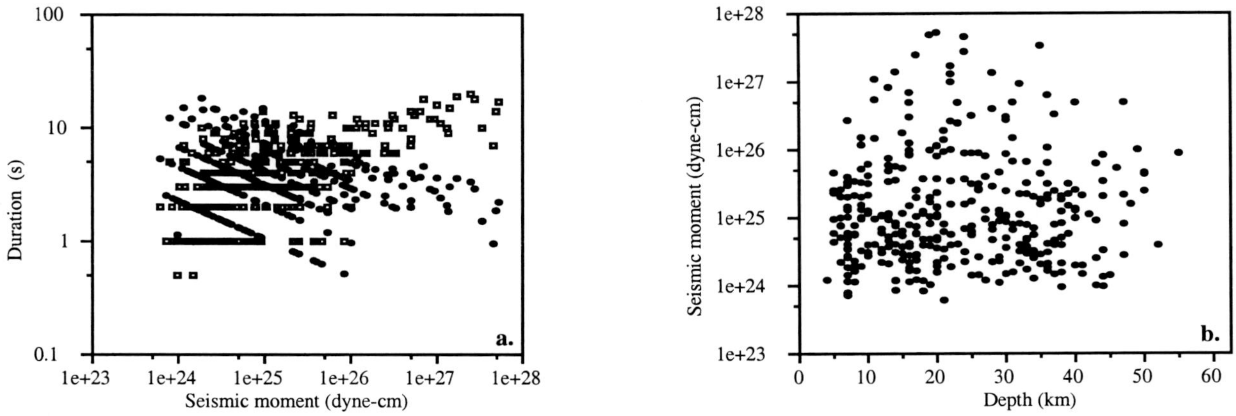

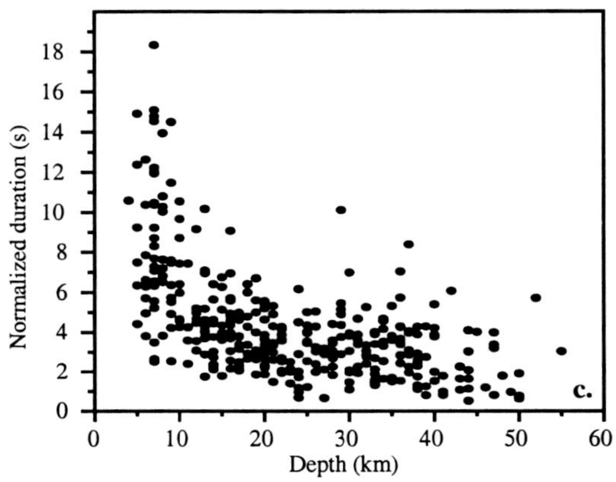

Figure 11. Relationship of source duration and depth with seismic moment. (a) Plot of source duration as a function of moment. Event moments are taken from the Harvard CMT catalog. Open squares represent raw durations, which show an expected increase with increasing moment. Linear pattern of symbols results from the fact that source duration is measured from the source time function in 1 s intervals. Solid circles represent normalized source duration as a function of moment. The increase in source duration with increasing moment has been removed from the data. (b) Seismic moment as a function of event depth. We see no trend in this plot, with moments scattered over our depth range. (c) Composite plot of normalized source duration as a function of depth. There is an overall trend of decreasing source duration with increasing depth, and wide scatter is likely due to the composite nature of the plot.

As data accumulate, it may prove viable to map out lateral variations in faulting complexity and coupling depth in greater detail, but the present results indicate that rather subtle variations will be found.

DISCUSSION

Further mapping out of lateral variations in behavior of earthquake ruptures within each zone and between zones remains a desirable goal, but for the remainder of this study we focus on the common depth-dependence apparent in Figure 11c. Seismological analysis reveals only gross attributes of the source energy release, and there is substantial non-uniqueness when interpreting source time function characteristics in terms of dynamical behavior. We will consider several simple end-member possibilities to frame the problem. One possible explanation for the observed variations in source duration is systematic variation with depth of rigidity of the material in the seismogenic zone. This is an extension of the notion that tsunami earthquakes have slow rupture velocities because the slip occurs in low rigidity sediments at very shallow depths. Rigidity is a key material property that can be related to source duration through its direct influence on shear wave velocity. For a simple unilateral rupture model, source duration is inversely related to rupture velocity. The rupture velocity (V_r) is empirically found to be approximately equal to $80\% \pm 10\%$ of the shear wave velocity (β) [*Scholz*, 1990],

$$\beta = \frac{\mu}{\rho} \tag{1}$$

where μ is the rigidity and ρ is the material density. If we assume a constant static stress drop model for scaling, variations in the duration can be associated with variations in rigidity. The constant stress drop model appears valid for a range of earthquake magnitudes [e.g., *Abercrombie*, 1995], although we do expect scatter to result from variations in stress drop. Thus we can use our source duration measurements to estimate volume-averaged (over the range of large strain accumulation) rigidity variations with depth in the seismogenic zone.

In order to calculate the rigidity, we use a constant density of 2.7 g/cm^3. It is clear that the density will

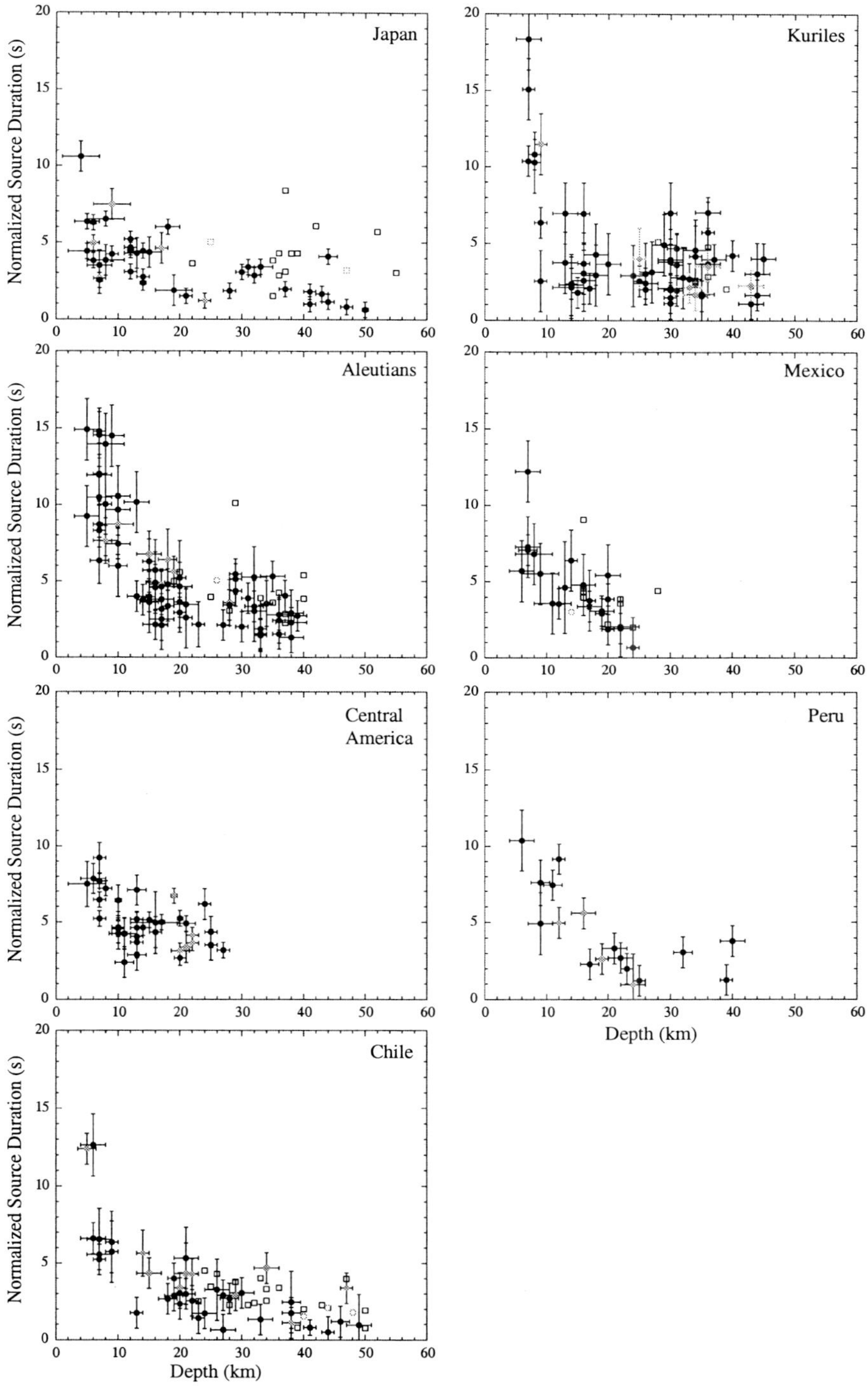

Figure 12. Normalized source duration as a function of source depth for each of the analyzed regions. Open squares indicate events analyzed in a similar fashion by *Tichelaar and Ruff* [1991, 1993]. All of the 354 event source durations have been scaled using the cube root of seismic moment, normalized to a M_w=6.0 event. Gray symbols indicate those thrust events satisfying our initial criteria, but have depth estimates suggesting that they did not occur at the plate interface.

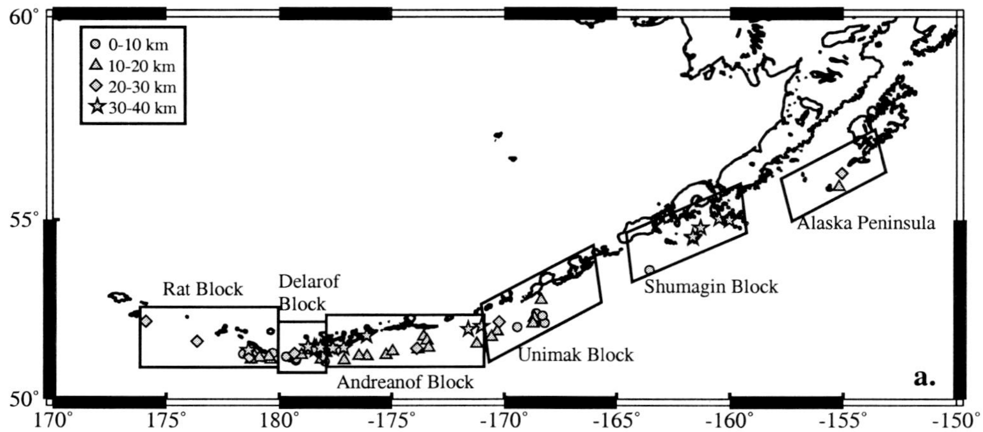

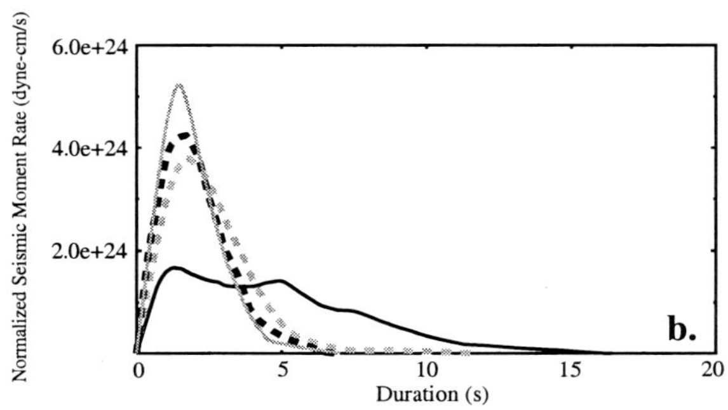

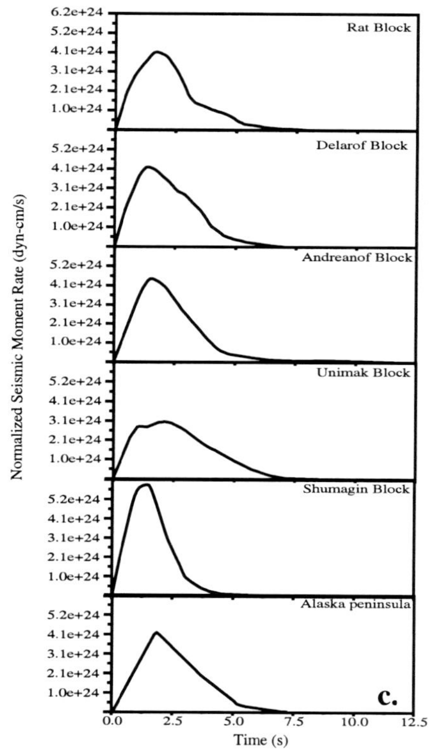

Figure 13. (a) Detailed map of Alaska-Aleutian Islands study area. Symbols show NEIC location of events with symbol shape denoting depth of event (see legend). Boxes indicate distinct tectonic blocks from *Geist et al.*[1988]. (b) Average normalized time functions for the depth bins 0-10 km (solid black), 10-20 km (dashed gray), 20-30 km (dashed black), and 30-40 km (solid gray). (c) Along strike bins of source time functions for blocks indicated in (a). Average source time functions were generated in same way as in (b), however only events with depths greater than 10 km were included in these bins.

change somewhat with depth, but we use a constant density for simplicity. The other parameter needed to estimate rigidity is a source dimension, as the source duration (τ) equals

$$\tau = \frac{\text{source dimension}}{V_r}. \tag{2}$$

We assume a uniform source dimension of 10 km for our moment-scaled rupture durations. This dimension is consistent with the 3-4 s duration for a typical M_w 6.0 event, with a 3.0-3.5 km/s rupture velocity. This choice of a rupture dimension affects the baseline of the rigidity values, but not any systematic depth variations. It is not presently possible to independently determine the rupture dimensions for all of our events using directivity analysis because of their small magnitude. Since it is unlikely that they all rupture precisely (scaled) 10 km dimensions, we expect significant scatter to remain.

Figure 14 shows the seismogenic zone rigidity estimates obtained for the seven subduction zones considered in this study. There is an order of magnitude

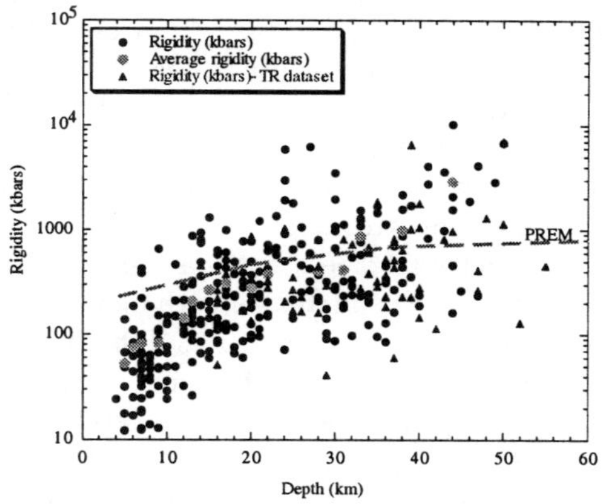

Figure 14. Plot of estimated rigidity variations along the megathrust for the entire dataset. Small black circles indicate rigidity values calculated from our measured source durations, triangles indicate rigidity estimates for the 68 events in the *Tichelaar and Ruff* [1991, 1993] datasets, and open circles are average rigidity values over a depth interval. The solid black line indicates the rigidity values estimated from PREM shear wave velocities and densities.

scatter in the data at all depths, but we find a general trend of increasing rigidity with increasing depth in the range of 5-50 km. Rigidity values estimated from PREM are included in the figure to provide a reference earth comparison for a layered crust/mantle intraplate oceanic environment. Our average values are very similar to PREM at depths of 20-40 km, but are lower than PREM by a factor of as much as 5 at depths shallower than 20 km. Given the likely contributions to error in the rigidity estimates from the factors described above, only the average trend should be considered, not the full range of values.

One concern inherent in these estimates is the use of Harvard CMT moments in scaling the source durations. These moments are calculated using the PREM structure for seismic wave excitation. If the rigidity variations in Figure 14 are correct, the excitation should be recomputed for a corresponding decrease of rigidity near the surface. Overestimating the rigidity in the modeling may have yielded seismic moments that are too small for shallow events, which would underestimate the corrections for moment scaling. A quantitative correction for this effect is very difficult, both because the rigidity is not independently known, and the excitation should be computed for a realistic three-dimensional model to obtain unbiased moments. We believe that the overall effect of such correction for moment scaling would be a slightly reduced range of rigidity values, but the general trend would be preserved.

Studies of tsunami earthquakes have inferred comparable (factor of 5 to 10) reductions of rigidity for the seismogenic zone necessary to reduce rupture velocities to about 1 km/s and to account for the enhanced slip needed to satisfy tsunami excitation given the observed seismic moments [*Kanamori*, 1972; *Pelayo and Wiens*, 1992; *Kanamori and Kikuchi*, 1993; *Satake*, 1994; *Heinrich et al.*, 1998]. Thus, these calculations suggest that all subduction zones may have shallow regions with low rigidity properties that could enable large tsunami earthquakes to take place. Whether they do or not is likely to be a consequence of the slip history of the deeper portions of the seismogenic zone, along with the myriad factors that control the transition from stable sliding to stick-slip instabilities.

If the rupture velocity is decreased because of low shear velocities, a fairly thick zone of reduced rigidity must be present to affect the volumetric strain release during rupture. The rigidity estimates in Figure 14 would then be averaged over the seismogenic zones, and local properties right at the plate interface may vary even more. The fact that our rigidity estimates do not display any seismic moment dependence (especially when very large tsunami event results are included), indicates that the scale of the region of low rigidity may be substantial. Various factors could produce a distributed zone of low rigidity material near the fault zone. A thick zone of sediments both above and within the seismogenic zone is a likely candidate for significant reduction of rigidity relative to hard rock values. Lower rigidities can also be associated with increased water content in the material, which is abetted by having high porosity sediments. Increased water content can also be the result of phase transitions such as smectite to illite, and even basalt to eclogite, which can release water from the hydrous phase.

Other factors may be important. Previous studies have examined the depth dependence of moment release [*Zhang and Schwartz*, 1992], maximum coupling depth [*Tichelaar and Ruff*, 1991; 1993], stress drop and source durations for deep earthquakes [e.g., *Ekström and Engdahl*, 1989; *Vidale and Houston*, 1993; *Bos et al.*, 1998; *Houston et al.*, 1998], and lateral variations in earthquake occurrence [*Tanioka et al.*, 1997]. These authors have invoked a number of possible factors for causing the variability in earthquake rupture, such as amount and types of sediment being subducted, thermal state, hydrologic effects, and changes in subducting plate roughness. Systematics in any of these factors is also a plausible cause for the depth dependence of the source duration and rigidity. However, with the current

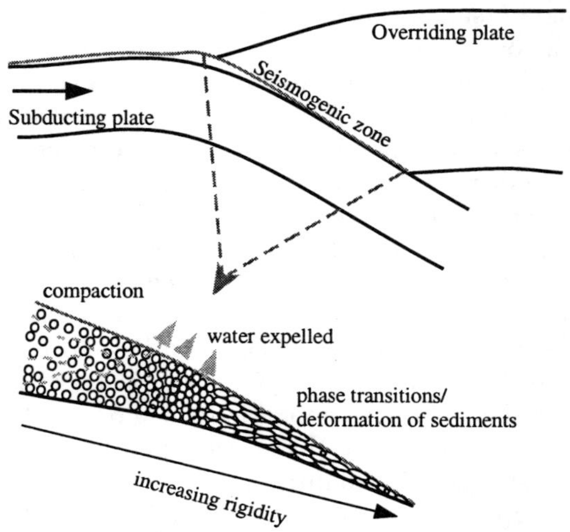

Figure 15. Generalized schematic of a subduction zone, with a detailed blow up of the seismogenic zone interface, modified from *Bilek and Lay* [1999b].

data set, it is difficult to determine which is the most important cause of the variations. Figure 15 shows a schematic of the seismogenic zone indicating some of the possible processes likely involved in changing the material rigidity.

The model of rigidity variations is non-unique. We assumed that the scaled rupture area of the events is constant with volumetric material properties controlling the variations in duration. Another end member model involves a constant rupture velocity with a varying rupture area for each event. Such a model implies static stress drop variations with depth. We lack independent estimates of the fault area for each event, which is needed to resolve the trade-off between rupture area and rupture velocity [*Vidale and Houston*, 1993]. Instead we relate static stress drop to seismic moment through its relationship to fault displacement and area. For a circular crack model, static stress drop is

$$\Delta\sigma = \frac{7\pi M_o}{16r^3} \tag{3}$$

where r is the radius of the circular fault area and M_o is the seismic moment [*Kanamori and Anderson*, 1975]. If we make a further assumption that the rupture velocity approximately equals the shear velocity, we can substitute the product of the shear velocity β and the measured source duration τ for the fault radius, leaving

$$\Delta\sigma = \frac{7\pi M_o}{16\beta^3\tau^3} \tag{4}$$

for the static stress drop.

Figure 16 shows corresponding estimates of the static stress drop as a function of depth for each of the studied regions. For each calculation, we use the Harvard CMT catalog moments and a constant $\beta = 3.5$ km/s. The shear wave velocity will likely vary with depth because of rigidity variations with depth, but we use a constant value for simplicity for these calculations. This model predicts that stress drop increases with increasing depth ranging over an order of magnitude. This is certainly plausible, however, a study of dynamic stress drop for recent large earthquakes shows that the stress drop estimated from source time functions is basically constant for the events once the change in shear velocity (and therefore rigidity variations) with depth and region is taken into account [*Ruff*, 1998]. Efforts to improve our ability to estimate rupture dimensions for moderate size events are essential if we are to resolve the trade-off between fault area and rupture velocity variations.

The discussion above invokes very simple notions of earthquake mechanics, either with rupture velocity being controlled by volumetrically averaged rigidity and shear velocity variations, or with variable static stress drop resulting from depth-dependent changes in frictional behavior. It is likely that such effects could be coupled, with low rigidity regions tending to have larger fault areas, so that the true explanation lies in between. However, perhaps the most plausible model is one in which the micromechanical properties of the fault zone influence the macroscopic earthquake rupture. In particular, porous sediments at shallow depths in the seismogenic zone may be fluid saturated, with fluids playing a critical role in earthquake slip. Such a model is considered by *Kanamori and Heaton* [1999], who especially

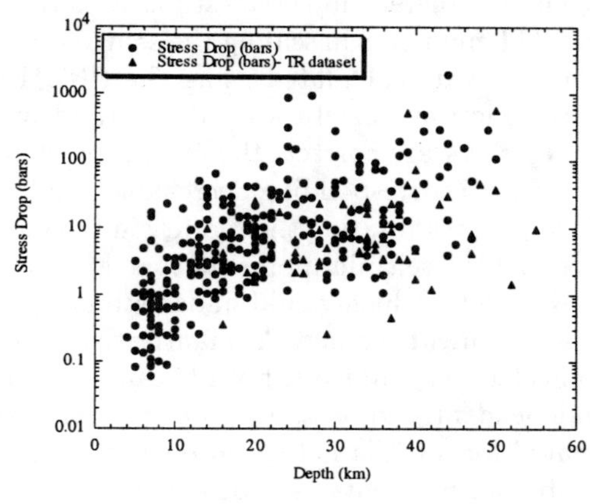

Figure 16. Plot of stress drop variations along the megathrust with depth. Small black circles indicate stress drop estimates calculated from our source durations and triangles indicate stress drop estimates calculated from the *Tichelaar and Ruff* [1991, 1993] datasets.

emphasize the microscale interactions of frictional heating and fluid pressurization as key factors in the macroscopic behavior of faulting. Essentially, it may not be necessary to have large volumes of low rigidity material if the very presence of fluid rich sediments in the seismogenic zone can directly reduce rupture velocity intrinsically. Such models need further elaboration, but our observations provide key targets for explanation.

CONCLUSIONS

Earthquake source time functions for a large number of events in seven circum-Pacific zones indicate that moment-normalized source rupture duration decreases with increasing depth along the seismogenic zone. There are minor regional differences in details of this relationship that may be related to the tectonic environment, but the depth dependence appears to be robust. The Alaska-Aleutian Islands region has the most complete dataset, and shows dramatic differences in the shape of the source time functions with depth, with minor changes in time function shape along strike of the trench, indicating that the depth dependence is more than any along-strike variations for this region. Two end member models of earthquake rupture process are considered to explain the observations, one with varying volumetrically-averaged source zone rigidity and the other with variable static stress drop. For the first model, we find that rigidity increases by a factor of 5 over the depth range of 5 to 20 km. Low rigidity of the shallow portion of the seismogenic zone may the result of sediments, high porosity, and weakly consolidated materials, all of which diminish in volume with depth. The variable stress drop model predicts an order of magnitude increase in stress drop with increasing depth. It is possible that the two scenarios are coupled, with material property variations influencing stress drop, but it may also be true that our observations reflect microscale influences of sediments and fluids in the fault zone on rupture propagation.

Acknowledgments. We made extensive use of the Harvard CMT catalog. GMT software [*Wessel and Smith*, 1991] was used for figure preparation. All data were obtained from the IRIS DMS. This work was supported by NSF EAR 9418643. This is contribution number 407 of the Institute of Tectonics, University of California, Santa Cruz.

REFERENCES

Abercrombie, R.E., Earthquake source scaling relationships from -1 to 5 M_L using seismograms recorded at 2.5 km depth, *J. Geophys. Res., 100*, 24015-24036, 1995.

Bilek, S.L., and Lay, T., Variation of interplate fault zone properties with depth in the Japan subduction zone, *Science, 281*, 1175-1178, 1998.

Bilek, S.L., and Lay, T., Comparison of depth dependent fault zone properties in the Japan trench and Middle America trench, *Pure Appl. Geophys., 154*, 433-456, 1999a.

Bilek, S.L., and Lay, T., Rigidity variations with depth along interplate megathrust faults in subduction zones, *Nature, 400*, 443-446, 1999b.

Bos, A.G., Nolet, G., Rubin, A., Houston, H., and Vidale, J.E., Duration of deep earthquakes determined by stacking of Global Seismograph Network seismograms, *J. Geophys. Res., 103*, 21059-21065, 1998.

Christensen, D.H., and Ruff, L.J., Analysis of the tradeoff between hypocentral depth and source time function, *Bull. Seism. Soc. Am., 75*, 1637-1656, 1985.

Dziewonski, A.M., and Anderson, D.L., Preliminary reference earth model, *Phys. Earth Planet. Inter., 25*, 297-356, 1981.

Ekström, G., and Engdahl, E.R., Earthquake source parameters and stress distribution in the Adak Island region of the central Aleutian Islands, *J. Geophys. Res., 94*, 15499-15519, 1989.

Geist, E.L., Childs, J.R., and Scholl, D.W., The origin of summit basins of the Aleutian ridge: Implications for block rotation of an arc massif, *Tectonics, 7*, 327-341, 1988.

Heinrich, P., Schindele, F., Guibourg, S., and Ihmlé, P.M., Modeling of the February 1996 Peruvian tsunami, *Geophys. Res. Lett., 25*, 2687-2690, 1998.

Heki, K., Miyazaki, S., and Tsuji, H., Silent fault slip following an interplate thrust earthquake at the Japan trench, *Nature, 386*, 595-598, 1997.

Houston, H., Benz, H.M., and Vidale, J.E., Time functions of deep earthquakes from broadband and short period stacks, *J. Geophys. Res., 103*, 29895-29913, 1998.

Hyndman, R.D., Yamano, M., and Oleskevich, D.A., The seismogenic zone of subduction thrust faults, *The Island Arc, 6*, 244-260, 1997.

Ihmlé, P.M., Gomez, J.-M., Heinrich, P., and Guibourg, S., The 1996 Peru tsunamigenic earthquake: Broadband source process, *Geophys. Res. Lett., 25*, 2691-2694, 1998.

Johnson, J.M., and Satake, K., Estimation of seismic moment and slip distribution of the April 1, 1946, Aleutian tsunami earthquake, *J. Geophys. Res., 102*, 11765-11779, 1997.

Kanamori, H., Mechanism of tsunami earthquakes, *Phys. Earth Planet. Int., 6*, 246-259, 1972.

Kanamori, H., Rupture process of subduction-zone earthquakes, *Ann. Rev. Earth Planet. Sci., 14*, 293-322, 1986.

Kanamori, H., and Anderson, D.L., Theoretical basis of some empirical relations in seismology, *Bull. Seism. Soc. Am., 65*, 1073-1095, 1975.

Kanamori, H., and Heaton, T.H., Microscopic and macroscopic physics of earthquakes, this volume, 1999.

Kanamori, H., and Kikuchi, M., The 1992 Nicaragua earthquake: A slow tsunami earthquake associated with subducted sediments, *Nature, 361*, 714-716, 1993.

Kawasaki, I., Asai, Y., Tamura, Y., Sagiya, T., Mikami, N., Okada, Y., Sakata, M., and Kasahara, M., The 1992 Sanriku-Oki, Japan, ultra-slow earthquake, *J. Phys. Earth, 43*, 105-116, 1995.

Lay, T., Kanamori, H., and Ruff, L., The asperity model and the nature of large subduction zone earthquakes, *Earthquake Pred. Res., 1*, 3-71, 1982.

Newman, A.V., and Okal, E.A., Teleseismic estimates of radiated seismic energy: The E/M_o discriminant for tsunami earthquakes, *J. Geophys. Res., 103*, 26885-26898, 1998.

Oleskevich, D.A., Hyndman, R.D., and Wang, K., The updip and downdip limits to great subduction earthquakes: Thermal and structural models of Cascadia, south Alaska, SW Japan, and Chile, *J. Geophys. Res., 104*, 14965-14991, 1999.

Pacheco, J.F., Sykes, L.R., and Scholz, C.H., Nature of seismic coupling along simple plate boundaries of the subduction type, *J. Geophys. Res., 98*, 14133-14159, 1993.

Peacock, S.M., Metamorphism, dehydration, and the importance of the blueschist-eclogite transition in subducting oceanic crust, *Geologic. Soc. of Amer. Bull. 105*, 684-694, 1993.

Pelayo, A.M., and Wiens, D.A., The November 20, 1960 Peru tsunami earthquake: Source mechanism of a slow earthquake, *Geophys. Res. Lett., 17*, 661-664, 1990.

Pelayo, A.M., and Wiens, D.A., Tsunami earthquakes: Slow thrust-faulting events in the accretionary wedge, *J. Geophys. Res., 97*, 15321-15337, 1992.

Ruff, L.J., Multi-trace deconvolution with unknown trace scale factors: Omnilinear inversion of P and S waves for source time functions, *Geophys. Res. Lett., 16*, 1043-1046, 1989a.

Ruff, L.J., Do trench sediments affect great earthquake occurrence in subduction zones? In: L. Ruff and H. Kanamori (Editors), Subduction Zones, II., *Pure Appl. Geophys., 129*, 263-282, 1989b.

Ruff, L.J., Asperity distributions and large earthquake occurrence in subduction zones, *Tectonophys., 211*, 61-83, 1992.

Ruff, L.J., Quasi-dynamic stress drop of recent large earthquakes, *EOS Trans., AGU, 79*, 657, 1998.

Ruff, L.J., and Kanamori, H., Seismicity and the subduction process, *Phys. Earth Planet. Inter., 23*, 240-252, 1980.

Ruff, L.J., and Kanamori, H., The rupture process and asperity distribution of three great earthquakes from long period diffracted P-waves, *Phys. Earth Planet. Int., 31*, 202-230, 1983.

Ruff, L.J., and Miller, A.D., Rupture process of large earthquakes in the northern Mexico subduction zone, *Pure Appl. Geophys., 142*, 101-171, 1994.

Satake, K., Mechanism of the 1992 Nicaragua tsunami earthquake, *Geophys. Res. Lett., 21*, 2519-2522, 1994.

Scholz, C., *The mechanics of earthquakes and faulting*, 449 pp., Cambridge University Press, Cambridge, U.K., 1990.

Shapiro, N.M., Singh, S.K., and Pacheco, J., A fast and simple diagnostic method for identifying tsunamigenic earthquakes, *Geophys. Res. Lett., 25*, 3911-3914, 1998.

Tanioka, Y., and Satake, K., Fault parameters of the 1896 Sanriku tsunami earthquake estimated from tsunami numerical modeling, *Geophys. Res. Lett., 23*, 1549-1552, 1996.

Tanioka, Y., Ruff, L., and Satake, K., What controls the lateral variation of large earthquake occurrence along the Japan trench?, *Island Arc, 6*, 261-266, 1996.

Tichelaar, B.W., and Ruff, L.J., Seismic coupling along the Chilean subduction zone, *J. Geophys. Res., 96* , 11997-12022, 1991.

Tichelaar, B.W., and Ruff, L.J., Depth of seismic coupling along subduction zones, ıJ. Geophys. Res., 98, 2017-2037, 1993.

Vidale, J.E., and Houston, H., The depth dependence of earthquake duration and implications for rupture mechanics, *Nature, 35*, 45-47, 1993.

Wessel, P., and Smith, W.H.F., Free software helps map and display data, *EOS, 72*, 441-445, 1991.

Zhang, Z., and Schwartz, S.Y., Depth distribution of moment release in underthrusting earthquakes at subduction zones, *J. Geophys. Res., 97*, 537-544, 1992.

S.L. Bilek and T. Lay, Earth Sciences Dept., Earth and Marine Sciences Bldg. A232, University of California, Santa Cruz, Santa Cruz, CA 95064) (e-mail: sbilek@es.ucsc.edu; thorne@es.ucsc.edu)

Effects of Loading Rate and Normal Stress on Stress Drop and Stick-slip Recurrence Interval

Stephen L. Karner and Chris Marone

Department of Earth, Atmospheric, & Planetary Sciences, Massachusetts Institute of Technology, Cambridge, Massachusetts

In this paper we report on investigations of frictional restrengthening during repetitive stick-slip on initially bare granite surfaces (nominal contact area 25 cm^2). The tests were conducted in a double-direct shear apparatus at room-temperature and humidity. Normal stress was held constant, and shear was induced by controlling the velocity of a loading piston. Samples exhibited quasi-periodic stick-slip instabilities. Appreciable yielding and precursory creep occur prior to failure. Stress drop amplitudes range from 0.1 to 3.1 MPa (or 4-49% of the failure strength). We study the effects of loading rate (0.5 to 300 μm/s) and normal load (5 and 10 MPa) on stress drop and stick-slip recurrence interval. At a given loading rate, there is a positive correlation between stress drop and recurrence interval, indicating healing rates of ~4 MPa per decade increase in recurrence time. However, the combined data from all velocities show a lower rate, suggesting an apparent healing rate of ~1 MPa per decade increase in recurrence time. We find a consistent scaling between different loading rates when stress drop is compared to inter-event loading displacement. The data for all velocities delineate a single trend and show larger stress drops for greater pre-failure loading displacement. Failure strength and post-slip stress levels converge with increasing velocity. Our data indicate that stress drop and post-seismic stress levels depend on loading rate and recurrence interval in a complex way.

1. INTRODUCTION

Estimates from seismic moment-magnitude data show that stress drop increases by 1-5 MPa per decade change in earthquake recurrence time [*Kanamori and Allen*, 1986; *Scholz et al.*, 1986; *Marone et al.*, 1995]. These results indicate that faults restrengthen, or heal, during the interseismic period. Fault healing is also inferred from seismic surveys of faults that participate in earthquake rupture. *Li et al.* [1998] found that fault-zone p- and s-wave velocities increased with time after the 1992 Landers earthquake (M_w 7.3), indicating healing and consolidation of the fault zone. These observations are consistent with data from laboratory experiments performed under a wide range of conditions, which show that static friction increases as a function of contact time [e.g. *Dieterich*, 1972, 1978; *Chester and Higgs*, 1992; *Fredrich and Evans*, 1992; *Marone and Kilgore*, 1993; *Beeler et al.*, 1994; *Karner et al.*, 1997; *Nakatani and Mochizuki*, 1996; *Marone*, 1998a; *Karner and Marone*, 1998; *Richardson and Marone*, 1999]. However, the relationship between seismic estimates of earthquake stress drop and laboratory measurements of healing and static friction is poorly understood. Hence, it is important to investigate the mechanics of healing and the relationship between healing and frictional instability. This may be achieved by analyzing stick-slip instabilities observed from laboratory experiments.

To investigate the repetitive stick-slip behavior of rock friction, and to study the effects of loading conditions on

GeoComplexity and the Physics of Earthquakes
Geophysical Monograph 120

fault healing, we have sheared samples of initially bare rock surfaces. We find that stress drop amplitude is dependent on velocity, normal stress, recurrence interval and effective stiffness. Our results are consistent with those of previous laboratory studies, in that stress drop increases with greater normal stress and slower loading rates. The frequency of stick-slip events depends inversely on velocity via a power-law relationship. When cast in terms of recurrence time, we observe two rates at which stress drop increases with interseismic interval. The first is given by inverse loading rate such that slower velocities yield longer recurrence times and larger stress drop. A second, higher rate occurs for a given velocity and is determined by stochastic variations in fault strength between instabilities coupled with the finite elastic stiffness of the sample assembly. We analyze our data in terms of rate- and state-dependent friction constitutive laws, and we find that the model predictions capture the major features. Our results highlight the importance of accounting for variations in earthquake recurrence interval in studies of the seismic cycle and earthquake source characteristics.

2. THE EXPERIMENTS

Experiments were conducted on samples of Westerly Granite, at room-temperature and humidity, in a hydraulically driven double-direct shear testing apparatus [described by *Karner and Marone*, 1998; *Mair and Marone*, 1999]. Samples consisted of a triple-block geometry where a central forcing block is sheared between two stationary side blocks (e.g. the inset shown in Figure 1a). Samples were machined to produce side blocks with dimensions of 5x5x3 cm^3, and a central sliding block of 5x8x4 cm^3. Hence, the nominal contact area on which shear occurred was 5x5 cm^2. The same forcing blocks were used for each test, and surface roughness was controlled by machining with a #100 grinding wheel (rms roughness of ~5μm).

For each experiment, normal stress was maintained constant by controlling the loadforce signal in a servo-controlled feedback loop (resolution better than 0.1 kN). Shear was induced by controlling the velocity of the piston loading the central block of the sample assembly (control step-size of 0.1 μm/s). We monitored the applied forces and displacements with sampling rates up to 1000 Hz. The parameters that were investigated included loading rate (0.5 to 300 μm/s) and normal stress (5 or 10 MPa). To study the effects of velocity on stick-slip behavior, we imposed step changes in loading rate throughout each test. Several tens of stick-slip cycles occurred for each velocity step (up to ~70 cycles), and between 400 to 500 instabilities were recorded for the 3 cm of slip accommodated by our sample dimensions. We have analyzed each stick-slip event, and we compare our data to both previous laboratory studies and numerical simulations using existing rate- and state-dependent friction laws.

3. RESULTS

3.1. Characteristics of Repetitive Stick-slip

We show a portion of data for one experiment, including the initial load-up from zero shear stress, in Figures 1 and 2. For this test, normal stress was held constant (10 MPa) and loadpoint velocity was varied from 10 to 300 μm/s. In Figure 1a, we present the data as a function of time in order to better distinguish the stick-slip cycles for different loading rates. For the conditions of our experiments, samples exhibited unstable sliding, with instabilities occurring repeatedly. We observe a clear effect of loading rate on the frequency and amplitude of stick-slip instability. The data indicate that stick-slip amplitude decreases, and frequency increases with increasing loading rate.

In Figure 1b we focus on several stick-slip cycles as a function of loadpoint displacement. The data show the characteristic stick-slip behavior observed by previous researchers [e.g. *Ohnaka*, 1973; *Engelder et al.*, 1975; *Engelder and Scholz*, 1976; *Johnson and Scholz*, 1976; *Shimamoto and Logan*, 1981; *Teufel and Logan*, 1978; *Wong and Zhao*, 1990]. That is, shear stress increases quasi-linearly as sample loading proceeds, indicating an elastic stiffness of the loading column of ~0.065 MPa/μm. As deformation continues, shear stress departs from the quasi-linear trend indicating premonitory slip. Shear stress continues to increase until it reaches a peak value (τ_{max}) that corresponds to the traditional definition of static friction. During an instability, shear stress decreases to a fraction of the peak strength (τ_{min}). We define stress drop ($\Delta\tau$) as the difference between the pre-slip peak stress (τ_{max}) and the post-slip minimum stress (τ_{min}). With subsequent loading, samples continue to slide in an unstable manner as can be seen by the repeated, quasi-periodic stick-slip cycles.

In Figure 2 we show sample slip for the same segment of data shown in Figure 1. Sample slip is calculated from the measured shear force and loadpoint displacement, using the elastic stiffness obtained from apparatus calibrations (k=0.065 MPa/μm). Calculating sample slip in this way is justified by recent tests where we monitor slip directly by using transducers mounted on the sample blocks. Over long time intervals, the cumulative slip rate resembles the background loading velocity (Figure 2a). When viewed in detail (Figure 2b), samples exhibit intervals of relatively little motion (stick), followed by accelerating slip that

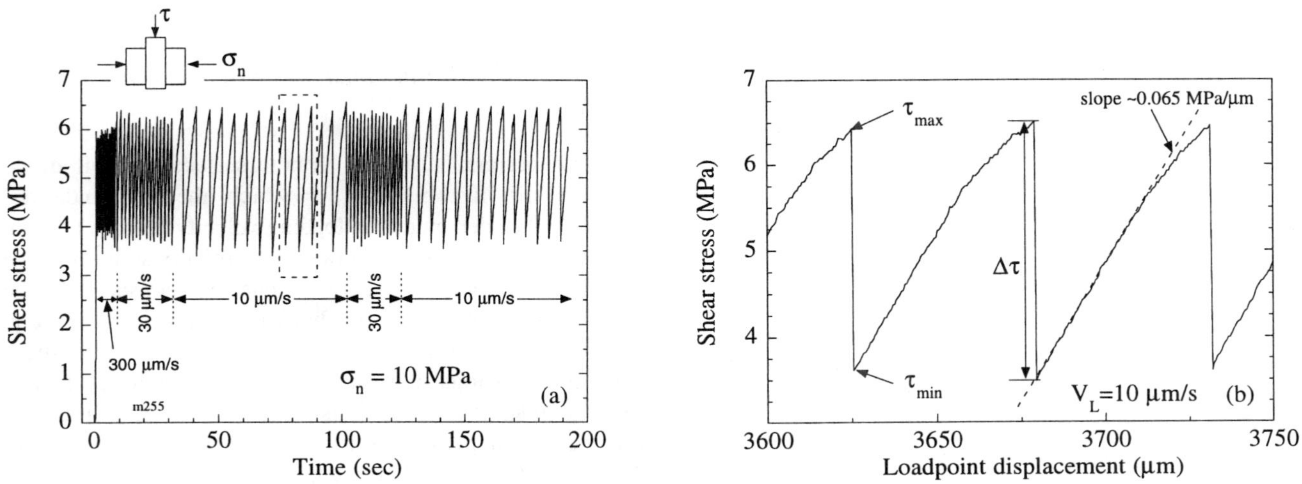

Figure 1. Shear stress data from a typical section of an experiment. **a)** Shear stress is plotted as a function of time. Loading rate was stepped between 10, 30, and 300 μm/s. Data show repeated stick-slip events, with velocity-dependent variations in stick-slip amplitude and frequency. **b)** An enlarged view of data from (a), plotted against loadpoint displacement. Loading rate for this section was 10 μm/s. For each instability, data show elastic loading with stiffness of ~0.065 MPa/μm (expressed as shear stress per unit loadpoint displacement), followed by yielding and failure.

culminates in failure and rapid motion (slip instability). We define premonitory slip as the amount of motion that occurs during yielding of the sample and prior to failure (i.e. departure from the linear elastic loading curve). The recurrence interval for a given slip instability is defined as the time since the previous slip event. The data shown in Figures 1 and 2 indicate an inverse scaling of recurrence time with loading rate, inasmuch as the frequency of instabilities is greater for faster velocities.

3.2. Relationship Between Velocity and Recurrence Interval

It is important to note that recurrence interval is not directly controlled in our experiments. However, we find that recurrence interval is dependent on loading rate, which is a control parameter. In Figure 3, we show the association between loading rate (V_L) and recurrence time (t_r) for stick-slip cycles where normal stress was 10 MPa. The imposed loading rate is the primary control for the length of time between slip events, with slower velocities resulting in longer recurrence intervals. The data indicate a relationship that is described by a power-law expression:

$$t_r = m\left(V_L\right)^n \tag{1}$$

where m is a scaling constant, and n is the power-law exponent (-1.15 for our tests). For a single velocity the data show a range of recurrence times, a feature that is discussed in section 4. In the following section we present our data in terms of recurrence interval, which allows for better comparison to seismic observations. We explore the dependence of stress drop and fault strength on recurrence interval.

3.3. Dependence of Shear Stress on Recurrence Interval

We measured the stress drop amplitude for each instability and plot these against recurrence interval in Figure 4. While the data show some scatter (discussed in more detail below), there are distinct correlations between stress drop and the loading conditions. Stress drops are generally larger for tests conducted at greater normal stress. Furthermore, we observe two distinctly different healing rates that depend on the imposed loading velocity. For a given normal stress and velocities from 0.5 to 300 μm/s, stress drop is greater for slower loading rates and, hence, longer recurrence intervals (β_1 ~ 0.75-1 MPa per decade change of recurrence time). For a given velocity there is a second, larger healing rate (β_2), of ~3.5 to 4 MPa per decade change in time. Thus, the restrengthening rates for a single velocity are quite different to the overall healing rates observed for a wide range of loading rates.

Peak and minimum stresses for all stick-slip cycles are shown as a function of recurrence time in Figure 5. The

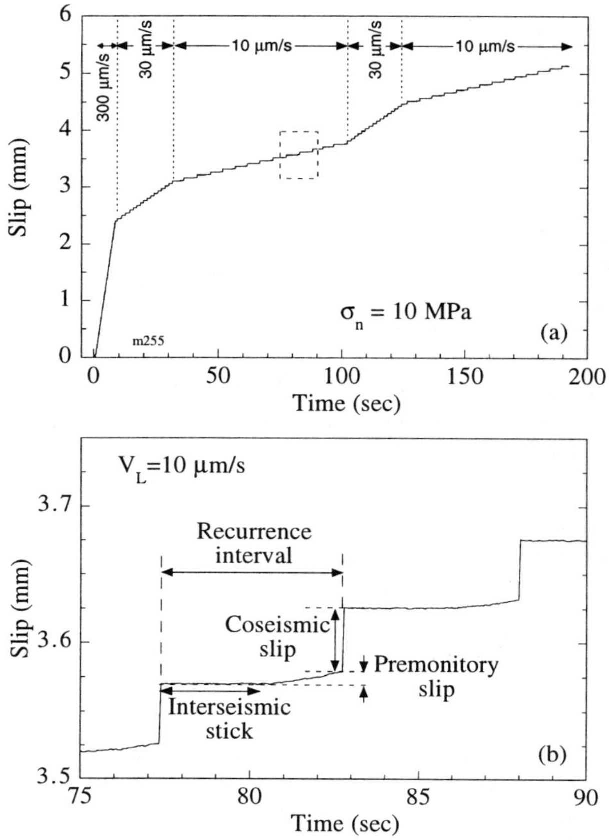

Figure 2. Cumulative slip across shear surface plotted as a function of experiment time (the same test shown in Figure 1). Slip is determined from the measured loadpoint displacement, loadforce, and elastic stiffness. **a)** The long-term slip rate closely matches the imposed loading velocity. **b)** Individual cycles show intervals of sample stick, followed by accelerating premonitory slip and instability (coseismic slip). Recurrence interval is measured from the point at which shear stress is minimum after an instability, to the point for which shear stress is maximum prior to the next instability.

data show systematic variations as a function of loading rate and normal stress. First, τ_{max} and τ_{min} are larger as normal stress increases. Second, there is a small dependence of τ_{max} and τ_{min} on loading rate. The values of τ_{max} are nearly constant, or may decrease slightly, with increasing velocity (shorter recurrence times). For a single velocity, τ_{max} increases slightly for longer recurrence intervals. Also, τ_{min} increases slightly for faster loading rates (Figure 5b). These observations are consistent with data from individual stick-slip cycles (as shown in Figure 1a), and measurements of stress drop (Figure 4).

4. DISCUSSION

4.1. Fault Healing Rates Inferred From Stress Drop Measurements

Our experiments on bare granite surfaces exhibit quasi-periodic slip instabilities, for which the amplitude and frequency are dependent on loading velocity (Figure 1). Our data show an inverse power-law relationship between loading rate and instability frequency, consistent with seismic observations of aftershock recurrence intervals [e.g. *Schaff et al.*, 1998]. Furthermore, we observe that stress drop increases logarithmically with recurrence interval, with healing rates that are consistent with earthquake data obtained from natural faults [*Kanamori and Allen*, 1986; *Cao and Aki*, 1986; *Marone et al.*, 1995]. If the healing rates observed from our experiments are taken as a measure of fault strength, then our results are comparable to previous work detailing the time-dependence of static friction [e.g. *Dieterich*, 1972, 1978; *Beeler et al.*, 1994; *Nakatani and Mochizuki*, 1996; *Marone*, 1998a; *Richardson and Marone*, 1999], and fault restrengthening inferred from seismic surveys [*Li et al.*, 1998]. However, the two different healing rates shown by our data (β_1 and β_2 in Figure 4) indicate that fault restrengthening may occur via an interplay of several processes. In order to better understand the different restrengthening rates, we first investigate the larger healing rate observed for individual velocities (β_2 in Figure 4).

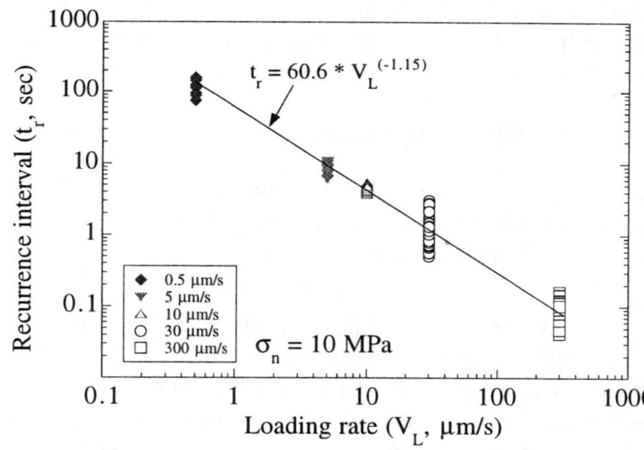

Figure 3. Recurrence interval plotted as a function of the imposed loading rate. Data are from stick-slip events occurring in three experiments for which σ_n=10 MPa and loading rates were varied from 0.5 to 300 µm/s. We observe a power law relationship between loading rate and recurrence interval.

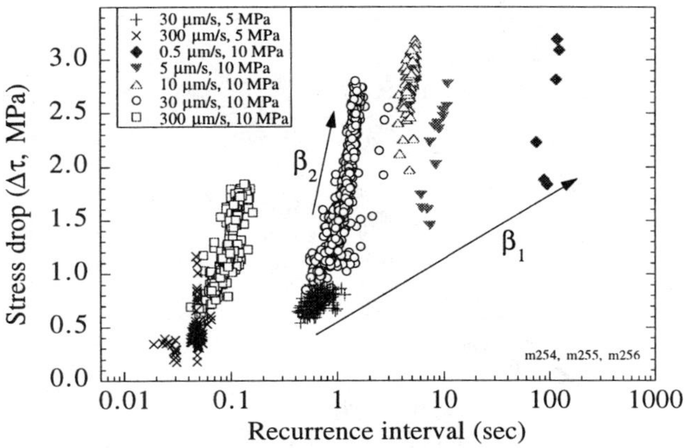

Figure 4. Measured stress drop values are shown as a function of recurrence time. Data are from the same stick-slip events shown in Figure 3, and from tests conducted at a normal stress of 5 MPa. Stress drop increases with normal load. The data show two distinct healing rates (β_1 and β_2). For a given normal load, stress drop increases by ~1 MPa per decade change in time (β_1). For a given loading rate, stress drop increases by ~4 MPa per decade change in recurrence interval (β_2). Data scatter reflects variations of material strength between stick-slip events occurring within a single experiment, and also reproducibility between experiments.

In Figure 6a, we directly compare stress drop data to recurrence interval for instabilities where V_L=30 μm/s and σ_n=10 MPa. We present data from three consecutive experiments to show data reproducibility. There is no systematic dependence of stress drop levels on the experiment order (m254, m255, and m256, consecutively). For a given experiment, the data display a range of stress drop values (~1.5 MPa, or ~75% of the range for all tests). Furthermore, we do not observe systematic variations in stress drop as a function of cumulative shear displacement or experiment time. Post-experiment observation of our samples revealed that only minor amounts of gouge had accumulated. This is important because previous studies have identified the production of wear material (gouge) as a mechanism responsible for displacement dependent friction evolution [e.g. *Byerlee*, 1967; *Marone*, 1998b]. Thus, the lack of a systematic evolution of stress drop amplitudes with displacement is consistent with the small amounts of gouge production during our tests.

The data shown in Figure 6a define a quasi-linear trend that is bounded by a line calculated from the imposed velocity and loading stiffness. This line corresponds to the loadpoint displacement between stick-slip events, and for zero precursory slip would equal the stored elastic energy.

To the first order, the trend shown in Figure 6a implies that healing rates for a single velocity (β_2) are determined by the stiffness of the sample assembly. To explore this further, we present stress drop data from all instabilities in Figure 6b. Stress drop is plotted as a function of the interseismic loadpoint displacement calculated for each instability ($x=V_L t_r$). The data for all stick-slip events describe a single quasi-linear trend consistent with the loading stiffness observed from individual stick-slip cycles (k=0.065 MPa/μm, as shown in Figure 1b). Hence, this indicates that stress drop can be described in terms of loading rate, recurrence interval, and stiffness (k) by the relation:

$$\Delta\tau = V_L t_r k \quad (2)$$

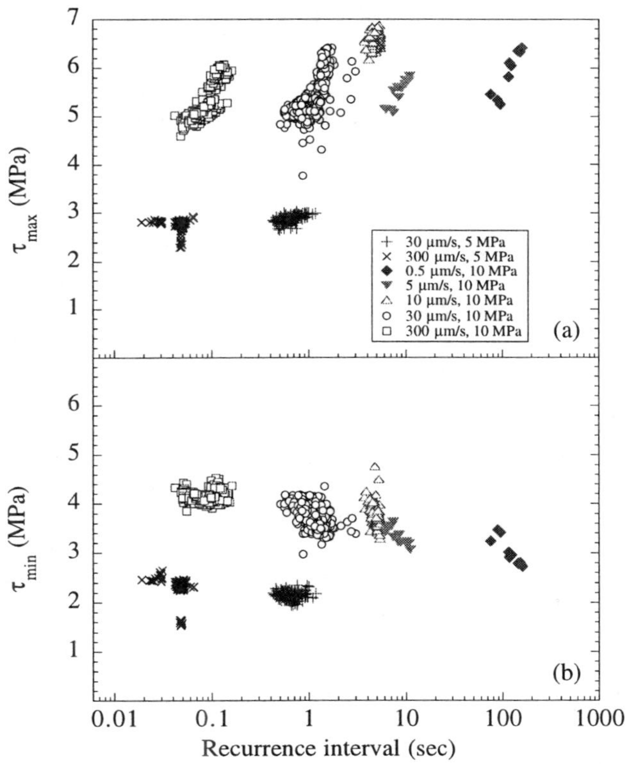

Figure 5. Shear stress data as a function of recurrence time. Data are from the same tests shown in Figure 4. **a)** Maximum shear stress prior to instability. For a given loading rate, peak stress increases with recurrence interval. For a constant normal load (i.e. including all loading rates), data show little or no dependence of peak stress on recurrence interval. **b)** Minimum stress following a slip event. Data are for the same stick-slip events shown in (a). Minimum stress increases with normal load and velocity.

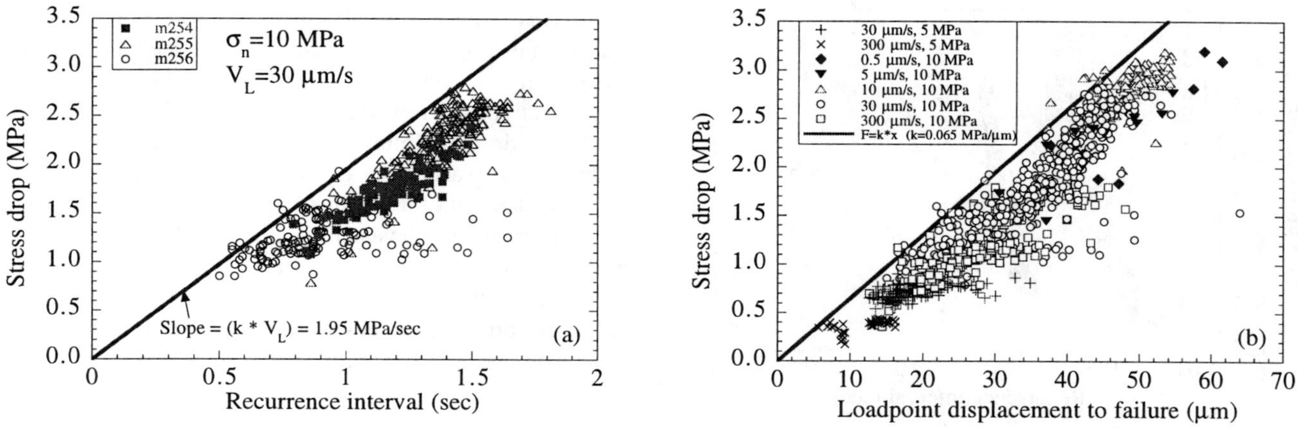

Figure 6. Stress drop data for the events shown in Figure 4. **a)** Stress drop is plotted against recurrence interval for events that occurred at the same loading conditions (i.e. σ_n=10 MPa, V_L=30 µm/s). Data from m254, m255 and m256 indicate experiment reproducibility. For a given experiment, data show a range of ~1.5 MPa in stress drop values. All data hug the calculated elastic loading curve (the line with slope kV_L, where k=0.065 MPa/µm). **b)** Stress drop data from all stick-slip events shown as a function of loadpoint displacement prior to instability. The data describe a single quasi-linear trend. For comparison to data, we show a line calculated for elastic loading of a material with stiffness of 0.065 MPa/µm. Our data hug the calculated loading curve, and consistently plot to the right of the line.

This expression can be derived from classical Coulomb theory (as shown by *Beeler and Wong*, 1999) and is consistent with data presented by other researchers [e.g. *Wong and Zhao*, 1990; *Gu and Wong*, 1991; *Berman et al.*, 1996; *Nasuno et al.*, 1998].

4.2. Second-order Variations and Data Scatter

While Equation 2 offers a first-order approximation to our data (Figure 6b), the majority of our data consistently plot to the right of the calculated linear trend. These second-order characteristics are partly due to reproducibility between experiments (as shown in Figure 6a). However, the lack of a dependence of stress drop on cumulative displacement or experiment time indicates that other factor(s) influence stick-slip behavior. For a given test, data scatter may be due to stochastic variations in strength of the sliding surfaces between slip events, or due to the effects of premonitory slip that occurs prior to instability. As the latter is consistent with our observations of yielding prior to sample failure (e.g. Figure 1b), we investigate whether stress drop is dependent on the amount of premonitory slip. We remove the linear elastic loading curve from the displacement data (see Figures 1b and 6b, Equation 2) and show the calculated values of premonitory slip in Figure 7. We do not observe any significant relationship between premonitory slip and stress drop.

The lack of a significant dependence of stress drop on premonitory slip indicates that data scatter may result from random variations in the properties of the slipping surface. As we do not control the state of the shearing surfaces between each instability, we suggest that differences in

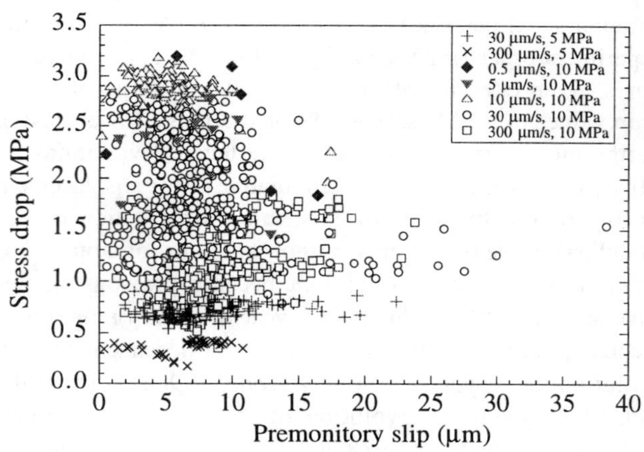

Figure 7. Stress drop shown as a function of the calculated premonitory slip to failure. Data are for the same events shown in Figure 6b. Slip is calculated as the difference between observed loadpoint displacements, and the theoretical loading curve for a stiffness of 0.065 MPa/µm. There is no systematic dependence of stress drop on pre-failure slip.

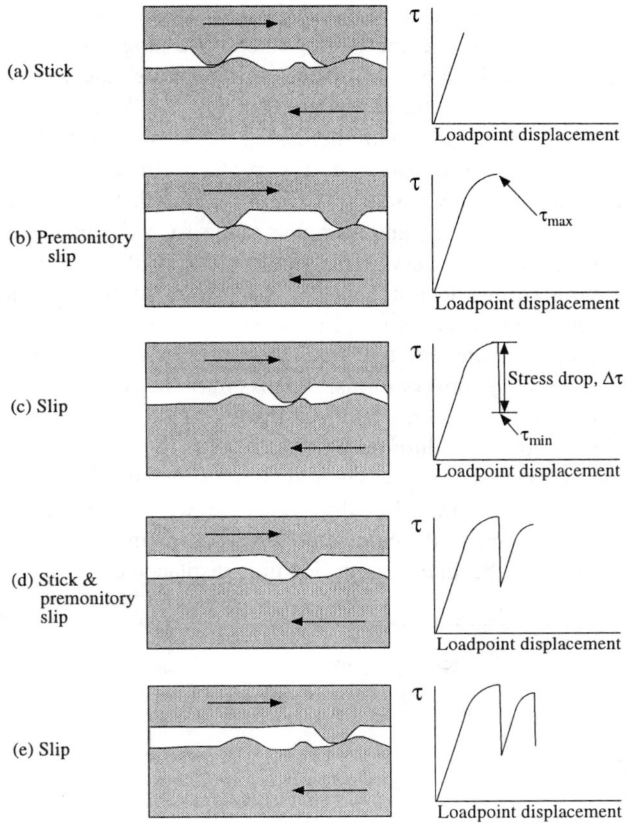

Figure 8. Schematic illustration showing how asperity contact size and population may affect shear loading behavior for stick-slip instabilities. Panels a) through e) represent consecutive snap-shots for shear on bare surfaces. The sample loads elastically with little slip (a), followed by yielding and premonitory slip (b). At failure, samples slide rapidly and exhibit a sharp reduction in shear load (c). Stick-slip cycles occur repeatedly (d-e). Peak strength and stress drop are affected by the characteristics of the asperity population.

surface evolution between stick-slip cycles may account for the stochastic character of our data. This may arise from slip and/or time dependent variations in the population and character (e.g. shape, strength) of contacting asperities, as shown schematically in Figure 8. For example, a population of strong asperities would raise the material failure strength, and inhibit premonitory slip (thus, samples would exhibit less yielding). If the number of strong asperities differed between cycles, then premonitory slip, failure strength and, perhaps, stress drop would also vary.

Other micromechanical processes may account for the variability within our data. For example, gradual deformation of contacting asperities during loading (e.g. by brittle failure) would relieve some of the imposed shear load and lead to greater premonitory slip and longer recurrence intervals. Alternately, during dynamic slip the characteristics of an asperity population may have an influence on the level of dynamic overshoot before slip is arrested. Detailed studies of surface interactions may determine which of these mechanisms is responsible for the second-order variations shown by our data. However, such work is beyond the scope of this paper.

4.3. Modeling of Data Using Rate- and State-dependent Friction Laws

We compare our data with results from forward models using existing rate- and state-dependent friction constitutive laws. For the simulations we used the Dieterich (or slowness) form of the friction evolution law [*Dieterich*, 1979], where friction (μ) at slip rate (V) is described by:

$$\mu = \mu_0 + a\ln\left(\frac{V}{V_0}\right) + b\ln\left(\frac{V_0\theta}{D_c}\right) \tag{3}$$

Here, μ_0 is steady-state sliding friction at a reference velocity (V_0), D_c is the critical slip distance, a and b are empirical scaling constants. The state variable (θ) evolves via the relation:

$$\frac{d\theta}{dt} = 1 - \left(\frac{V\theta}{D_c}\right) \tag{4}$$

Equations (3) and (4) are coupled with an expression describing the elastic interaction between the sliding surface and the testing apparatus: $d\mu/dt = k(V_l - V)$, where k is the system stiffness and V_l is the velocity of the loading piston.

The numerical simulations have been conducted for the boundary conditions of our experiments ($0.5 < V_L < 300$ μm/s), and using friction parameters (a, b, and D_c) such that the critical stiffness equals the system stiffness (0.0065 change in friction per μm, determined from stiffness value shown in Figure 1b and σ_n=10 MPa) via the relation:

$$k = K_c = \frac{(b-a)}{D_c} \tag{5}$$

To model oscillating instabilities, we perturb the numerical system by including a small slide-hold-slide (hold time of 0.0002 s) early in the simulations. The model results show frictional behavior similar to the stick-slip

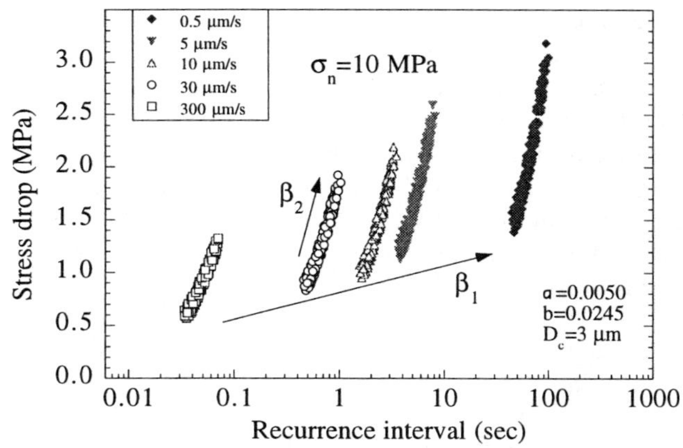

Figure 9. Results from numerical simulations using rate- and state- dependent friction laws. Simulations are forward models using the Dieterich (or slowness) law with constitutive parameters as shown. To simulate physical changes of the contacting surfaces, the state variable was randomly perturbed by a small amount (±2.5%) in the middle of each slip instability. Stress drop amplitudes are plotted as a function of recurrence interval. Results display two rates of increase in stress drop amplitudes as a function of recurrence time, consistent with our laboratory observations.

instabilities observed in our experiments. However, the simulations predict perfectly periodic behavior where stress drop amplitudes and recurrence intervals are constant with increasing slip. This is because state evolution is the same for each stick-slip cycle, which may not be the case in our experiments. Hence, to investigate the relationship between stress drop and recurrence interval we randomly perturb the state variable by a small amount (between ±2.5% of the value of state at τ_{max}) in the middle of each slip instability. This procedure is justified by physical models in which the nature of the sliding surfaces varies during stick-slip (e.g. due to wear and asperity fracturing). In such models, the greatest variability in surface characteristics would likely occur during dynamic slip. A key feature of the perturbation in state is that it yields a range of stick-slip recurrence intervals and stress drops, as we observe experimentally.

Results from these forward models are presented in Figures 9-11. We show the calculated stress drop amplitudes for instabilities in Figure 9, and the associated τ_{max} and τ_{min} values in Figure 10. The model predictions capture the first order observations from our experiments. That is, the frequency of instability (i.e. recurrence interval) is strongly dependent on the imposed loading rate (similar to Figure 3, and Equation 1). The calculated stress drop amplitudes show two distinct restrengthening rates (Figure 9), consistent with our laboratory data (Figure 4). The simulations show greater healing rates for a single velocity than for all velocities combined. Furthermore, the model predictions show little dependence of τ_{max} on loading rate and that τ_{min} increases with increasing velocity (Figure 10), consistent with our data (i.e. Figure 5). However, for a given loading rate we note that the model predictions show a greater range of τ_{min} and a smaller range for τ_{max} than that observed from our data. This may be due to the fact that the simulations do not consider the effects of inertia and neglect radiated energy. Furthermore, the simulations predict less scatter for τ_{max} than for τ_{min} which differs from our laboratory observations. This may be due to the dependence of τ_{max} on system stiffness, or on second-order variations in pre-failure fault strength.

As is the case for our laboratory data, we find that the numerically determined healing rates are well described by Equation 2. When the calculated stress drop amplitudes are compared to the interseismic loading displacement (Figure

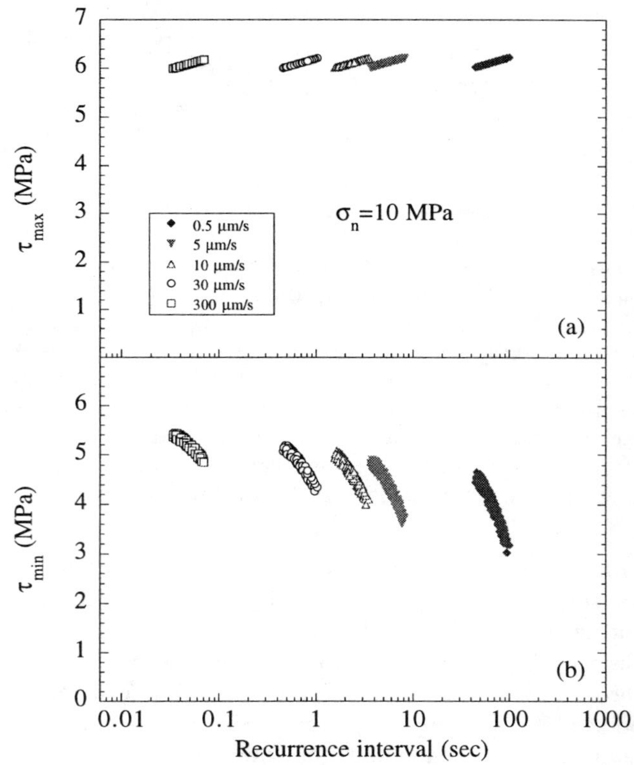

Figure 10. Calculations of τ_{max} and τ_{min} for the same numerical simulations shown in Figure 9, plotted as a function of recurrence interval. **a)** For a given loading rate, peak stress levels (prior to instability) increase slightly with recurrence time, while for **b)** minimum stress levels (after instability) decrease with recurrence interval.

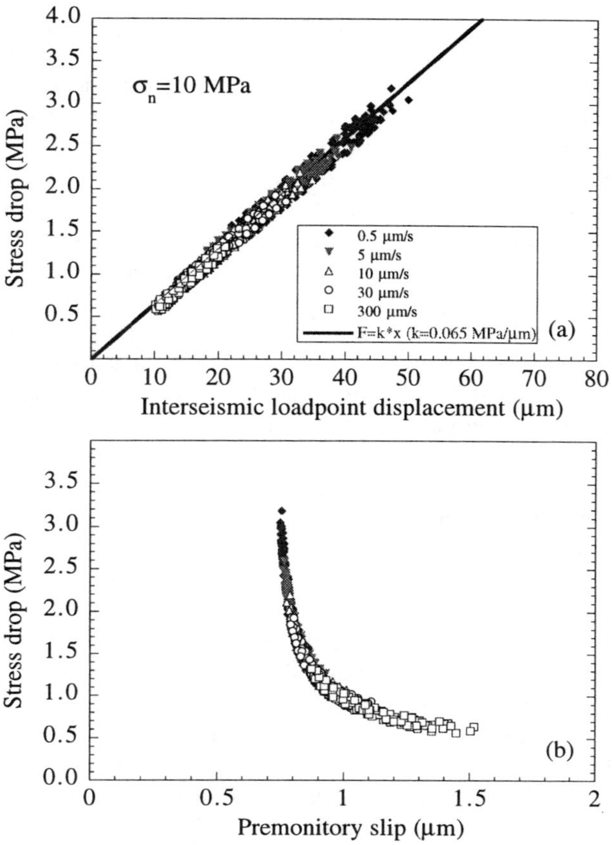

Figure 11. **a)** For comparison to our experiment results, we show stress drop amplitudes from simulations (same as those presented in Figure 9) plotted against loadpoint displacement prior to failure. Stress drop amplitudes plot as a single line that is well described by an elastic loading curve. **b)** Model calculations of stress drop plotted as a function of premonitory slip. Results show an inverse relationship between stress drop and premonitory slip, and that premonitory slip is small compared to our experiment data.

11a), the predicted values describe a single line with a slope given by the system stiffness (compare with Figure 6b). The scatter predicted by the models is consistent with data from our experiments, suggesting that stochastic evolution of state is a plausible explanation of our experimental observations. However, we note that the amount of scatter predicted by the models is less than that shown by our laboratory data. This may arise from differences between the observed and modeled amounts of premonitory slip. Model calculations of premonitory slip are shown in Figure 11b. The simulations show less premonitory slip than observed from our experiments, and an inverse relationship between slip and stress drop. While our data do not show a dependence of stress drop on premonitory slip (compare Figures 11b and 7), it may be that such a relationship is masked by the data scatter. The larger values of premonitory slip and the greater amount of scatter observed from our laboratory measurements may result from stochastic variations of the state of the sliding surface during the yielding phase prior to failure.

4.4. Comparison to Previous Studies

4.4.1 Seismic Observations. Our results are consistent with seismic observations, inasmuch as inter-event recurrence interval scales inversely with loading rate [e.g. *Schaff et al.*, 1998], and that stress drop increases with longer recurrence intervals [*Kanamori and Allen*, 1986; *Marone et al.*, 1995] and with slower loading rates [*Cao and Aki*, 1986]. Furthermore, our data show a range of recurrence intervals and stress drop values for a given velocity, also consistent with seismic observations [*Cao and Aki*, 1986]. By reanalyzing the data of *Kanamori and Allen* [1986] in terms of velocity, *Cao and Aki* [1986] found that stress drop values exhibited considerable scatter for a given loading rate. From this they reasoned that fault strength is affected by heterogeneities in the number of strong asperities. If variations in asperity geometry and strength are also considered, then such a micromechanical model would account for the stochastic nature of our room temperature data.

We recognize that the conditions at hypocentral depths introduce other factors that could influence fault healing rates and stress drop levels. Laboratory studies have explored several of these factors - including temperature, fluid-rock interactions, and variations in stress state [e.g. *Chester and Higgs*, 1992; *Linker and Dieterich*, 1992; *Fredrich and Evans*, 1992; *Karner et al.*, 1997; *Nakatani*, 1998; *Karner and Marone*, 1998; *Richardson and Marone*, 1999]. As it is unlikely for every fault to experience the same physico-chemical conditions, variations in these parameters may account for some scatter observed from seismic data. However, the conditions at hypocentral depths provide added complexity which must be separated from the purely physical mechanisms that effect fault healing (such as loading rate).

4.4.2. Laboratory Data. Our data show that stress drop decreases with increasing loading rates (i.e. lower recurrence times, Figures 1 and 4), in agreement with results from previous studies of frictional instability for geologic materials [*Ohnaka*, 1973; *Engelder et al.*, 1975; *Engelder and Scholz*, 1976; *Johnson and Scholz*, 1976; *Shimamoto and Logan*, 1981; *Teufel and Logan*, 1978; *Wong and Zhao*,

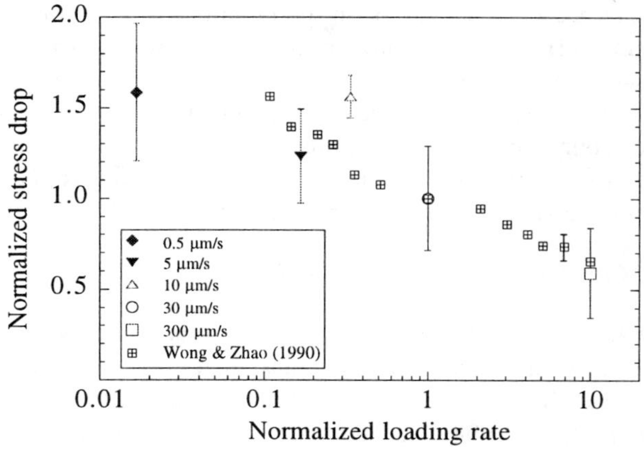

Figure 12. Our stress drop data are plotted with data from *Wong and Zhao* [1990]. For comparison, we normalize our data by the mean value for a loading rate of 30 μm/s (1.8 MPa). Velocities are normalized by 30 μm/s. We present our data as mean and standard deviation for each normalized velocity, and we show the maximum uncertainty presented by *Wong and Zhao* [1990]. Our results are consistent with those of *Wong and Zhao* [1990], with a similar inverse scaling between stress drop and loading rate.

1990], for metals [*Rabinowicz*, 1958], and for thin molecular films separating sheets of mica [*Berman et al.*, 1996]. *Wong and Zhao* [1990] presented stick-slip data from experiments conducted in a triaxial deformation apparatus at normal stresses from 25 to 250 MPa. They performed velocity-stepping tests on fine-grained quartz gouge sandwiched between Westerly granite forcing blocks. They averaged the data for each velocity step, and presented the results as the ratio between consecutive velocity groups. Hence, their data were essentially normalized by reference values of loading velocity, stress drop, and recurrence interval. To remain consistent with *Wong and Zhao* [1990] and to assist with comparison, we have normalized our stress drop and loading rate data by reference values (corresponding to the mean stress drop of ~1.8 MPa at a velocity of 30 μm/s). We present a comparison between our data, and those of *Wong and Zhao* [1990] in Figure 12. The inverse relationship that we observe between stress drop and loading rate are in agreement with the data of *Wong and Zhao* [1990]. Furthermore, *Wong and Zhao* [1990] observed that their normalized stress drop levels increased with normalized recurrence interval, which is consistent with the results of this study.

Berman et al. [1996] also investigated the effects of loading rate on stick-slip behavior from experiments conducted on thin molecular films sandwiched between mica surfaces. As a function of increasing velocity, they observed that static friction levels decreased and post-failure minimum friction increased. Their data define a critical velocity (V_c) above which stick-slip motion transitioned to continuous stable sliding ($V_c \sim 7$ to 8 μm/s for their tests). A similar critical velocity has been observed from dry friction experiments on bristol board and also on paper [*Baumberger et al.*, 1994; *Heslot et al.*, 1994]. These researchers observed a transition from stick-slip to stable sliding that depends on loading rate and stiffness. Their data showed that stable sliding is promoted by both increased stiffness and increased velocity. In the vicinity of the bifurcation of slip behavior, they also found that frictional characteristics differed significantly for low and high loading rates. From these observations, they identified that the low velocity regime was creep dominated and could be defined by a characteristic length scale, while the high velocity regime was inertia dominated and has a characteristic time scale. Hence, the combined results of *Berman et al.* [1996], *Baumberger et al.* [1994], and *Heslot et al.* [1994] indicate that stiffness and loading rate can effect the behavior of stick-slip motion and the transition to stable sliding.

To compare with the study of *Berman et al.* [1996], we show in Figure 13b the levels of peak stress (τ_{max}) and post-slip minimum stress (τ_{min}) from our tests as a function of loading rate, and for a constant normal stress of 10 MPa. The data show that τ_{max} decreases slightly with loading rate, while τ_{min} increases with velocity. Hence, τ_{max} and τ_{min} converge with faster loading rates, consistent with the observations of *Berman et al.* [1996]. For the conditions of our experiments, and in light of the results of *Berman et al.* [1996], *Baumberger et al.* [1994], and *Heslot et al.* [1994], our data suggest that a critical velocity exists above which stable sliding may be achieved. This would correspond to a Hopf bifurcation between unstable and stable sliding at a fast loading rate, similar to those reported previously [*Klein et al.*, 1997; *Scholz*, 1998]. Given that this critical velocity would be large (>1000 μm/s), we suggest that the transition from stick-slip to stable sliding would occur in an inertia-dominated regime. To date, we have not observed this bifurcation, which may occur at a velocity greater than we can presently achieve with our testing apparatus. However, we have observed a similar velocity-dependent convergence of τ_{max} and τ_{min} from shear tests on layers of soda-lime glass beads [*K. Frye*, personal communication]. The existence of this transition in sliding behavior may have implications for studies of earthquake mechanics and models of rupture dynamics.

5. SUMMARY

To study stick-slip behavior of bare rock surfaces, we have conducted experiments on Westerly granite in a biaxial

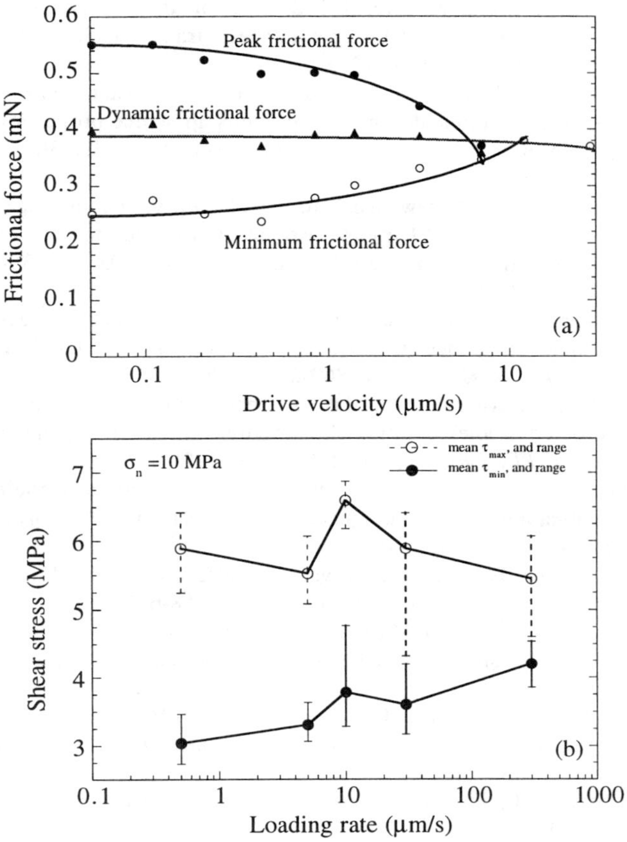

Figure 13. Peak and minimum stress levels shown as a function of loading rate. **a)** Data from *Berman et al.* [1996] showing measured peak, dynamic, and minimum frictional force from stick-slip instabilities for tests on thin molecular films between mica surfaces. Their data show convergence of peak and minimum stress levels as a function of loading rate. Above a critical velocity (~8 μm/s), they observe a transition to stable sliding. **b)** Shear stress data from our experiments as a function of loading rate. For clarity, we show the mean value and range of shear stress data for each velocity. For faster velocities, τ_{min} increases and τ_{max} decreases slightly. Thus, peak and minimum stress levels converge as velocity increases similar to results from *Berman et al.* [1996].

apparatus at room-temperature and humidity. We have analyzed over 1500 instabilities, and the data show several interesting features. We observe that stress drop amplitude increases with greater normal stress, consistent with previous studies. We also observe two apparent rates of restrengthening. For a single loading rate, stress drop increases by ~4 MPa per decade in recurrence time. The combined data from all velocities show that stress drop increases by ~1 MPa per decade in time. For the conditions of our tests, this discrepancy in restrengthening rates can be characterized in terms of stiffness of the loading column, loading rate, and recurrence interval. After removing elastic effects, we do not observe any systematic dependence of stress drop on premonitory slip. Measurements of pre-failure peak stress, and post-failure minimum stress, show convergence with faster loading rates. This is consistent with previous studies that identify a critical velocity marking a transition between stick-slip and stable sliding, and that this transition lies in an inertia-dominated regime. Our results are important for studies of the seismic cycle and modeling of earthquake source characteristics, as stress drop depends not only on loading rate but also on stochastic variations in earthquake recurrence times.

Acknowledgments. This research was supported by NSF Grant EAR-9627895 and by USGS Grant 99HQGR0003. We thank the anonymous reviewers for their constructive comments and thoughtful suggestions about this manuscript. We are also grateful to K. Frye, K. Mair, U. Mok, E. Richardson and J. Renner for their stimulating discussions about this work.

REFERENCES

Baumberger, T., Heslot, F., and Perrin, B., Crossover from creep to inertial motion in friction dynamics, *Nature*, 367, 544-546, 1994.

Beeler, N.M., Hickman, S.H., and Wong, T-f., Earthquake stress drop and laboratory inferred interseismic strength recovery, *Manuscript in preparation.*

Beeler, N.M., Tullis, T.E., and Weeks, J.D, The roles of time and displacement in the evolution effect in rock friction, *Geophys. Res. Lett.*, 21, 1987-1990, 1994.

Berman, A.D., Ducker, W.A., and Israelachvili, J.N., Experimental and theoretical investigations of stick-slip friction mechanisms, In *Physics of Sliding Friction*, Edited by B.N.J. Perrson and E. Tossatti, Kluwer Academic Publ., Netherlands, 1996.

Byerlee, J.D., Frictional characteristics of granite under high confining pressure, *J. Geophys. Res.*, 72, 3639-3648, 1967.

Cao, T., and Aki, K., Effect of slip rate on stress drop, *Pageoph.*, 515-529, 1986.

Chester, F.M., and Higgs, N.G., Multimechanism friction constitutive model for ultrafine quartz gouge at hypocentral conditions, *J. Geophys. Res.*, 97, 1857-1870, 1992.

Dieterich, J.H., Time-dependent friction in rocks, *J. Geophys. Res.*, 77, 3690-3697, 1972.

Dieterich, J.H., Time-dependent friction and the mechanics of stick-slip, *Pure Appl. Geophys.*, 116, 790-805, 1978.

Dieterich, J.H., Modeling of rock friction 1: Experimental results and constitutive equations, *J. Geophys. Res.*, 84, 2161-2168, 1979.

Engelder, J.T., Logan, J.M., and Handin, J., The sliding characteristics of sandstone on quartz fault-gouge, *Pure Appl. Geophys.*, 113, 69-86, 1975.

Engelder, J.T., and Scholz, C.H., The role of asperity indentation and ploughing in rock friction II, *Int. J. Rock Mech. Mining Sci. Geomech. Abstr., 13, 155-163, 1976.*, 113, 69-86, 1976.

Fredrich, J.T., and Evans, B., Strength recovery along simulated faults by solution transfer processes, *Proc. 33rd Natl. Rock Mech. Symp.*, ed. W. Wawersik, 121-130, Rotterdam: Balkema, 1992.

Gu, Y. and Wong, T-f., Effects of loading velocity, stiffness and inertia on the dynamics of a single degree of freedom spring-slider system, *J. Geophys. Res.*, 96, 21677-21691, 1991.

Heslot, F., Baumberger, T., Perrin, B., Caroli, B., and Caroli, C., Creep, stick-slip, and dry-friction dynamics: Experiments and a heuristic model, *Phys. Rev. E*, 49, 4973-4988, 1994.

Johnson, T.L., and Scholz, C.H., Dynamic properties of stick-slip friction of rocks, *J. Geophys. Res.*, 81, 881-889, 1976.

Kanamori, H., and Allen, C.R., Earthquake repeat time and average stress drop, In *AGU Geophys. Mono. Vol. 37*, Edited by S. Das, J. Boatwright & C.H. Scholz, 227-236, 1986.

Karner, S.L., and Marone, C., The effect of shear load on frictional healing in simulated fault gouge, *Geophys. Res. Lett.*, 25, 4561-4564, 1998.

Karner, S.L., Marone, C., and Evans, B., Laboratory study of fault healing and lithification in simulated fault gouge under hydrothermal conditions, *Tectonophysics*, 277, 41-55, 1997.

Klein, W., Rundle, J.B., and Ferguson, C.D., Scaling and nucleation in models of earthquake faults, *Phys. Rev. Lett.*, 78, 3793-3796, 1997.

Li, Y-g, Vidale, J.G, Aki, K., Xu, F., and Burdette, T., Evidence of shallow fault zone strengthening after the 1992 M7.5 Landers, California, earthquake, *Science*, 279, 217-220, 1998.

Linker, M.F., and Dieterich, J.H., Effects of variable normal stress on rock friction: observations and constitutive equations, *J. Geophys. Res.*, 97, 4923-4940, 1992.

Mair, K., and Marone, C., Friction of simulated fault gouge at a wide range of velocities and variable normal stresses, *J. Geophys. Res.*, 1999, In press.

Marone, C., The effect of loading rate on static friction and the rate of fault healing during the earthquake cycle, *Nature*, 391, 69-72, 1998a.

Marone, C., Laboratory-derived friction laws and their application to seismic faulting, *Annu. Rev. Earth Planet. Sci.*, 26, 643-696, 1998b.

Marone, C., and Kilgore, B., Scaling of the critical slip distance for seismic faulting with shear strain in fault zones, *Nature*, 362, 618-621, 1993.

Marone, C., Vidale, J.E., and Ellsworth, W., Fault healing inferred from time dependent variations in source properties of repeating earthquakes, *Geophys. Res. Lett.*, 22, 3095-3098, 1995.

Nakatani, M., A new mechanism of slip-weakening and strength recovery of friction associated with the mechanical consolidation of gouge, *J. Geophys. Res.*, 103, 27239-27256, 1998.

Nakatani, M., and Mochizuki, H., Effects of shear stress applied to surface in stationary contact on rock friction, *Geophys. Res. Lett.*, 23, 869-872, 1996.

Nasuno, S., Kudrolli, A., Bak, A., and Gollub, J.P., Time-resolved studies of stick-slip friction in sheared granular layers, *Phys. Rev. E*, 58, 2161-2171, 1998.

Ohnaka, M., Experimental studies of stick-slip and their application to the earthquake source mechanism, *J. Phys. Earth*, 21, 285-303, 1973.

Rabinowicz, E., The intrinsic variables affecting the stick-slip process, *Proc. Phys. Soc. London*, 71, 668-675, 1958.

Richardson, E., and Marone, C., Effects of normal stress vibrations on frictional healing, *J. Geophys. Res.*, 1999, In press.

Schaff, D.P., Beroza, G.C., and Shaw, B.E., Postseismic response of repeating aftershocks, *Geophys. Res. Lett.*, 25, 4549-4552, 1998.

Scholz, C.H., Earthquakes and friction laws, *Nature*, 391, 37-42, 1998.

Scholz, C.H., Aviles, C.A., and Wesnousky, S.G., Scaling differences between large interplate and intraplate earthquakes, *Bull. Seismol. Soc. Am.*, 76, 65-70, 1986.

Shimamoto, T., and Logan, J.M., Effects of simulated fault gouge on the sliding behavior of Tennessee sandstone: nonclay gouges, *J. Geophys. Res.*, 86, 2902-2914, 1981.

Teufel, L.W., and Logan, J.M., Effect of displacement rate on the real area of contact and temperature generated during frictional sliding of Tennessee sandstone, *Pure Appl. Geophys.*, 116, 840-872, 1978.

Wong, T-f., and Zhao, Y., Effects of loadpoint velocity on frictional instability behavior, *Tectonophys.*, 175, 177-195, 1990.

S.L. Karner (corresponding author) and C. Marone, Department of Earth, Atmospheric and Planetary Science (54-720), Massachusetts Institute of Technology, 77 Massachusetts Ave., Cambridge, MA, USA, 02139

Changes in earthquake size-frequency distributions underlying accelerating seismic moment/energy release

Steven C. Jaumé

Queensland University Advanced Centre for Earthquake Studies, The University of Queensland, Brisbane, Queensland, Australia

A considerable number of moderate to great earthquakes have been preceded by an increase in the rate of smaller events in the surrounding region, resulting in an acceleration in the rate of seismic energy/moment release as the time of the mainshock is approached. The so-called "critical point" models for this behavior postulate that a growing correlation length in the earth's crust underlies this phenomenon. In these models, the correlation length controls the maximum size of events in the earthquake population. As it grows it allows progressively larger earthquakes to occur, and thus changes the earthquake size-frequency distribution at large magnitudes. Here I test this hypothesis by examining changes in the earthquake size-frequency distribution of 17 known cases of accelerating seismic energy/moment release for which the space-time dimensions of this behavior have been defined and adequate earthquake catalogs exist. I find that for 15 of these 17 cases, observed changes in the earthquake size-frequency distribution are consistent with the predictions of the critical point hypothesis. For the other two cases, an increase in the rate of seismicity at all magnitudes appears to have occurred. These results suggest that critical point behavior underlies most but not all cases of accelerating seismic energy/moment release.

1. INTRODUCTION

There are now a sizeable number of cases where moderate to great earthquakes are known to be preceded by an increase in the rate of smaller earthquakes within a broad surrounding region [e.g., *Bowman et al.*, 1998; *Brehm and Braile*, 1988, 1999; *Bufe et al.*, 1994; *Sykes and Jaumé*, 1990]. The size of this region is also known to scale with the size of the oncoming mainshock [*Bowman et al.*, 1998; *Brehm and Braile*, 1998; *Sykes and Jaumé*, 1990] and the increase in seismicity rate is confined to events with magnitudes within 2.0 to 3.0 units of the mainshock magnitude [e.g., *Bufe and Varnes*, 1993; *Brehm and Braile*, 1998; 1999; *Ellsworth et al.*, 1981]. In many cases the cumulative seismic moment or Benioff strain release in these regions can be modeled using a power-law time-to-failure relationship that gives reasonable predictions of the time and magnitude of the oncoming mainshock [e.g., *Brehm and Braile*, 1998; 1999; *Bufe and Varnes*, 1993; *Bufe et al.*, 1994].

The recognition of an increasing number of these cases now provides an adequate set of data to begin testing predictions of the various models put forward to explain this behavior. Some authors have already conducted tests of model predictions for this behavior [e.g., *Bowman et al.*, 1998; *Gross and Rundle*, 1998]. My goal in this paper is to test a specific prediction of the so-called "critical point" models [e.g., *Saleur et al.*, 1996;

GeoComplexity and the Physics of Earthquakes
Geophysical Monograph 120

Sornette and Sammis, 1995]; i.e., that the physical process underlying accelerating seismic energy sequences is the growth of long-range correlations in a crustal fault system that allow for progressively larger earthquakes to occur in a region. *Jaumé and Sykes* [1999] conducted preliminary tests of this hypothesis based upon a small number of cases; here I examine a larger and more diverse set of data.

In the body of this paper I first review models for accelerating seismic moment/energy release (hereafter ASR) and their predictions, and describe the specific prediction being tested in this paper. Next I describe the data used in this study and the methodology for analysis of the data. I introduce a simple graphical method for displaying differences between earthquake size-frequency distributions that allows for comparison between distributions with different mainshock magnitudes. I then use this method to compare changes in earthquake size-frequency distributions with time for a number of published cases of ASR.

2. MODELS FOR ACCELERATING SEISMIC MOMENT/ENERGY RELEASE

Early models put forward to explain ASR noted the similarity between the accelerating seismicity sequences and acceleration in other measures of material deformation seen before failure in many different types of natural and man-made materials [*Sykes and Jaumé*, 1990; *Bufe and Varnes*, 1993]. *Bufe and Varnes* [1993] specifically reviewed crack nucleation/propagation and damage mechanics models when deriving the power-law time-to-failure relationship they used to model ASR in the San Francisco Bay region.

A number of problems arise when comparing crack nucleation/damage mechanics models to natural cases of ASR. While an acceleration in energy release (or other measure of deformation) before failure is a clear feature of these models, there are a number of other features of the observed sequences that do not appear to be predicted. In their review of ASR, *Jaumé and Sykes* [1999] pointed out, that besides the acceleration in seismic energy/moment release, there are three additional features that characterize these sequences: 1) the earthquakes involved in ASR almost exclusively fall outside the rupture zone of the mainshock [*Sykes and Jaumé*, 1990], 2) the increased rate of earthquake occurrence is limited to events within about 2.0 magnitude units of the mainshock [*Ellsworth et al.*, 1981], and 3) the size of the region involved in the ASR sequence scales with the size of the oncoming mainshock [*Sykes and Jaumé*, 1990; *Bowman et al.*, 1998; *Brehm and Braile*, 1998].

More recently a group of authors have proposed what is called the "critical point" model to explain the observations of ASR [*Sornette and Sammis*, 1995; *Sammis et al.*, 1996; *Saleur et al.*, 1996]. Conceptually, these models consider a large earthquake to be analogous to a critical point in a fluid or magnet. In such models a regional fault system progressively orders itself under the influence of strain accumulation and small earthquakes; a large earthquake can only occur when large scale correlations exist within the system. In these models ASR is a consequence of this ordering phenomena, which allows for progressively larger earthquakes to occur.

There are several predictions of the critical point model which match the observational evidence reviewed by *Jaumé and Sykes* [1999]. One feature of the critical point model is that an earthquake can both be considered a failure of a particular localized volume and also part of the failure process that leads to even larger events (up to the size of the system). This predicts not only the observed scaling of the region size with mainshock magnitude (see above), but also that ASR should happen prior to mainshocks with a wide variety of magnitudes. The latter observation has been seen in both computer models [*Huang et al.*, 1998] and observational data [*Bowman et al.*, 1998; *Brehm and Braile*, 1998; 1999].

A central concept in the critical point model is that there is a growing correlation length of the regional stress field prior to the occurrence of the mainshock event [*Sornette and Sammis*, 1995]. Indeed it is only at the critical point itself that correlations exist at all scales and the occurrence of the largest earthquake in a region becomes possible. In this model it is the correlation length that sets a limit on the largest possible earthquake size; it is the growth of the correlation length with time which produces ASR. Based upon an analysis of coarse grained models and simulations that capture the basic physics of the earthquake process, *Rundle et al.* [1999] have developed explicit relationships between the correlation length, the time to the mainshock, and the size-frequency distribution of the earthquake population. Following *Rundle et al.* [1999], the correlation length ξ_c can be defined as:

$$\xi_c = \gamma(K_L V |\delta t|)^{-1/4} \quad (1)$$

where K_L is an spring (elastic) constant, V is the driving velocity, $|\delta t| = |t_{sp} - t|$ (t_{sp} is the time of the "spinodal" or mainshock in the model of *Rundle et al.* [1999])

and γ is a constant. Thus the correlation length grows as the time of the mainshock approaches. The correlation length ξ_c controls a "cutoff area" A^* by:

$$A^* = \frac{(\xi_c \gamma)^{4/\sigma_f}}{k} \qquad (2)$$

where k is another constant and σ_f is the "surface exponent" which generally has a value of 1 [*Rundle et al.*, 1997]. A^* puts a limit on the size of the largest event in the non-cumulative earthquake size-frequency distribution (defined by the size of the failed area A):

$$n(A, \delta t) = \frac{n_o}{A^{\tau-1}} \exp[-A/A^*] \qquad (3)$$

Rundle et al. [1999] note that *Kagan* [1994] finds just such a distribution for the global earthquake size-frequency distribution in terms of seismic moment, M_o, with a size scaling exponent of 1.5 (i.e., $\tau = 2.5$) in equation (3).

To illustrate the effect of the growth in ξ_c on the event size-frequency distribution as the mainshock time approaches, in Figure 1 I plot cumulative event size-frequency distributions for two values of δt. The constants in the equations above where chosen so that the b-value in the Gutenberg-Richter relation is 1.0 at small magnitudes. For large δt the influence of the cutoff area A^* limits the size of the largest events. When δt is small the size-frequency scaling remains the same up to the largest event in the distribution.

Thus a prediction of the critical point model is that the earthquake size-frequency relationship should change as the time of the mainshock grows closer, and that this change should have its greatest effect on the large magnitude range of the distribution (i.e., consistent with observations). *Jaumé and Sykes* [1999] compared earthquake size-frequency distributions during the 1st and 2nd halves of four known accelerating seismic energy sequences and found that three of four cases showed changes consistent with this model. In this paper I test a far greater number of cases to see if they are consistent with this prediction.

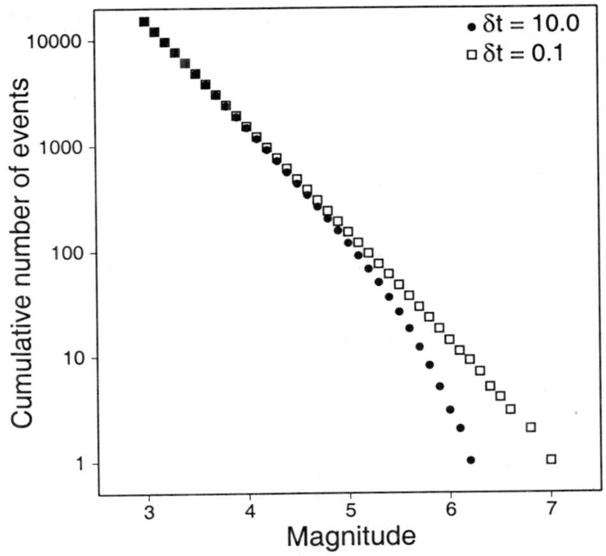

Figure 1. Illustration of changes in the earthquake size-frequency distribution as a function of time remaining until the critical point, based upon equation (3).

3. DATA

As stated above, the main purpose of this work is to expand upon the observations of *Jaumé and Sykes* [1999] and more fully test a prediction of the critical point hypothesis, i.e., that it is a change in the underlying earthquake size-frequency distribution with time that produces the acceleration in seismic energy release. Fortunately, in recent years there have been a number of studies of ASR before moderate to great earthquakes in different tectonic regimes. I have endeavored to gather earthquake catalogs for as many published cases as possible.

In selecting the earthquake data for use in examing changes in the size-frequency distribution, I have restricted the time and space windows of the catalogs to that used by the authors of the individual studies. This has caused some complications, specifically with regard to southern California earthquakes examined both by *Bowman et al.* [1998] and *Brehm and Braile* [1999]. The authors of these studies examine several large earthquakes in common, but use different methodologies for determining "optimal" time and space windows for the ASR sequences. Therefore I examine these cases using the time and space windows of both *Bowman et al.* [1998] and *Brehm and Braile* [1999].

A number of hurdles were encountered in selecting cases of ASR for this study. The first is that not all studies of ASR clearly define both the space and time windows used in selecting earthquake data for study. I have in some cases been able to get additional information on these sequences [e.g., southern California space-time windows used by *Brehm and Braile*, 1999; D. Brehm, pers. comm.], but not in all cases. Another hurdle encountered were limitations in the earthquake catalogs in the different regions. The most serious limitation was changing levels of completeness of small magnitude

Table 1. Earthquakes Used in This Study Together With Space-time Parameters Defining Associated Accelerating Energy Release Sequences

Date	Location	Magnitude	Critical Radius	Time (year)	Source
1996/6/10	Aleutians	7.9	210[a]	1980.46-1996.45	BNV94[b]
1992/8/28a	California	7.3	150	1970.50-1992.49	B&O98[c]
1992/8/28b	California	7.3	120	1979.50-1992.49	BB99[d]
1989/10/18	California	7.0	200	1943.81-1989.80[e]	B&O98
1995/2/05	New Zealand	7.0	167	1964.03-1995.10	R99[f]
1983/5/2a	California	6.7	175	1980.35-1983.34	B&O98
1983/5/2b	California	6.7	40	1976.35-1983.34	BB99
1994/1/17a	California	6.7	73	1992.76-1994.05	B&O98
1994/1/17b	California	6.7	155	1971.56-1994.05	BB99
1994/6/18	New Zealand	6.7	139	1964.43-1994.46	R99
1971/2/9	California	6.6	100	1967.12-1970.11	B&O98
1987/11/24	California	6.6	275	1982.91-1982.90	B&O98
1968/4/8	California	6.5	240	1958.28-1968.27	B&O98
1980/5/25	California	6.1	240	1970.40-1980.40	BB99
1986/7/8a	California	5.6	40	1985.40-1986.52	B&O98
1986/7/8b	California	5.6	15	1969.53-1986.52	BB99
1980/2/25	California	5.5	20	1968.15-1980.15	BB99
1982/10/25	California	5.4	70	1975.33-1982.82	BB99
1997/3/5	Australia	5.1	70	1972.19-1997.18	JMB99[g]
1980/2/14	Virgin Islands	4.8	24	1979.29-1980.12	B&O98,VB96[h]
1989/4/27	Central USA	4.3	5	1981.13-1989.33	BB98[i]

[a]This is the along-strike length of the Delarof Islands segment as defined by *Nishenko and Jacob* [1990].
[b]*Bufe et al.* [1994].
[c]*Bowman et al.* [1998].
[d]*Robinson* [1999].
[e]Time window of *Jaumé and Sykes* [1999].
[f]*Brehm and Braile* [1999].
[g]*Jaumé et al.* [1999].
[h]*Varnes and Bufe* [1996].
[i]*Brehm and Braile* [1998].

events with time. As seen in Figure 1, the critical point hypothesis predicts that the number of events of small magnitude should remain constant and that changes in frequency should only occur at higher magnitudes. A number of cases had to be rejected from this study because of incompleteness at small magnitudes during the early part of the sequence. Another problem, seen for the 1952 M = 7.5 Kern County, California earthquake, was a change in the earthquake magnitude determination procedure with time [*Hutton and Jones*, 1993]. A smaller number of cases were rejected from consideration for this reason.

The remaining cases used in this study are listed in Table 1, together with the time and space windows used to define the accelerating sequence and the source for this information. Additional cases studied by *Bowman et al.* [1998], *Brehm and Braile* [1998; 1999] and *Robinson* [1999] were not used in this study, generally due to incompleteness in the earthquake catalogs.

4. METHOD

Jaumé and Sykes [1999] conducted a preliminary test of the prediction that the underlying earthquake size-frequency distribution changes as the time of the mainshock approaches. There they took earthquakes occurring within the time-space windows defined by ASR sequences and divided the resulting earthquake catalogs into two halves based upon time, and compared the resulting frequency-magnitude distributions. They found that in three of four cases there was a greater frequency

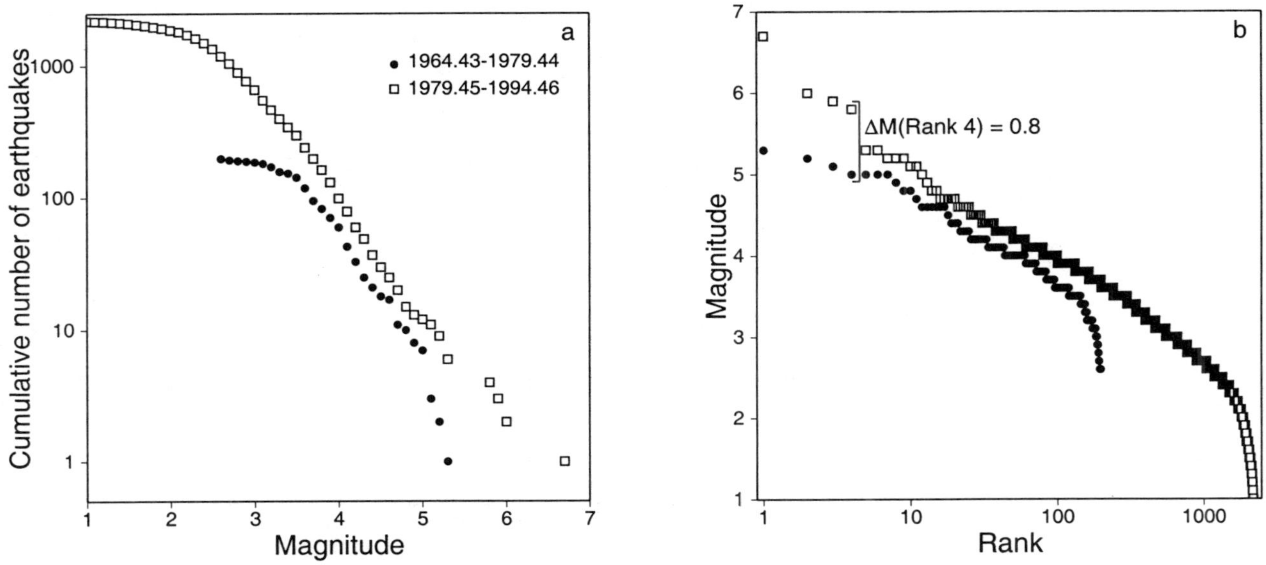

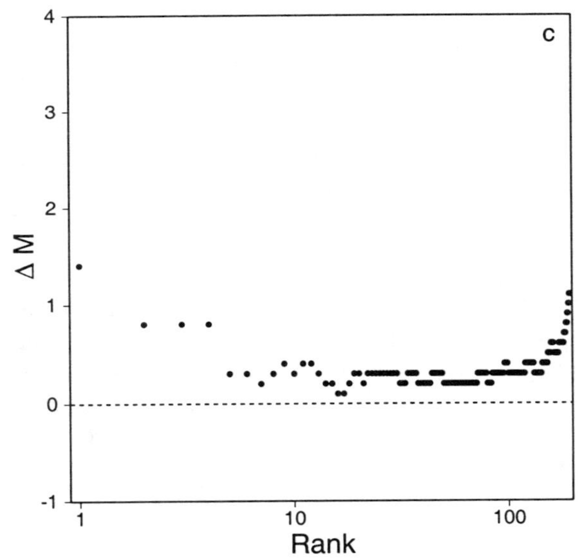

Figure 2. Example of magnitude-frequency, rank-ordering, and magnitude difference graph, using the data from the ASR sequence associated with the 1994 Arthur's Pass, New Zealand earthquake. (a) Traditional frequency-magnitude distributions, with the time periods used to define the two distributions shown. (b) Rank-ordering graph of the same data. Notice that all the data is plotted, and not grouped into magnitude bins as in A. The value of ΔM at rank 4 is also shown. (c) The *magnitude difference* graph. This is simply the difference in magnitude (ΔM), as a function of rank, between the two sequences. Note that it is defined up to highest rank in whichever distribution has the smallest total number of events.

of larger events during the second half of the ASR sequence, but little or no change in the rate of smaller events.

Although the results of *Jaumé and Sykes* [1999] are encouraging, there are difficulties with simply comparing frequency-magnitude distributions. The first, noted by *Jaumé and Sykes* [1999], is that although three of the four cases showed evidence for changes in frequency-magnitude distributions, the overall frequency-magnitude distributions in each case were different. Another difficulty, more apparent in this work which includes a wider range of mainshocks, is how to compare results from cases with different mainshock magnitudes.

A possible solution to these problems is to plot the magnitude-frequency distributions not in the traditional manner but using the rank-ordering technique [*Gumbel*, 1958] (Figure 2). Unlike the traditional Gutenberg-Richter formula, rank-ordering statistics emphasize the large event tail of the distribution. In seismology, rank-ordering statistics have been used to estimate earthquake recurrence times in eastern North America [*Howell*, 1993; 1994] and to test for changes in the scaling of

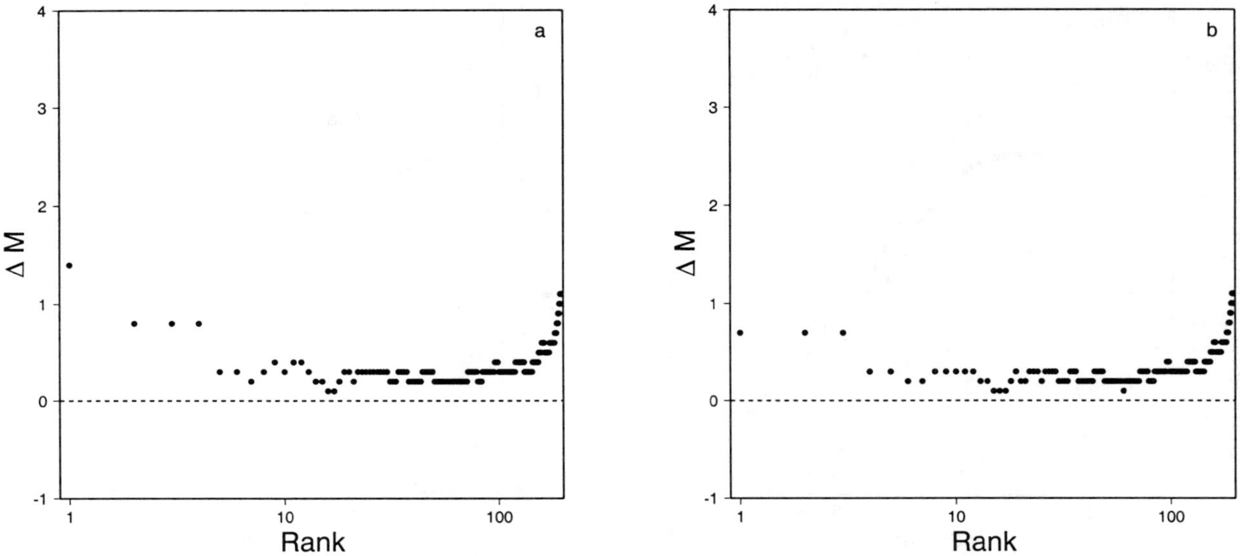

Figure 3. Magnitude difference graphs for the 1994 Arthur's Pass, New Zealand earthquake sequence, both (a) with the mainshock included in the distribution, and (b) without the mainshock.

earthquake populations at large magnitude [*Sornette et al.*, 1996].

This rank-ordering method is, in effect, simply a case of switching the **x** and **y** axes on the traditional magnitude-frequency graph. The largest event in the distribution is assigned to rank 1, the second largest event to rank 2, etc., until the smallest event is reached (Figure 2b). The utility of the method in this case is that one can now simply and clearly define the *difference* in size-frequency distributions as a function of event rank. This allows for any *changes* in the size-frequency distributions for ASR sequences to be easily compared. Figure 2 is an example of this procedure, using the case of the 1994 M = 6.7 Arthur's Pass, New Zealand earthquake [*Robinson*, 1999], going from a traditional magnitude-frequency graph (Figure 2a), to a rank-ordering graph (Figure 2b), to what I call here a *magnitude difference* graph (Figure 2c).

The magnitude difference graph for the Arthur's Pass earthquake (Figure 2c) shows many features in common with the other cases in Table 1. First and foremost, an increase in the rate of larger earthquakes shows up as a large ΔM at small rank. ΔM decreases to a smaller value as rank increases and remains there for a large range in rank, indicating that the two earthquake distributions have similar size-frequency scaling (i.e., the same b-value in the Gutenberg-Richter relationship) in this range. At large rank ΔM is seen to increase again, this time as a consequence of increased catalog completeness for small events with time. This "uplifted tail" is seen for nearly all sequences with long time windows.

As noted by *Jaumé and Sykes* [1999], it is not clear whether or not the mainshock that ends the accelerating energy release sequence should be included when comparing the two distributions. Thus I produce magnitude difference graphs both with and without the mainshock included (Figure 3). The major difference between the two is that the larger values of ΔM at small rank are reduced, but still remain above the "background" level seen at higher ranks.

5. RESULTS

As mentioned above, during the initial review of ASR sequences it was noted that many earthquake catalogs have large changes in completeness through time. In some cases these changes seriously interfere with the interpretation of the magnitude differences. An example of this is the ASR sequence associated with the 1981 M = 5.5 Santa Barbara, California earthquake [*Brehm and Braile*, 1999], shown in Figure 4. In this case it is unclear whether or not ΔM on the magnitude difference graph reaches a stable minimum before rising again due to the incompleteness of the earthquake catalog during the earlier time period. Such cases, which were most common for the smaller and/or historically

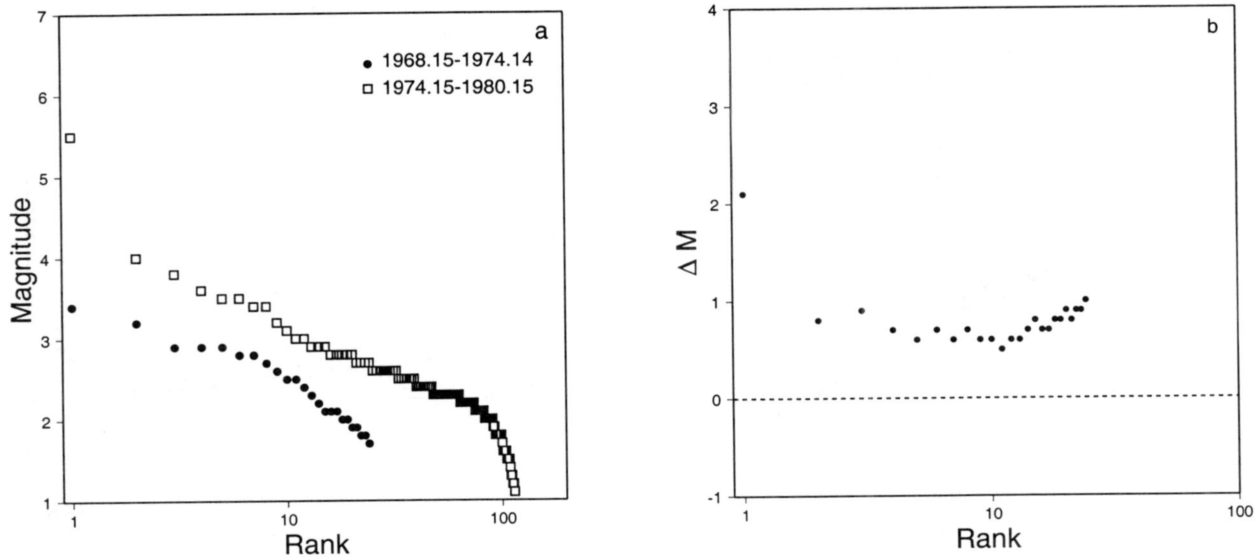

Figure 4. (a) Rank-ordering and (b) magnitude difference graphs for the M = 5.5 1981 Santa Barbara, California earthquake sequence.

earlier mainshocks, were discarded from further consideration and are not included on Table 1.

The most common results are those similar to the 1994 M = 6.7 Arthur's Pass, New Zealand earthquake (Figure 2). An example from a different location is the ASR sequence before the M = 7.3 1992 Landers, California earthquake (Figure 5). The magnitude difference monotonically decreases from a maximum at rank 1 to about rank 10, then is stable at a level $\Delta M \sim 0.1$-0.3 from rank 10 to nearly rank 10000, before increasing again due to changes in the catalog completeness level. Note that, based on equation (3) and Figure 1, the critical point hypothesis predicts ΔM should become zero at high rank. Therefore, I call these cases where ΔM reaches a stable minimum but clearly does not reach zero "modified critical point behavior", to distinguish them from the prediction of equation (3).

In some cases the minimum magnitude difference does indeed lie at zero for high ranks. This is most common for relatively short ASR sequences. An example of this is the ASR sequence associated with the 1994 M = 6.7 1994 Northridge, California earthquake (Figure 6), using the space-time window of *Bowman et al.* [1998].

While most cases examined are for large mainshock magnitudes and therefore large space-time windows, there are cases with relatively small mainshocks (and therefore small space-time windows) that compare favorably with their larger brethren. An example of this is the ASR sequence associated with the 1980 M = 4.8 Virgin Islands earthquake (Figure 7). Although the total data set is small (the magnitude difference can only be defined up to rank 14), it shows the main features in common with the others, i.e., a magnitude difference monotonically decreasing with rank to a lower level at which it remains. There is no apparent change in the catalog completeness in this case; this is most likely due to the short time span of the sequence.

A small number of cases examined do not show the features predicted for critical point behavior. An example is the 1996 M = 7.9 Delarof Islands earthquake (Figure 8), previously discussed by *Jaumé and Sykes* [1999]. As in that work, I find the difference between the two earthquake size-frequency distributions can best be described as an increase in the rate of events at all sizes (seen as a nearly constant ΔM in Figure 8), not an increase in rate of the larger events in the distribution. I also note here that one could obtain a similar magnitude difference graph because of an artificial magnitude shift in the earthquake catalog [*Habermann*, 1987], in addition to the apparent increase in catalog completeness at small magnitudes.

6. DISCUSSION

In Table 2 I summarize the results of this study by classifying the changes in the earthquake size-frequency distributions during the ASR sequences with respect to

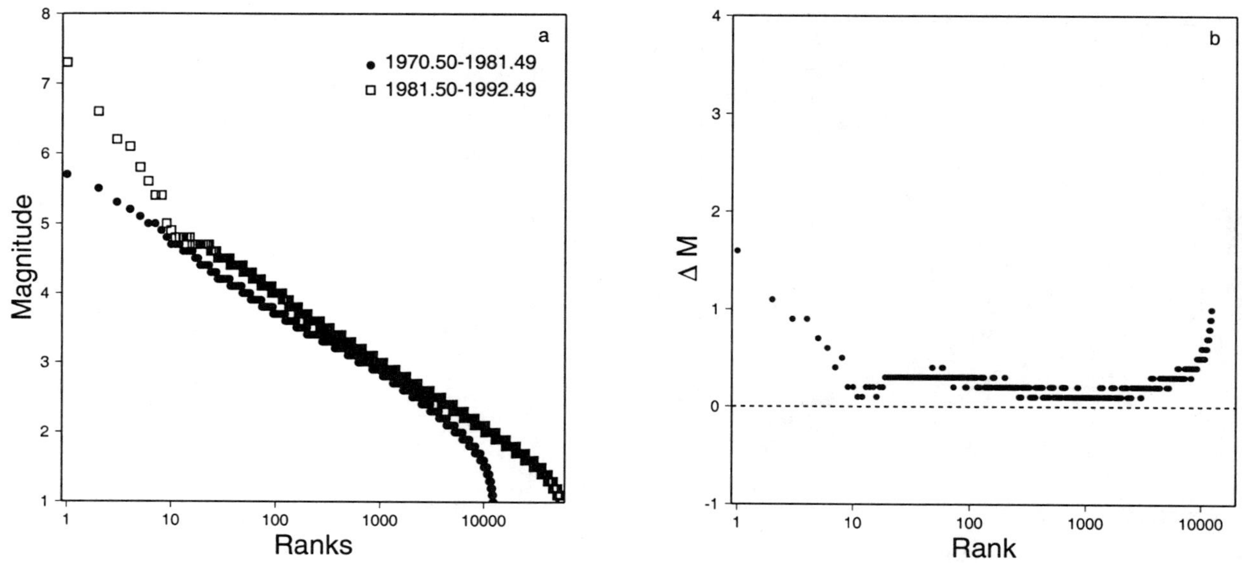

Figure 5. (a) Rank-ordering and (b) magnitude difference graphs for the 1992 M = 7.3 Landers, California earthquake sequence. The time-space dimensions of Bowman *et al.* (1998) are used to construct this figure.

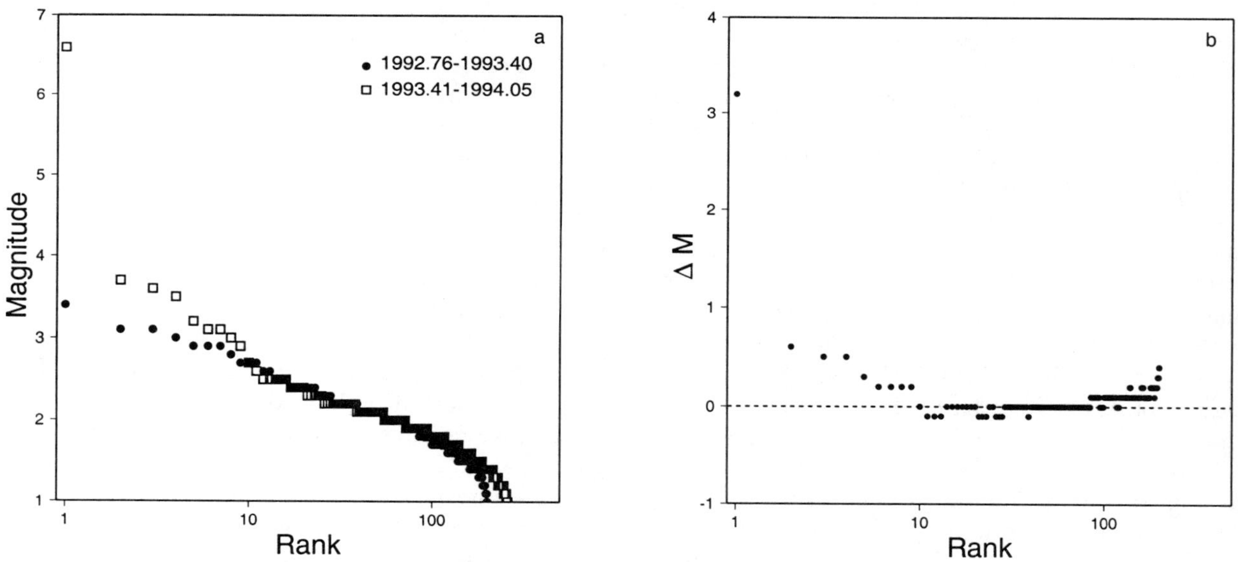

Figure 6. (a) Rank-ordering and (b) magnitude difference graphs for the M = 6.7 1994 Northridge, California earthquake sequence. The time-space dimensions of Bowman *et al.* (1998) are used to construct this figure.

their consistency with the critical point hypothesis. Of the 17 sequences studied here, 15 were found to be consistent (with some modification) with the critical point hypothesis. The other two sequences were found to be more consistent with either an overall increase in seismicity rate or perhaps a man-made shift in the magnitudes in the earthquake catalog.

As noted in Table 2, only three of the sequences are perfectly consistent with the predictions of the critical point hypothesis, as described by equation (3). Of the rest 10 of 15 are better described by what I call "modified critical point behavior". The final two of the 15 cases show either critical point or modified critical point behavior, depending upon the space-time windows used.

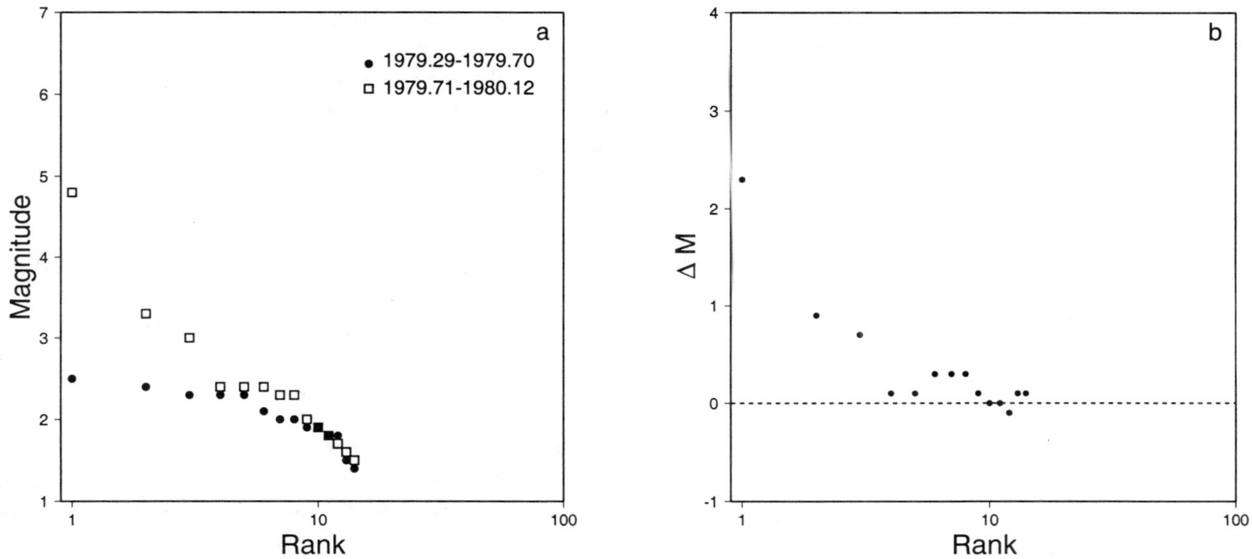

Figure 7. (a) Rank-ordering and (b) magnitude difference graphs for the 1980 M = 4.8 Virgin Islands earthquake sequence.

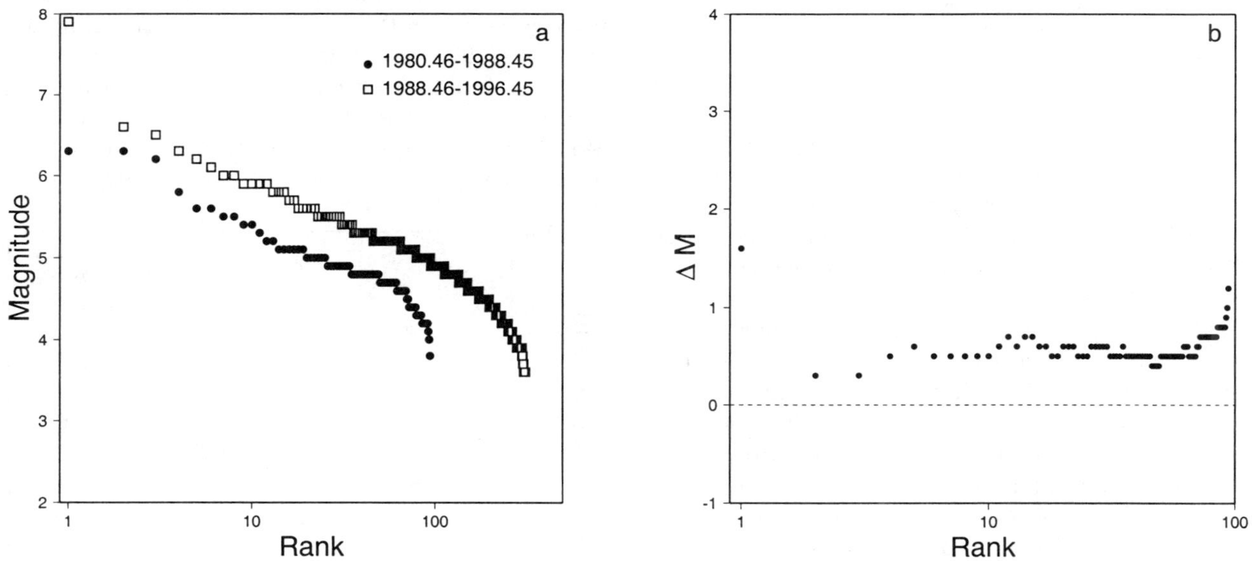

Figure 8. (a) Rank-ordering and (b) magnitude difference graphs for the 1996 M = 7.9 Delarof Islands, Alaska earthquake sequence.

I decided to examine the last two cases more closely, to find out if there is any correlation between aspects of the space-time windows and the type of behavior displayed. I find that the shorter time windows (1.3 and 3 years respectively) correlate with critical point behavior and the longer time windows (7 and 22.5 years) with modified critical point behavior. However, upon looking at the larger set of 15 cases, there is no clear correlation between behavior and either the time or space windows used (Figure 9).

A possible explanation for the majority of cases having modified critical point behavior is the prescence of aftershocks in the earthquake catalogs I used. As has been noted by others, aftershocks can make a substantial fraction (up to 70%) of a regional earthquake catalog [e.g., *Reasenberg*, 1985; *Knopoff and Lee*, 1999].

Table 2. Classification of Accelerating Energy Release Sequences With Respect to Critical Point Behavior

Date	Location	Magnitude	Behavior
1996/6/10	Aleutians	7.9	OI[a]
1992/8/28	California	7.3	MCP[b]
1989/10/18	California	7.0	MCP
1995/2/05	New Zealand	7.0	MCP
1983/5/2	California	6.7	CP?MCP
1994/1/17	California	6.7	CP/MCP
1994/6/18	New Zealand	6.7	MCP
1971/2/9	California	6.6	MCP
1987/11/24	California	6.6	MCP
1968/4/8	California	6.5	MCP
1980/5/25	California	6.1	OI
1986/7/8	California	5.6	CP
1980/2/25	California	5.5	MCP
1982/10/25	California	5.4	MCP
1997/3/5	Australia	5.1	CP
1980/2/14	Virgin Islands	4.8	CP
1989/4/27	Central USA	4.3	MCP

[a]Overall seismicity rate increase.
[b]Modified critical point behavior.
[c]Critical point behavior.

Given that larger events generally produce greater numbers of aftershocks and that the critical point hypothesis predicts a growth in the maximum earthquake size as the critical point is approached, something akin to the modified critical point behavior described here could be expected. If this is the case, the prescence or absence of modified critical point behavior has more to do with the contribution of aftershocks to the total number of events in the earthquake catalog than with any fundamental difference in physical behavior (although differences in the relative abundance of aftershocks may imply other aspects of earthquake physics). This concept can potentially be tested by repeating the analysis presented here using "declustered" earthquake catalogs; that work will be left for a future study.

Finally, I wish to ask the question "How well do the different critical point/modified critical point sequences compare?" If all of these cases are the result of a similar physical processes, expressed in different tectonic settings, it may be expected that the changes in the size-frequency distributions should overlap. Figure 10 presents a composite magnitude difference graph using the 15 cases of critical point/modified critical point behavior. As noted in the Methods section, the magnitude difference method presented here removes the absolute magnitude difference between the sequences and allows them to be easily compared.

Although there is scatter in the data, the overall behavior of the 15 cases presented in Figure 10 is remarkably consistent. For the cases in which the mainshock is retained in the distribution, the magnitude difference ΔM decreases monotonically from rank 1 to about rank 10, beyond which it remains essentially constant. When the mainshock is removed the distribution, the difference in ΔM at low versus high rank is more subdued and in one case (i.e., the 1997 Burra, South Australia earthquake) ΔM at rank 1 is less than zero. However, the general trend, ΔM rising above the "background" at small rank, is preserved.

While the trends in Figure 10 are clear, the scatter in ΔM ranges from 0.5 to over 1.0 at small ranks. How much scatter should be expected in this data, and what significance does it have? One possible avenue of investigation is to look at the results of earthquake simulation models that also show this behavior, and analyze the changes in the model earthquake size-frequency distributions in a similar manner. I note here that preliminary results in this direction are encouraging; *Mora and Place* [1999] have found accelerating energy release sequences in a simulation model of an evolving granular zone. They examined changes in earthquake size-frequency distributions during ASR in their model using the method in *Jaumé and Sykes* [1999], and find

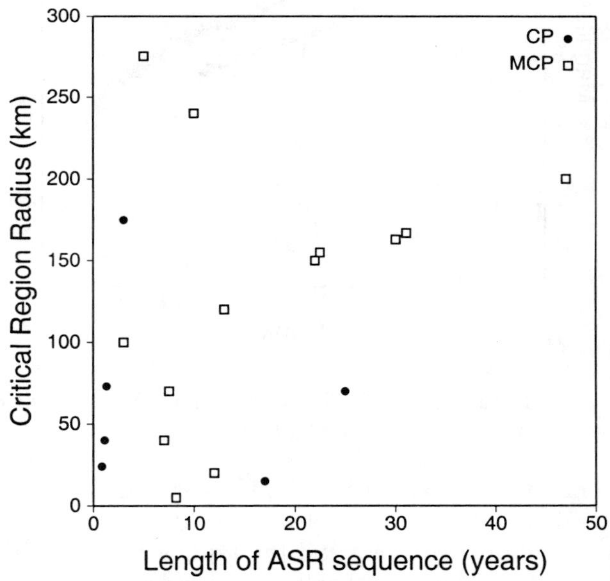

Figure 9. Critical point (filled circles) and modified critical point (open squares) behavior as a function of space-time window of the ASR sequence.

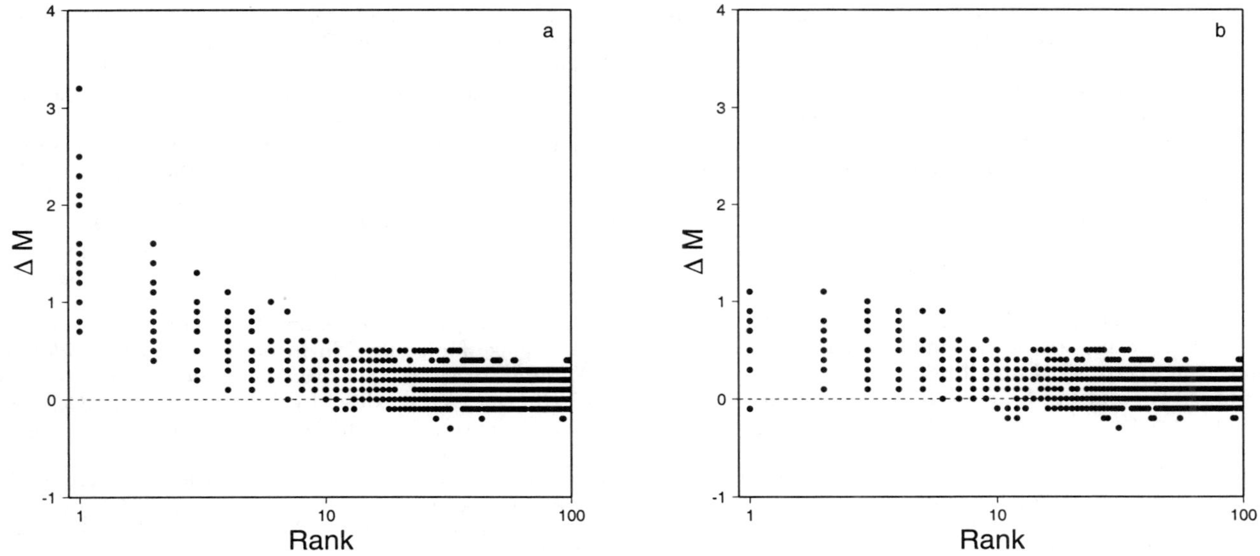

Figure 10. Composite magnitude differences graphs for the 15 cases (filled circles) that show critical point behavior, both (a) including and (b) excluding the mainshock in the distribution. Only the first 100 ranks are shown.

changes in the size-frequency distributions similar to what is reported here. Other simulation models of the earthquake process, particularly those involving multiple faults, should be examined for this behavior and compared to observational results.

7. CONCLUSIONS

Of 17 cases where accelerating seismic energy release is known to have occurred before a moderate-to-great earthquake, 15 cases show changes in the earthquake size-frequency distribution during the lead-up to the mainshock that are consistent with that predicted by the critical point hypothesis. Specifically, the distributions show evidence for growth in the size of the largest earthquakes as the time of the mainshock is approached. For the remaining two cases, the changes in the size-frequency distributions are more consistent with an overall increase in the rate of seismicity at all magnitudes. This suggests that critical point behavior underlies most, but not all, observed cases of accelerating seismic energy release.

Acknowledgments. Discussions with P. Mora, D. Weatherley, R. Cuthbertson, and M. Winter helped clarify my thinking during the course of this work. I especially thank P. Mora, D. Weatherley and two anonymous reviewers for critical reviews of the manuscript. I also thank D. Weatherley for reminding me about rank-ordering statistics and D. Brehm for pre-publication materials. D. Love (PIRSA), K. McCue (AGSO) and R. Robinson (IGNS) provided earthquake catalogs for Australia and New Zealand. Other earthquake catalogs were downloaded from the Council of the National Seismic System and Center for Earthquake Research and Information web pages. The GMT system software [*Wessel and Smith*, 1991] was used in the production of the figures.

REFERENCES

Bowman, D. D., G. Ouillon, C. G. Sammis, D. Sornette, and A. Sornette, An observational test of the critical earthquake concept, *J. Geophys. Res., 103*, 24,359-24,372, 1998.

Brehm, D. J., and L. W. Braile, Intermediate-term prediction using precursory events in the New Madrid Seismic Zone, *Bull. Seismol. Soc. Am., 88*, 564–580, 1998.

Brehm, D. J., and L. W. Braile, Intermediate-term earthquake prediction using using the modified time-to-failure method in southern California, *Bull. Seismol. Soc. Am., 89*, 275-293, 1999.

Bufe, C. G., and D. J. Varnes, Predictive modeling of the seismic cycle in the greater San Francisco Bay region, *J. Geophys. Res., 98*, 9871-9983, 1993.

Bufe, C. G., S. P. Nishenko and D. J. Varnes, Seismicity trends and potential for large earthquakes in the Alaska-Aleutian region, *Pure Appl. Geophys., 142*, 83-99, 1994.

Ellsworth, W. L., A. G. Lindh, W. H. Prescott, and D. G. Herd, The 1906 San Francisco earthquake and the seismic cycle, in *Earthquake Prediction: An international review,* edited by D. W. Simpson and P. G. Richards, pp. 126–140, AGU, Washington, D. C., 1981.

Gumbel, E. J. *Statistics of extremes*, 375 pp., Columbia University Press, New York, 1958.

Gross, S., and J. Rundle, A systematic test of time-to-failure analysis, *Geophys. J. Int.*, *133*, 57-64, 1998.

Habermann, R. E., Man-made changes in seismicity rates, *Bull. Seismol. Soc. Am.*, *77*, 141-159, 1987.

Howell, B. F., Jr., Recurrence expectation for earthquakes in eastern North America south of 50° latitude, *Seism. Res. Lett.*, *64*, 139-147, 1993.

Howell, B. F., Jr., Earthquake recurrence rates in the central Atlantic United States, *Seism. Res. Lett.*, *65*, 149-156, 1994.

Huang, Y., H. Saleur, C. Sammis, and D. Sornette, Precursors, aftershocks, criticality and self-organized criticality, *Europhys. Lett.*, *41*, 43-48, 1998.

Hutton, L. K., and L. M. Jones, Local magnitudes and apparent variations in seismicity rates in Southern California, *Bull. Seismol. Soc. Am.*, *83*, 313-329, 1993.

Jaumé, S. C., and L. R. Sykes, Evolving towards a critical point: a review of accelerating seismic moment/energy release prior to large and great earthquakes, *Pure Appl. Geophys.*, *155*, 279-305, 1999.

Jaumé, S. C., P. Mora and C. G. Bufe, Accelerating seismic energy release before moderate to large earthquakes: three Australian examples (extended abstract), *Proc. 1999 Australian Earthquake Engineering Society Conf.*, paper 16, 1999.

Kagan, Y. Y., Observational evidence for earthquakes as a nonlinear dynamical process, *Physica D*, *77*, 160-192, 1994.

Knopoff, L., and M. W. Lee, The self-organization of aftershocks, *Proc. 1st ACES Workshop*, 463-465, 1999.

Nishenko, S. P., and K. Jacob, Seismic potential of the Queen Charlotte-Alaska-Aleutian seismic zone, *J. Geophys. Res.*, *95*, 2511-2532, 1990.

Mora, P., and D. Place, Accelerating energy release prior to large events in simulated earthquake cycles: implications for earthquake forecasting, *Proc. 1st ACES Workshop*, 513-520, 1999.

Reasenberg, P., Second-order moment of Central California seismicity, *J. Geophys. Res.*, *90*, 5479-5495, 1985.

Robinson, R., A test of the precursory accelerating moment release model on some recent New Zealand earthquakes, *Geophys. J. Int.*, in press, 1999.

Rundle, J. B., W. Klein, S. Gross, and C. D. Ferguson, Traveling density wave models for earthquakes and driven threshold systems, *Phys. Rev. E*, *56*, 293-307, 1997.

Rundle, J. B., W. Klein, and S. Gross, Physical basis for statistical patterns in complex earthquake populations: models, predictions, and tests, *Pure Appl. Geophys.*, *155*, 575-607, 1999.

Sammis, C. G., D. Sornette, and H. Saleur, Complexity and earthquake forecasting, in *Reduction and Predictability of Natural Disasters, SFI Studies in the Science of Complexity, vol. XXV*, edited by J. B. Rundle, W. Klein, and D. L. Turcotte, Addison-Wesley, Reading, Mass., 143-156, 1996.

Saleur, H., C. G. Sammis, and D. Sornette, Discrete scale invariance, complex fractal dimension, and log-periodic fluctuations in seismicity, *J. Geophys. Res.*, *101*, 17661-17677, 1996.

Sornette, D., and C. G. Sammis, Complex critical exponents from renormalization group theory of earthquakes: implications for earthquake predictions, *J. Phys. I France*, *5*, 607-619, 1995.

Sornette, D., L. Knopoff, Y. Y. Kagan, and C. Vanneste, Rank-ordering statistics of extreme events: application to the distribution of large earthquakes, *J. Geophys. Res.*, *101*, 13,883-13,893, 1996.

Sykes, L. R., and S. C. Jaumé, Seismic activity on neighboring faults as a long-term precursor to large earthquakes in the San Francisco Bay region, *Nature*, *348*, 595-599, 1990.

Varnes, D. J., and C. G. Bufe, The cyclic and fractal seismic series preceding an m_b 4.8 earthquake on 1980 February 14 near the Virgin Islands, *Geophys. J. Int.*, *124*, 149-158, 1996.

Wessel, P., and W. H. F. Smith, Free software helps map and display data, *Eos Trans. AGU*, *72*, 445–446, 1991.

Observation of Systematic Variations in Non-Local Seismicity Patterns from Southern California

K. F. Tiampo, J. B. Rundle, S. McGinnis, and S. J. Gross

Colorado Center for Chaos and Complexity, Cooperative Institute for Research in Environmental Sciences, University of Colorado, Boulder, CO

W. Klein

Department of Physics, Boston University, Boston, MA

We have discovered systematic space-time variations in the seismicity from southern California using a new technique. Our procedure is based upon the idea that seismic activity corresponds geometrically to the rotation of a pattern state vector in the high-dimensional correlation space spanned by the eigenvectors of a correlation operator. Using our technique it is possible to isolate emergent regions of coherent, correlated seismicity. Analysis of data taken only up to December 31, 1991 reveals that the appearance of the coherent correlated regions is often associated with the future occurrence of major earthquakes in the same areas. These major earthquakes include the 1992 Landers, the 1994 Northridge, and the 1999 Hector Mine events.

1. INTRODUCTION

Earthquakes strike without warning, causing great destruction and loss of life. A poignant example is the recent Izmit, Turkey, earthquake of August 17, 1999, which resulted in the deaths of over 17,000 persons. Other recent large events include the M ~ 7.6 Taiwan earthquake of September 20, 1999, whose death toll now exceeds 2000 persons, the M ~ 7.3 Landers, California, event of June 28, 1992, and the M ~ 7.1 Hector Mine, California, earthquake of October 16, 1999. Many similar examples have been documented over the course of time [*Richter,* 1958; *Scholz,* 1990].

While a long-sought goal of earthquake research has been the reliable forecasting of these great events, very little progress has been made in developing a successful, consistent methodology [*Geller et al.,* 1997; *Kanamori,* 1981]. Despite the fact that the largest of these events span distances of more than 500 km, no reliable precursors have ever been detected. It is difficult for most scientists to understand why events of this magnitude are not preceded by at least some causal process, which would presumably imply the existence of premonitory signals. In the past, the search for such signals understandably focused on local regions near the earthquake source. Many of these techniques require intensive and expensive monitoring efforts [*Geller et al.,* 1997]. Various patterns of seismic activity centered on the source region have been proposed, including phenomena such as characteristic earthquakes [*Schwartz, et al.,* 1981; *Ellison and Cole,* 1997], Mogi donuts [*Mogi,* 1969; *Mogi,* 1977], seismic gaps [*Haberman,* 1981; *House et al.,* 1981], precursory quiescence [*Knopoff and Yamashita,* 1988; *Wyss and Haberman,* 1988; *Wyss et al.,* 1996; *Kato et al.,* 1997], precursory activation [*Evison,* 1977; *Shaw et al.,* 1992; *Dodge et al.,* 1996], Time-to-Failure and Log-Periodic precursory distributions [*Bufe et al.,* 1993; *Saleur et al.,*

GeoComplexity and the Physics of Earthquakes
Geophysical Monograph 120

1996; *Gross and Rundle,* 1998], temporal clustering [*Frolich,* 1987; *Dodge et al.,* 1995; *Rundle et al.,* 1997], and earthquake triggering over large distances [*Hill et al.,* 1993; *King et al.,* 1994; *Pollitz and Sacks,* 1997]. Since these hypothesized patterns are localized on the eventual source region, the fact that one must know or suspect where the event will occur before they can be applied is a major drawback to their implementation.

Recent observational evidence has suggested that earthquakes can be characterized by strongly correlated dynamics [*Bufe and Varnes,* 1993; *Press and Allen,* 1995; *Knopoff et al.,* 1996; *Bowman et al.,* 1998; *Brehm and Braile,* 1998; *Gross and Rundle,* 1998; *Brehm and Braile,* 1999]. Realistic numerical simulations of earthquakes also suggest that space-time pattern structures are non-local in character, another consequence of strong correlations in the underlying dynamics [*Rundle,* 1988; *Rundle et al.,* 2000]. Variables in many of these dynamical systems can be characterized by a phase function that involves both an amplitude and a phase angle. The simulations have suggested that seismicity can be described by phase dynamics [*Mori and Kuramoto,* 1998; *Rundle et al. 2000*]. Here, the important changes in seismicity are associated primarily with rotations of the vector phase function in a high-dimensional correlation space [*Fukanaga,* 1970; *Holmes et al.,* 1996]. Changes in the amplitude of the phase function are unimportant, or not relevant. The most familiar examples of these are quantum mechanical systems, but examples also exist in the macroscopic world, including weak turbulence in fluids and reaction-diffusion systems [*Mori and Kuramoto,* 1998].

These results suggest that space-time patterns of seismic activity directly reflect the existence of space-time correlations in the underlying stress and strain fields. Previous research has indicated that the development of correlations in the stress field is a necessary precondition for the occurrence of large earthquakes [*Rundle et al.,* 2000]. The correlation patterns, which represent emergent space-time structures, evidently form and evolve over time intervals of years preceding the main shock. Longer time intervals and larger correlated areas are associated with larger main shocks. The probability for observing such an anomalous correlation can be computed directly from the simulated seismicity data using the square of the anomalous pattern state vector [*Rundle et al.,2000*]. These are the methods that we use in the present analysis of data from southern California.

We test the hypothesis that anomalous, non-local space-time patterns and correlations associated with recorded events can be detected in real seismicity data years prior to the main shock. The seismicity data employed in our analysis is taken from existing observations in southern California between the years 1932 and the present. Using only a subset of this data covering the period from January 1, 1980 through December 31, 1991, we compute the probability for finding an anomalous spatial correlation at all sites in southern California over several intervals preceding December 31, 1991. We then superimpose on this map the locations of main shocks larger than 5.0 that occurred between January 1, 1992 and November 1, 1999, that is, the ~ 8 years following the time interval from which we computed the probabilities. We observe a striking correspondence between regions of increased probability and the location of the recent main shocks, tending to support the results first observed in our simulations. In particular, we note that the epicenter of the recent October 16, 1999, M ~ 7.0 Hector Mine earthquake in southern California occurred at a location that is identified as one of the high probability locations. From the size of the candidate source regions, the magnitude of the possible events can be estimated as well. A likelihood ratio test of the method on both the real southern California seismicity catalog, and a second catalog in which times of events had been reassigned randomly, indicates that the method does find coherent correlated structures in the data.

2. METHOD

As mentioned above, our method is based primarily on the idea that the time evolution of seismicity can be described by phase dynamics. We therefore define a real-valued seismic phase function, $\hat{S}(x_i,t_0,t)$, a unit vector whose tail is fixed at the origin, and whose head is constrained to move on the unit sphere in an N-dimensional (N large) correlation space. $\hat{S}(x_i,t_0,t)$ is a non-local function, and is the mathematical embodiment of the idea that earthquake fault systems are characterized by strongly correlated dynamics. Geometrically, time evolution of $\hat{S}(x_i,t_0,t)$ corresponds to rotations about the origin in a series of correlated random walk increments through small solid angles on the unit sphere. Formation of an emergent correlated pattern in seismic activity over a time interval Δt is associated with rotation of the $\hat{\mathbf{S}}$-vector in a persistent direction. When these persistent directions are examined, previously undetectable, systematic variations in seismicity become evident, as described in *Rundle et al., 2000.*

For our analysis, the phase function $\hat{S}(x_i,t_0,t)$

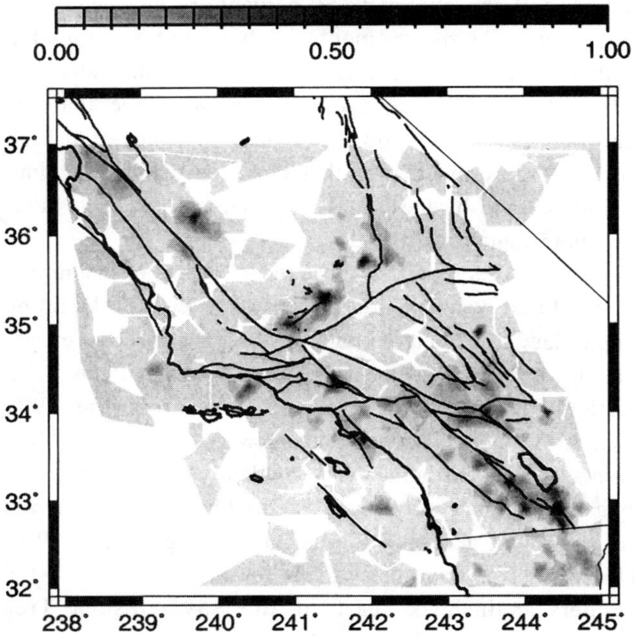

Figure 1. Normalized seismicity function **S**(1932,1991), M ≥ 3.0, for southern California.

characterizes the seismic activity in southern California between 32° and 37° latitude, and -115° to -122° longitude. It should also be noted that while our initial choice for total area was relatively arbitrary, we have varied the region size in recent analyses, by as much as a factor of two, and have found it to make little difference in the final results. In addition, the catalog was not declustered, as it is the correlations in the data set, the best known examples of which are local patterns of seismic activity or quiescence, that are identified by this method. Since it is well known that seismicity in active regions is a noisy function [*Kanamori,* 1981], we work with temporal averages of seismic activity. The geographic area is partitioned into N square regions approximately 11 km on a side, centered on a point x_i. Within each box, a time series is defined using the Caltech seismic catalog obtained from the online SCEC database. For southern California, the instrumental data begins in 1932 and extends to the present. The instrumental coverage was sparse in the early years, and is substantially more complete today. In general, the seismicity catalog is considered complete for magnitudes M ≥ 3.

We define the activity rate $\psi_{obs}(x_i, t)$ as the number of earthquakes per unit time, of any size, within the box centered at x_i at time t. The geographic region that $\hat{S}(x_i, t_0, t)$ represents is taken large enough so that seismic activity can be considered an incoherent superposition of phase functions. The seismicity function $S(x_i, t_0, t)$ is then defined as the time average at x_i of $\psi_{obs}(x_i, t)$ over the period (t_0, t),

$$S(x_i, t_0, t) = \frac{1}{(t - t_0)} \int_{t_0}^{t} \psi(x, t) dt.$$

Events included in $\psi_{obs}(x_i, t)$ or $S(x_i, t_0, t)$ are restricted to those for which the magnitude M ≥ 3, so as to ameliorate sensitivity to changes in detection thresholds through time.

Considered as a function of the N locations x_i, $S(x_i, t_0, t)$ represents a vector in N-dimensional correlation space with its tail fixed at the origin. The vector space is spanned by the eigenvectors, or eigenpatterns, of an N-dimensional, Karhunen-Loeve correlation matrix $C(x_i, x_j)$. The elements of $C(x_i, x_j)$ are obtained by cross-correlating a set of N seismic activity time series associated with each box x_i [*Rundle et al., 2000]*. As an aside, we note that purely random processes are characterized only by an amplitude in correlation space, not by any preferred direction. This follows from the fact that if the time series defining $C(x_i, x_j)$ were uncorrelated, $C(x_i, x_j)$ would be the identity matrix.

Figure 1 shows one example of **S**(1932,1991) superimposed on a map of southern California. It is clear that **S**(1932,1991) is an unremarkable function, and appears to show little evidence of any phenomena precursory to the M ~ 7.3 Landers, California event that occurred on June 28, 1972.

In the past [*Scholz,* 1990], investigators have generally focused on attempts to detect systematic variations in the both the amplitude and phase of $S(x_i, t_0, t)$, or alternatively the amplitude and phase variations in the corresponding rate of seismic activity $R(x_i, t)$. Following our assumption that seismicity is characterized by phase dynamics, we define $\hat{S}(x_i, t_0, t)$ as the unit vector pointing in the direction of $\hat{S}(x_i, t_0, t)$ using an L2 norm. As an incoherent superposition of functions must have zero mean, we remove the spatial mean of $\hat{S}(x_i, t_0, t)$ and create a unit vector by dividing by the standard deviation. Therefore,

$$\hat{S}(x_i, t_0, t) = \frac{[S(x_i, t_0, t) - \overline{S}(x_i, t_0, t)]}{\sigma},$$

$$\text{where } \overline{S}(x_i, t_0, t) = \frac{1}{N} \int_{all x} S(x_i, t_0, t) dx,$$

the spatial mean of $S(x_i, t_0, t)$, and

$$\sigma^2 = \frac{1}{N} \int_{all x} \left[S(x_i, t_0, t) - \overline{S}(x_i, t_0, t) \right]^2 dx,$$

the variance of $S(x_i, t_0, t)$.

Under the phase dynamics assumption, the important changes in seismicity $\Delta\hat{S}(x_i,t_1,t_2) = \hat{S}(x_i,t_0,t_2) - \hat{S}(x_i,t_0,t_1)$ over the time interval (t_1, t_2) correspond to rotations in correlation space about the origin.

Thus the important observable is this difference $\Delta\hat{S}(x_i,t_1,t_2)$. Recall that $S(x_i, t_0, t)$ is a spatially local function. However, due to the normalization above, which involves information from the entire active region of N boxes, $\hat{S}(x_i,t_0,t)$ is a spatially non-local function.

We now compute the increase in probability $\Delta P(x_i, t_1, t_2)$ associated with formation of a spatial correlation at location x_i over the time interval $\Delta t = (t_2 - t_1)$. Because a correlation function can be interpreted as a probability, the eigenvectors of the correlation operator $C(x_i, x_j)$ are effectively the square root of a probability. Any vector such as $\hat{S}(x_i,t_0,t)$ can be written as a linear expansion of such a complete set of eigenvectors which span the correlation vector space. Thus, the increase in probability $\Delta P(x_i, t_1, t_2)$ is related to the square of $\Delta\hat{S}(x_i,t_1,t_2)$. In addition, as the principle of conservation of probability implies that the integral over all space of $\Delta P(x_i, t_1, t_2)$ is equal to zero, we find that $\Delta P(x_i,t_1,t_2) = |\Delta\hat{S}(x_i,t_1,t_2)|^2 - \mu_P$, where μ_P is the spatial mean of $|\Delta\hat{S}(x_i,t_1,t_2)|^2$.

3. RESULTS AND DISCUSSION

Plate 1 shows plots of all $\Delta P > 0$, using only existing seismicity data acquired prior to January 1, 1992, six months before the June 28, 1992 occurrence of the M ~ 7.3,Landers, CA, earthquake. Recall that the increase in ΔP above the background level as measured by μ_P should be interpreted as a tendency to form a spatially correlated region of seismic activity, and that such regions evidently must be present for larger earthquakes to occur. An increase in ΔP appears to represent an increased chance of an earthquake occurring near that location. The color coding on Figure 2 is scaled to the largest value of ΔP on any of Figures 2 a, b, c or d. The largest 30 percent of points is represented by red, approximately 20 percent by yellow, and approximately 50 percent are between white and green. At the moment, we have not yet found a method to convert these relative numbers into absolute values of probability change, however, we expect that this may be possible with further study. The inverted blue triangles represent events that occurred during the time period covered by the plot, to indicate colored boxes that need not be analyzed further.

Blue circles represent more recent events of magnitude M > 5.0 that occurred after January 1, 1992. It should be emphasized again that no data for these more recent events was used in constructing the colored boxes in Plate 1. In particular, we include circles representing the 1992 Landers sequence and the recent M ~ 7.1 Hector Mine events. These earthquakes are evidently associated with a long-lived arcuate structure of colored boxes cutting across the local fault geometry that began forming prior to 1980. This structure continues down to the southeast of the Landers mainshock, east of the 1992 Joshua Tree earthquake, and the lack of subsequent activity to date may indicate this site as a potential rupture zone in the near future.

Visual examination of Plate 1 indicates that recent large events (blue circles) that occurred after January 1, 1992 are clearly associated with detectable locations of positive ΔP that formed prior to January 1, 1992. However there is clearly some variability, particularly for smaller events, depending on the choice of time interval (t_1, t_2). Larger events tend to be associated with larger colored regions that form earlier and persist longer after the event. Since earthquake fault dynamics are now believed to be associated with critical phenomena [*Rundle and Klein,* 1995; *Klein et al.,* 1997; *Gell-Mann et al.,* 2000; *Rundle et al.,* 2000], we hypothesize that there may be a scaling relation between the area A of the correlated region and the time interval t prior to the main shock at which the correlation begins to form such that $t \propto A^{\eta}$, where η is a critical exponent near 1. Since the linear size of our boxes is approximately 11 km, one should not expect events significantly smaller than M ~ 6, whose characteristic linear source dimension is 10 km, to be well resolved by our procedure. Yet even the smaller circles associated with M ~ 5 - 6 events seem to occur in proximity to colored boxes, albeit those that appear only a short time interval before the main shock.

To test the hypothesis that the formation of correlated regions identified by this method are indeed coherent

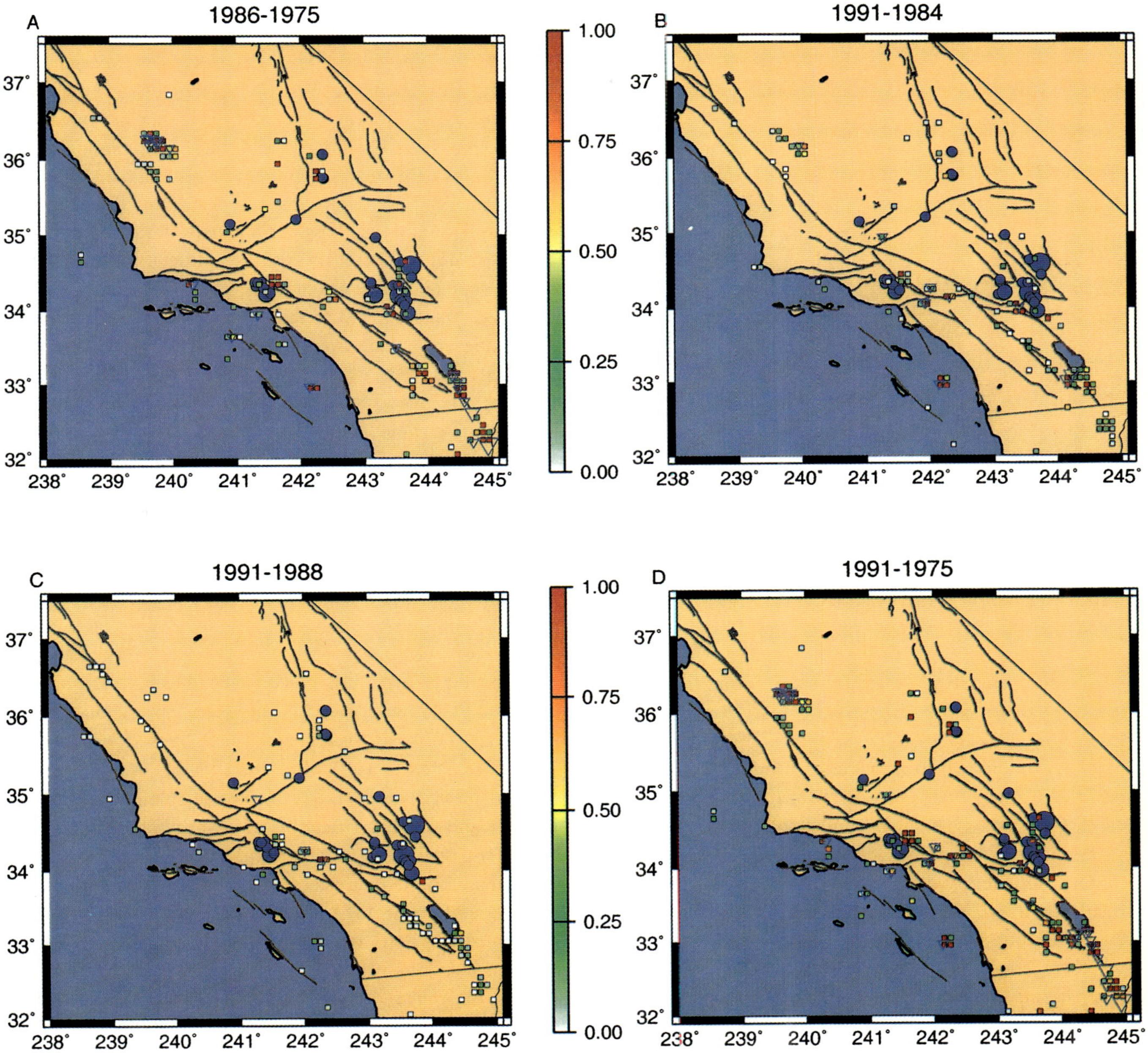

Plate 1. Maps of ΔP for the time intervals a) 1986-1975, b) 1991-1984, c) 1991-1988 and d) 1991-1975. In constructing these maps, no data is used from time periods after December 31, 1991. Inverted triangles represent events that occurred during the indicated time periods, with three sizes corresponding to magnitudes M of: $5 < M < 6$, $6 \leq M < 7$, $7 \leq M$. Filled circles represent events that occur after January 1, 1992. Three increasing circle sizes again correspond to the same magnitude ranges as for the inverted triangles.

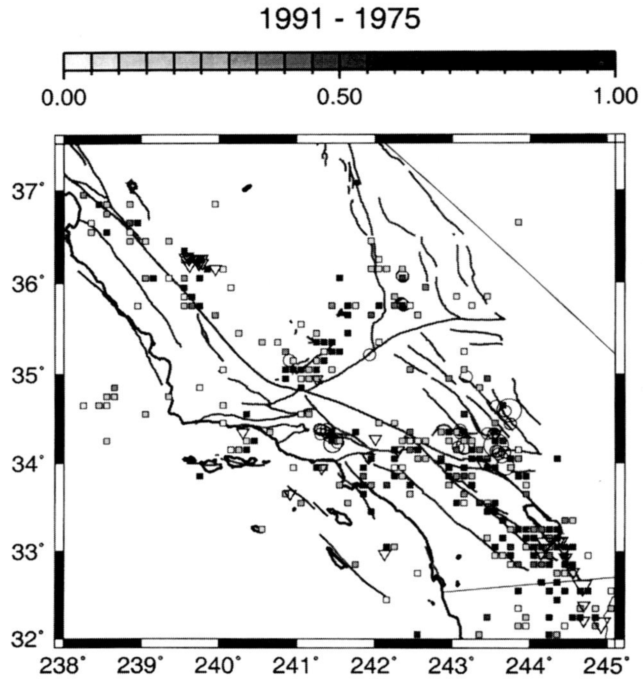

Figure 2. Map of $\Delta P > 0$ for the time interval 1991-1975, for the random catalog generated by applying a uniform random distribution of times between 1932 and 1991 to the same locations as the actual catalog.

space-time structures that are related to future large events, we carried out a likelihood ratio test [*Bevington and Robinson,* 1992; *Gross and Rundle,* 1998] on our model using a comparison to a random seismicity catalog. The random catalog was constructed from the instrumental catalog by using the same number of events at the same locations, but assigning occurrence times drawn from a uniform probability distribution over the years 1932-1991, resulting in a Poisson distribution of interevent times. Randomizing the catalog in this way should destroy whatever coherent space-time structure exists, effectively declustering the catalog. We applied our method to this random catalog and obtained the colored boxes shown in Figure 2, which corresponds to the same time period as Plate 1d, 1991-1975. One can see that there are many more colored boxes in Figure 2 than in 2d, and that the boxes are more broadly distributed in space.

To apply the likelihood ratio test to both Figures 2d and 3, we assumed a probability density function for each box, colored or not, with a Gaussian distribution, whose peak value is $\Delta P(x_i, t_1, t_2) + \mu_P$, since probabilities in a likelihood test must all be positive, and whose width is that of the colored boxes, approximately 11 km. We then calculated the log likelihoods for the blue circles in both Figures 2d and 3, and found values of -14.5439 for Plate 1d and –17.4239 for Figure 2. These values correspond to a likelihood ratio of $e^{2.88} \sim 17.8142$, indicating that the colored boxes obtained from the actual instrumental catalog are more likely to be associated with the locations of the blue circles than the colored boxes obtained from the random catalog. The physical reason for this large ratio is that the likelihood test invokes a penalty for colored boxes that are not sufficiently near to blue circles (``false positives"), and there are many more such boxes in Figure 2 than in Plate 1d. From this test, we conclude that there are coherent space-time correlation structures in the instrumental catalog that our method identifies. This supports the theory that earthquake fault systems contain space-time correlations which are effectively destroyed by declustering the catalog.

In analyzing the meaning of Plate 1 we emphasize that while our method may identify higher risk areas, there is no certainty at this time that every box will be located near the site of a future large earthquake. There are a number of examples in Plate 1 where a box appears during one time period, then disappears over a longer time period without the occurrence of a major earthquake ("false negatives"). One example of this is the colored boxes which appear near 34.7° latitude, 238.6° longitude, during the period 1975 to 1986, but disappear during the periods 1986 to 1991 and 1988 to 1991. Further attempts at optimization of the method must focus on better spatial location of events and the identification of a minimum number of both false positives and false negatives.

In addition, it appears that, as coherent space-time structures form in the underlying physics which drive the earthquake system, our method attempts to fit the nearest known locations of previous seismicity. One example of this is the four red boxes in the location of the 1971 San Fernando earthquake, as shown in Plate 1d. We hypothesize that these locations are simply the nearest locations available to the method for applying the increased probability associated with the adjacent 1994 Northridge event. In examining the time periods shown, we note that the method described above effectively subtracts out the effects of any San Fernando aftershocks remaining in the catalog in 1975. To test the assumption that the changes in probability are not exclusively associated with the location of aftershock sequences, we plotted the change in $\hat{S}(x_i,t_0,t)$, $\Delta\hat{S}(x_i,t_1,t_2)$. The results show that the four squares adjacent to the 1994 Northridge event denote an area of *decreased* $\hat{S}(x_i,t_0,t)$, i.e. precursory quiescence, not aftershocks due to the 1971 San Fernando earthquake. Events which go off during this time period, such as the 1983 Coalinga earthquake and the 1979 Imperial Valley event, display seismic activation, while the Landers sequence is a complicated mix of positive and negative $\Delta\hat{S}(x_i,t_1,t_2)$. These findings support our conclusions from both the numerical simulations and theoretical analysis that this technique does not simply identify areas associated with past events and their aftershock sequences.

4. CONCLUSIONS

In summary, we conclude that we have observed systematic variations in seismicity prior to recent southern California earthquakes. Our method employs data from existing seismic monitoring networks as well as a theoretical understanding obtained from numerical computer simulations to identify coherent space-time structures in seismicity. These space-time patterns in the seismic activity directly reflect the existence of correlated structure in the underlying stress and strain fields, a necessary precondition for the occurrence of large earthquakes. Depending on the nature of future seismic activity in the region, as well as future modifications and extensions of the theory and technique, this procedure may prove useful in analysis of future trends in seismic activity.

Acknowledgements. We would like to thank J. Bernard Minster and Andrea Donnellan for their helpful review and comments during the drafting of this paper. We would also like to acknowledge useful discussions with J. Perez-Mercader and M. Gell-Mann. Work carried out by K.F.T. was supported under NASA Fellowship No. NGT5-30025 to the Cooperative Institute for Research in Environmental Sciences (CIRES) at the University of Colorado, and by a CIRES Fellowship funded under NOAA Grant No. NA67RJ0153. Work by J.B.R. was supported by U.S. Dept of Energy Grant No. DE-FG03-95ER14499 to CIRES and NASA Grant No. NAG5-5168 to CIRES. The work of S.M. was supported by NASA Fellowship No. ESS-97-0110 to CIRES, and the work of W.K. was supported under U.S. Dept of Energy Grant No. DE-FG-2-95ER14498 to the Physics Dept. and Center for Computational Science at Boston University.

REFERENCES

Bevington, P. R., D. K. Robinson, *Data Reduction and Error Analysis for the Physical Sciences,* McGraw-Hill, N.Y., 1992.

Bowman, D. D., G. Ouillon, C. G. Sammis, A. Sornette and D. Sornette, *J. Geophys. Res., 103,* 24,359, 1998.

Brehm D. J., and L. W. Braile, *Bull. Seis. Soc. Am., 88,* 564, 1998.

Brehm D. J., and L. W. Braile, *Bull. Seis. Soc. Am., 89,* 275, 1999.

Bufe, C. G., and D. J. Varnes, *J. Geophys. Res., 98,* 9871, 1993.

Dodge, D. A., G. C. Beroza, W. L. Ellsworth, *J. Geophys. Res., 100,* 9865, 1995.

Dodge, D. A., G. C. Beroza, W. L. Ellsworth, *J. Geophys. Res., 101,* 22371, 1996.

Ellsworth, W. I., and A. T. Cole, *Seis. Res. Lett., 68,* 298, 1997.

K. Mogi, *Bull. Earthquake Res. Inst. Tokyo Univ., 47,* pp. 395 417, 1969.

Evison, F. F., *Nature, 266,* 710, 1977.

Frohlich, C., *J. Geophys. Res., 92,* 13,944, 1987.

Fukunaga, K., *Introduction to Statistical Pattern Recognition,* Academic Press, N.Y., 1970.

Geller, R. J., D. D. Jackson, Y. Y. Kagan, F. Mulargia, *Science, 275,* 1616 (1997).

Gell-Mann, M., J. Perez-Mercader, J. B. Rundle, *in The Physics of Earthquakes*, edited by J. B. Rundle, W. Klein and D. L. Turcotte, AGU, Washington, D.C., 2000.

Gross, S., and J. Rundle, *Geophys. J. Int., 133,* 57, 1998.

Haberman, R. E., *in Earthquake Prediction: an International Review*, edited by D. W. Simpson, II, and P. G. Richards, pp. 29-42, AGU, Washington, D. C., 1981.

Hill, D. P., et al., *Science, 260,* 1617, 1993.

Holmes, P., J. L. Lumley, G. Berkooz, *Turbulence, Coherent Structures, Dynamical Systems and Symmetry,* Cambridge University Press, U.K., 1996.

House, L. S., L. R. Sykes, J. N. Davies, K. H. Jacob, *in Earthquake Prediction: an International Review*, edited by D. W. Simpson, II, and P. G. Richards, pp. 81-92, AGU, Washington, D. C., 1981.

Kanamori, H., *in Earthquake Prediction: an International Review*, edited by D. W. Simpson, II, and P. G. Richards, pp. 1-19, AGU, Washington, D. C., 1981.

Kato, N., M. Ohtake, T. Hirasawa, *Pure Appl. Geophys., 150,* 249, 1997.

King, G. C. P., R. S. Stein, J. Lin, *Bull. Seis. Soc. Am., 84,* 935, 1994.

Klein, W., J. B. Rundle, C. D. Ferguson, *Phys. Rev. Letters, 78,* n. 19, 1997.

Knopoff, L., and T. Yamashita, *Terra Cognita, 8,* 118, 1988.

Knopoff, L., T. Levshina, V. I. Keilis-Borok and C. Mattoni, *J. Geophys. Res., 101,* 5779, 1996.

Mogi, K., *Bull. Earthquake Res. Inst., Tokyo Univ., 47,* pp. 395-417, 1969.

Mogi, K., *Proc. Symp. on Earthquake Prediction, Seis. Soc. Japan,* pp. 203-214, 1977.

Mori H., and Y. Kuramoto, *Dissipative Structures and Chaos,* Springer-Verlag, Berlin, 1998.

Pollitz, F. F., and I. S. Sacks, *Bull. Seis. Soc. Am., 87,* 1, 1997.

Press F., and C. R. Allen, *J. Geophys. Res., 100,* 6421, 1995.

Richter, C. F., *Elementary Seismology,* Freeman, San Francisco, 1958.

Rundle, J. B., *J. Geophys. Res., 93,* 6255, 1988.

Rundle, J. B., and W. Klein, *Rev. Geophys. Space Phys., Suppl.,* July 1995.

Rundle, J. B., S. Gross, W. Klein, C. Ferguson, D. L. Turcotte, *Tectonophysics, 277,* 147, 1997.

Rundle, J. B., W. Klein, K. F. Tiampo and S. Gross, *Phys. Rev. E,* in press, March 2000.

Saleur, H., C. G. Sammis, and D. Sornette, *J. Geophys. Res., 101,* 17,661, 1996.

Scholz, C. H., *The Mechanics of Earthquakes and Faulting,* Cambridge University Press, Cambridge, U.K., 1990.

Schwartz, D. P., et al., *Earthquake Notes, 52,* 71, 1981.

Shaw, B. E., J. M. Carlson, J. S. Langer, *J. Geophys. Res., 97,* 479, 1992.

Wyss, M., and R. E. Haberman, *Pure Appl. Geophys., 126,* 319, 1988.

Wyss, M., K. Shimaziki, T. Urabe, *Geophys. Jour. Int., 127,* 735, 1996.

K.F. Tiampo, J.B. Rundle, S. McGinnis and S.J. Gross, Colorado Center for Chaos and Complexity, Cooperative Institute for Research in Environmental Sciences, University of Colorado, Boulder, CO 80309-0216.

W. Klein, Dept. of Physics, Boston University, Boston, MA 02215.

Introducing a New Paradigm for Computational Earth Science: A Web-Object-Based Approach to Earthquake Simulations

Geoffrey C. Fox

School for Computational Science and Information Technology and Department of Computer Science, Florida State University, Dirac Science Library, Tallahassee, Florida

Ken Hurst, Andrea Donnellan, and Jay Parker

Jet Propulsion Laboratory/California Institute of Technology, Pasadena California

Computer simulations will be key to substantial gains in understanding the earthquake process. Emerging information technologies make possible a major change in the way computers are used and data is accessed. An outline of a realizable computational infrastructure includes standardization of data accessibility, harnessing high-performance computing algorithms, and packaging simulation elements as distributed objects across wide networks. These advances promise to reduce dramatically the frustration and cost of doing earthquake science as they transform the fragmentary nature of the field into one of integration and community.

1. INTRODUCTION

Earthquakes in urban centers are capable of causing enormous damage. The recent January 16, 1995 Kobe, Japan earthquake was only a magnitude 6.9 event and yet produced an estimated $200 billion loss. Despite an active earthquake prediction program in Japan, this event was a complete surprise. The 1989 Loma Prieta and 1994 Northridge earthquakes were also unexpected and caused billions of dollars of damage as well as loss of lives. Partly as a result of these events the volume of earthquake related data being collected is rapidly increasing. Simulations are the necessary next step in order to understand earthquakes and examine the broad parameter space related to them.

The overarching goal of earthquake physics is to characterize and predict the behavior of systems of earthquake faults. One way to address this question is to take advantage of the currently available computational power and ever increasing wealth of data to construct realistic models of the earthquake process. These models can then be run for hundreds of thousands of virtual years to span several cycles of the entire system, and a statistical mechanics approach applied to look for patterns in both the synthetic and real seismicity and crustal deformation catalogues.

Recent work in earthquake physics has focused largely on understanding dynamic rupture processes, such as how ruptures grow into large earthquakes and how faults heal themselves. Other work has focused on analyzing observed seismicity in an attempt to look for precursory activity. Much more recently investigators have begun studying earthquakes using a systems approach in which individual faults interact with an entire system of faults.

In parallel efforts, a great wealth of geophysical data pertaining directly to the earthquake problem is being collected. The traditional seismic networks are being improved with broadband and strong ground motion instruments. GPS networks have been expanding globally and have virtually replaced the more traditional trilateration and triangulation networks. Concerted efforts in paleoseismology have added a wealth of data on the surface characteristics of major faults, particularly in southern California.

GeoComplexity and the Physics of Earthquakes
Geophysical Monograph 120

Rapidly expanding datasets and the advent of object broker technologies make it possible to pursue the development of complex, sophisticated models for predicting the behaviors of fault systems. Surface geodetic, seismicity, strong motion, and other data provide the necessary constraints for carrying out realistic simulations of fault interactions. Information technology is providing the means for defined and accessible data formats and code protocols that allow for communicating distributed components, the basis for the next wave of advanced simulations. The emerging technologies of IT enable frameworks for documentation of codes and standards as well as visualization and data analysis tools that will make for true community-based efforts. Without such tools it will be impossible to construct the more complex models and simulations necessary to further our understanding of earthquake physics.

Simulations are critical to understanding the behavior of fault systems, a major problem in earth science, because earthquakes occur in the real earth at irregular intervals on timescales of hundreds to thousands of years. Simulations generate arbitrarily long seismicity catalogs and provide a numerical laboratory in which the physics of earthquakes can be investigated from a systems viewpoint.

Simulations also provide invaluable feedback for the planning and design of future data collection efforts. Emerging information technology applied to the data and analysis will result in a revolutionary change in the manner in which earth scientists explore and analyze data . The cost of data access will plummet while the usefulness of data will multiply. Forging the feedback loop between the simulations and design of data collection efforts will further enhance the value and utility of the data.

Recent research indicates that the phenomena associated with earthquakes occur over many scales of space and time. Understanding the dynamic processes responsible for these events will require not only a commitment to develop the necessary observational datasets, but also the technology required to use these data in the development of sophisticated, state-of-the-art numerical simulations and models. The models can then be used to develop an analytical and predictive understanding of these large and damaging events, thus moving beyond the current, more descriptive approaches routinely employed. Future approaches emphasizing the development of predictive models and simulations for earthquakes will be similar to methods now used to understand global climate change, the onset of the El Niño-Southern Oscillation events, and the evolution of the polar ozone depletion zones.

1.1. Current Problems in Earthquake Physics

The goal of the earthquake physics working group of the Southern California Earthquake Center (SCEC) is to model the evolution of seismic histories and space-time evolution of the stress field in southern California in order to understand space-time clustering of earthquakes. The focus has been on modeling two-dimensional networks of faults and on simulating the occurrence of large and intermediate-magnitude earthquakes to compare with contemporary earthquake catalogs.

It is not surprising that until recently the major focus in earthquake physics has been on the rupture process since seismic data represent the largest data set and the catalogue extends approximately half a century back from present. Since seismographs record earthquakes they give us insight into the elastic response of the earth's crust and into the rupture process. In contrast, new data types such as crustal deformation data provide us with information about the processes leading up to failure as well as information on the static response from earthquakes. Surface deformation data can be used as boundary conditions to continuum models of the earthquake cycle.

The current state of the science of earthquake physics is rather disjoint. Several investigators have constructed complex and realistic models of a single facet of earthquake processes, while others have performed statistical analysis of seismicity. Computer performance is now such that these facets can be joined together into comprehensive models of the entire earthquake process.

Simulations are the only comprehensive means to study earthquake fault systems because earthquakes occur on timescales of decades to thousands of years. Creating physically realistic simulations requires an understanding of the mechanical properties of faults and the bulk material surrounding the faults, perhaps inclusive of the entire lithosphere. Building such understanding requires a concentrated modeling effort. Thus earthquake simulations of increasing realism will help aid planning efforts by predicting the level of shaking from a hypothetical event, and will also help focus attention on areas and phenomena that need additional measurements or theoretical development, leading to yet more realism in the simulations.

1.2. Computational Overview

There is substantial international interest in the use of large-scale computation in the earthquake field including an activity in Japan where major computational resources are being deployed and an effort among several Asia-Pacific nations including USA (the so-called APEC initiative). Here we will focus on an American activity known as GEM for its goal to produce a "General Earthquake Model" (http://www.npac.syr.edu/projects/gem, http://milhouse.jpl.nasa.gov/gem).

There are currently no approaches to earthquake forecasting that are uniformly reliable. The field uses phenomenological approaches, which attempt to forecast individual events or more reliably statistical analyses giving probabilistic predictions. The development of these methods has been complicated by the fact that large events responsible for the greatest damage repeat at irregular intervals of hundreds to thousands of years, and so the limited historical record has frustrated phenomenological studies. Direct numerical simulation has not been extensively pursued due to the complexity of the problem and the (presumed) sensitivity of the occurrence of large events to detailed understanding of earth constituent make up, myriad initial conditions, and the relevant micro-scale physics which determines the underlying friction laws.

This field is different from many other physical sciences such as climate and weather, as it so far has made little use of parallel computing and only now is starting its own "Grand Challenges". It is thus not known how important large-scale simulations will be in Earthquake science. Nevertheless it is essentially certain that they can provide a numerical laboratory of semi-realistic earthquakes which will enable development and testing of other more phenomenological methods based on pattern recognition. Also, simulations provide a powerful way of integrating data into statistical and other such forecasting methods as has been demonstrated in the use of data assimilation techniques in other fields.

This field has some very challenging individual simulations but it has only just started to use high performance computers. Therefore the most promising computations at this stage involve either scaling up existing simulations to large system sizes with modern algorithms or integrating several computational components with assimilated data, thus creating prototype full fault-system simulations. The latter has important real-world applications in the area of crisis response and planning. For example, one may carry the computations through from initial sensing of stress build up through the structural simulation of building and civil infrastructure responding to propagating waves. This is discussed briefly in Sections 5.3 and 5.4.

Earthquake fault modeling exhibits many different types of codes discussed in Section 3, which eventually could be linked together to support either real time response to a crisis or fundamental scientific studies. The computational framework GEMCI introduced in Section 2 has been carefully designed to support multiple types of component-component linkage including rich user front ends, termed problem solving environments.

Indeed, such a comprehensive modeling environment must be established to enable full exploitation of the new and expanding types of geophysical data relevent to both earthquake science and engineering. Codes must be able to communicate to allow for cross-validation between models. The wealth of geophysical data and modeling codes now being collected and developed needs to be standardized to allow rapid and easy sharing of information, retrieval of data, and model development and validation. At present, researchers laboriously transform the data formats into ones useable by their own individual programs. Current practices that are frustrating and wasteful include FTP of data sets larger than needed, parsing and reformatting of data, and transcription of information directly from printed papers.

Existing modeling and simulation programs are fragmented; however, most of the pieces are in place to construct powerful integrated simulations. We can envision this approach providing the key to crustal data assimilation, and successfully addressing the self-consistent use of physics at multiple scales. For example, kinematic models use surface deformation data, and digital elevation models (DEMs) to estimate plate and microplate motions. Elastic models use surface observations to estimate fault geometries and slip rates. Viscoelastic calculations incorporate surface deformation, DEMs, and lab, heat-flow, stress, seismic and geologic data into models that produce estimates of crustal rheology and structure, fault geometry, fault slip and stress rates, heat flow, gravity, and refined estimates of surface deformation. These provide the inputs for quasi-static dynamic models of single faults and systems of faults. The quasi-static dynamic models provide space-time patterns, correlations, and information about fault interactions.

As the models become more complex and as the data volume continues to increase we must harness high-performance techniques that allow fast data incorporation and running more and more complex models in shorter and shorter times. Some promising new techniques include fast multipole methods (descibed in Section 4), pattern recognition, statistical mechanics, and adaptive meshing techniques. Tasks easily carried out for simple problems need to be automated for complex problems such as southern California.

A comprehensive computational infrastructure would result in numerous investigators utilizing the solid earth data and models produced. Rather than current practices which rarely go beyond sharing results, these technologies will make it easy (when desired) to share and probe mathematical assumptions and cross-check the simulations and analysis methods while a particular topic is hot (for example during the data collection and analysis phase that always follows major earthquakes). The cost per scientific output would drop dramatically. Instead of spending long hours manipulating data, investigators will be exploring and interpreting the data.

The value will further increase because more sophisticated models and tools will be created and tested speedily within this comprehensive environment. For example, data assimilation tools will allow ingestion and comparison of large volumes of data rather than small subsets; three-

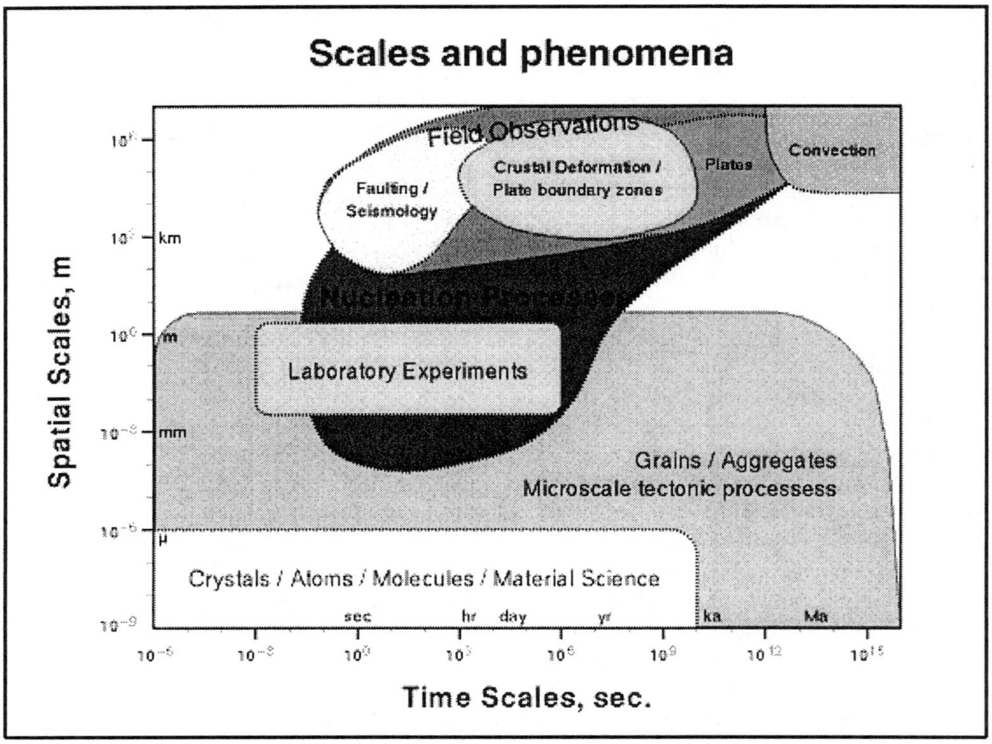

Figure 1. Spatial and temporal scales in earthquake science.

dimensional adaptive meshing technology for constructing finite element meshes makes three-dimensional finite element modeling of complex interacting fault systems practicable.

1.3. Geoscience Overview

Earthquake science embodies a richness present in many physical sciences as there are effects present spread over more than ten orders of magnitude in spatial and temporal scales (Figure 1).

Success requires linking numerical expertise with the physical insight needed to coarse grain or average the science at a fine scale to be used phenomenologically in simulations at a given resolution of relevance to the questions to be addressed. Again the nonlinear fault systems exhibit a wealth of emergent, dynamical phenomena over a large range of spatial and temporal scales, including space-time clustering of events, self-organization and scaling. An earthquake can be modeled as a clustering of slipped fault segments as seen in studies of critical phenomena. As in the latter field, one finds (empirically) scaling laws that include the well-known Gutenberg-Richter magnitude-frequency relation, and the Omori law for aftershocks (and foreshocks). Some of the spatial scales for physical fault geometries include:

The microscopic scale (~ 10^{-6} m to 10^{-1} m) associated with static and dynamic friction (the primary nonlinearities associated with the earthquake process).

The fault-zone scale (~ 10^{-1} m to 10^{2} m) that features complex structures containing multiple fractures and crushed rock.

The fault-system scale (10^{2} m to 10^{4} m), in which faults are seen to be neither straight nor simply connected, but in which bends, offsetting jogs and sub-parallel strands are common and known to have important mechanical consequences during fault slip.

The regional fault-network scale (10^{4} m to 10^{5} m), where seismicity on an individual fault cannot be understood in isolation from the seismicity on the entire regional network of surrounding faults. Here concepts such as "correlation length" and "critical state" borrowed from statistical physics have led to new approaches to understanding regional seismicity.

The tectonic plate-boundary scale (10^{5} m to 10^{7} m), at which Planetary Scale boundaries between plates can be approximated as thin shear zones and the motion is uniform at long time scales.

2. GEM COMPUTATIONAL INFRASTRUCTURE (GEMCI)

2.1. Introduction

The components of GEMCI can be divided into eight areas.

1) Overall Framework including agreement to use appropriate "commodity industry standards" such as XML (a language for metadata) and CORBA (a distributed object access standard and broker), as well as more specialized high performance computing standards like MPI (Message Passing Interface).

2) Use of GEMCI to construct multiple Problem Solving Environments(PSE's) to address different scenarios.

3) Web-based User Interface to each PSE

4) Simulation engines built in terms of the GEMCI framework

5) Geophysical-specific libraries such as modules to estimate local physics and friction. These would also use the GEMCI framework which would already include generic libraries

6) Data analysis and Visualization

7) Data Storage, indexing and access for experimental and computational information

8) Complex Systems and Pattern Dynamics Interactive Rapid Prototyping Environment (RPE) for developing new phenomenological models -- RPE includes analysis and visualization aspects and would be largely on the client (the local lightweight workstation). In contrast, the large simulations in 4) above, are naturally thought of as distributed server side computational objects.

In the remainder of this Section, we describe the overall GEMCI framework and how it can be constructed in terms of components built according to emerging distributed object and Web standards and technologies. This describes the "coarse grain" (program level) structure of the GEMCI environment. There are myriad important details inside each module (or grain), which could be a finite element simulation code, data streaming from a sensor, a visualization subsystem, a Java eigensolver used in the client RPE or field data archived in a web-linked database. In Section 3, we describe some of the existing simulation modules available to the GEM collaboration while Section 4 goes into one case in detail. This is the use of fast multipole methods in large-scale Green's function computations. Section 5 illustrates how these ideas can be integrated together into a variety of different scenarios. These essentially correspond to different problem solving environments that can be built by using the same GEMCI framework to link GEM components in various ways. In Section 6, we make brief remarks on other parts of the GEMCI framework and speculate a little on the future.

There are several important activities which have pioneered the use of object based techniques in computational science. *Legion* has developed a sophisticated object model optimized for computing (http://www.cs.viginia.eu/~legion/) and such a framework could be integrated into GEMCI. Currently we are focussing on broad capabilities available in all important distributed object approaches and we can refine this later as we develop more infrastructure. *Nile* developed the use of CORBA for experimental data analysis (http://www.nile.utexas.edu/) but we need a broader functionality. *POOMA* is an interesting technology developed at Los Alamos (http://www.acl.lanl.gov/PoomaFramework/) aimed at object oriented methods for finer grain objects used to build libraries as discussed in Section 2.5. GEMCI could use modules produced by POOMA as part of its repository of coarse grain distributed components.

2.2. Distributed Objects and the Web

The nature of the web demands that any computing environment based in it must be based on distributed object models. In the following sections we discuss several aspects of the GEMCI in light of the demands of the web environment.

2.2.1. Multi-Tier Architectures. Modern information systems are built with a multi-tier architecture that generalizes the traditional client-server to become a client-broker-service model. This is seen in its simplest realization with the classic web access which involves 3 tiers – the Browser runs on a client; the middle-tier is a Web server; the final tier or backend is the file system containing the Web page (Figure 2). One could combine the Web server and file system into a single entity and return to the client-server model. However the 3-tier view is better as it also captures the increasingly common cases where the Web Server does not access a file system but rather the Web Page is generated dynamically from a backend database or from a computer program invoked by a CGI script. More generally the middle tier can act as an intermediary or broker that allows many clients to share and choose between many different backend resources. This is illustrated in Figure 3, which shows this architecture with the two interfaces separating the layers. As we will discuss later the specific needs and resources of the Earthquake community will be expressed by metadata at these interfaces using the new XML technology.

This 3-tier architecture (often generalized to a multi-tier system with several server layers) captures several information systems which generalize both the web-page access model of Figure 2 and the remote computer program invocation model of Figure 4.

The architecture builds on modern distributed object technology and this approach underlies the "Object Web"

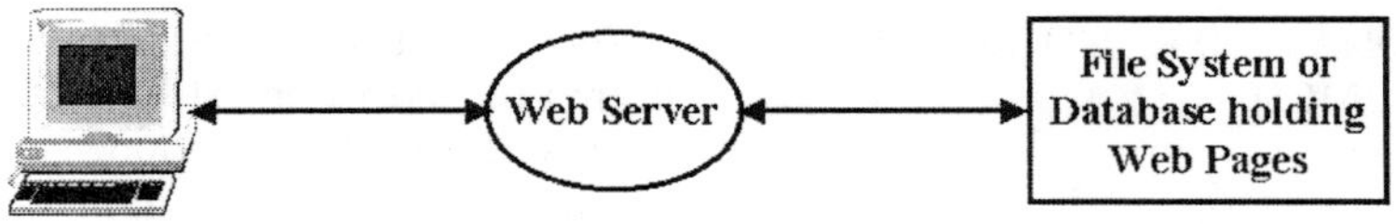

Figure 2. Three-tier computing model illustrated by basic web access.

approach to building distributed systems. Before describing this, let us define and discuss several computing concepts.

2.2.2. Clients, Servers and Objects. A *server* is a free standing computer program that is typically multi-user and in its most simplistic definition, accepts one or more inputs and produces one or more outputs. This capability could be implemented completely by software on the server machine or require access to one or more (super)computers, databases or other information resources such as (seismic) instruments.

A *client* is typically single-user and provides the interface for user input and output. In a distributed system, multiple servers and clients, which are in general geographically distributed, are linked together. The clients and servers communicate with *messages* for which there are several different standard formats.

An (electronic) *object* is essentially any artifact in our computer system. Two examples among many are displayed in Figure 4. One kind is simply a computer program. The best-known distributed object is a web page.

In the example of Figure 2, a *Web Server* accepts an HTTP request and returns a web page. HTTP is a simple but universal protocol (i.e. format for control information) for messages specifying how Web Clients can access a particular distributed object – the web page. A Database Server accepts an SQL request and returns records selected from the database. SQL defines a particular "Request for Service" in the interface of Figure 3 but usually the messages containing these requests use a proprietary format. New standards such as JDBC (The Java Database Connectivity) imply a universal message format for database access with vendor dependent bridges converting between universal and proprietary formats. An Object Broker as in Figure 4 uses the industry CORBA standard IIOP message protocol to control the invocation of methods of a distributed object (e.g. run a program). IIOP and HTTP are two standard but different protocols for communicating between distributed objects.

2.2.3. Services in a Distributed System. Distributed objects are the units of information in our architecture and we need to provide certain critical operating system *services* to support them. Services include general capabilities such as "persistence" (store information unit as a disk file, in a database or otherwise), and "security" as well as more specialized capabilities for science such as visualization. One critical set of services is associated with the unique labeling of objects and their look-up. We are familiar with this with Domain Name Servers and Yellow Page services for computers on the Internet. Web pages with URL's build on this technology and Web search engines like Alta Vista provide

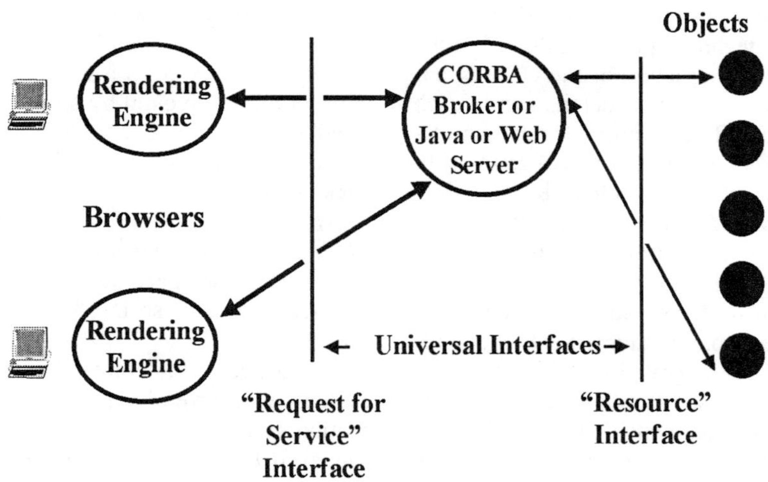

Figure 3. General three-tier architecture.

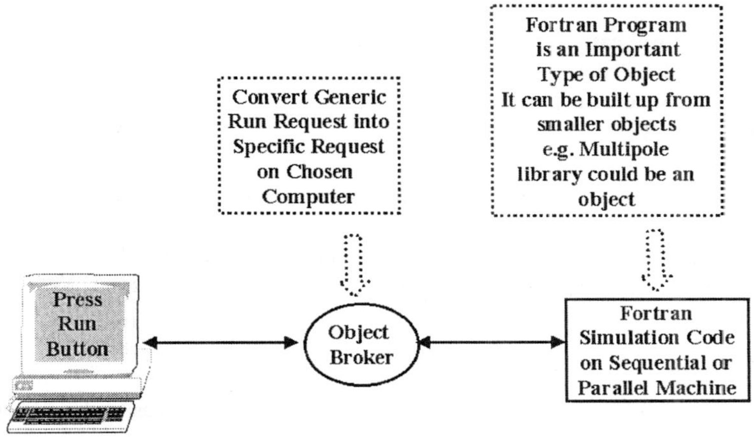

Figure 4. Simple simulation formulated as a "distributed computational object."

a sophisticated look-up service. More general objects can use natural variants of this approach with a possibly arcane URL linking to a database, supercomputer or similar resource. The "resource" interface in Figure 3 defines the properties of back-end resources and how to access them. In particular it defines the equivalent of a URL for each object. The set of these resource specifications forms a database, which defines a distributed object repository.

2.2.4. XML Extended Markup Language. The new XML technology is used to specify all resources in the GEMCI. A good overview of the use and importance of XML in Science can be found in *Bosak and Bray* (1999, http://www.sciam.com/1999/0599issue/0599bosak.html) and we illustrate it in Figure 5, which specifies a computer program used in a prototype GEM problem solving environment described later and shown in Figure 15.

Readers who are familiar with HTML will recognize that XML has a similar syntax with elements defined by nested tags such as <application > .... </application>. This information is further refined by attribute specification as in the string id=disloc, which naturally indicates that this is the disloc application code. XML is a very intuitive way of specifying the structure of digital objects as simple ASCII byte streams. One could equally well specify the same information by appropriate tables in an object-relational database like Oracle, and indeed XML files can easily be stored in such a database if you require their powerful data access and management capabilities. Correspondingly relational database contents can easily be exported to XML format. XML is obviously more easily written and read than a complex database. Further there are growing set of powerful tools which can process XML – these include parsers, editors and the Version 5 browsers from Microsoft and Netscape which can render XML under the control of a powerful language XSL which specifies the mapping of tags to the display device. In the multi-tier generalization of Figure 3, one has multiple linked servers in the middle tier. These servers exchange information, which is expressed as a stream of objects. These objects need to be "serialized" so as to be transportable between servers. There are many ways of converting complex data-structures into a stream of bytes but XML is one of the most general and often the best choice.

2.2.5. Dynamic Resources. Traditionally one labels a computer with an IP address that reflects a particular physical domain, or addresses a web page with a URL, which reflects a particular server and file system. This approach is appropriate for a fixed resource but not well designed for mobile or dynamic resources such as palm top devices and the growing number of Internet enabled consumer products. These resources are often transient and one cannot assume their continuous availability. There is substantial academic and commercial interest in new object registration, look-up and connection approaches supporting fault tolerance and dynamic clients and servers. Objects must be self defining and able to announce themselves universally to a network of registration servers. Jini from Sun Microsystems (http://www.sun.com/jini/) and the Ninja system from UC Berkeley (http://ninja.cs.berkeley.edu/) are well known examples of new approaches to dynamic objects. This technique can be used for both programs and data sets as objects. Thereby either new compute capabilities or new information (raw from sensors or processed) can be dynamically incorporated into simulation systems.

2.2.6. The Object Web. The Object Web signifies the merger of distributed object and web technologies, which is implicitly embodied in the discussion above. There are four rather distinct but important object standards.

CORBA is the Industry Standard supporting objects in any language on any platform. New features in CORBA

```
<?xml version="1.0"?>
<!DOCTYPE application SYSTEM "ApplDescV2.dtd">
<application id="disloc">
<target id="osprey4.npac.syr.edu">
 <status installed="Yes"/>
 <installed>
  <CmdLine command="/npac/home/webflow/GEM/JAY/dis2loc" />
  <input>
   <inFile Path="/npac/home/webflow/GEM/JAY/" Name="disloc.output"/>
   <source Host="osprey4.npac.syr.edu" Path="/npac/home/Jigsaw/WWW/tmp"
      Name="disloc.out"/ >
  </input>
  <output>
   <outFile Path="/npac/home/webflow/GEM/JAY/" Name="simplex.input" />
   <dest Host="osprey4.npac.syr.edu"
      Path="/npac/home/webflow/GEM/JAY/simplex/" Name="s.in" />
  </output>
  <stdout Host="aga.npac.syr.edu" Path="/npac/home/haupt/webflow/history/"
      Name="job2001.out" >
  <stderr Host="aga.npac.syr.edu" Path="/tmp/" Name="haupt_job2001.err" >
 </installed>
</target>
</application>
```

Figure 5. XML Used to describe the computational problem defined in Figure 14.

tend to be deployed relatively slowly as they have a cumbersome approval process and must satisfy complex constraints.

COM is the Microsoft standard, which is confined to PC's but broadly deployed there and high performance.

Java is the software infrastructure of the web and although single language, the same object can be used on any computer supporting the Java VM.

XML comes from the Web Consortium and has been briefly described above. It can be used to define general objects in an intuitive format illustrated in Figure 5.

The *Pragmatic Object Web* implies that there is no outright winner in the distributed object field and one should mix and match approaches as needed. For instance, CORBA Objects can use interfaces (as in Figure 3) defined in XML, Clients and Servers programmed in Java, with rendered displays using COM.

2.3. Architecture of the GEMCI Problem Solving Environment

A Problem Solving Environment or PSE is an application that integrates access to the data, computers and tools needed for a particular computational science area. There is general agreement that Object Web technology is the natural software infrastructure for building PSE's, for the entities one needs to integrate can be considered as distributed objects. With this choice, there is a growing trend to term web-based PSE's as portals in analogy to the term used to describe environments built commercially to allow access to personal or business information. Commercial portals allow both administrative and user customizability from a suite of objects and services supporting them.

GEMCI illustrated in Figure 6, is an Object Web PSE and has the classic three-tier structure discussed above.

In GEMCI, everything is a "distributed object" whether it be a simulation on a supercomputer, the basic GEM Web pages, the notes from a field trip entered on a palm top, CNN real-time coverage of the latest earthquake or the data streaming in from sensors. GEMCI provides an integrated view of these diverse resources with XML definitions for the raw objects themselves and the data they produce. The services shown in Figure 6, from collaboration, security, object discovery, visualization and computer access, are generic to all computing portals. Building GEMCI using the same approach and tools as other portals ensures the availability of these services. They will require customization as for instance there are many different visualization packages and each requires non-trivial work to include in such a portal. Again, collaboration corresponds to sharing distributed objects, and this can currently only be automated for some objects. Many web pages can be shared using generic

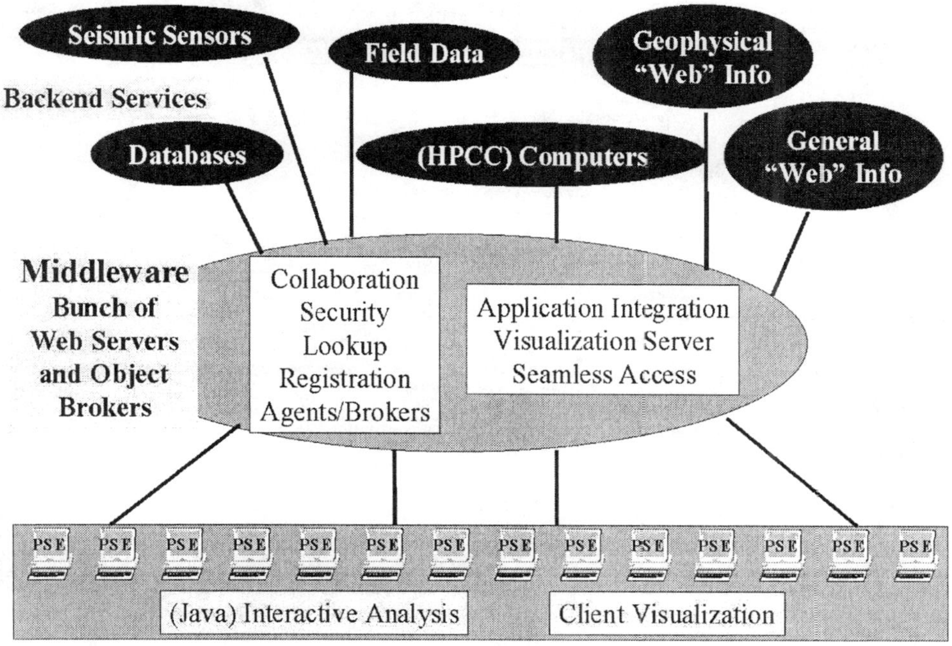

Figure 6. Three-tier architecture of a GEM Problem Solving Environment.

techniques illustrated in Figure 5 but sharing say the control and output of a general simulation can require quite a lot of custom modifications.

Most importantly, to use GEMCI shown in Figure 6, one needs to define the entities in the GEM environment as distributed objects. For computer programs this implies a rather arcane process termed "wrapping the program as a distributed object". Operationally this implies allowing a middle-tier server (the CORBA object broker or Java application Server) to be able to run the program on one or more machines, specify the input files and either specify output files or access them as streams of data in the fashion of UNIX pipes. Each distributed object technology has a rather different approach to this using what is termed an IDL or Interface Definition Language and specialized Java and C++ code to implement the wrapping.

We can use the concept of the Pragmatic Object Web to simplify this process. Our strategy is to define all relevant properties of computer programs in XML as illustrated in Figure 5. These properties are used to generate the needed object wrappers, either statically or dynamically. This approach requires the user specify what they know – the properties of their program – while the filter copes with the obscure syntax of each object model. Obviously this also allows one to support all object models – COM CORBA, and Java – by changing the filter. In this way one can adapt to changes in the commercial infrastructure used in the middle tier.

One must apply the XML object definition strategy to all entities in GEMCI; programs, instruments and other data sources and repositories. This gives the metadata defining macroscopically the object structure. In addition, one needs to look at the data stored in, produced by or exchanged between these objects. This data is itself a typically a stream of objects, each an array, a table or more complex data structure. One could choose to treat the data at some level as an unspecified (binary) "blob" with XML defining the overall structure but detailed input and output filters used for the data blobs. As an example, consider the approach that an electronic news organization could take for their data. The text of news flashes would be defined in XML but the high volume multimedia data (JPEG images and MPEG movies) would be stored in binary fashion with XML used to specify <IMAGEOBJECT> or <MOVIEOBJECT> metadata.

As explained earlier, systematic use of XML allows use of a growing number of tools to search for, manipulate, persistently store and render the information. It facilitates the linkage of general and specific tools/data sources/programs with clearly defined interfaces. This will help the distributed GEM collaborators to separately develop programs or generate data, which will be easily able to interoperate.

More generally XML standards will be defined hierarchically starting with distributed information systems, then general scientific computing and finally application specific object specifications.

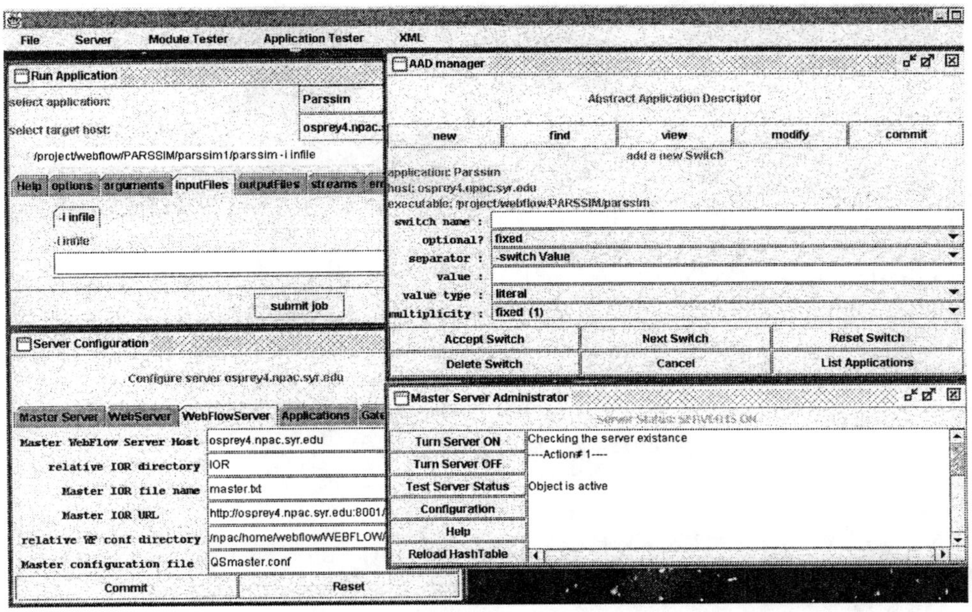

Figure 7. This illustrates several components of the WebFlow system from NPAC. (Counter-clockwise from bottom-left): The Server Configuration and Master Server Administrator are WebFlow administrative tools that allow one to configure and monitor the middle tier servers. AAD manager (Abstract Application Descriptor) is a tool to incorporate new applications to the system: the interface shown simplifies creation of an XML definition of the application. This definition can be used for dynamical creation of front-end application interfaces (window "Run Application").

For example GEMCI would develop its own syntax for seismic data sensors but could build on general frameworks like the XSIL scientific data framework developed by Roy Williams at Caltech (http://www.cacr.caltech.edu/SDA/xsil/index.html). XSIL supports natural scientific data structures like arrays and the necessary multi-level storage specification.

Another example is MathML which provides XML support for the display and formulation of Mathematics. We can expect MathML to be supported by tools like Web Browsers and white boards in collaborative scientific notebooks and allow one to enhance theoretical collaboration in GEM. There will for instance be modules that can be inserted into applications for parsing MathML or providing graphical user specification of mathematical formulae. This could be used in sophisticated implementations of the Complex Systems and Pattern Dynamics Interactive Rapid Prototyping Environment with scripted client side specification of new analysis methods. One can also use MathML in high level tools allowing specification of basic differential equations that are translated into numerical code. This has been demonstrated in prototype problem solving environments like PDELab but have so far not had much practical application (*Houstis et al.*, 1998 and http://www.cs.purdue.edu/research/cse/pdelab/pdelab.html). Greater availability of standards like MathML should eventually allow more powerful interchangeable tools of this type.

Finally we can mention a set of graphical XML standards such as X3D (3 dimensional objects) and VML which is a vector graphics standard, which can be expected to be important as basis of application specific plot and drawing systems.

2.4. Building the GEMCI Problem Solving Environment

We have described above some aspects of the base strategy of defining GEM components as distributed objects. Here we discuss some existing experience from NPAC at Syracuse in integrating such objects and tools that manipulate them into an overall environment. The NPAC team has built several exemplar problem solving environments for both the NSF and DoD HPCMO (High Performance Computing Modernization Office) supercomputer centers. In Figure 7, we show some useful tools including a first cut at a "wizard" that helps produce the distributed object wrappers described above.

This AAD (Abstract Application Descriptor) can be extended to allow specification of all needed input parameters of an application, with an automatic generation of input forms respecting default values and allowed value ranges.

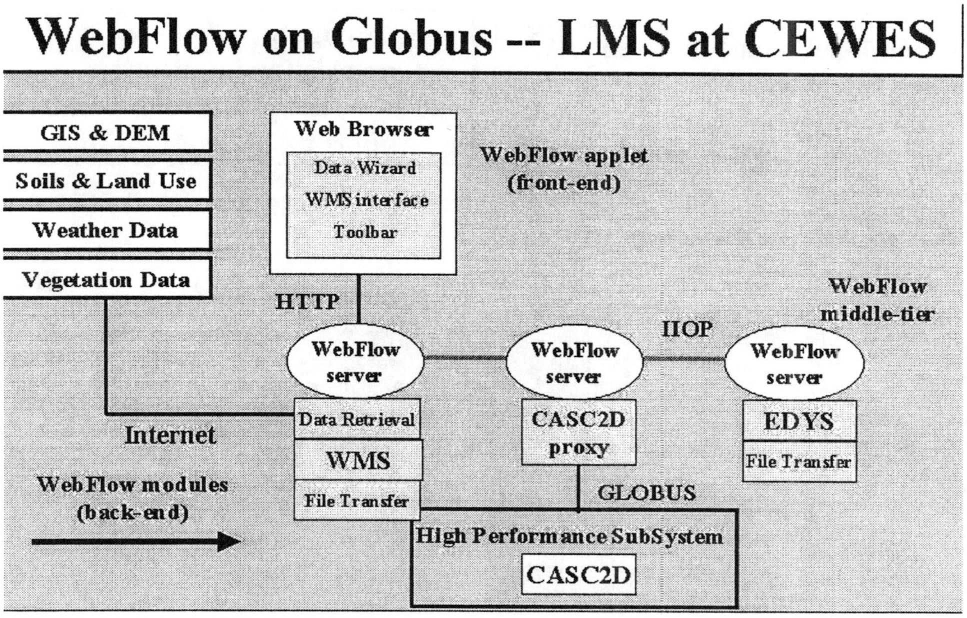

Figure 8. Three-Tier Architecture of DoD "Land Management" application whose front end is shown in Figure 9.

The object wrappers of an application should not only invoke the code and allow parameter specification but have built in description/help systems.

One major NPAC Problem Solving Environment was built for the Landscape Management System (LMS) project at the U.S. Army Corps of Engineers Waterways Experiment Station (ERDC) Major Shared Resource Center (MSRC) at Vicksburg, MS, under the DoD HPC Modernization Program, Programming Environment and Training (PET). The application can be idealized as follows.

A decision maker (the end user of the system) wants to evaluate changes in vegetation in some geographical region over a long time period caused by some short term disturbances such as a fire or human activities. One of the critical parameters of the vegetation model is soil condition at the time of the disturbance. This in turn is dominated by rainfall that possibly occurs at that time. Consequently as shown in Figure 8, the implementation of this project requires:

1) Data retrieval from remote sources including DEM (data elevation models) data, land use maps, soil textures, dominating flora species, and their growing characteristics, to name a few. The data are available from many different sources, for example from public services such as USGS web servers, or from proprietary databases. The data come in different formats, and with different spatial resolutions.

2) Data preprocessing to prune and convert the raw data to a format expected by the simulation software. This preprocessing is performed interactively using WMS (Watershed Modeling System) package.

3) Execution of two simulation programs: EDYS for vegetation simulation including the disturbances and CASC2D for watershed simulations during rainfalls. The latter results in generating maps of the soil condition after the rainfall. The initial conditions for CASC2D are set by EDYS just before the rainfall event, and the output of CASC2D after the event is used to update parameters of EDYS and the data transfer between the two codes had to be performed several times during one simulation. EDYS is not CPU demanding, and it is implemented only for Windows95/98/NT systems. On the other hand, CASC2D is very computationally intensive and typically is run on powerful backend supercomputer systems.

4) Visualization of the results of the simulation. Again, WMS is used for this purpose.

One requirement of this project was to demonstrate the feasibility of implementing a system that would allow launching and controlling the complete simulation from a networked laptop. We successfully implemented it using WebFlow middle-tier servers with WMS and EDYS encapsulated as WebFlow modules running locally on the laptop and CASC2D executed by WebFlow on remote hosts. Further the applications involved showed a typical mix of supercomputer and computationally less demanding personal computer codes. LMS was originally built using specialized Java Servers but these are now being replaced by

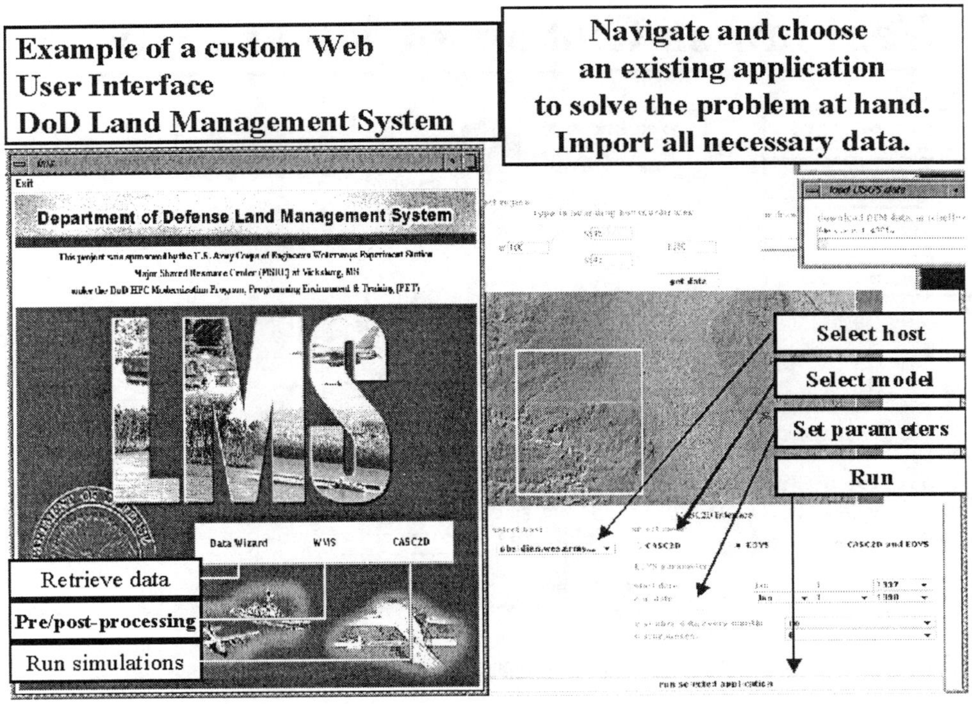

Figure 9. Example of a Web Interface for a "Land Management Problem Solving Environment" built by NPAC for the Department of Defense ERDC Laboratory in Vicksburg, MS.

commercial CORBA object brokers but in either case the architecture of Figure 8 is consistent with the general structure of Figures 6 and 3.

For this project we developed a custom front-end shown in Figure 9, that allows the user to interactively select the region of interest by drawing a rectangle on a map. Then one could select the data type to be retrieved, launch WMS to preprocess the data and make visualizations, and finally launch the simulation with CASC2D running on a host of choice.

Our use of XML standards at the two interfaces in Figure 3, allows us to change front end and middle-tier independently. This allowed us the middle tier upgrade described above which will bring security and "seamless access" capability to our PSE's.

Seamless access is an important service, which is aimed at allowing applications to be run on an arbitrary appropriate backend. Most users wish their job to just run and usually do not mind what machine is used. Such seamless capability is rather natural in the architecture of Figure 3. Essentially the front end defines the "abstract" job passing its XML specification to a middle tier server. This acts as a broker and instantiates the abstract job on one of the available backend computers. The middle-tier backend XML interface is used to specify both the available machines sent to the servers and the "instantiated job" sent from the servers. We build the back end mapping of jobs to machine using the important Globus technology (http://www.globus.org). Globus is a distributed or meta-computing toolkit providing important services such as resource look-up, security, message passing etc.

One can view WebFlow (Figures 7 and 8) as a "kit" of services, which can be used in different ways in different applications. In Figure 10, we show another capability of WebFlow, which supports the ability to compose complex problems by linking different applications together with data flowing between different modules chosen from a palette. This WebFlow service is supported by a Java front end and a middle tier service matching abstract modules to back end computers and supporting the piping of data between the modules.

2.5. Libraries or Distributed Components?

Since the beginning of (computing) time, large programs have been built from components that are assembled together into a complete job. One can do this in many ways and perhaps the best known is the use of libraries linked together. Modern "object-oriented" approaches can be used in this fashion with method calls. Design frameworks can be used to establish a systematic methodology with say specific interfaces ("calling sequences") to be followed by for

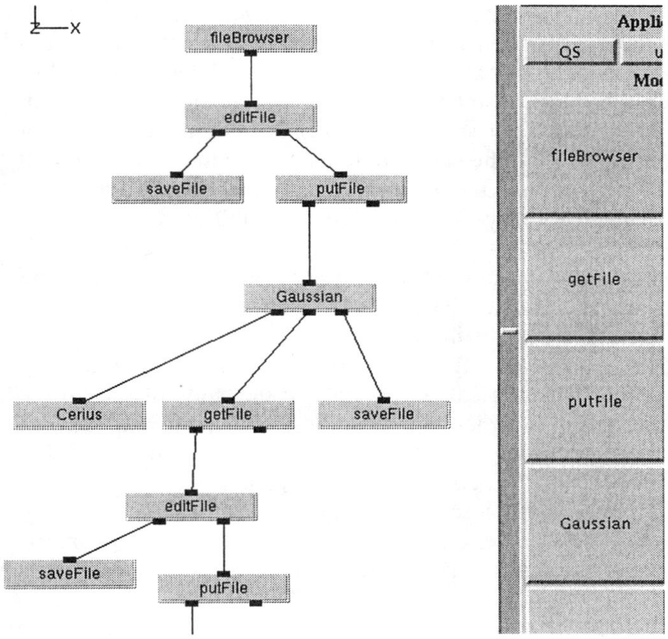

Figure 10. Fragment of WebFlow Composition Tool linking modules in a quantum simulation (chemistry) application.

instance different implementations of friction laws. The discussion in Section 2.4 above describes a rather different component model where possibly distributed modules are linked together in a dynamic fashion. This is more flexible than the library approach and further as explained above, this supports distributed components. The component approach is not surprisingly less efficient than the library approach as component linkage is typically obtained through explicit exchange of messages rather than efficient compiled generated parameter passing as for library methods. The inefficiency is particularly serious for small components as the component mechanisms come with large start up (latency) overheads. Thus we adopt a hybrid approach with small objects using the traditional library mechanism within a set of agreed interfaces and design frameworks to promote easier interchange of modules. On the other hand, large objects (roughly "complete programs") use a component approach within a distributed object framework, which is defined in XML.

3. CURRENT GEM COMPUTATIONAL COMPONENTS

The first step in building the GEMCI has been to inventory the computer codes currently available in the community. A summary of the results of this inventory is maintained at http://milhouse.jpl.nasa.gov/gem/gemcodes.html and the status as of the time of writing is given below.

The categories covered by the existing codes include codes for understanding dislocations: elastic models - both half-space and layered, viscoelastic models, finite element and boundary element codes, and inversion codes. There are a set of codes for understanding how faults interact: cellular automata, finite element and boundary element codes, fault friction models. Some codes are designed to understand the dynamics of the earthquake rupture itself. Finally there are codes for displaying the results of various simulations. We follow with short descriptions of the codes grouped by type.

Elastic

3D-DEF – Performs elastic dislocation boundary-element calculations.

coulomb – Computes 3D elastic dislocation and 2D boundary element stress and strain.

DYNELF – Models 3D elastodynamic finite difference with frictional faults.

faultpatch – Generates earthquake sequences, given fault geometries and loading rates using .

FLTSLP – Inverts groups of focal mechanism solutions or slickenline data for orientation and relative magnitudes of principal strain rates and for relative micropolar vorticity.

GNStress – Model stresses induced by faulting, for studying fault interaction.

layer – Calculates surface displacements and strains for vertical strike-slip point source in horizontal layer above half-space.

RNGCHN – Calculate surface displacements and strains in elastic half space.

scoot – 2D elastodynamic finite difference with frictional fault.

simplex – 3D inversion of geodetic data for displacement on faults.

Viscoelastic

DYNELF – Models 3D elastodynamic finite difference with frictional faults.

FLTGRV and FLTGRH – Compute 3 vector components of surface displacement from slip on a dipping thrust fault contained within an elastic layer overlying a viscoelastic-gravitational half space.

STRGRH and STRGRV – Computes 3 vector components of surface displacement from slip on a dipping strike slip fault contained within an elastic layer overlying a viscoelastic-gravitational half space.

Virtual_California – Realistic cellular automata (CA) viscoelastic earthquake simulator.

VISCO1D – Computes viscoelastic spherical deformation due to faulting or dike emplacement.

Cellular Automata

faultpatch – Generates earthquake sequences, given fault geometry's and loading rates using Cellular automata methods.

Virtual_California – Realistic cellular automata (CA) viscoelastic earthquake simulator.

Dynamic

DYNELF – Models 3D elastodynamic finite difference with frictional faults

scoot – 2D elastodynamic finite difference with frictional fault

Inverse

FLTSLP – Inverts groups of focal mechanism solutions or slickenline data for orientation and relative magnitudes of principal strain rates and for relative micropolar vorticity.

simplex – 3D inversion of geodetic data for displacement on faults.

qoca – optimal combining of various geodetic measurments resulting in a consistent deformation field.

Visualization

VelMap – Aid for visualizing displacement and velocity of crust.

4. LARGE-SCALE SIMULATIONS IN GEMCI

4.1. Overview of Earthquake Fault System Simulations

With reasonable approximation, one may model the long-term evolution of stresses and strains on interacting fault segments with a Green's function approach. As in other fields this method leads to a boundary (the faults) value formulation, which looks numerically, like the long-range force problem. The faults are paneled with segments (with area of some 100 m^2 in definitive computations) which interact as though they were dipoles. The original calculations of this model used the basic $O(N^2)$ algorithm but a new set of codes will be using the "fast multipole" method adapted from gravitational-particle astrophysics. There are interesting differences between the earthquake and gravitational applications. In gravity we get wide ranges in density and dynamical effects from the natural clustering of the gravitating particles. Earthquake "particles" are essentially fixed on complex fault geometries and their interactions fall off faster than those in the astrophysical problem.

Several variants of this model have been explored including approximations, which only keep interactions between nearby fault segments. These "cellular automata" or "slider-block" models look very like statistical physics with an earthquake corresponding to clusters of particles slipping together when the correlation length gets long near a critical point. The full Green's function approach should parallelize straightforwardly in either $O(N^2)$ or multipole formulation. However cellular automata models will be harder to parallelize, as we know from experience with the corresponding statistical physics case where clustering models have been extensively studied.

An interesting aspect of these simulations is that they give the "numerical laboratory" for the study of space-time patterns in seismicity information. This type of analysis was used successfully in the climate field to aid the prediction of "El Nino" phenomena. These pattern analyses may or may not need large computational resources although they can involve determination of eigensolutions of large matrices which is potentially time consuming.

4.2. Green's Function Formulation and Approximations

4.2.1. Basic Equations. Let us consider in more detail the problem addressed by the *Virtual_California* code described in Section 3 and introduced in *Rundle* 1988a. If one is given a network of faults embedded in an Earth with a given rheology, subject to loading by distant stresses, and neglecting elastic waves (see discussion below), the evolution of the state of slip $\mathbf{s}(\mathbf{x}, t)$ on a fault at $(\mathbf{x}, t)$ is determined from the equilibrium of stresses according to Newton's Laws:

$$\partial \mathbf{s}(\mathbf{x}, t) / \partial t = \Phi\{ \Sigma_i (\sigma_i) \} \qquad (1)$$

where $\Phi\{\}$ is a nonlinear functional, and $\Sigma_i(\sigma_i)$ represents the sum of all stresses acting within the system. These stresses include

1) The interaction stress $\sigma_{int}[\mathbf{x},t; \mathbf{s}(\mathbf{x}',t'); \mathbf{p}]$ provided by transmission of stress through the Earth's crust arising from background tractions $\mathbf{p}$, as well as stresses due to slip on other faults at other sites $\mathbf{x}'$ at times t';

2) The cohesive fault frictional stress $\sigma_f[\mathbf{x},t; \mathbf{s}(\mathbf{x},t)]$ at the site $(\mathbf{x},t)$ associated with the state of slip $\mathbf{s}(\mathbf{x},t)$; and

3) Other stresses such as those due to dynamic stress transmission and inertia.

The transmission of stress through the Earth's crust involves both dynamic effects arising from the transient propagation of seismic waves, and from static effects that persist after wave motion has ceased. Rheologic models

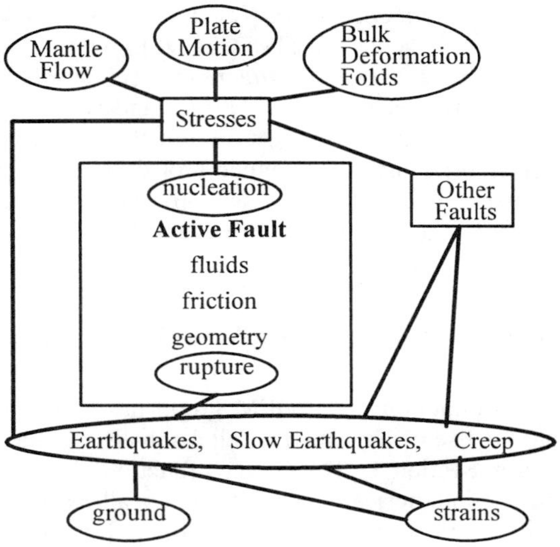

Figure 11. Logical structure for model of Section 4.

typically used for the Earth's crust between faults are all linear (e.g., *Rundle and Turcotte,* 1993) and include

1) A purely elastic material on both long and short time scales;

2) A material whose instantaneous response is elastic but whose long term deformation involves bulk flow (viscoelastic); and

3) A material that is again elastic over short times, but whose long term response involves stress changes due to the flow of pore fluids through the rock matrix (poroelastic).

In Figure 11, we show the basic conceptual "wiring diagram" for the model, which indicates the interplay between loading stresses, rupture, interactions with other faults, and relaxation processes following a major earthquake.

4.2.2. The Green's Function Approximation. Considering GEM models that assume a linear interaction rheology between the faults implies that the interaction stress can be expressed as a spatial and temporal convolution of a stress Green's function $\mathbf{T}_{ij}^{kl}(\mathbf{x} - \mathbf{x}', t - t')$ with the slip deficit variable $\phi(\mathbf{x}, t) = \mathbf{s}(\mathbf{x}, t) - \mathbf{V}t$, where $\mathbf{V}$ is the long term rate of offset on the fault. Once the slip deficit is known, the displacement Green's function $\mathbf{G}_{ij}^{k}(\mathbf{x} - \mathbf{x}', t - t')$ can be used to compute, again by convolution, the deformation anywhere in the surrounding medium exterior to the fault surfaces (e.g. *Rundle* 1988a). In the first implementation of GEM models, we expect to specialize in the case of quasistatic interactions, even during the slip events. Observational evidence supports the hypothesis that simulations carried out without including inertia and waves will have substantial physical meaning. *Kanamori and Anderson* (1975) estimated that the seismic efficiency η, which measures the fraction of energy in the earthquake lost to seismic radiation, is less than 5%-10%. This implies that inertial effects in the dynamical evolution of slip in studying large populations of earthquakes will be of lesser importance for initial calculations. Elastic waves will be included in later simulations when errors arising from other effects are reduced to the 5%-10% level. At present, inclusion of these effects is severely limited by available computational capability, so we anticipate that it may be practical to include only the longest wavelengths or largest spatial scales.

4.2.3. Inelastic Rheologies. In quasistatic interactions, the time dependence of the Green's function typically enters only implicitly through time dependence of the elastic moduli (e.g., *Lee*, 1955). Because of linearity, the fundamental problem is reduced to that of calculating the stress and deformation Green's function for the rheology of interest. For materials that are homogenous within horizontal layers, numerical methods to compute these Green's functions are well known (e.g., *Okada*, 1985, 1992; *Rundle*, 1982a,b, 1988; *Rice and Cleary*, 1976; *Cleary*, 1977; *Burridge and Varga*, 1979; *Maruyama*, 1994). Problems in heterogeneous media, especially media with a distribution of cracks too small and too numerous to model individually, are often solved by using effective medium approaches, self-consistency assumptions (*Hill*, 1965; *Berryman and Milton*, 1985; *Ivins*, 1995a,b), or damage models *Lyakovsky et al.* (1997). Suffice to say that a considerable amount of effort has gone into constructing quasistatic Green's functions for these types of media, and while the computational problems present certain challenges, the methods are straightforward as long as the problems are linear. In the initial work, we expect to focus on elastic (with possible incorporation of damage parameters) and layered viscoelastic models only.

4.3. Friction Models:

Friction models to determine the slip condition must augment the overall differential equations. At the present time, six basic classes of friction laws have been incorporated into computational models. They fall into three families as follows.

1) Two basic classes of friction models arise from laboratory experiments:

Slip Weakening. This friction law (*Rabinowicz*, 1965; *Bowdon and Tabor*, 1950; *Beeler et al.*, 1996; *Li*, 1987; *Rice*, 1993; *Stuart and Tullis*, 1995) assumes that the frictional stress at a site on the fault $\sigma_f = \sigma_f[\mathbf{s}(\mathbf{x}, t)]$ is a functional of the state of slip. In general, $\sigma_f[\mathbf{s}(\mathbf{x}, t)]$ is peaked at regular intervals. The current state of the system is found from enforcing the equality $\sigma_f[\mathbf{s}(\mathbf{x}, t)] = \sigma_{int}[\mathbf{x}, t; \mathbf{s}(\mathbf{x}', t'); \mathbf{p}]$ prior to, and just after, a sliding event.

Rate and State. These friction laws are based on laboratory sliding experiments in which two frictional surfaces are slid over each other at varying velocities, usually without experiencing arrest (*Dieterich*, 1972; 1978; 1981; *Ruina*, 1983; *Rice and Ruina,* 1983; *Ben Zion and Rice*, 1993; 1995; 1997; *Rice*, 1993; *Rice and Ben Zion*, 1996). In these experiments, the laboratory apparatus is arranged so as to be much "stiffer" than the experimental "fault" surfaces. The rate dependence of these friction laws refers to a dependence on logarithm of sliding velocity, and the state dependence to one or more state variables $\theta_i(t)$, each of which follows an independent relaxation equation.

2) Two classes of models have been developed and used that are based on laboratory observations, but are computationally simpler.

Coulomb-Automata. These are widely used because they are so simple (e.g., *Rundle and Jackson,* 1977; *Nakanishi*, 1991; *Brown et al*., 1991; *Rundle and Brown*, 1991; *Rundle and Klein*, 1992; *Ben Zion and Rice*, 1993, 1995, 1997). A static failure threshold, or equivalently a coefficient of static friction μ^S is prescribed, along with a residual strength, or equivalently a dynamic coefficient of friction μ^D. When the stress at a site increases, either gradually or suddenly, to equal or exceed the static value, a sudden jump in slip (change of state) occurs, that takes the local stress down to the residual value. These models naturally lend themselves to a Cellular Automaton (CA) method of implementation.

Velocity Weakening. This model (*Burridge and Knopoff*, 1967; *Carlson and Langer*, 1989) is based on the observation that frictional strength diminishes as sliding proceeds. A constant static strength $\sigma_f = \sigma^F$ is used as above, after which the assumption is made that during sliding, frictional resistance must be inversely proportional to sliding velocity.

3) Two classes of models are based on the use of statistical mechanics involving the physical variables that characterize stress accumulation and failure. Their basic goal is to construct a series of nonlinear stochastic equations whose solutions can be approached by numerical means:

Traveling Density Wave. These models (*Rundle et al*., 1996) are based on the slip weakening model. The principle of evolution towards maximum stability is used to obtain a kinetic equation in which the rate of change of slip depends on the functional derivative of a Lyapunov functional potential. This model can be expected only to apply in the mean field regime of long range interactions, which is the regime of interest for elasticity in the Earth. Other models in this class include those of *Fisher et al.* (1997) and *Dahmen et al.* (1997).

Hierarchical Statistical Models. Examples include the models by *Allègre et al.* (1982, 1996); *Smalley et al.* (1985); *Blanter et al.* (1996); *Allègre and Le Mouel* (1994); *Heimpel* (1996); *Newman et al.* (1996); and *Gross* (1996). These are probabilistic models in which hierarchies of blocks or asperity sites are assigned probabilities of failure. As the level of external stress rises, probabilities of failure increase, and as a site fails, it influences the probability of failure of nearby sites.

4.4. Multipole Methods and Fast Numerical Simulation

4.4.1. Introduction. The GEM group has started to investigate the use of advanced parallel solvers for the codes described in Section 3. The Green's function approach can be formulated numerically as a long-range all-pairs interaction problem and this can be straightforwardly parallelized using well-known algorithms. However one cannot reach the required level of resolution without switching from an order $O(N^2)$ to one of the $O(N)$ or $O(N \log N)$ approaches. As in other fields, this can be achieved by dropping or approximating the long-range components and implementing a neighbor-list based algorithm. However it is more attractive to formulate the problem as interacting dipoles and adapt existing fast-multipole technology developed for particle dynamics problems. We have already produced a prototype general purpose "fast multipole template code" by adapting the very successful work of *Salmon and Warren* (1997). These codes have already simulated over 300 million gravitating bodies on a large distributed memory system (a 4500-processor subset of the ASCI "Red" machine), so we expect these parallel algorithms to scale efficiently up to the problem sizes needed by GEM.

4.4.2. Multipolar Representation of Fault Systems. A primary key to a successful implementation of GEM models of faults systems will be to utilize computationally efficient algorithms for updating the interactions between fault segments. Converting the Green's function integrals to sums, without truncation or approximation, would require $O(N^2)$ operations between earthquakes, and possibly more for segments of faults experiencing an earthquake. For quasistatic interactions, the Green's functions $\mathbf{T}_{ij}{}^{kl}$ and $\mathbf{G}_{ij}{}^{k}$ for linear elasticity have a simple time dependence. Moreover, the Green's functions for linear viscoelasticity and for linear poroelasticity can be obtained from the elastic Green's functions using the correspondence principle (e.g., *Lee*, 1955; *Rundle* 1982a,b). These simplifications strongly suggest that multipole expansions (*Goil*, 1994; *Goil and Ranka*, 1995) will be computationally efficient algorithms.

The stress and displacement Green's functions $\mathbf{T}_{ij}{}^{kl}$ and $\mathbf{G}_{ij}{}^{k}$ represent the tensor stress and vector displacement at $\mathbf{x}$

due to a point double couple located at **x**' (*Steketee*, 1958). The orientation at **x**' of the equivalent fault surface normal vector, and of the vector displacement on that fault surface, are described by the indices *i* and *j*. Displacement and stress indices at the field point **x** are described by indices *k* and *l*. Integration of $\mathbf{T}_{ij}{}^{kl}$ and $\mathbf{G}_{ij}{}^{k}$ over the fault surface then corresponds to a distribution of double couples. For that reason, representation of the stress over segments of fault in terms of a multipole expansion is the natural basis to use for the GEM computational problem. In fact, the use of multipolar expansions to represent source fields in earthquake and explosion seismology was introduced by *Archambeau* (1968) *and Archambeau* and *Minster* (1978), and later revisited from a different perspective by *Backus* and *Mulcahy* (1976). *Minster* (1985) gives a review of these early representations.

4.4.3. Application of Fast Multipole Methods to GEM. In the gravitational N-body problem, each body interacts with every other one in the system according to the familiar law of gravitational attraction. Simply computing all pairs of interactions requires $N(N\text{-}1)/2$ separate evaluations of the interaction law. This formulation of the problem has some important advantages: it is easy to code, it is easy to vectorize and parallelize, it is readily expressible in High Performance Fortran, and it is even amenable to special-purpose hardware [e.g. GRAPE]. Nevertheless, even today's fastest special-purpose systems, running in a dedicated mode for extended times at rates of nearly 1 TERAFLOP, cannot simulate systems larger than about 100,000 bodies.

Tremendous computational savings may be realized by combining bodies into "cells" and approximating their external field with a truncated multipole expansion. When this idea is applied systematically, the number of interactions may be reduced to O(N log$N$) (*Appel*, 1985; *Barnes and Hut*, 1986) or O(N) (*Greengard and Rokhlin*, 1987; *Anderson*, 1992). The cells are generally arranged in a tree, with the root of the tree representing the entire system, and descendants representing successively smaller regions of space. *Salmon and Warren* (1997) have demonstrated that such codes can run in parallel on thousands of processors and have simulated highly irregular cosmological systems of over 300 million bodies using ASCI facilities.

The basic idea of fast multipole methods is shown in the three versions of Figure 12. In 12(a), we show the simple O(N^2) approach with a single particle on the left interacting with all the individual particles in a box on the right. The naïve computational simplification shown in Figure 12(b) is to replace (in the gravity case) the particles by their center of mass and use a single force calculation. However this approximation may not be accurate enough and so in Figure 12(c) we show the original box broken into 8 sub-boxes with the total summed over the individual sub-box contri-

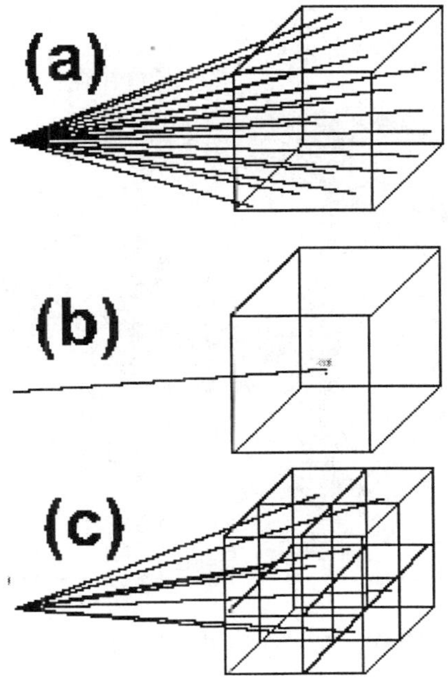

Figure 12. Three ways of calculating long range forces.

butions. This idea can be automated and implemented in a clever way so that the boxes used get smaller as you move closer to a given particle or if there is significant variation in particle density. Further there are subtle but very efficient parallel versions of the method to build the set of hierarchical sub-boxes (which form a tree so that these are often termed tree codes) which in the case of gravity is rebuilt each time step. In Figure 13 below, we illustrate the adaptive division generated for 10,000 particles clustered in a two-dimensional disk.

There is a direct analogy between the bodies in an astrophysical N-body system and the fault segments in a GEM. In both cases, there exists a pair-wise interaction that seems to require O(N^2) interactions. But if we represent the distribution of sources in a region by a multipole expansion, the external field generated by a large number of bodies can be computed to any desired degree of accuracy in constant time. Thus, the GEM problem can also be reduced to O(Nlog$N$) or O(N) total interactions, so that large calculations are tractable. On the other hand, although multipole methods can deliver large performance gains, they also require a considerable infrastructure. This is especially true of efficient parallel implementations. We intend to develop the multipole version of GEM using a library that has been abstracted from Salmon and Warren's successful astrophysical N-body codes. This new library is:

Modular - The "physics" is cleanly separated from the "computer science", so that in principle, alternative physics

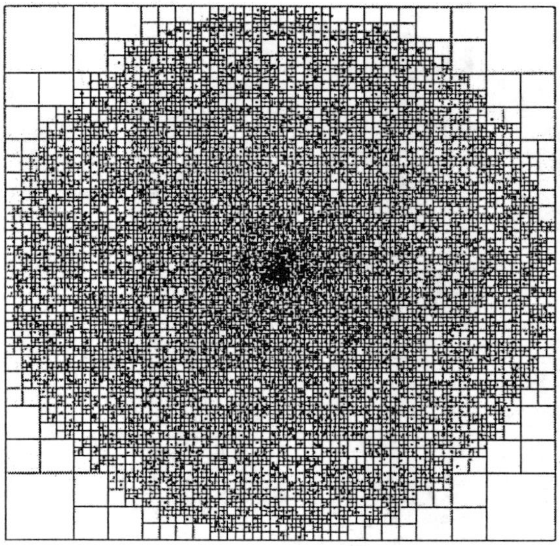

Figure 13. Adaptive division of disk of particles into cells.

modules such as the evaluation of the GEM Green's functions, can simply be "plugged in". The first non-gravitational demonstration was a vortex dynamics code written by *Winckelmans et al.* (1995). The interface to the physics modules is extremely flexible. A general decision-making function tells the treecode whether or not a multipole, or any other approximation, is adequate for a given field evaluation. Short-range interactions, which vanish outside a given radius, can be handled as well.

Tunable - Careful attention to analytical error bounds has led to significant speed-ups of the astrophysical codes, while retaining the same level of accuracy. Analytic error bounds may be characterized as quantifying the fact that the multipole formalism is more accurate when the interaction is weak: when the analytic form of the fundamental interaction is well-approximated by its lower derivatives; when the sources are distributed over a small region; when the field is evaluated near the center of a "local expansion"; when more terms in the multipole expansion are used, and when the truncated multipole moments are small. These issues are primarily the concern of the "physics" modules, but the library provides a sufficiently powerful interface to make these parameters adjustable. The formulation is general enough that the same library can be used to support evaluation of $O(N)$, $O(N \log N)$ and $O(N^2)$ approximation strategies, simply by changing the decision criteria and interaction functions.

Adaptive - The tree automatically adapts to local variations in the density of sources. This can be important for GEM as it is expected that large earthquakes are the result of phenomena occurring over a wide range of length and time scales.

Scalable - The library has been successfully used on thousands of processors, and has sustained 170 Gflops aggregate performance on a distributed system of 4096 200Mhz PentiumPro processors.

Out of core - The library can construct trees, and facilitates use of data sets that do not fit in primary storage. This can allow one to invest hardware resources into processing rather than memory, resulting in more computations at constant resources.

Dynamically load balanced - The tree data structure can be dynamically load-balanced extremely rapidly by sorting bodies and cells according to an easily computed key.

Portable - The library uses a minimal set of MPI primitives and is written entirely in ANSI C. It has been ported to a wide variety of distributed memory systems - both 32-bit and 64-bit. Shared memory systems are, of course, also supported simply by use of an MPI library tuned to the shared memory environment.

Versatile - Early versions of the library have already been applied outside the astrophysics and molecular dynamics area. In particular the Caltech and Los Alamos groups have successfully used it for the vortex method in Computational Fluid Dynamics.

In the full GEM implementation, we have a situation similar to the conventional $O(N^2)$ N-body problem but there are many important differences. As the most obvious difference, the GEM case corresponds to double-couple interactions, which correspond to a different force law between "particles" from the gravitational case. Further the critical dynamics -- namely earthquakes -- are found by examining the stresses at each time step to see if the friction law implies that a slip event will occur. As discussed above, many different versions of the friction law have been proposed, and the computational system needs be flexible so we can compare results from different laws. Noting that earthquakes correspond to large-scale space-time correlations including up to perhaps a million 10-to-100 meter segments slipping together shows analogies with statistical physics. As in critical phenomena, clustering occurs at all length scales and we need to examine this effect computationally. However, we find differences with the classical molecular dynamics N-body problems not only in the dynamical criteria of importance but also in the dependence of the Green's function (i.e. "force" potential) on the independent variables. Another area of importance, which is still not well understood in current applications, will include use of spatially dependent time steps (with smaller values needed in active earthquake regions). An important difference between true particles and GEM is that in the latter case, fault positions are essentially fixed in space.

Thus the N-body gravitational problem involves particles whose properties are time-invariant but whose positions change with time, while GEM involves "particles" whose positions are fixed in time, but whose properties change with the surrounding environment. Of course a major challenge in both cases is the issue of time-dependent "clustering" of "particles." It may be possible to exploit this in the case of GEM -- for example by incrementally improving parallel load balancing.

4.5. Computational Complexity

Current evidence suggests that forecasting earthquakes of magnitude ~6 and greater will depend upon understanding the space-time patterns displayed by smaller events; i.e., the magnitude 3's, 4's and 5's. With at least 40,000 km^2 of fault area in southern California, as many as 10^8 grid sites (10-meter segment size) will be needed to accommodate events down to magnitude 3. Extrapolations based upon existing calculations indicate that using time steps of ~100 seconds implies ~10^8 time steps will be required to simulate several earthquake cycles. This leads to the need for teraflop class computers in this as in many physical simulations.

Here we make the conservative assumption that the GEM dipole-dipole Green's function evaluations are ten times as computationally expensive as the Newtonian Green's functions evaluated in Salmon and Warren's code. At this stage, we cannot guess how far the teraflop class of computer will take us and the systems needed to support research, crisis managers or insurance companies assessing possible earthquake risk, may require much higher performance.

4.6. Integration of Simulation Modules into GEMCI

As we described in Section 2.5, we will adopt a hierarchical model for the simulation components used in GEMCI. The full Greens function programs would be macroscopic distributed objects which can be manipulated by systems like WebFlow and linked with other coarse grain components such as sensor data, other programs such as finite-element solvers and visualization systems. These components are made up of smaller objects used in library fashion. Here we have the friction modules and different solver engines such as the conventional $O(N^2)$ and fast multipole methods with each having sequential and parallel versions. The latter would often exist in multiple forms including an openMP version for shared memory and MPI version for distributed memory.

5. TYPICAL COMPUTATIONAL PROBLEMS

Here we give three sample scenarios, which would correspond to distinct problem solving environments built using GEMCI.

5.1. Seismicity Models and Data Assimilation

Our first example of how the GEMCI might be used is drawn from an attempt to create a computer model of the seismicity of California or other seismically active region. Such a model, to the extent that it is realistic, could be quite useful in guiding intuition about the earthquake process, and suggesting new measurements or lines of inquiry.

Making such a model realistic requires many different types of data and there are for instance already substantial archives accumulated by the Southern California Earthquake Center (SCEC), as well as the Seismological Laboratory of the California Institute of Technology, and the Pasadena field office of the United States Geological Survey. The relevant data includes:

1) Broadband seismic data from the TERRASCOPE array.

2) Continuous (SCIGN) and "campaign style" geodetic data.

3) Paleoseismic data collected on the major faults of southern California.

4) Near field strong motion accelerograms of recent earthquakes.

5) Field structural geology of major active faults.

6) Other data including GPS data, leveling data, pore fluid pressure, *in situ* stress, heat flow, downhole seismic data, multi-channel seismic data, laboratory measurements of mechanical properties of various rocks, 3-D geologic structure, gravity data, magneto-telluric data, hydrology data, and ocean tide data (to constrain coastal uplift).

These will be used, for example, to update the fault geometry models used by GEM, and to update fault slip histories used to validate earthquake models.

A new and extremely promising type of geodetic data is *Synthetic Aperture Radar Interferometry* (InSAR), which permits "stress analysis of the Earth." A number of SAR missions are currently acquiring data over southern California, including the C-band (5.8 cm) European ERS 1/2 satellites and the L-band Japanese JERS satellite. These missions have already produced revolutionary images of the complete deformation fields associated with earthquakes in the United States and Japan (e.g., *Massonnet et al.,* 1993). These techniques rely on radar interferograms that represent the deformation field at a pixel resolution of a few tens of meters over areas of tens of thousands of square kilometers, and over time intervals of weeks to years. We are now essentially able to see the complete surface deformation field due to an earthquake, with excellent prospects for similar views of deformation due to interseismic strain-accumulation processes.

There are productive scientists who have built entire careers on each individual type of data mentioned above. No individual scientist at this time has access to complete collections of all these data types, nor the detailed expertise that would be required to use all these data in a computational model. Consequently, the traditional models are based on some subset of the data, with which we can try to predict other sets. However a more powerful approach can be based around an environment in which people can collaborate to construct and test models that span multiple data sets and which can then be embellished by other investigators.

To maximize the utility of any given data set, the people who collect and care about the data must be the ones who archive it and make it available. Attempts to construct monolithic databases of diverse earth science data using dedicated people and facilities have met with limited success. It appears to work better if the people or community who "own" the data are the ones who archive it. Thus, we need an environment, which will allow easy collaboration, and access to diverse distributed data sets. To return to our example of modeling the seismicity of California, an investigator might start with a map of faults in the region, a constitutive law for determining the slip on those faults derived from laboratory measurements, and some sort of loading criteria perhaps derived from GPS measurements. From these inputs, a model is constructed which reproduces the statistics of rupture as shown by paleoseismicity data and earthquake catalogs. Subsequently, another investigator takes that model and adds pore fluid pressure modeling in order to investigate the effects of fluids on the nucleation of earthquakes and the migration of seismicity. This process is repeated with other investigators collaborating and developing competing and/or complementary models. The models can be benchmarked against other data sets and against each other. This kind of an environment will enable us to make the most rapid possible progress at understanding earthquake processes.

5.2. Future Modeling Directions

The basic method for constructing simulations is to compile and process data to obtain estimates of fault geometry and properties, including the properties of the bulk material surrounding the fault. Forward models can be run for hundreds of thousands of years using the above parameters as inputs in order to establish dynamic patterns of the system. Pattern analysis can then performed on the calculated deformation and seismicity to establish temporal and spatial correlations in the simulated data. The simulations can be compared to real geophysical data, which sample only a short timespan of the entire system. In parallel, pattern analysis can be performed on actual geophysical data to search for temporal anomalies. By using a statistical mechanical approach we can address whether statistical patterns form in the system and if so, characterize them. We can also address whether earthquake fault systems can be parameterized without a detailed knowledge of rheological details and assess how the dynamics of space-time patterns relate to the dynamics of fault systems.

There are several questions in earthquake physics that can be addressed with crustal deformation data, a relatively new data set. Some of the questions identified at the Plate Boundary Observatory workshop in Snowbird, Utah, 1999 follow. What processes occur on faults that cause strain localization? What is the architecture of fault zones below the seismogenic zone? What is the relation between strain and fault processes? How do faults interact? What is the nucleation process of earthquakes, or how do they initiate and stop? And what are the conditions (temperature, pressure, stress, . . .) within active fault zones? These questions can be answered by combining high temporal resolution data (seismic, strain, and GPS) and spatially detailed data (InSAR) into models of the deformation process. By answering these questions through the construction of detailed models we can then address how space-time patterns relate to the dynamics of fault systems.

Combining the broad range of models from kinematic to quasi-static dynamic will result in a physically realistic model of earthquake systems based on observations and data. One major question of earthquake physics is whether broad scale stress correlations build up that produce large events or clusters of events. If such correlations exist, are they detectable either in seismicity patterns or in geodetic data? We describe below a means of developing models to test for this without looking at the details of rupture processes.

Geologic, geodetic, and seismic data can all be used to establish fault geometry and slip rates. Paleoseismologic data provide details on slip per event, slip rate and timing of past events on surface rupturing faults. Seismic imaging data as well as borehole data provide details of geometry at depth. Structural analysis of retro-deformed cross sections adds information on slip rate and fault evolution. The geodetic data provide information on fault locations, including blind faults, as well as slip rate. Both strain and rotation rates can be determined from the full velocity gradient tensor provided by GPS data. Strain rates indicate the style of faulting expected from various faults (strike-slip, thrust, oblique) and spatial variations in strain identify fault bounded blocks.

The fault geometries and rates must be coupled with parameters of the mechanical properties of the crust to generate realistic quasi-static models in which stress is transferred and diffused between faults and in the crust. Crustal rigidity structure can be determined from seismic velocity models [e.g. Hauksson, 1997]. These data can be combined with surface deformation observations into viscoelastic models to estimate rheological parameters of the crust as well as place added constraints on slip rates [e.g. Hager et al., 1999].

Fairly abrupt changes of velocity have been observed across the Ventura and Los Angeles basins [Donnellan et al., 1993; Argus et al., 1999]. These velocity profiles can be incorporated into viscoelastic models. The surface deformation data are particularly useful when combined with paleoseismologic data because viscoelastic models indicate that the velocity profile across faults should change with time as stress relaxation occurs in the lower crust following earthquakes. When available, paleoseismologic data make it possible to determine (or at least estimate) the point in the earthquake cycle that the geodetic data have been collected, thus improving estimates of lower crustal viscosity.

Crustal viscosity or relaxation time determines how well the upper crust is coupled to the mantle, which drives the processes that occur at the surface. A weak lower crust tends to generate broader features and effectively results in far-field or side driving stresses, while a strong lower crust results in strong coupling between the crust and the mantle, localizing effects and amplifying the importance of driving stresses from below. These two styles of driving forces result in widely different observed patterns of deformation at the surface. Furthermore, a strong lower crust results in near-fault strain localizations even late in the earthquake cycle, while a weak lower crust rapidly dissipates stress.

The surface deformation profile is further complicated by fault zone properties. Recent geodetic results indicate that afterslip on the rupture plane can occur for up to two years following moderate to large earthquakes [*Pollitz, et al.*, 1998; *Donnellan and Lyzenga,* 1998]. Results from the Northridge earthquake are consistent with rate and state parameters determined in the lab, suggesting that faults are weaker than the surrounding material throughout the earthquake cycle. Future detailed measurements following large earthquakes will clarify behavior of both faults and the lower crust.

Until recently faults have been treated fairly independently from neighboring faults in models. Computer power is available now to model systems of faults. *Rundle et al.* [1997] model how stress is transferred to other faults following earthquakes. The models do not allow a viscous lower crust, but future calculations that allow for both elastic stress transfer and viscous dissipation will elucidate whether or how stress correlations build up in the crust. Running models of interacting fault systems that are constrained by observations will be the first step in applying pattern recognition and statistical mechanics to the earthquake problem.

5.3. Response to an Earthquake in Southern California

An illustrative scenario is shown in the first part of Figure 14 which links multiple datasets, modeling codes, research centers and scientists for real-time earthquake response. A prototype interface supporting this scenario is shown in Figure 15. Other data types and models (eg., seismic and field data) would be included in any operational system, but are omitted here for purposes of exposition. The goal is to form a rapid consensus among researchers concerning the characterization of the deformation field and the location, size, and direction of slip on a fault following an earthquake. This consensus can be used to guide decisions on both civil and scientific responses to the quake.

This initial analysis is followed at the bottom of Figure 14 by detailed simulations of the type described in Section 4. We describe the whole process in detail below.

Following an earthquake in Southern California, the location and magnitude are calculated based on seismic data within minutes by Caltech/USGS, and currently are broadcast to several users via email and pagers. The information on location and magnitude could then be automatically used to define an area wherein instruments might be expected to record a signal (the program *disp*). Data from these stations would be given priority in retrieval and analysis. In this example we will assume that the data in question is GPS data from the Southern California Integrated Geodetic Network (SCIGN) array. Retrieval in this case is done by telephone modem. As soon as the list of possibly affected stations has been generated, the database at the USGS is checked. If any of the stations on the list have not had data downloaded since the quake, computers at the USGS begin dialing the selected stations and retrieving the data.

Data from these stations would then be processed for rapid analysis to determine the measured displacements of the stations (program *GIPSY*). If the measured displacements are large enough, emergency and scientific personnel are notified via email and pager. These displacements are then automatically fed into an inversion routine (program *simplex*) which solves for the best fit single fault displacement. This single fault displacement is in turn fed back into a forward elastic half space model, which yields a preliminary map of displacements over the whole area (program *disloc*).

At this point this map is shared between various scientists and emergency personnel, using systems like Tango Interactive described in Section 6 that allow the collaboration and interaction of multiple people viewing and manipulating the same data set over the Internet. The emergency personnel can use the preliminary map in combination with a Geographical Information System data about utilities, lifelines, etc. to help assign resources to various areas. The scientists will use the preliminary map to help design a strategy for collecting additional measurements. They can also collaborate on refining the single fault model, possibly breaking the single preliminary fault into several segments, introducing more realistic material properties, or including more data, before rerunning the inversion.

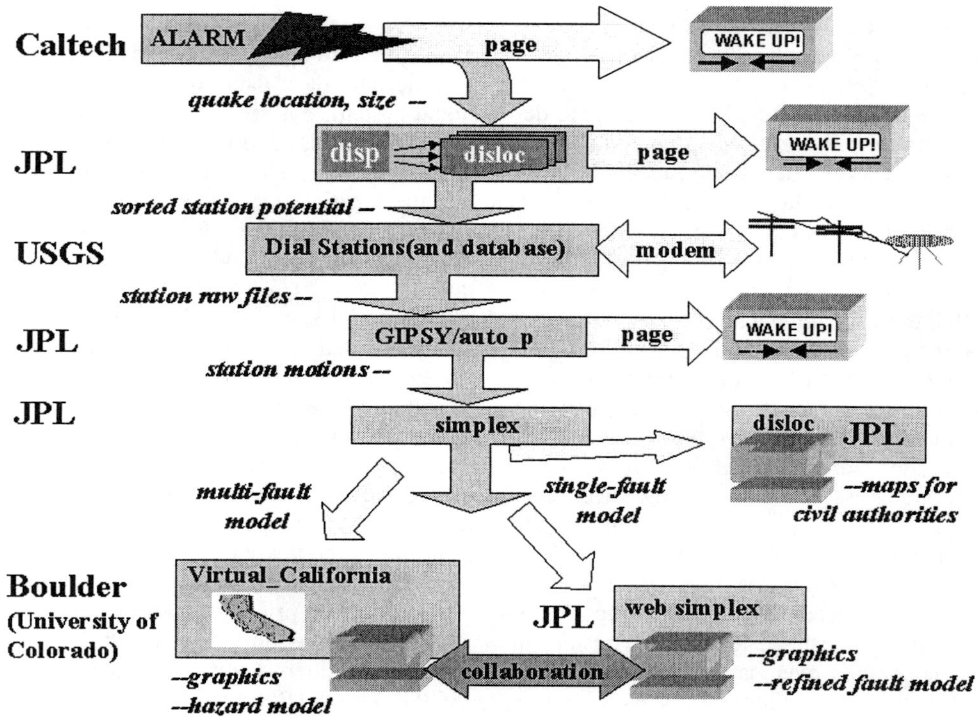

Figure 14. Simulations used in Response to an Earthquake.

This environment permits the rapid determination and dissemination of preliminary information about the earthquake and the collaborative refining of that information following an event. The rapid dissemination of information can greatly aid both the civil and scientific response to the quake. Resources can be more efficiently allocated to the areas where they are needed, and scientific measurements can be focused to provide information critical to refining our understanding of the earthquake system. Once an acceptable model of the earthquake has been determined, various models can be used to estimate the updated earthquake hazard for adjacent areas. Since there are currently several competing models for this, consensus will undoubtedly involve multiple runs of multiple models and significant discussion among scientific colleagues.

Each of these models as well as the various pieces of the automated processes described above have been developed by different people under different assumptions, and is developed, run, and maintained on computers under the control of the developer. The technologies such as CORBA and Enterprise Javabeans described in Section 2, allow appropriate access and security mechanisms in this complex evolving distributed system.

The 1999 Izmit Turkey earthquake provided a recent example of how a system like this could have been useful. Following that earthquake, many geoscientists got together in a series of conference telephone calls to try to piece together what had happened, and what was an appropriate response. Some participants initially only knew what had been reported in the media. Others knew of specific pieces of data concerning the earthquake or of actions being taken by various groups and individuals. It is safe to say that no one had a complete picture. Much of the conference call was devoted to informing everyone about all the pieces of data and all the various initiatives that people were pursuing or might pursue. Similar calls and emails occurred after the 1992 Landers and 1994 Northridge earthquakes. Having a system such as has been described above wherein participants could share maps, descriptions, programs, data sets, and graphs and wherein they could interactively and collaboratively manipulate the data and programs, both synchronously and asynchronously, would immeasurably aid the rapid and accurate diagnosis of what has happened and what should be done next.

5.4. Fundamental Computational Science Studies in Earthquake Forecasting

Above we described a relatively complicated real-time analysis scenario. Another important problem-solving environment supports that process typically used in computational science. Here the main steps are development of simulation codes, refinement of analysis techniques, initiating runs with multiple parameter values, and comparison

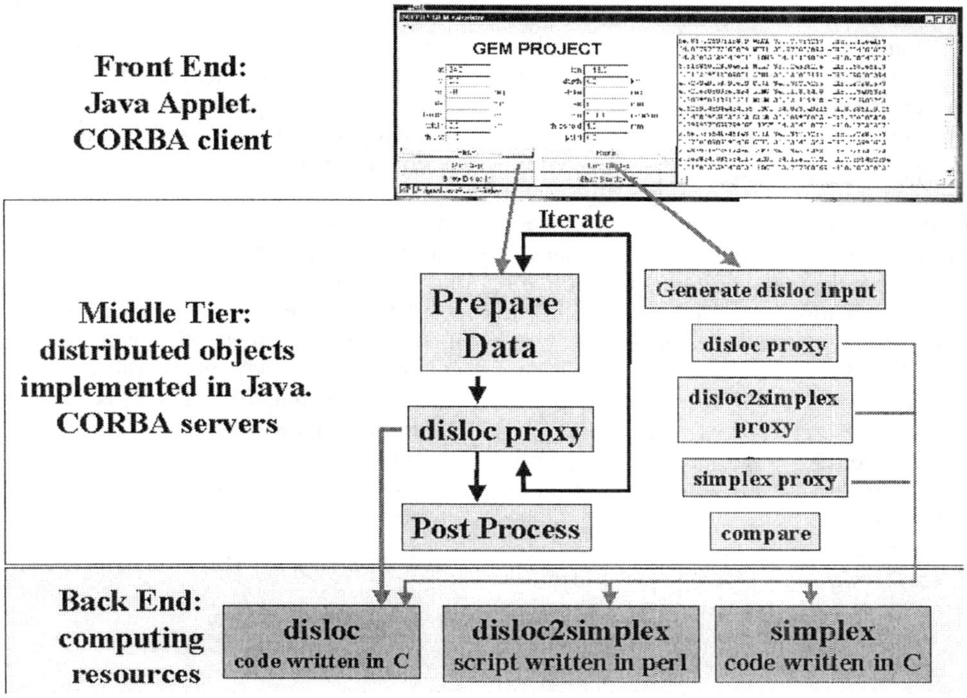

Figure 15. Computational structure of a fragment of the scenario in Figure 14, which was implemented by NPAC as a prototype GEM Problem Solving Environment.

of results with multiple data sets. This type of analysis has similarities with the previous two scenarios but now the collaboration for instance would not emphasize real-time issues so much but extend over many months or years. Technically this implies the need for powerful asynchronous tools linking over greater spans of time and space than the tightly coupled scenario of Section 5.3. The tools needed here include the interactive rapid prototyping environment supporting pattern analysis and visualization. This aspect entails somewhat different trade-offs than the core simulations, in that interactivity is perhaps more critical than performance. As discussed in Section 6, this could suggest that this part of the problem-solving environment would be implemented client-side using interpretative languages. We would also need to experiment with many different ways of linking programs together and so we would have to support both program development and execution. The real-time constraints of Section 5.3 would on the other hand emphasize execution of pre determined program modules.

This problem-solving environment would naturally link with that in Section 5.1 as GPS, InSAR and broadband seismic (TERRASCOPE) data, together with archived and newly developed paleoseismic data can be used in conjunction with the simulation capabilities to establish the relevant model parameters. These parameters include the current geometry of faults, slip rates on any given segment, recurrence intervals and historic variations in slip during earthquakes (leading to estimates of frictional parameters), deformation data (leading to estimates of elastic plate thickness and sub-crustal stress), relaxation times, poroelastic stress, relaxation in the crust following earthquakes (leading to estimates of drained and undrained elastic moduli), and variations in seismicity (leading to estimates of the variable properties of friction and fault geometry at depth). One must support the fitting of models to data with a suite of techniques (e.g., *Menke*, 1989), including least squares, evolutionary programming, and simulated annealing (*Michalewicz*, 1996; *Holland*, 1975; *Rawlins*, 1991), among others. In addition, one can expect to develop new methods so as to adapt models to assimilate new data as that becomes available, a concept that has served meteorological and climate studies extremely well. Self-adaptation techniques can be based on the same kinds of back-propagation methods that have been useful in analysis of neural network models (*Hertz et al.*, 1991).

5.5. Seismic Waves and Earthquake Engineering

The most mature computations in the earthquake field are perhaps those used to calculate seismic waves and the corresponding response of buildings to them. In fact Clayton in the Caltech Geophysics department performed one of the very first Caltech Cosmic Cube computations to simulate the motion of earthquake waves in the Los Angeles Basin. This wave motion can in principle be generated from the earthquake "events" calculated in the Virtual_California simulations described in Sections 4 and 5.4. These are large-scale finite element problems with complex grids and a recent NSF Grand Challenge project at CMU was very successful in this area (http://www.cs.cmu.edu/~quake/). The wave motion can be used as a forcing function for structural dynamics computations of buildings, roads and other civil infrastructure (http://www.curee.org/).

6. SUMMARY OF GEMCI

We introduced in Section 2.1 a computational environment GEMCI and described its overall framework in terms of distributed objects supported by commodity technologies like CORBA COM Java and XML. In the following sections we have discussed some especially important modules and how they are integrated into overall systems aimed at particular problem areas. Although we have not discussed them in detail, we will use some basic services, in particular security, collaboration and visualization. Security does not need much customization for computational science from the capabilities available from the commodity technologies for electronic commerce. Both Globus and Legion have developed good security approaches for high performance computing and GEMCI will be adopt such solutions. There is much experience in visualization in many projects although it is not clearly understood how to formulate this generally in systems like GEMCI. We can however expect that the substantial efforts by both Department of Energy and NSF will lead to visualization approaches appropriate for GEMCI. Collaboration corresponds to sharing of objects in GEMCI and this could correspond to shared rendering of data or the results of simulations. This technology is less mature than that for security and visualization although the Web has enabled powerful "asynchronous" collaboration where one posts information on web pages and your collaborators access your information at a later time. Synchronous collaboration can be important – especially in scenarios like those of Section 5.2. Here approaches like Tango Interactive from NPAC at Syracuse (http://www.npac.syr.edu/tango) are early prototypes that are expected to mature into powerful synchronous sharing for web pages and related data sets. An important area that we have not yet discussed in detail is the handling of data and its linkage to simulation. Here one of the leaders is the NSF NPACI partnership (http://www.npaci.edu/DICE/) whose data-intensive activity is led by Reagan Moore. This approach makes extensive use of the same object standards advocated in GEMCI and we expect to be able to use the best available hardware and software to link data and simulations.

In summary, we believe that GEMCI has been designed to take advantages of the best available high performance and commodity information and simulation resources. We cannot predict the future of this rapidly changing field but GEMCI has made choices that should be able to track and exploit advances.

The earthquake community is uniquely positioned to take advantage of technology advances on many fronts. The data volume and quality is rapidly increasing as geodetic and seismic networks become denser and broader band. Recent advances in information technology will enable modeling environments to be constructed in which data and parameters are easily incorporated into and passed between models. Computational power is such that complex three-dimensional fault system problems can now be approached, and algorithms developed in other fields (plasma physics, astrophysics) are well-suited to be applied to the earthquake problem.

Acknowledgements. The research described in this paper was partially performed by the Jet Propulsion Laboratory, California Institute of Technology, under a contract with the National Aeronautics and Space Administration.

REFERENCES

Allègre, C., J.L. Le Mouel and A. Provost, Scaling rules in rock fracture and possible implications for earthquake prediction, Nature, 297, 47-49, 1982.

Allègre, C.J. and J.L. Le Mouel, Introduction of scaling techniques in brittle fracture of rocks, *Phys. Earth Planet Int.*, 87, 85-93, 1994.

Allègre, C.J., J.L. Le Mouel, H.D. Chau and C. Narteau, Scaling organization of fracture tectonics and earthquake mechanism, *Phys. Earth Planet Int.*, 92, 215-233 1996.

Anderson, C.R., SIAM J. Sci. Stat. Comp., 13, 923, 1992.

Appel, A W. , An efficient Program for Many-Body Simulation, SIAM J. Sci. Stat. Comp., 6, 85, 1985.

Appel, A.W.,SIAM Journal on Computing., 6, 85, 1985.

Archambeau, C.B., General theory of elastodynamic source fields, *Rev. Geophys. Space Phys.*, 16, 241-288, 1968.

Archambeau, C.B. and J.B. Minster, Dynamics in prestressed media with moving phase boundarires: A continuum theory of failure in solids, *Geophys. J. R. Astr.* , 52, 65-96, 1978.

Argus, D. F., M. B. Heflin, A. Donnellan, F. Webb, D. Dong, K. Hurst, D. Jefferson, G. Lyzenga, M. M. Watkins, J. F. Zumberge, Shortening and thickening of metropolitan Los Angeles measured and inferred by using geodesy, *Geology,* 27, 703-706, 1999.

Backus, G and M. Mulcahy, Moment tensors and other phenomenological descriptions of seismic sources - I: Continuous displacements, *Geophys. J. Roy. Astr. Soc.*, 46, 341-361, 1976.

Barnes, J. and Hut, P., Nature, 324, 446, 1986.

Barnes, J.E. and P Hut, A Hierarchical (O (N log$N$)) Force-Calculation Algorithm, Nature, 324, 446-449, 1986.

Beeler, N.M., T.E.Tullis, M.L. Blanpied and J.D. Weeks, Frictional behavior of large displacement experimental faults, J. Geophys. Res., 101, 8697-8715, 1996.

Beeler, N. and TE Tullis, Self healing slip pulses in dynamic rupture models due to velocity dependent strength, *Bull. Seism. Soc*, 1996.

Ben-Zion, Y. and J. R. Rice, earthquake failure sequences along a cellular fault zone in a three-dimensional elastic solid containing asperity and nonasperity regions, J. Geophys. Res., 98, 14109-14131, 1993.

Ben-Zion, Y. and J. R. Rice, Slip patterns and earthquake populations along different classes of faults in elastic solids, J. Geophys. Res., 100, 12959-12983, 1995.

Ben-Zion, Y. and J. R. Rice, Dynamic simulations of slip on a smooth fault in an elastic solid, J. Geophys. Res., 102, 17771-17784, 1997.

Berryman, J.G. and G.W. Milton, Normalization constraint for variational bounds on fluid permeability, *J. Chem. Phys.*, 83, 754-760, 1985.

Blanter, EM, MG Schnirman, JL Le Mouel and C.J. All□gre, Scaling laws in block dynamics and dynamic self-organized criticality, *Phys. Earth. Planet. Int., 99, 295-307*, 1997.

Bosak, J. and T. Bray, XML and the Second-Generation Web, *Scientific American, 280(5)*, 89-93, 1999.

Brown, S.R, C.J. Scholz and J.B. Rundle, A simplified spring-block model of earthquakes, *Geophys. Res. Lett., 18,* 2, 215-218, 1991.

Burridge, R and L. Knopoff, Model and theoretical seismicity, Bull. Seism. Soc. Am., 57, 341-371, 1967

Burridge, R. and C.A. Varga, The fundamental solution in dynamic poroelasticity, *Geophys. J. Roy. astr. Soc.*, 58, 61-90, 1979.

Carlson, J.M and J.S. Langer, Properties of earthquakes generated by fault dynamics, *Phys. Rev. Lett.*, 62, 2632-2635, 1989.

Cleary, M.P., Fundamental solutions for a fluid-saturated porous solid, *Int. J. Solids Str.*, 13, 785-806, 1977.

Dahmen K, Ertas D, Ben-Zion Y, Gutenberg-Richter and characteristic earthquake behavior in simple mean-field models of heterogeneous faults, *Phys Rev E.* 58, 1494-1501 1998

Dieterich, J.H., Time dependent friction in rocks, *J. Geophys. Res.*, 77, 3690-3697, 1972

Dieterich, J.H., Time dependent friction and the mechanics of stick slip, *Pure Appl. Geophys.*, 116, 790-806, 1978.

Dieterich, J.H., Constitutive properties of rock with simulated gouge, In *Mechanical Behavior of Crustal Rocks. AGU Geophys. Monograph 24*, Washington, DC, American Geophysical Union, 103-120, 1981.

Donnellan A., BH Hager, RW King, and TA Herring, Geodetic Measurement of Deformation in the Ventura-Basin-Region, Southern California, J Geophys Res 98, 21727-21739, 1993.

Donnellan A., Lyzenga G.A., GPS observations of fault afterslip and upper crustal deformation following the Northridge earthquake J. Geophys Res 103 21285-21297 1998 .

Fisher, D. S., K. Dahmen, S. Ramanathan and Y. Ben-Zion, Statistics of earthquakes in Simple Models of Heterogeneous Faults, Phys. Rev. Lett., 78, 4885-4888, 1997.

Goil, S., Primatives for problems using heirarchical algorithms on distributed memory machines, NPAC Tech Rept. SCCS-687, http:/www.npac.syr.edu/techreports/index.html, 1994.

Goil, S. and S. Ranka, Parallelization of heirarchically structured applications on distributed memory machines, NPAC Tech Rept. SCCE-688, http:/www.npac.syr.edu/techreports/index.html, 1995.

Greengard, L. and Rokhlin, V., A Fast Algorithm for Particle Simulations, J. Comp. Phys, 73, 325-348, 1987.

Gross, S., On the scaling of average slip with rupture dimensions of earthquakes, pp.157-166 in JB Rundle, D. Turcotte and W. Klein, eds., *Reduction and Predictability of Natural Disasters,* vol XXV in the Santa Fe Inst. series on the Sciences of Complexity, Addison Wesley, 1996.

Heimpel, M., earthquake size-frequency relations from an analytical stochastic rupture model, *J. Geophys. Res.*, 101, 22435-22448 1996.

Hauksson, E., and J. Hasse. Three-dimensional V-P and V-P/V-S velocity models of the Los Angeles basin and central Transverse Ranges, California, *J. Geophys Res.* 102: (B3) 5423-5453, 1997

Hertz, J. , A. Krogh and R.G. Palmer, *Introduction to the Theory of Neural Computation*, Lecture Notes I, Santa Fe Inst. Addison Wesley, Reading, MA, (1991).

Hill, R. A self consistent mechanics of composite materials, *J. Mech.. Phys. Sol.*, 13, 213-222, 1965.

Holland, J.H., *Adaptation in natural and artificial systems*, University of Michigan Press, Ann Arbor, 1975.

Houstis, E. N., J. R. Rice, S. Weerawarana, A. C. Catlin, P. Papachiou, K. Y. Wang, and M. Gaitatzes, PELLPACK: A Problem Solving Environment for PDE Based Applications on Multicomputer Platforms, *ACM Trans. Math. Software, 24,* 30-73, 1998.

Ivins, E.R. and C.G. Sammis, On lateral viscosity contrast in the mantle and the rheology of low frequency geodynamics, *Geophys. J. Int*, 123, 305-322 1995b.

Ivins, E.R., *Composite Constitutive Theory for a Solid Earth of Strong Lateral Viscosity Contrast*, Ph.D. Dissertation, Univ. Southern Calif., 1995a.

Kanamori, H. and D.L. Anderson, Theoretical basis for some empirical relations in seismology, *Bull. Seism. Soc. Am.*, 65, 1073-1095, 1975.

Li, V.C., Mechanics of shear rupture applied to earthquake zones, in *Fracture Mechanics of Rock*, ed. B.K. Atkinson, Academic Press, London, 1987, pp. 351-428.

Lyakhovsky, V., Y. Ben-Zion and A. Agnon, Distributed Damage, Faulting, and Friction, J. Geophys. Res., 102, 27635-2764, 1997.

Maruyama, T., Fluid pressure responses of a water-saturated porous elastic multilayered half space to water pumping and atmospheric loading, *J. Phys. Earth*, 42, 331-375, 1994.

Massonnet, D., M Rossi, C. Carmona, F. Adragna, G. Peltzer, K. Feigle, T Rabaute, The Displacement Field of the Landers Earthquake Mapped by Radar Interferometry, *Nature* 364, 138-142 1993.

Menke, W., Geophysical Data Analysis: Discrete Inverse Theory, Academic Press, 289p 1989

Michalewicz, Z., *Genetic Algorithms+Data Structures = Evolution Programs*, Springer-Verlag, New York, 1996.

Minster, J.-B., Twenty five years of sesimic source theory, in *The VELA Program, A Twenty-Five Year Review of Basic Research*, ed. A.U. Kerr, Defense Advanced Research Projects Agency, pp. 67-116, 1985.

Newman, W.I., D.L. Turcotte and A. Gabrielov, A hierarchical model for precursory seismic activation, pp. 243-260, in J.B. Rundle, D.L. Turcotte and W. Klein, *Reduction and Predictability of Natural Disasters*, vol. XXV in the Santa Fe Institute series on complexity, Addison Wesley, 1996.

Okada, Y., Surface deformation duei to shear and tensile faults in a half-space, *Bull. Seism. Soc. Am.*, 75, 1135-1154, 1985.

Okada, Y., Internal deformation due to shear and tensile faults in a half-space, *Bull. Seism. Soc. Am.*, 82, 1018-1040, 1992.

Pollitz FF, Burgmann R, Segall P., Joint estimation of afterslip rate and postseismic relaxation following the 1989 Loma Prieta earthquake, J Geophys. Res. 103, 26975-26992 1998.

Rabinowicz, E., *Friction and Wear of Materials*, John Wiley, New York, 1965.

Rawlins, G.J., *Foundations of Genetic Algorithms*, Morgan-Kaufman publishers, San Francisco, 1991.

Rice, J.R. and M.Cleary, Some basic stress diffusion solutions for fluid saturated elastic porous media with compressible constituents, *Rev. Geophys. Space Phys.*, 14, 227-241, 1976.

Rice, J.R., Spatio-temporal complexity of slip on a fault, *J. Geophys. Res.*, *98*, 9885-907, 1993.

Rice J. R. and Y. Ben-Zion, Slip complexity in earthquake fault models, Proc. Natl. Acad. Sci. U.S.A, 93, 3811-3818, 1996.

Rice, J.R. and A.L. Ruina, Stability of steady frictional slipping, *J. Appl. Mech.*, 105, 343-349, 1983.

Ruina, A, Slip Instability and State Variable Friction Laws, J. *Geophys. Res.* 88, 359-370, 1983.

Rundle, J.B. and D.D. Jackson, Numerical simulation of earthquake sequences, *Bull. Seism. Soc. Am.*, 67, 1363 - 1378, 1977.

Rundle, J.B., Viscoelastic - gravitational deformation by a rectangular thrust fault in a layered earth, *J. Geophys. Res.*, 87, 7787 - 7796, 1982.

Rundle, J.B., Some solutions for static and pseudo - static deformation in layered, nonisothermal media, *J. Phys. Earth*, 30, 421 - 440, 1983.

Rundle, J.B., A physical model for earthquakes: 1, Fluctuations and interactions, *J. Geophys. Res.*, 93, 6237-6254, 1988.

Rundle, J.B. and S.R. Brown, Origin of rate dependence in frictional sliding, *J. Stat. Phys.*, 65, 403-412, 1991.

Rundle, J.B. and W. Klein, Nonlinear dynamical models for earthquakes and frictional sliding: An overview, *Proc. 33rd Symposium Rock Mech.*, A.A. Balkema, Rotterdam, 1992.

Rundle, J.B. and D.L. Turcotte, New directions in theoretical studies of tectonic deformation: A survey of recent progress, in *Contributions of Space Geodesy to Geodynamics: Crustal Dynamics*, ed. D.E. Smith and D.L. Turcotte, Geodynamics Ser. Vol. 23, AGU, Washington, DC, 1993.

Rundle, J.B., W. Klein and S. Gross, Dynamics of a traveling density wave model for earthquakes, Phys. Rev. Lett., 76, 4285-4288, 1996.

Rundle, J.B., Salmon, J.K and M.S. Warren, Parallel supercomputing with comodity components, *Proc. Intl. Conf. Parallel Distributed Processing Tech. Appl. (PDPTA '97)*, ed. M.S. Warren, D.J. Becker, M.P. Goda, J.K. Salmon and T. Sterling, Las Vegas, NV, June 30-July 3, 1997.

Smalley, R.F., D.L. Turcotte and S.A. Solla, A renormalization group approach to the stick slip behavior of faults, *J. Geophys. Res.*, 90, 1884-1900, 1985.

Stein RS, The role of stress transfer in earthquake occurrence, *Nature,* 402 605-609 1999 .

Stuart, W.D. and T.E. Tullis, Fault model for preseismic deformation at Parkfield, California, *J. Geophys. Res.*, 100, 24079-24101, 1995.

Winckelmans, G.S., JK Salmon, MS Warren, A Leonard and B Jodoin, Application of fast parallel and sequential tree codes to computing three-dimensional flows with vortex element and boundary element methods, in *Vortex Flows and Related Numerical Methods,* Montreal, 1995.

Andrea Donnellan, Jet Propulsion Laboratory, MS 238-600, 4800 Oak Grove Drive, Pasadena, CA 91109-8099.

Geoffrey C. Fox, School for Computational Science and Information Technology and Department of Computer Science, Florida State University, Dirac Science Library, Tallahassee, Florida.

Ken Hurst, Jet Propulsion Laboratory, MS 238-600, 4800 Oak Grove Drive, Pasadena, CA 91109-8099.

Jay Parker, Jet Propulsion Laboratory, MS 238-600, 4800 Oak Grove Drive, Pasadena, CA 91109-8099.

Modeling Complex Crustal Processes

William J. Bosl

Geophysics Department, Stanford University

Geophysical processes in the Earth, such as earthquakes, are complex processes that exhibit emergent behavior. That is, fundamental physical and chemical processes coupled together result in phenomena that are not apparent in isolated laboratory experiments or theoretical studies of uncoupled, linear processes. Rigorous quantitative investigation of complex earth processes requires the development of appropriate quantitative tools for expressing and exploring the laws that govern complex processes. Computer-based models allow investigation of the consequences of theories that cannot be explored in any other way. As the need to include more processes in coupled models and more detailed theories about earthquake cycles grows, so does the need to work at higher levels of software abstraction. This can be accomplished through the development of a standard, extensible set of geophysical software components. The basic mechanical, fluid, and chemical processes that, when coupled together in their most general forms, are thought to control the earthquake cycle, are presented here, with particular emphasis on the continuum mechanics seismogenic upper crust as a chemically evolving, thermoporoelastic medium. A detailed finite element discretization for the thermoporoelastic continuum equations is presented, as well as references for numerical modeling of other important processes. A framework for geoscience software components is discussed, the purpose of which is to integrate the complicated numerical models required for realistic quantitative investigations of earthquake physics and to provide a standard quantitative approach to coordinate a community modeling effort.

1. INTRODUCTION

Geology in this century has often been considered primarily a descriptive science because geologic processes were thought to be too complex and random to be submitted to rigorous mathematical analysis. Mathematical methods in geophysics are well developed for processing and analyzing data that has been gathered using geophysical methods. Geophysics certainly uses classical and modern physical methods to conduct measurements of the earth. But science is more than just the gathering of data. It also includes theoretical constructs which make sense out of the data. Geophysics must also concern itself with the formulation of quantitative physical models (laws) that give meaning to the patterns observed in data. At its most fundamental level, science is concerned with making models that allow scientific prediction. We cannot claim to "understand" a phenomenon unless we can prescribe a rule, that is, an equation or a computational algorithm, that allows us to say something about the future behavior of a system, given information about its present state. Quantitative

GeoComplexity and the Physics of Earthquakes
Geophysical Monograph 120

models or laws of nature not only connect data points on a graph, but tell us how to extend the plot beyond the data points.

In this paper, I propose the idea that the complex processes of the earth, and in particular earthquakes, cannot be understood in terms of analytically tractable differential equations. In order to quantify earthquake physics, the approach of classical physics in this century, whereby the essence of a process is found by reduction to a single rule, such as Schroedinger's equation, or a Hamiltonian formulation of mechanics, will fail. Earthquakes are an emergent process, one that manifests itself only when many complex, coupled processes operate simultaneously. Such processes can be represented by algorithmic laws, whose consequences can be investigated only through the use of computer technology. In fact, the new science of complex systems would not have developed without the concurrent evolution of modern electronic computers.

2. SCIENTIFIC LAWS AND PHYSICAL PROCESSES

A scientific law is a rule that tells us how to make future predictions about a system or variable of interest. A predictive rule is not required to encapsulate all aspects of a system in order to attain the status of a scientific law, but only certain aspects which are of interest. Thus, Newton's First Law, when applied to the motion of the Earth in the Sun's gravitational field, doesn't deal with the momentum imparted to the Earth by solar radiation as it travels through the space (though this must be included when calculating the orbital decay of artificial satellites orbiting the Earth), or with the internal tidal deformations resulting from the Earth's non-rigid material constitution. The law describes a certain aspect of the dynamical behavior of the Earth which has been abstracted from the real system of processes which make up the Earth. A physical law is a quantitive expression of some aspect of a system that allows us to make predictions about that aspect of its future state.

A predictive rule formulated as an algorithmic statement can aquire the status of a scientific law just as readily as an elegant differential equation. *Wolfram* [1984] defined scientific laws as algorithms or procedures for determining how systems behave. Any physical process, any quantitative scientific law, can be represented as a computational process. "When the computer program is executed, it causes the numbers and symbols to be modified in the way specified by the scientific laws. It thereby allows the consequences of the laws to be deduced" [*Wolfram*, 1984, p. 439]. The equation $F = \frac{d(mv)}{dt}$, is a compact, symbolically rich way of expressing a computational process that allows us to determine the future dynamical state of massive bodies.

Toffoli and Margulis [1987] argue that the mathematics of any age reflects the computational tools that are available. Since Newton, the emphasis has been on models (formulations of scientific laws) that are continuous, linear, and involve few lumped parameter variables:

> This emphasis does not reflect a preference of nature, but rather the fact that the human brain, aided only by pencil and paper, performs best when it handles a small number of symbolic tokens having substantial conceptual depth (e.g. real numbers, differential operators).... The advent of digital computers has shifted the region of optimum performance" [*Toffoli and Margolus*, 1987].

This is not to imply that differential equations are not useful for modeling the earth. Rather, discrete implementations of coupled sets of nonlinear differential equations, perhaps with heterogeneous coefficients that cannot be summarized in functional form, together with algorithmic rules to describe processes such as fault rupture, may be the simplest quantitative representation of certain earth systems that properly model the phenomenon of interest.

A quantitative law or rule that cannot be simplified further is called *computationally irreducible* by *Wolfram* [1984]. Predictions about the future state of physical processes that are computationally irreducible systems can be made only by evolving the model step by step in time; there is no shortcut. Wolfram contends that computationally irreducible processes may be the norm rather than the exception in nature. Earth processes, which exhibit a great degree of complexity and apparent randomness, seem to support this claim.

For example, the orbit of a single planet around the sun is a computationally reducible system. If the perturbations of other planets are ignored, the orbit can be described for all time by a single equation. Predictions about any future state can be computed rather easily. A model for diagenesis of granular rocks is an example of a computationally irreducible given in Bosl [1998]. In this model, the deposition of small particles within the spaces between grains during early burial is modelled as follows: Flow through a granualar rock is computed using a lattice Boltzmann method. Small particles, the diagenetic material, are deposited at a small, specified fraction of the pore volume where flow is fastest. The fraction is intended to represent the finite size of the deposited particle; this is not a continuous process. But the diagenetic process then alters the flow field, which is recomputed. The new flow field determines where deposition will next occur, and so on. This process of depositing particles where flow is highest is simple to describe algorithmically. But there is no way to write an equation that describes the time evolution of the complex pore geometry. The pattern of pore evolution can be calculated only by evolving the whole system through each step in time.

The earthquake process may be a computationally irreducible process. The equations that describe the porous, fluid-saturated, chemically evolving, thermoelastic crust and the viscoelastic asthenosphere cannot be solved analytically when realistic coefficients are present, but discrete forms of these equations can be solved with computers. The laws that govern the faulting process are not yet known, but those that are proposed follow discontinuous, algorithmic rules. For example, rupture occurs if some condition is met, otherwise nothing happens. The condition which determines whether or not rupture should occur may involve a complicated combination of friction laws, stress evolution, diagenetic evolution of fault properties, boundary conditions, and so on. The computer, as a means for exploring the consequences of the scientists' ideas about how nature works, may open up a golden age of discovery for the earth sciences, whose processes are complex and cannot be adequately described by the continuous, linear equations of classical physics.

Yet the computer is often a great hindrance to scientific exploration and creativity. The difficulty of implementing scientists' ideas in the form of a computer code that can be evolved forward in time prohibits the use of the computer as a laboratory for testing complex ideas. There is a great need for tools that will shorten the time and effort needed to take an abstract, yet quantitatively precise, scientific idea from its conception to a working simulation that accurately represents the idea and allows its consequences to be explored. It is at this point that science begins; the preliminary software construction, though difficult, is merely technical detail from a scientific viewpoint. A software framework specifically tailored for exploration of processes in the solid earth is needed to transform the computer into a canvas on which the geophysicist can explore the consequences of his or her creative ideas.

3. COMPLEX SYSTEMS IN NATURE AND SOFTWARE

The motivating assumption behind this paper is that the complexity typically observed in the earthquake process, and Earth systems in general, is fundamentally due to the coupling of many physical processes that, when considered alone, are well-known in terms of classical physics and chemistry and can be described algorithmically and quantitatively. The primary goal of computational experimentation is to be able to test the consequences of the various theories which geoscientists conceive to see if in fact they have predictive value. Quantitative description and, eventually, prediction are thus reasonable expections. This is the goal of all physical theory and is in fact the essence of science. However, the fundamental physical processes involved in such complex phenomena as earthquakes must be adequately modeled in simulations before we can begin to understand the relative importance of each factor in controlling the total complex system. By this, it is implied that all of the relevant physical processes must be included, they must be properly coupled, and the evolution of controlling parameters or material properties must be modeled. It is not sufficient to use a simple diffusion equation with constant coefficients, for example, to discuss the role of fluids in crustal processes if this model is not the correct description of how fluids flow in the thermoporoelastic, chemically evolving material that makes up the crust.

In order to understand the physics of earthquakes, it is not sufficient to find formulas or rules that merely reproduce the statitics of earthquake patterns in space and time. We know that mechanical and chemical processes, which are known fundamental processes, control crustal dynamics. We do not yet understand how these fundamental processes work together to cause the complex phenomenon known as earthquakes. When we truly understand earth dynamics, the complex, chaotic, fractal patterns that are observed in nature will emerge from the basic physics. They must, or else we will not be able to claim to understand the physics of earthquakes. Pattern matching may be a first step in science, but it does not constitute scientific understanding.

There is a great temptation to simplify mathematical equations to make them amenable to solution by the methods which are available to us. Simplification is desirable only when insignificant terms can be excluded. A general rule for earth systems is that when seemingly simple processes occur together in media which vary spatially and temporally, the resulting behavior may be wholly unlike the dynamics of simple, uncoupled processes. The tectonic and hydrologic forces that drive the dynamics of the solid earth are not independent of each other and must be considered as complex systems. Earth systems exhibit several characteristics that contribute to the overall complexity of crustal processes:

TABLE 1. Examples of complex dynamics in the crust.

Type	Variables involved	Physical example
Coupled processes	p, σ (mean stress)	Mandel-Cryer effect [a]
Spatial heterogeneity	p, σ, k	Anomolously high Pp[b]
Nonlinear parameters	k = k (p)	Piston-like pressure profile [c]
Chemical alteration	p, ϕ	Abnormal p compartments [d]
Faulting, earthquakes	Friction and above variables	Earthquake cycle [e]

[a]Cryer [1962]
[b]Bosl [1999]
[c]Yilmaz, et al. [1994]
[d]Walder and Nur [1984]
[e]Byerlee [1996]

coupling of two or more processes, spatially heterogeneous material properties which evolve in time, nonlinear terms in the governing equations, and rule-based discontinuous processes. Some examples of each of these and their potential dramatic effects are briefly discussed now. Table 1 gives some examples in summary form.

3.1. Coupled Processes

It is possible for two simple linear processes to exhibit nonlinear, unpredictable behavior when coupled together. A striking example of this is the coupling of fluid flow through an elastic porous medium. A comparison between simple fluid diffusion and coupled poroelastic diffusion was done by *Cryer* [1963], which illustrates how surprising behavior can arise from simple linear processes that are dynamically coupled. Cryer considered a fluid saturated poroelastic sphere, with initial internal pore pressure of $p = 1$ and boundary pressure of 0. Mean stress on the boundaries was held constant with a value of -1 (compressive), so as to exactly balance the fluid pressure and had an initial internal value of -1 also. Uncoupled porous medium flow theory predicts that fluid pressure inside the sphere will decay monotonically from its initial value to zero. Cryer found, remarkably, that in a fully coupled poroelastic medium, fluid pressure inside the sphere initially *rises* before decaying to the steady-state pressure. This is an illustration of how coupling two or more linear processes, even with homogeneous properties, can lead to quite surprising behavior.

3.2. Heterogeneous Properties

The prediction of contaminant movement and the flow of fluids in a petroleum reservoir rely heavily today on the use of geostatistical methods to quantify the role of heterogeneities in controlling fluid flow in the crust [*Tompson, et al.*,1998]. Measurements of the hydrologic properties of faults have revealed their heterogeneous character (see especially the collection of articles edited by *Hickman, et al.* [1994]. These studies demonstrate not only that permeability is quite heterogeneous in and near fault zones, but is also strongly anisotropic. In some cases, heterogeneous permeability may combine with poroelastic effects to cause unexpected abnormally high pore pressures. This may have important implications for fluids trapped in low permeability fault zones [*Bosl*, 1999].

3.3. Nonlinearities

Nonlinear behavior is one of the predominant characteristics of complex dynamical systems. Yilmaz, et al. [1994] showed that even fairly mild permeability nonlinearities can cause significant deviations from expected linear behavior when permeability is a nonlinear function of pressure. This may be an appropriate physical model for flow in materials where the permeability is due primarily to fractures. Fractures tend to be much more compliant than pore spaces in granular rocks and will tend to open or close in response to the pressure of the fluid that is flowing through them. The regions around many major faults may be composed largely of micro-fractured rock. When the permeability depends strongly on the fluid pressure, a shock-wave type of pressure front may result. This is similar to nonlinear thermal waves that occur in high-temperature plasmas when radiation heat conduction is the dominant heat transfer mechanism [*Zeldovich*, 1966]. Thus, qualitatively accurate simulations of the poroelastic response of fractured materials may require that nonlinear permeability be considered in simulations.

Mantle convection, which is the driving force of tectonic plate movement, can be represented by coupled equations for heat convection and fluid flow. The equations are highly nonlinear because of temperature and pressure-dependent viscosity and nonlinear stress-strain relationships. The dynamics of these processes are complex and the consequences of different theories about mantle properties can only be explored through simulation [*Christensen*, 1989].

3.4. Time Evolution of Material Properties

The properties of a porous material change over relatively short timescales. The primary agent in the rapid alteration of rocks in the earth is moving water. The time scale for chemical alteration of the flow properties of crustal rocks may be short enough to play a significant role in a number of geological processes in the shallow crust, including earthquakes. *Byerlee* [1996], for example, presents a model for the earthquake cycle which depends upon evolving permeability and porosity in and near the fault zone. Computer simulations to test such hypotheses must include the effects of evolving material properties.

3.5. Scientific and Engineering Computing

To understand the behavior of complex systems we must be able to carry out quantitative computer experiments and manipulate each of the parameters of the model so as to determine their effect on the whole system under various conditions. The use of the computer as an experimental tool to explore the consequences of competing physical models is often referred to as *computational science.* This activity may be distinguished from the use of the computer to compute the answer to a problem where the physical model is known. We may refer to this latter activity as *engineering computing.* Both of these kinds of numerical computing can be mathematically challenging, both can stretch the capabilities of the largest supercomputers to their limit, and both are very important to the advancement of pure and applied science. Earthquake science is not ready to carry out engineering calculations. The most important processes are not yet understood. But earthquake science needs to carry out computer experiments with various physical models in order to determine the consequences of these models in coupled, complicated, realistic settings.

Engineering calculations require that the underlying physical models, hence the mathematical equations that are being solved, be known and relatively stable. Commercial software packages generally offer the best means by which to solve such problems, because the codes required to solve complex engineering problems can represent many programmer-years of effort to build. Scientific codes sometimes require as much effort to build, yet scientists, graduate students, and industrial researchers often build their own codes from scratch, perhaps with the help of math libraries. There are several reasons for this, including the desire to "own" the code and perhaps a distrust of code written by others. But the main reasons many scientists feel the need to write new code may be found in the task itself: scientific codes (as defined here) are used to experiment with various physical and mathematical models, compare different constitutive laws, and test the consequences of alternative theoretical ideas. It is necessary for the scientist to be able to change fundamental aspects of the code in order to carry out the numerical experiments. This is why so much effort is expended in scientific circles to write and rewrite numerical codes. Yet there is much repeated and wasted effort in this approach. We can do better.

The advanced mathematical methods and new quantitative models, such as cellular automata models of physical systems being developed by computer scientists and mathematicians are not easily used for exploration of earth systems until they are implemented in software. (See, for example, *Chopard and Droz*, 1998; *Wolfram*, 1994; *Toffoli and Margolus*, 1994). The earth science community needs a set of software components which represent the dynamics of various physical processes and that are designed to be easily coupled together, extended, and modified. Earth scientists want to explore the consequences of physical processes; computational tools are thus needed which allow geophysicists to work at the level of physical processes, not at the level of numerical math and low-level optimization.

4. SOLID EARTH SUBSYSTEMS

Complex systems in the earth may be studied computationally by modeling fundamental processes and subsystems, then coupling these together. This approach allows the computer to be used as an experimental laboratory to investigate the effects of single processes on the entire system. Complex dynamics can emerge from the complicated interactions and nonlinear response of basic processes. This approach requires that software components be constructed in a way that mimics nature. Complex processes such as earthquakes may be thought of as a hierarchy of subprocesses: the fundamental laws of solid and fluid mechanics, thermodynamics, and chemistry govern the behavior of the brittle upper crust, the asthenosphere, faults, and the mantle. In some cases, the relevant physical processes are well-understood. In others, it is not clear which basic

TABLE 2. The most common dynamical variables which describe the physical state of the crust.

Dynamical variable	Governing equations	Controlling parameters
Displacements, u_1, u_2, u_3	Navier equations	elastic moduli
Displacement rates	Viscoplastoelastic	moduli, rock viscosity
Pore pressure	Hydraulic diffusion	permeability, viscosity
Temperature	Thermal advection/diffusion	thermal conductivity
Porosity	Solid phase fraction	mineral fractions
Mineral species	Geochemical system	activity coefficients

physical processes (which are known) are most important, nor is it understood what happens when several processes are coupled.

The primary subsystems which are important for earthquake physics are described here. Processes in the upper crust are discussed in detail, since it is often considered to be an elastic body for earthquake studies. There is considerable evidence that thermal and fluid processes also play a significant role in earthquake physics, so the full thermoporoelastic equations are developed. Asthenospheric and mantle dynamics are not described in detail because references to these equations are readily available. Faults are discontinuities in the (continuum) upper crust. A brief discussion of how they may be incorporated into an evolving, complex thermoporoelastic medium is given. The primary purpose of this section is to describe several of the most important subsystems in the earth can be reflected in a hierarchical system of software components which can be used, modified, and combined to investigate complex earth systems.

4.1. Processes in the Upper Crust

The brittle crust is often modeled as an elastic solid for purposes of studying seismic waves, fault mechanics, and rock physics. Hydrologists and geochemists, who are generally interested in fluid dynamics or aqueous chemistry, treat the crust as an immobile porous medium through which fluids flow. Faults can greatly complicate matters for both the mechanical and hydrologic aspects of crustal models. Of course, physical and chemical processes occur simultaneously and in a coupled fashion, resulting in dynamical behavior that is sometimes counter-intuitive, may even appear random, and is difficult to predict.

First, a general set of equations for thermal, mechanical, and hydrologic processes is derived in a form that is intended for computer simulation. That is, general assumptions about coefficients (material properties) are retained. More than just presenting a set of equations, a strategy is given for implementing numerical software that is designed for experimentation. General assumptions about the spatial and temporal variability of material properties are made so as to not dismiss heterogeneous coefficients and nonlinearities that might be important in some circumstances. Though geochemical modeling will not be considered in any detail here, the role of geochemical models in the general modeling framework is indicated. Though the framework outlined is appropriate for general simulations on many scales, the specific intention is to build a simulation framework that can support computational experimentation with fault models and earthquake source mechanics embedded in a complex continuum model of the crust. Detailed presentation of the thermoporoelastic equations are given here because they are often absent from discussions of earthquake physics.

Table 2 lists some of the most common fundamental variables of interest when describing the physical state of the earth's crust. Along with each variable are the simplest (linear, uncoupled) equations which are used to model the variable and the most important material properties that effect its dynamics. Geochemistry is a vast subject and the modeling of geochemical systems is a complex undertaking in itself. Of primary interest for modeling the physical state and evolution of the crust is aqueous geochemistry as an agent for modifying the material properties of the crust, such as permeability and elastic moduli. Porosity is usually considered to be a material property rather than a dynamical variable. However, when the rock material that makes up the crust has evolving properties due to chemical changes, porosity can be considered a fundamental variable which depends primarily on the total solid phase fraction in any given cell volume. Key material properties, such as permeability, thermal diffusivity, and elastic moduli can be derived from porosity, given other parameters which define the quantitative relationship between porosity and other properties. The Kozeny-Carmen rule [*Dullien*, 1992] for computing permeability from porosity is an example. To use this rule, an exponent and a constant parameter must be specified in addition to porosity.

There are many equations which can be specified to govern the evolution of the relatively small set of dynamical variables shown in the preceding table. The basic governing equations listed in the table may be coupled into systems of equations. Or, nonlinear terms may be added to the simplest linear equations. In some situations, the material parameters may be nonlinear functions of the dynamical variables. For some materials, anisotropic elastic moduli may be appropriate. One of the primary tasks that must be accomplished by computer simulation is to determine which equations model the observed behavior of the crust by experimenting with different fundamental equations, couplings between equations, and spatial distributions of material properties. Complex dynamics can emerge from a combination of simple processes, each of which is understood in isolation. Complex fault models do not introduce any new independent variables; rather, faults represent discontinuities in the displacement field. Fault models may be embedded in a complex continuum model. Ideally, one would like to simply specify the particular continuum model at run time, as well as the particular fault model, including friction laws and so on. The means for doing this will be discussed after the continuum equations have been developed.

One of the most important aspects of the crust which is not governed directly by any of the above equations is faulting. Fault movement may alter some or all of the primary variables listed above, as well as the material properties which govern the evolution of the primary variables. From a modeling point of view, faults may be considered geometrical regions that surround a set of surface discontinuities in the continuum. Movement along the fault surfaces are governed by laws distinct from the continuum equations. For example, displacement along boundary points which make up opposing faces of a fault might be displaced by applying a friction law when shear stresses exceed some threshhold. At the same time, thermal conductivity equations might be unaltered by the presence of the fault, except that a heat source is added during sliding. The inclusion of faults in a simulation of the crust will necessarily involve complicated gridding and decision algorithms which will not be discussed in this paper. Generic fault components can, however, be easily included in the general framework presented here and this will be indicated where appropriate. It is important to emphasize that the thermoporoelastic continuum can be treated as a subsystem distinct from, yet tightly coupled to, a fault subsystem. Models designed from this viewpoint allow numerical experimentation with complex friction laws and "standard" continuum models by one group of scientists while another group experiments with evolving continuum properties and its role in the earthquake cycle while assuming simple fault dynamics. This is the flexibility that is sought by the construction of a set of standard earth science software components.

In the following development, we will assume that the material properties are functions of space, time and the state of the system, such as pressure, temperature and stress. Permeability is a tensor, rather than a scalar, since flow anisotropy in fault zones may be significant. Of course, inclusion of these complexities in the equations is certainly not exhaustive. However, the framework presented here is intended to allow for expansion and development as understanding of crustal physics grows.

4.1.1. Mass Balance Equations. A rigorous derivation of equations that can be found in other papers and texts will not be given here. (*Rice and Cleary*, [1976], is a classic paper, for example; *Charlez*, [1998], gives a particularly thorough development of thermoporoelasticity; *Detournay* [1993], and *Wang* [1998], treat isothermal poroelasticity). Rather, a review of those aspects of the derivations that differ from standard treatments is given. This primarily involves coefficients (permeability, porosity, elastic moduli) or state variables (temperature, pore pressure) which are usually assumed to be constant in space and time, restrictive assumptions which will not be made here. Only the case of a single fluid phase which saturates the porous material will be considered in the present study since this is of greatest importance for earthquake studies. Multiphase flow may be of importance in some regions, such as near geothermal sources and perhaps volcanoes.

Darcy's law states that the flux of fluid in the x_i direction is

$$q_i = \phi v_i = -\frac{k_{ij}}{\mu}\left(\frac{\partial p}{\partial x_j} + \rho g_j\right) \tag{1}$$

where q_i is the volumetric flux per unit area, ϕ is the porosity, v_i is the fluid velocity, ρ is the fluid density, g_j is the acceleration vector due to gravity, k_{ij} is the absolute permeability tensor, μ is the viscosity, and p is the fluid pressure. Fluid mass conservation is expressed as the divergence equation:

$$\frac{\partial m}{\partial t} - \rho\frac{\partial q_i}{\partial x_i} = 0 \tag{2}$$

Substituting equation 1 into equation 2 yields an equation for flow in porous media:

$$\frac{\partial m}{\partial t}=\rho\frac{\partial}{\partial x_i}\left(\frac{k_{ij}}{\mu}\left(\frac{\partial p}{\partial x_j}+\rho g_j\right)\right) \quad (3)$$

Viscosity, μ, may depend significantly on the temperature of the fluid, which is spatially variable in general. Hence, the viscosity term must remain inside the spatial derivative.

The time derivative of the fluid mass requires careful evaluation. The fluid mass per unit volume is the product of the fluid density and the local porosity of the material. Since our intent is to simulate crustal behavior on a macroscopic scale, grid blocks, which define elementary volumes in simulations, will be assumed to be large enough to treat porosity and other material properties as well-defined continuum variables.

Fluid density will be assumed to depend on pressure and temperature. It may also depend on the concentration of dissolved species, but in this paper chemical species concentrations are treated as given quantities, either specified constant values or derived from an unspecified simulator.

Porosity will depend on the fundamental state variables pressure, temperature, and hydrostatic stress. It can also vary with time due to inelastic pore volume changes associated with material deposition and dissolution or plastic deformation. In functional notation,

$$\rho=\rho(p,T) \quad (4)$$

$$\phi=\phi(p,T,\sigma,t)=\tilde{\phi}(p,T,\sigma)\,\phi_0(t) \quad (5)$$

Note that the pore volume has been separated into a time-dependent porosity ϕ_0,which is assumed to vary by chemical changes and inelastic deformation and a multiplicative scaling porosity that is dependent upon pressure, temperature, and hydrostatic stress. In the relaxed state, that is, when $p-p_{ref}$, $T-T_{ref}$, and $\sigma-\sigma_{ref}$ are zero, $\tilde{\phi}$ will be equal to one.

The mass derivative with respect to time may be written

$$\frac{\partial m}{\partial t}=\frac{\partial(\rho\phi)}{\partial t}=\phi\frac{\partial\rho}{\partial t}+\rho\left(\phi_o\frac{\partial\tilde{\phi}}{\partial t}+\tilde{\phi}\frac{\partial\phi_0}{\partial t}\right)$$

$$=\phi\frac{\partial\rho}{\partial t}+\rho\phi_o\frac{\partial\tilde{\phi}}{\partial t}+\rho\tilde{\phi}\left(\frac{\partial\phi_0}{\partial t}\right)_{inelastic} \quad (6)$$

The pore volume scaling factor, $\tilde{\phi}(p,T,\sigma)$, will generally be close to one for matrix porosity, but may vary over a considerable range in fractured media. In fact, if the anisotropic compressibility of fractures is to be considered, the full stress tensor rather than just the hydrostatic stress must be considered in order to evaluate the porosity change due to stress. In this case, the porosity (and consequently the permeability) will depend on the orientation of the stress field. There are many situations in which this will be significant for crustal dynamics [*Zoback*, 1992]. In this situation, quantities derived from porosity, for example permeability, will depend on other parameters besides porosity, such as the stress orientation, pore pressure, mean fracture orientation, and so on. This level of complexity can be incorporated by making the material properties (coefficients in the equations) functions of the state variables. The role of inelastic porosity changes in generating abnormally high pore pressure and its consequences for tectonic processes was explored by *Walder and Nur* [1984].

Fluid density depends only on the pressure and temperature of the fluid; note that temperature and pressure are time-dependent, however. When dissolved minerals are present, the density may also be a function of aqueous species concentration. It will be assumed here that chemical concentrations are low enough that their effect on fluid density can be ignored. There are many cases when this is not true, for instance, when sea water intrudes a freshwater aquifer. But this is not important for tectonic processes. Thus, the time derivative for fluid density is:

$$\frac{\partial\rho}{\partial t}=\left(\frac{\partial\rho}{\partial p}\right)\frac{\partial p}{\partial t}+\left(\frac{\partial\rho}{\partial T}\right)\frac{\partial T}{\partial t} \quad (7)$$

If the coefficients in parentheses are taken to be constants we have

$$\frac{\partial\rho}{\partial p}=\beta_\rho\rho \quad (8)$$

$$\frac{\partial\rho}{\partial T}=\alpha_\rho\rho \quad (9)$$

Similarly, the time derivatives for ϕ may be expanded as

$$\frac{\partial\phi}{\partial t}=\left(\frac{\partial\phi}{\partial p}\right)\frac{\partial p}{\partial t}+\left(\frac{\partial\phi}{\partial T}\right)\frac{\partial T}{\partial t}+\left(\frac{\partial\phi}{\partial\sigma}\right)\frac{\partial\sigma}{\partial t}+\frac{\partial\phi_0}{\partial t} \quad (10)$$

In many situations, it is correct to model the variation of porosity with pressure, temperature, and hydrostatic as constants also:

$$\frac{\partial\phi}{\partial p}=\beta_\phi\phi \quad (11)$$

$$\frac{\partial \phi}{\partial T} = \alpha_\phi \phi \quad (12)$$

$$\frac{\partial \phi}{\partial \sigma} = \gamma \phi \quad (13)$$

The elastic constants in equations 11 through 13 may be used in equation 6 and equation 10 to yield

$$\frac{\partial \phi}{\partial t} = \phi \left(\beta_\phi \frac{\partial p}{\partial t} + \gamma \frac{\partial \sigma}{\partial t} + \alpha_\phi \frac{\partial T}{\partial t} \right) + \widetilde{\phi} \left(\frac{\partial \phi_0}{\partial t} \right)_{in} \quad (14)$$

Combining these expressions, we get:

$$\frac{\partial m}{\partial t} = \rho \phi \left(b \frac{\partial p}{\partial t} + \gamma \frac{\partial \sigma}{\partial t} - (\alpha_\phi + \alpha_\rho) \frac{\partial T}{\partial t} \right) + \rho \left(\frac{\partial \phi}{\partial t} \right)_{in} \quad (15)$$

where $b = \beta_\phi + \beta_\rho$. The value of these parameters in terms of commonly measured elastic and hydraulic parameters will be examined later. Note that the variation of porosity due to inelastic pore space evolution in equation 15 uses the fact that $\widetilde{\phi} \frac{\partial (\phi_0)}{\partial t} = \frac{\partial (\phi)}{\partial t}$.

Following the derivation in *Charlez* [1991] , chapter 8, the coefficients in equation 15 can be identified in terms of common material parameters. Let

$$c = \frac{9\mu (\nu_u - \nu)}{2GB^2 (1 + \nu)(1 + \nu_u)} \quad (16)$$

Substituting equation 1 and equation 15 into equation 2 yields

$$c \left(\frac{\partial p}{\partial t} + B \frac{\partial \sigma}{\partial t} \right) - a \frac{\partial T}{\partial t} = \frac{\partial}{\partial x_i} \left(\frac{k_{ij}}{\mu} \left(\frac{\partial p}{\partial x_j} + \rho g_j \right) \right) - BK_u \left(\frac{\partial \phi}{\partial t} \right)_{in} \quad (17)$$

where $a = (1 - \nu)(\alpha_d - \alpha_u) / (1 + \nu_u)$, α_d and α_u are the drained and undrained thermal expansion coefficients, respectively, K_u is the undrained bulk modulus, and other parameters are as before. The bulk modulus can be written in terms of the shear modulus and Poisson ratio, parameters that have already been introduced:

$$K_u = \frac{2G (1 - \nu_u)}{3 (1 - 2\nu_u)} \quad (18)$$

To allow for evolution of the pore space to modify the coefficients, a useful form for the parameters c is that given in Charlez (1997):

$$c = \phi \{ C_f + C_r \} \quad (19)$$

where C_f and C_r are the fluid and rock compressibilities, respectively, and a in *McTigue* [1986]:

$$a = \phi (\alpha_f - \alpha_s) \quad (20)$$

where α_f and α_s are the fluid and rock thermal expansion coefficients, respectively. The fluid compressibility can be assumed to be constant. For small porosity changes, the rock compressibility may also be assumed constant, though an accurate representation requires that C_r will vary over time as the rock evolves inelastically.

Two more equations are needed to resolve the system. These are derived from the strain compatibility conditions and from energy balance (that is, thermal advection-diffusion equation). It should be especially noted that we have not assumed that the permeability and elastic coefficients are constant. In the most general case, these coefficients may be spatially variable, may be nonlinear functions of pressure, temperature, and stress, and may change over time due to chemical alteration of the porous structure. The two additional equations required are derived below.

4.1.2. Heat Transport. A full treatment of the thermodynamics of fluids should include internal energy changes, mechanical work performed by the fluid, and work against friction arising from the viscosity of the fluid. For many tectonic problems, it is sufficient to represent some of the material properites by appropriate averages between solid (mineral) and fluid phases. The energy balance equation for thermal advection-diffusion of a single phase fluid is

$$\frac{c_e}{\rho c_p} \frac{\partial T}{\partial t} - \rho_f c_{pf} q_i \frac{\partial T}{\partial x_i} - \frac{\partial}{\partial x_i} \left(D_{ij} (\mathbf{x}) \frac{\partial T}{\partial x_j} \right) = Q_T (\mathbf{x}) \quad (21)$$

where c_e, ρ, c_p are the *effective* thermal conductivity, density, and specific heat, respectively, which may be taken to be the volume-fraction weighted averages of fluid and mineral properties. ρ_f is the fluid density, c_f is the specific heat of the fluid, D_{ij} is the effective thermal diffusivity tensor of the saturated porous medium, $Q_T (\mathbf{x})$ is a heat source and q_i is the Darcy velocity which may be found from equation 1. The latter term couples thermal advection to equation 17. More precise descriptions of the thermal diffusivity are possible if the medium is considered to be a multiphase system where solid and fluid phases are present. Each phase, solid or fluid, then has a distinct thermal diffusivity and the mass fraction or saturation of each phase is a dependent variable that changes with time as flow proceeds and as chemical reactions alter the solid matrix. The thermal diffusivity will, in general, be spatially variable. Thus, it is important to keep the diffusivity parameter within the outer spatial derivative in equation 21.

To complete the system of partial differential equations, we need to introduce a constraint on the elastic deformation that ensures that the strains will be continuous across the domain. The derivation of this equation follows from the compatibility equations [*Rice and Cleary*, 1976; *Charlez*, 1991; *Charlez*, 1998] and has the form of a Poisson equation:

$$\frac{\partial^2}{\partial x_j^2}\left(\sigma + c_1 p + c_2 T\right) = 3g\left(\phi\rho + (1-\phi)\,\rho_r\right) \tag{22}$$

where the constants c_1 and c_2 may be expressed in terms of commonly-known material parameters by:

$$c_1 = \frac{2\left(\nu_u - \nu\right)}{B\left(1-\nu\right)\left(1+\nu_u\right)} \tag{23}$$

$$c_2 = \frac{4G\left(1+\nu\right)}{9\left(1-\nu\right)}\alpha_T \tag{24}$$

α_T is the effective bulk thermal expansion coefficient for the fluid-saturated medium in the drained case. The body force term on the right side of the equation is the gravitational force on the mean density of rock mass and fluid mass. ρ is the fluid density and ρ_r is the rock density. The latter may be spatially variable.

4.1.3. Elastic Deformation. Elastic deformation may be computed from the stress equilibrium equations with pore pressure and temperature as applied forces (equation 26). If the fully coupled poroelastic diffusion equations are used, as described above, the mean stress derived from solving the equilibrium or Navier equations (equations 26) will be exactly equal to the mean stress that is evolved in the coupled poroelastic equations, equations 17 and 22. It is computationally efficient to solve the coupled poroelastic equations and the stress equilibrium equations in this decoupled manner. However, this approach assumes that boundary and initial conditions can be assigned to p, σ and T. If boundary conditions for the mean stress cannot be assigned, or if displacement boundary conditions are required at some grid points, such as explicitly modeled fault surfaces, then the poroelastic equations for p and T must be solved simultaneously with the three equations for displacement in 26. The mean stress in this case becomes a derived parameter:

$$\sigma = \frac{K_B}{3}\varepsilon_{kk} = \frac{K_B}{3}\frac{\partial u_k}{\partial x_k} = \left(\lambda + \frac{2}{3}G\right)\frac{\partial u_k}{\partial x_k} \tag{25}$$

where K_B is the bulk modulus of the medium.

The three equations for the displacements in terms of pore pressure and temperature are:

$$\frac{\partial}{\partial x_i}\left(\left(\lambda + G\right)u_{k,k}\right) + \frac{\partial}{\partial x_k}\left(G u_{i,k}\right) = \\ \frac{\partial}{\partial x_i}\alpha\left(p - p_{ref}\right) - \frac{\partial}{\partial x_i}\alpha_T\left(T - T_{ref}\right) + \rho_r\left(\mathbf{x}\right)g_i \tag{26}$$

where u_i are the three components of the displacement vector, λ is the Lame coefficient, and G is the shear modulus. Note that both of these may be spatially variable. The Biot-Willis parameter, α, is a function of Skempton's coefficient and the drained and undrained Poisson's ratios: $\alpha = 3\left(\nu_u - \nu\right) / \left(B\left(1-2\nu\right)\left(1+\nu_u\right)\right)$. α_T is the effective bulk thermal expansion coefficient for the fluid-saturated medium as above. The right side of equation 26 is expressed in terms of the deviation of the pore pressure from a reference value, p_{ref}. The gravitational force vector, g_i, will have only a z-component. ρ_r is the density of the fluid-saturated rock at $\mathbf{x}$. If the independent variables are *deviations* from a reference value the gravitational force terms will be zero. Note that the elastic moduli are inside of the spatial derviatives.

Only quasi-static deformation has been discussed. For studies of crustal stress evolution on scales greater than an hour the quasi-static equations are sufficient. When dynamic effects are important, such as during fault rupture, a time derivative term must be included in equations 26 and these must be fully coupled to equations 29 and 31; the equilibrium equation for mean stress, equation 30, cannot be used because this was derived from an assumption of static stress equilibrium.

4.1.4. Summary of Thermoporoelastic Equations. The set of general equations for a thermoporoelastic material with six unknowns, p, σ, T, and the three components of displacement u_i, are as follows:

$$c\left(\frac{\partial p}{\partial t} + B\frac{\partial \sigma}{\partial t}\right) - a\frac{\partial T}{\partial t} = \\ \frac{\partial}{\partial x_i}\left(\frac{k_{ij}}{\mu}\left(\frac{\partial p}{\partial x_j} + \rho g_j\right)\right) - BK_u\left(\frac{\partial \phi}{\partial t}\right)_{in} \tag{27}$$

$$\frac{c_e}{\rho_e c_{pe}}\frac{\partial T}{\partial t} = \rho c_{pf} q_i \frac{\partial T}{\partial x_i} + \frac{\partial}{\partial x_i}\left(D_{ij}(\mathbf{x})\frac{\partial T}{\partial x_j}\right) + Q_T \quad (28)$$

$$\frac{\partial^2}{\partial x_j^2}\left[\sigma + \left(\frac{2(\nu_u - \nu)}{B(1-\nu)(1+\nu_u)}\right)p\right] + \frac{\partial^2}{\partial x_j^2}\left[\left(\frac{4G(1+\nu)}{9(1-\nu)}\alpha_T\right)T\right] = 3g(\phi\rho + (1-\phi)\rho_r) \quad (29)$$

$$\frac{\partial}{\partial x_i}((\lambda + G)u_{k,k}) + \frac{\partial}{\partial x_k}(Gu_{i,k}) = \frac{\partial}{\partial x_i}\alpha(p - p_{ref}) - \frac{\partial}{\partial x_i}\alpha_T(T - T_{ref}) + \rho_r g_i \quad (30)$$

where

$$q_i = -\frac{k_{ij}}{\mu}\left(\frac{\partial p}{\partial x_j} + \rho g_j\right) \quad (31)$$

in equation 28. These may be solved as two systems of equations, equations 27-29 as one set and equations 30 as another, each with three unknown variables. The strain tensor and stress tensor may be computed as derived quantities from the above equations using

$$\varepsilon_{ij} = \frac{1}{2}\left(\frac{\partial u_i}{\partial x_j} + \frac{\partial u_j}{\partial x_i}\right) \quad (32)$$

$$\sigma_{ij} = 2G\varepsilon_{ij} + \left(\left(K_B - \frac{2G}{3}\right)\frac{\partial u_k}{\partial x_k}\right)\delta_{ij} \quad (33)$$

Alternatively, when displacement boundary conditions are required, a single system of equations with five unknowns, equations 27, 28, and 30 are solved, with equation 25 used to compute the derived parameter σ instead of equation 29.

The full set of coupled partial differential equations may only be solved in its present general formulation by numerical approximation on a computer. Various assumptions allow simpler formulations of the thermoporoelastic problem which admit analytical solutions or straight-forward numerical solutions analogous to heat diffusion problems. Because the permeability is not assumed to be constant, equations 27 and 29 cannot be combined in the manner done by *Rice and Cleary* [1976] and others cited previously to give an evolution equation in a single variable $m = p + B\sigma$.

4.1.5. Material Properties. The mechanical and hydrologic properties of porous rocks may be modified quite rapidly on a geologic time scale by the flow of water through the pores. Evolution of rock properties is invoked as an essential process in some theories of the earthquake cycle [*Byerlee*, 1996; *Rice*, 1992]. Porosity is the primary variable in that is modified, with other variables properly determined from the evolving pore structure: permeability, the elastic modulii, and most of the other parameters are functions of the porosity of the material. Temporally and spatially variable material properties can be computed as a separate component in the simulation and updated in the system of continuum equations and the fault model. In this way, the different pore evolution models can be implemented by different researchers using standard continuum and fault components. This will be illustrated in more detail below.

4.2. Numerical Simulation of the Upper Crust

A thorough treatment of all of the numerical techniques needed to solve a complex set of coupled partial differential equations is beyond the scope of this paper. A brief overview of a finite element approach to the solution of the thermoporoelastic equations given above, equations 29-31, is given here. The Navier equations for static elastic deformation, equations 32, are standard and can be found in many texts, such as *Pande, et al.* [1990]. *Langtangen* [1999] contains an excellent discussion of finite element discretization and related numerical mathematics issues for scientists and engineers who are primarily interested in using the methods rather than proving theorems about the methods.

4.2.1. Finite Element Spatial Discretization. We first write the general set of equations in the following form:

$$a_1\frac{\partial(P + B\sigma + a_2T)}{\partial t} = \frac{\partial}{\partial x_i}\left[\frac{k_{ij}}{\mu}\frac{\partial}{\partial x_j}P\right] + f \quad (34)$$

$$\frac{\partial^2}{\partial x_i^2}(b_1P + \sigma + b_2T) = 0 \quad (35)$$

$$c\frac{\partial T}{\partial t} = q_i\frac{\partial T}{\partial x_i} + \frac{\partial}{\partial x_i}\left(d_{ij}\frac{\partial T}{\partial x_j}\right) \quad (36)$$

where a_1, a_2, f, b_1, b_2, c, and d_{ij} are simply shorthand notations for the coefficients in equations 27, 28, and 29.

The finite element discrete form of these equations is derived by approximating the unknown functions P, σ and T by a linear sum of basis functions. Here we use the method of weighted residuals or Petrov-Galerkin formulation *Helmig* [1997], where the weighting functions W_i and the basis functions N_i may be different. Petrov-Galerkin elements are used primarily to control numerical dispersion, which may be useful in the thermal advection terms. In the standard Galerkin finite element formulation, $N_i = W_i$.

$$p \approx \sum_{i=1}^{n} p_i(t) W_i \tag{37}$$

$$\sigma \approx \sum_{i=1}^{n} s_i(t) W_i \tag{38}$$

$$T \approx \sum_{i=1}^{n} \tau_i(t) W_i \tag{39}$$

where n is the number of nodes, W_i are the finite element weighting functions and the p_i, s_i, τ_i are unknown values to be determined. A spatially discrete weak form of the system of equations is given by

$$a_1 \mathbf{M}_{ij} \frac{d(p_j + Bs_j + a_2\tau_j)}{dt} + \mathbf{K}_{ij} p_j = d_i \tag{40}$$

$$\mathbf{B}_{ij}(b_1 p_j + s_j + b_2 \tau_j) = \gamma_i \tag{41}$$

$$c\mathbf{M}_{ij} \frac{d\tau_j}{dt} + \mathbf{D}_{ij}\tau_j = Q_i \tag{42}$$

where

$$\mathbf{M}_{ij} = \int_\Omega W_i N_j d\Omega \tag{43}$$

$$\mathbf{K}_{ij} = \int_\Omega \sum_{l=1}^{3}\sum_{k=1}^{3} \left(k_{kl}(\mathbf{x}) \frac{\partial W_i}{\partial x_k}\frac{\partial N_j}{\partial x_l} \right) d\Omega \tag{44}$$

$$\mathbf{D}_{ij} = \int_\Omega \sum_{k=1}^{3} \left[\left(\sum_{l=1}^{3} d_{kl}(\mathbf{x}) \frac{\partial W_i}{\partial x_k}\frac{\partial N_j}{\partial x_l} \right) + W_i q_k \frac{\partial N_j}{\partial x_k} \right] d\Omega \tag{45}$$

$$d_i = \int_\Omega f W_i d\Omega + \int_{\partial\Omega} g W_i d\Gamma \tag{46}$$

$$Q_i = \int_\Omega Q_T W_i d\Omega + \int_{\partial\Omega} q_T W_i d\Gamma \tag{47}$$

$$\mathbf{B}_{ij} = \int_\Omega \left(\frac{\partial W_i}{\partial x_k}\frac{\partial N_j}{\partial x_l} \right) d\Omega \tag{48}$$

$$\gamma_i = \int_\Omega \rho g W_i d\Omega \tag{49}$$

and the functions N_i are the finite element trial functions. Prescribed flux boundary conditions are the functions represented by g and g_T for pressure and temperature flux conditions. Mathematically, a fluid flux boundary condition, in the direction of $\hat{n}$, is

$$q_n = \sum_{l=1}^{3}\sum_{k=1}^{3} \left(k_{kl} n_k \frac{\partial p}{\partial x_l} \right) \tag{50}$$

The prescribed heat flux has similar definition. In a computer code, g and q_T are defined flux values at the boundaries or at internal source points.

4.2.2. Time Discretization. Temporal discretization is accomplished by the *theta* method:

$$\frac{u^r - u^{r-1}}{\Delta t} = \theta g^r + (1-\theta) g^{r-1} \tag{51}$$

When $\theta = 1$, for example, we recover the backward Euler approximation. $\theta = \frac{1}{2}$ gives the Crank-Nicholson scheme. The theta method can formally be applied to equation 40 by writing it in the form

$$\frac{d(p_j + Bs_j + a_2\tau_j)}{dt} = \frac{1}{a_1}\mathbf{M}^{-1}(\mathbf{d} - \mathbf{Kp}) \tag{52}$$

4.2.3. Linear and Nonlinear Solution Methods. Applying the theta scheme, we obtain a linear system for the unknowns at the time level r:

$$\mathbf{A}\mathbf{u}^r = \mathbf{c} \tag{53}$$

where

$$A_{ij} = \begin{pmatrix} a_1\mathbf{M}_{ij} + \theta\Delta t\mathbf{K}_{ij} & Ba_1\mathbf{M}_{ij} & a_1a_2\mathbf{M}_{ij} \\ b_1\mathbf{B}_{ij} & \mathbf{B}_{ij} & b_2\mathbf{B}_{ij} \\ 0 & 0 & c\mathbf{M}_{ij} + \theta\Delta t\mathbf{D}_{ij} \end{pmatrix} \tag{54}$$

$$u_j^r = \begin{pmatrix} p_j^r \\ s_j^r \\ T_j^r \end{pmatrix} \quad (55)$$

$$c_i = \begin{pmatrix} a_1 \mathbf{M}_{ij} u_j^{r-1} + (\theta - 1)\, \Delta t \mathbf{K}_{ij} p_j^{r-1} + \Delta t d_i \\ \gamma_i \\ c\mathbf{M}_{ij} u_j^{r-1} + (\theta - 1)\, \Delta t \mathbf{D}_{ij} p_j^{r-1} + \Delta t Q_i \end{pmatrix} \quad (56)$$

The linear system in 53 becomes nonlinear when the coefficients, such as the permeability tensor, depend on pressure, stress, or temperature, or evolve in time. The system of nonlinear equations may be solved using a Newton-Raphson iteration. In this method, the quantity $\mathbf{c} - \mathbf{A}\mathbf{u}^r$ is minimized using the the Jacobian (derivative) matrix of $\mathbf{c} - \mathbf{A}\mathbf{u}^r$. Details can be found in texts on numerical optimization such as *Dennis and Schnabel* [1983] and *Langtangen* [1999].

Equation 30 can be discretized in a similar manner. The resulting linear system is

$$\mathbf{A}\mathbf{x} = \mathbf{b} \quad (57)$$

where $\mathbf{x}$ is the vector of displacements, x = (u_1, v_1, w_1, u_2, v_2, w_2, ..., u_n, v_n, w_n). $\mathbf{A}$ is a 3n by 3n nonsymmetric matrix, which can be partitioned into n blocks of size 3×3 each. The general formula for block $\mathbf{A}_{ij}$, which represents the coupling between node i and j, is

$$\mathbf{A}_{ij} = \begin{matrix} \mathbf{a}_{ij}^{11} & \cdots & \mathbf{a}_{ij}^{13} \\ \vdots & \ddots & \vdots \\ \mathbf{a}_{ij}^{31} & \cdots & \mathbf{a}_{ij}^{33} \end{matrix} \quad (58)$$

and

$$\mathbf{a}_{ij}^{rs} = \int_\Omega \left[G \left(\sum_k N_{i,k} N_{j,k} \right) N_{i,s} N_{j,r} + \lambda N_{i,r} N_{j,s} \right] d\Omega \quad (59)$$

The corresponding partitioning for $\mathbf{b}$ is

$$\mathbf{b}_j = - \begin{pmatrix} \beta_j^1 \\ \beta_j^2 \\ \beta_j^3 \end{pmatrix} \quad (60)$$

where

$$\beta_j^r = \int_\Omega \left[(2G + 3\lambda) + \alpha_T \left(T - T_{ref} \right) N_{j,r} + \alpha \left(p - p_{ref} \right) \right] d\Omega \quad (61)$$

4.3. Faulting Processes

Frictional processes that control slip on fault segments are not well understood. The physical processes associated with faulting occur on many timescales, from the long-term evolution of fault zone state variables (stress, fluid pressure, temperature) and material properties, to frictional processes which control dynamic rupture on a scales of seconds. Many simplified models of fault behavior have been proposed and studied. Many of these are variations of the slider-block model originally developed by *Burridge* [1967] or some form of interacting point-faults [*Rundle*, 1998]. The interaction between faults is governed by the intervening continuum, which may be modeled as a simple homogeneous elastic material or as an evolving, heterogeneous material as discussed previously. Similarly, faults may be modeled as simple point sources, or as sliding surfaces governed by complicated friction laws that include state variables (stress, pore pressure, temperature). For computational experimentation, it will be important to abstract the essential features of a fault model from the details of a particular fault model.

The study of earthquakes might be reduced to two basic problems, one a long-time process and one a short-time processes: (1) what are the long-term forces and processes that lead to seismic rupture of faults; and (2) once a fault begins to rupture, what governs the rupture length and how do waves propagate in the surrounding complex medium. This paper is primarily concerned with developing models to study the first problem, so our brief discussion of faults is in this context.

4.3.1. Fault Models. There are two primary methods for modeling the interaction of faults embedded in a continuum background: semi-analytic or Green's function or Boundary Element methods, and finite element discretization with the fault modeled as an internal boundary. Semi-analytic or boundary element methods may be grouped together because they are mathematically very similar. The essence of these methods is the representation of faults as point sources embedded in an infinite domain. The influence at any location in the domain of all of the point sources can be determined by linear superposition of each source. The advantage of these methods is two-fold. First, faults are easily defined by their location and orientation in space. Second, if the goal of a computation is to calculate stress changes at only a limited number of points, such as at the locations of all other faults, there is an apparent savings in computational effort since values do not have to be computed at extraneous grid points.

The drawback of boundary element methods is that fundamental analytical solutions (Green's functions or influence functions) must be determined for the medium. This makes it impossible to model complex thermoporoelastic media with evolving material properties, since the influence functions cannot be determined.

Finite element methods require that faults be explicitly modeled by grid points. The advantage of this approach is that the medium in which the faults are embedded can include realistic complexity. Heterogeneous and evolving material properties, pore fluid flow, and heat dissapation can all be included. Complicated friction laws to define the interaction between the planes that make up the fault segment surfaces can be devised and studied. Realistic moving (plate) boundaries can be simulated. The cost of this greater realism is much greater computational effort due to the need for many more grid points to represent the domain and embedded faults.

4.3.2. Computational Effort. When many faults are involved, it is not clear that a semi-analytic method gives numerical advantages. The solution of a boundary element problem where, for example, nf faults are represent by a single point each, requires that a dense linear system of n_f equations in nf unknowns be solved. The solution of such a system requires $O\left(n_f^2\right)$ operations. New approaches (such as the fast multipole method used for astrodynamical problems) may reduce the computational effort to $O(n_f \log(n_f))$. Although the number of grid points for a finite element discretization may be larger, this can be minimized by using variable meshes. The matrices that result from finite element discretizations are sparse. Methods for sparse systems can converge in perhaps $O(N \log N)$ operations, where N is the number of grid points. Multigrid methods scale linearly; that is, they converge in $O(N)$ operations. Because finite element methods allow the seismogenic upper crust to be treated as a more complex material than boundary element methods, the development of appropriate solvers for thermoporoelastic problems may make this approach much more attractive for studies of earthquake physics.

It should be mentioned that mathematical methods for the types of problems that will arise in earthquake simulations may not yet exist. Multigrid methods are not generic solvers as, for example, iterative methods such as conjugate gradients. One of the problems with the irregular grids that must be employed for explicit fault modeling is that the notion of grid refinement is not clearly defined. The composite grid model discussed here may introduce a method for defining grid refinement. The background continuum grid, which may be assumed to be rectangular, and the embedded fault zones may each be refined separately before combining. Such mathematical details require further research by numerical mathematicians and may be of great benefit to computational geoscience research.

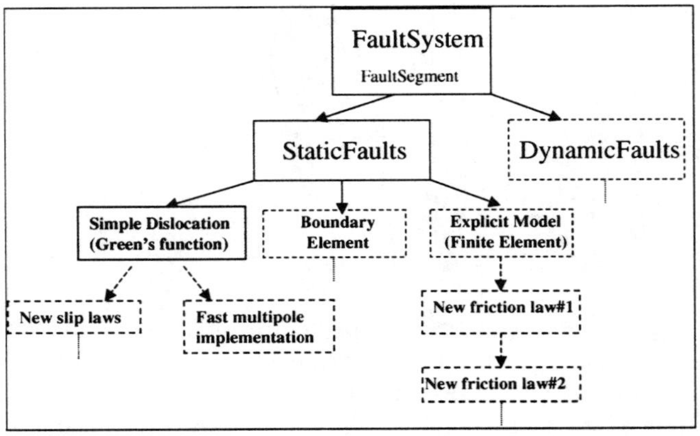

Figure 1. Schematic diagram of a fault system hierarchy.

4.3.3. Software Implementation. A hierarchical object-oriented paradigm is particularly well-suited to computational experimentation with many fault models, on many timescales, by many different researchers. Pre-existing faults may be treated as an array of planar segments having a particular orientation and location in space, embedded in a continuum thermoporoelastic background material. Each planar segment may be referred to as a *fault segment*. A *fault* is then simply an array of fault segments. A *fault system* is the total collection of faults in the entire simulation region. On timescales much longer than the rupture time of an earthquake event, the essential feature of fault behavior is that its strength and stress state evolve until a rupture event. The rupture event causes an instantaneous change in the stress state of the surrounding medium. This is the behavior that must be captured in a general earthquake model. In figure 1 a class hierarchy is illustrated for modeling faults in a continuum. By implementing generic fault behavior in the base class, a mechanism is available for experimenting with and comparing many different fault models in the context of different lithospheric models.

Implementation of fault dynamics must include the dual nature of the faults as objects with their own dynamics and as part of a continuum. That is, faults exhibit both continuum dynamics, including fluid flow,

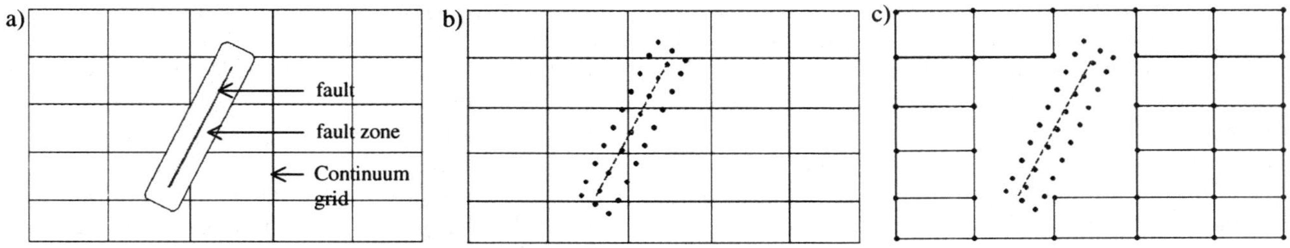

Figure 2. Uniform continuum grid with embedded fault segment. a) background grid and fault; b) gridded fault segment and surrounding fault zone imposed upon background grid; c) resulting composite grid. Dashed line in b) and c) denotes internal boundary points that make up the fault segment.

heat transport, and non-continuum dynamics. The latter primarily involves the rupture process, which may be modeled as movement along planar segments according to prescribed friction laws or other models that govern slip movement. In general, the timescale for continuum behavior is much longer than the timescale for fault rupture. This separation of scales may be exploited in the computer simulation of the earthquake cycle.

For semi-analytic fault models, continuum state variables are defined explicitly for fault zones, but may be interpolated from the background continuum grid. When a rupture event occurs on a fault segment, it's effect on the regional state variables (particularly stress) is computed and added to the pre-event state variables.

The continuum aspects of a fault system may be incorporated into a simulation by adding grid points to the simulation domain. Fault segments can be represented by any number of grid points. Addition of these grid points to the continuum domain can be accomplished by forming a composite grid. *Fremming* [1997] discusses the details of building composite grids for modeling faults. Figure 2 illustrates the idea of a composite grid for fault modeling.

Fault rupture occurs when the shear stress along a segment of the fault exceeds the frictional forces that prevent slip. The most common such "slip criterion" is the Coulomb criterion:

$$\tau_c = \tau_s + \mu_f (\sigma_n + p) \tag{62}$$

where τ_c is the shear stress on the fault, μ_f is the coefficient of internal friction or simply the frictional strength, σ_n is the normal stress on the fault (compression negative), and p is the pore pressure in the fault. This criterion can be made as simple or as complex as desired by the researcher. The general design presented here is intended to abstract general behavior into generic software components from which custom plug-and-play components may be easily derived. In this way, the scientist may concentrate on the consequences of the physical model, not code implementation. For a general model of faults it is not important which slip criterion is specified. Each fault segment must be able to compute from state variables and parameters whether or not its slip criterion has been met. For example, as the regional continuum model evolves in time, at each time step the relevant state variables are given to each fault segment and the segment makes a determination as to whether or not its slip criterion is exceeded. If not, the fault segment does nothing and the continuum model continues to evolve. The slip criterion may be different for each fault or even each fault segment. When the slip criterion for a segment is exceeded, the particular model for that fault segment takes control of the simulation.

Slip on a defined fault segment may completely ignore the details of the rupture process and only compute the stress perturbation that results from the rupture. For this case, an elastic dislocation algorithm, using an undrained Poisson ratio to account for the presence of fluids, may be used to compute changes to the regional stress and pore pressure fields. The perturbed stress field and other variables may be used then to check the slip criterion on other fault segments, and so on until all fault rupture ends. Small earthquakes or large earthquakes may result on any particular fault (that is, an array of fault segments) depending on the friction law, pore pressure and stress changes, and other factors that are included in the fault slip model. The number of segments in a single fault that rupture sequentially determines the size of the earthquake event. Computational experimentation may be used to examine the consequences of choosing various physical models.

After a single earthquake event, which may involve a complicated simulation in itself, control then passes back to the continuum field and it continues to evolve. The effects of slip on a defined fault segment can also be computed by solving the continuum equations with the fault segment composite grid points treated as boundary points with a fixed displacement. Both method for computing stress transfer during slip events may be implemented and either chosen at run time.

The regional stress field can also be updated dynamically as as fault segment ruptures. In this case, the only difference from the previous case is that the time from rupture onset to completion on a single segment is divided into many time increments. A continuum model which includes dynamic terms must also be used. The time-stepping is controlled by the fault segment during rupture. At each time step, an updated stress field is computed. The status of other fault segments can be checked at any time in the time stepping. Methods for computing stress changes during dynamic rupture, such as Green's function methods and fast multipole methods for implementing Green's function methods can be implemented as subclasses of the fault segment class and selected at runtime.

The frictional forces that control slip on fault segments may be best represented by microscale models. Models that are similar to molecular dynamics methods used in materials science [*Pechenik et al.*, 1999] have been proposed to study heat generation and fault surface interaction during rupture [See also *Place and Mora*, 1996]. Methods have been developed to allow coupling of molecular dynamics simulations to finite element methods for crack propagation problems [*Abraham*, 1998]. Such hybrid methods may represent the next generation of realistic fault models.

In summary, embedding faults into a continuum can be accomplished in a way that does not imply any particular fault rupture model. It provides a method for numerical experimentation with various fault models and friction laws within the context of a complex continuum. A fault system can be modeled as a collection of faults, each of which is an array of fault segments. Fundamental friction laws, slip criteria and methods for updating stress due to fault slip are contained in the various fault segment models that are implemented. Using object-oriented methods to build virtual fault classes will enable researchers to implement their own physical models quickly as subclasses of the fault class and use them in the context of a complex faulted continuum simulation model.

4.4. Geochemical Processes

Geochemistry is a vast subject and will not be dealt with here, though the component design of a general crustal dynamics model presented here can readily be extended to include coupling to a reactive chemistry model. If a geochemistry software component was available, it would take as input from the hydromechanical continuum component the state variables (fluid pressure, temperature, stress) at a given time and integrate chemical concentrations ahead to the next time step. The output from the chemistry model is taken as input to the thermoporoelastic model, most likely as a new solid phase fraction which modifies the porosity. Pressure, mean stress, and temperature are then integrated forward in time, and so on. Porosity and fluid velocity values may be shared by each component as the whole system evolves in time. If a geochemical code is available, sequential coupling to the hydromechanical model in the manner presented here should be straightforward. Extensive modeling capabilities for the evolution of crustal properties by hydrothermal geochemistry have been developed by many groups, for example, *Bolton, et al.* [1996]. These capabilities may be incorporated into the framework described.

4.5. Viscoelasticity and Mantle Processes

Earthquakes are primarily a phenomenon of the brittle crust. The brittle plates are driven at their lower boundary by movement of the viscous mantle. A viscoelastic component can be built using the same object technology outlined above and integrated into a general crustal dynamics simulation. Though it is not clear that a computer yet exists that could run a full mantle convection code coupled to a faulted poroelastic crust with geochemical processes, the design presented here allows for easy extension and plug-and-play numerical experimentation. Equations for modeling mantle convection are presented in several articles in *Peltier* [1989]. Numerical methods for discretizing viscoelastic equations are covered in *Pande, et al.* [1990] and *Langtangen* [1999].

5. A FRAMEWORK FOR EARTH MODELING

Scientific computing involves three aspects, all of which are important when the computer is used as an experimental research tool. These are mathematical formulation of the physical processes, selection of appropriate numerical methods, and robust, maintainable, and extensible software implementation of the mathematical and numerical concepts. Too often the latter aspect is not given enough attention. The result is that, at best, much time is wasted in trying to maintain or adapt complex codes. Even worse, erroneous results creep in due to unmanagable code complexity and undetected bugs. New software design methodologies are developing to meet these needs. Though they are well-developed in some applications, the new ideas have been slower to take hold in scientific and engineering codes. An attempt is made here to sketch out an object-oriented design of a set of software components that may be flexibly used for complex crustal simulation experiments.

5.1. Software Organization for Geocomputing

One of the greatest benefits to be realized by adopting a software component design strategy that the work of many researchers working in scattered locations can be coordinated and synchronized. Once a design is adopted by the community for a particular class, anyone who wants to incorporate a new physical model for that class need only wrap his or her code in the standard interface to make it available to the entire system. For example, a researcher who wants to test the consequences of a new friction law on faults embedded in a poroelastic continuum need only copy a fault segment component, give it a new name, and replace the internal algorithms with the new one. From the viewpoint of the simulation package, the new fault segment behaves just like any other fault segment; only the hidden, internal methods for computing fault segment variables is different.

5.1.1. Hierarchical Design Mimics Nature. The system presented here is designed to allow a hierarchical development of different models in each of the primary simulation components. Each hierarchy, such as the PorousMedium class hierarchy shown above, represents a set of different models that may be specified very easily by an input parameter. For example, one could choose to build a simulator which combines a PorousMedium component and a FaultSystem component. One could choose a coupled poroelastic model for the porous medium component, and test various fault models when coupled to or embedded in a porous medium. This enables a researcher who is, for example, primarily interested in fault mechanics to utilize rather complicated software components built by others whose expertise is in poromechanics. This organization into software components allows the complexity of large software systems to be hidden and managed.

5.1.2. Proposal for a General Earthquake Model. The organization of coupled systems of variables and equations into class hierarchies that was described above for thermoporoelasticity, faults, and so on may be extended to more general systems in a direct manner. Each software component should resemble an analogous physical system in terms of state variables and dynamical behavior. By building software components which can readily be coupled together with other components and altered to test new ideas, some of the advantages of commercial or publically available codes that are already built, tested, and working gained be gained, whilst retaining the scientific need for adaptability, flexibility, and extensibility. The goal is to develop a framework into which physical theories, encapsulated as software components, can be implemented with as little effort as possible, in a standard fashion. The framework will provide a convenient means for representing physical theories in software, so that these computationally irreducible "laws" may be used by other scientists.

The goal of a general earthquake model is to provide a framework for numerically experimenting with various physical models for crustal dynamics (including the upper mantle) in order to determine likely physical mechanisms. This requires more than just a single complicated computer code. Rather, a set of software components which can be coupled together in various ways and from which new components can be derived. Such a set of components can be used by researchers and expanded in a way that makes new theoretical ideas - implemented in software - available to a wide community.

From a general viewpoint, software components must derive from general classes that evolve certain state variables, such as stress, strain, displacements, pore pressure, and temperature, according to prescribed physical laws. The hierarchy of physical models organized into software components provides a scientific computing system that will enable new physical models and equations to be implemented. Figure 3 is an example of how higher level, more complex classes can be constructed by assembling more basic components. Like electronic components, new models implemented as software components can be used in conjunction with other, already tested, components in order to explore dynamical behavior of new theories in a larger, more complex context.

5.1.3. The Digital Earth Project: EarthObjects^{TM}. The ideas presented in this paper are being implemented in a prototype system. Some components have been implemented already. Figure 4 gives a schematic representation of the major components in the EarthObjects^{TM} package. Several components have been used to build a poroelastic simulator, SimQuake, that was used to study postseismic deformation and aftershocks (*Bosl and Nur*, 1998; *Bosl*, 1999). The manner in which components were used to build SimQuake is an exam-

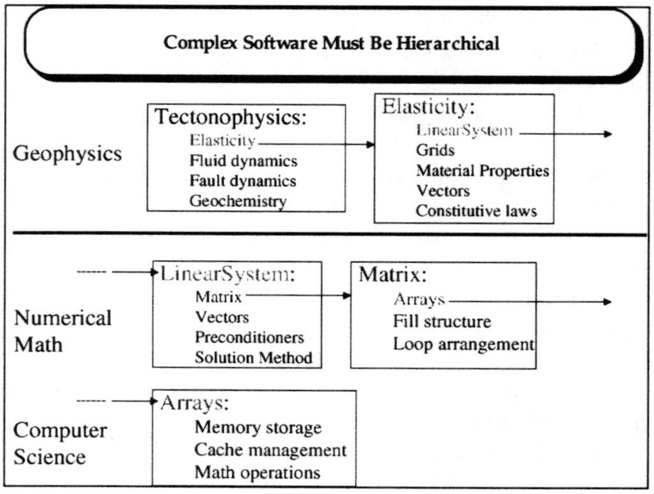

Figure 3. Complex software must be hierarchical in order to insulate the physical scientist from having to master complicated details of mathematical code optimization. As much as possible, the computational geoscientist must work at the level above the solid line in the diagram. Mathematicians must provide for computational scientists the numerical software tools that hide many of the details shown below the line.

ple of how EarthObjects^{TM} components may be used to quickly put together complicated general earthquake models. A simple simulator, which has been developed and used for studying postseismic deformation and aftershocks [*Bosl*, 1999] and an example of a more complex simulator are illustrated in figure 5.

5.2. Parallel Computers and Future Hardware

This paper is purposely focused on software rather than hardware. The task of the scientist is to build simulation software that models natural systems. The scientist has little control over hardware. He or she can merely use whatever computers are available at the time. It is true that physicists at national labs, for example, once had considerable influence over the design of supercomputers, but that influence is significantly reduced. Supercomputers are economically viable only if they rely on commodity hardware components and attempt to aim at a broader scientific market. Software that is to be maintainable and extensible in the future cannot be built to any particular hardware specifications.

Computer hardware, especially in the more specialized supercomputer arena, can be expected to change rather rapidly in the future. Complex software for specific applications, such as modeling earth processes, must have a lifespan that is longer than that of any particular computer design. Otherwise, geophysicists will forever be writing and updating codes for new machines, taking away from time spent actually running simulations and thinking about the implications for geophysics. We already assume that standard software tools, such as operating systems, compilers and windowing software, will be available on all platforms. These software packages, which are absolutely necessary for using computers, are built by experts in their domain. Though scientists might complain about a particular operating systems or windowing system, no scientist (who still wants to do science) would try to rewrite a new operating system or windowing system, even though these software systems can be written at the level of C code. It would take too much time away from his or her real interest. As standard methods for efficient numerical computing become more sophisticated and complicated to implement, it is only natural to allow scientific computing to evolve to a level higher than implementation of C or Fortran code. Solving sparse linear systems on parallel, distributed processors, for example, is a highly specialized task. Yet, from the point of view of earthquake physics, it is merely a technical detail that is irrelevant to physics and best left to specialists in that field. Straightforward Gaussian elimination will give the same numerical results; it just happens to be far too slow and memory consuming on current computers.

Supercomputers are defined as the fastest computers available at any particular time. Advanced research in computational science is often associated with supercomputers. This viewpoint is evident in books such as "Supercomputers and the Transformation of Science"

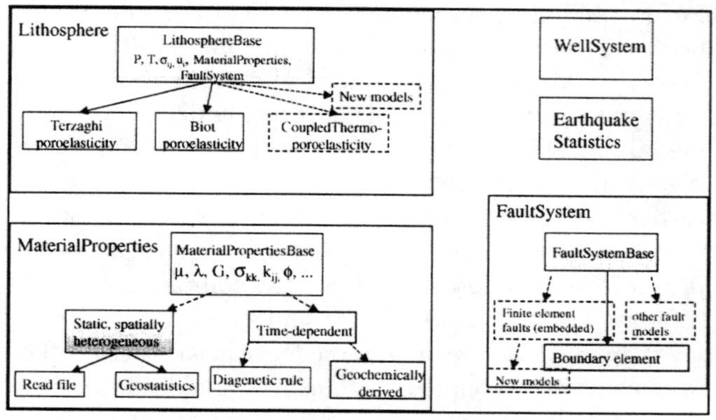

Figure 4. Schematic diagram of a component software system for crustal physics research. Shaded classes have been implemented.

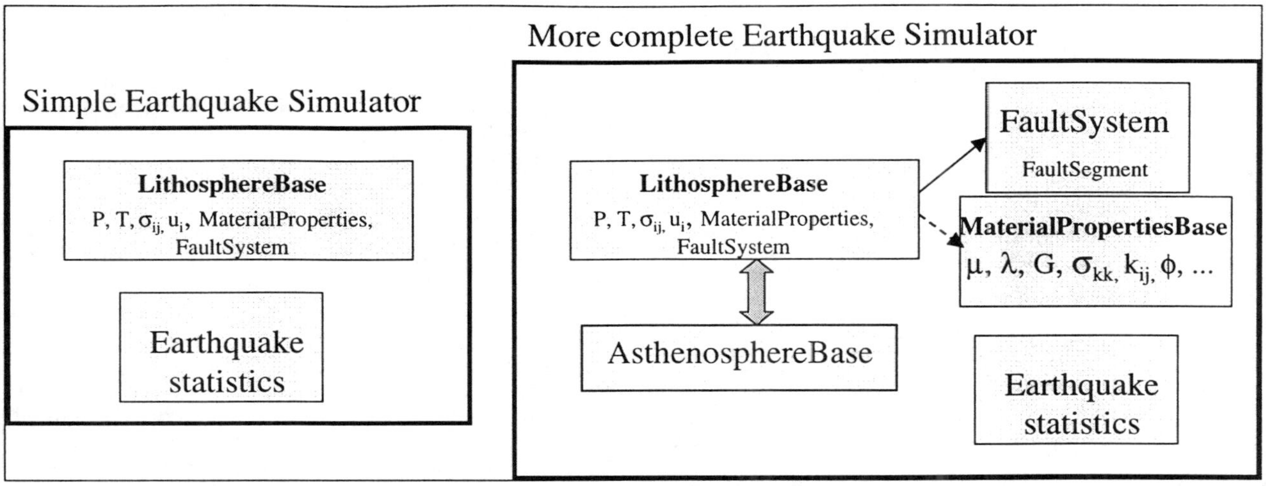

Figure 5. Building with EarthObjects component software. This simulator illustrates how component software can be used to build custom simulators for computational experimentation.

by *Kaufmann and Smarr* [1993]. It is certainly true that some scientific insights gained from computational experimentation would not have been possible without the very fastest computers available at that time. But today's supercomputer is tomorrow's desktop workstation. The association of computer hardware with scientific research really misses the point of computational science; the complexity is in the software. The dynamical processes instantiated in the software constitute the algorithmic laws of nature. As B. Woolley explains:

"One of the great technological hopes among experimental scientists of the late 1980s was the creation of a 'teraflop' computer, a computer capable of calculating a million million sums per second.... One could argue that this is simply another excursion in sciences's mission to discover ever more expensive instrumentation to make ever less relevant observations" [*Woolley*, 1994]. .

The greatest advances in our understanding of dynamical processes in the earth will result not because future computers will run larger simulations faster, however useful faster hardware may be. Scientific discoveries will come because future simulations (that is, software) will allow us to investigate and test novel, more complex ideas about physical processes in the earth. The software is the concrete representation of a scientist's ideas. Fast computers will never be a substitute for brilliant and creative scientific thinking.

The low-level complexity of computer codes that can use massively parallel computers efficiently is complicated and requires quite specialized knowledge. This situation is expected to continue. Paul Dubois, a computational scientist at Lawrence Livermore National Laboratory, mentions that architectures with 7 to 9 levels of memory hierarchies are now being designed . "There is no language, no software, no way of talking about any of that that anybody's invented. If they build those machines without thinking about how to program them, it's going to be really rough" [*Holmes*, 1997]. Simulation of complex earth systems, such earthquakes, must focus on experimentation with new physical ideas, implemented as quantitative algorithmic laws. As scientists, it is becoming too difficult to keep up with programming to these advanced architectures at the level of native C or Fortran and MPI calls. If we attempt to do this, scientific creativity will grind to a halt as we take on the role of technicians simply trying to keep up with efficiently porting our codes to newer machines. The only way to work around this situation is division of labor. Geophysicists must evolve to a higher level of simulation development.

Geoscientists need to build complex simulations from components which carry out tasks that are well-known already. There is no glory and nothing new to be learned from writing yet another finite element elasticity code. Software must be organized in a hierarchical fashion. Mathematical software components, which are optimized for efficiency on various computer architectures, must form the basis for higher level software components that implement physical theories. We would like to see a set of earth modeling components, an Earth Simulation Toolkit, built from a set of publically available "Math Objects", which can be maintained by experts in numerical computation and which insulate the geoscientist from the details of efficient, low-level numerical computing on sophisticated modern computer architectures. In this way, mathematical packages can be ported to new computer architectures without impacting the physical components which use them. Mathematical codes which are accomplishing this are now becoming available (see, for example, *Dahlen and Tveito*, 1997). The Diffpack^{TM} package (www.nobjects.com) is one of the most complete and well designed object-oriented numerical libraries) and will make possible a division of labor in which mathematicians and scientific programmers can focus on their specialties, while physical scientists pursue questions of concern to them. Indeed, *Dubois* [1997] predicts that in the near future (ten years from when his article was published or 2007) "discipline-specific class libraries and frameworks [will] form the basis for most developments".

6. SUMMARY

A quantitative formulation of the laws that govern complex earth systems, such as the earthquake cycle, cannot be written in the traditional manner as a closed set of differential equations. Though differential equations may be used to express some aspects of crustal dynamics that are a part of the earthquake cycle, earth materials are heterogeneous, nonlinear, and evolve in time, making discrete analogs of the equations necessary. Furthermore, some processes, such as faulting and diagenesis of fault properties, may be represented quantitatively as algorithmic laws, but not as equations. Computer simulation is the only means to experiment with different theories about the dynamics of complex processes. From this viewpoint, software that implements the various quantitative ideas becomes a new form of scientific law for the complex earth. In order to facilitate the advance of a quantitative understanding of the earth, and earthquakes in particular, the geoscience community must develop a common framework in which to build and use software components that represent the dynamical theories for various processes. Using a hierarchical component design which imitates nature's hierarchy of processes will allow a concrete, common body of quantitative "laws" that can be used and extended by geoscientists. There is no simpler way to express quantitative models of our understanding of nature.

Acknowledgment. This work was supported by USGS / NEHERP Grant #1434-HQ-97-GR-03 and DOE Basic Science Program Grant #FG03-86ER13601.

REFERENCES

Abraham, F. F., Spanning the length scales in dynamics simulation, *Computers in Physics*, *12*, 538–546, 1998.

Bolton, E. W., A. C. Lasaga, and D. M. Rye, A model for the kinetic control of quartz dissolution and precipitation in porous media flow with spatially variable permeability: Formulation and examples of thermal convection, *Journal of Geophysical Research*, *101*, 22157–22187, 1996.

Bosl, W. J., Computational studies of crustal fluids and earthquakes, Ph.D. thesis, Stanford University, 1999.

Bosl, W. J., and A. Nur, Numerical simulation of postseismic deformation due to pore fluid diffusion, in *Poromechanics: A Tribute to Maurice A. Biot*, edited by J.-F. Thimus, pp. 23–28, A. A. Balkema, Rotterdam, 1998.

Bosl, W. J., J. Dvorkin, and A. Nur, A study of porosity and permeability using a lattice boltzmann simulation, *Geophys. Res. Lett.*, *25*, 1475–1478, 1998.

Burridge, R., and L. Knopf, Model and theoretical seismicity, *Bulletin of the Seismological Society of America*, *57*, 341–371, 1967.

Byerlee, J., Model for episodic flow of high-pressure water in fault zones before earthquakes, *Geology*, *21*, 303–306, 1996.

Charlez, P. A., *Rock Mechanics Volume 1: Theoretical Fundamentals*, Editions Technip, Paris, 1991.

Charlez, P. A., *Rock Mechanics Volume 2: Petroleum Applications*, Editions Technip, Paris, 1997.

Chopard, B., and M. Droz, *Cellular Automata Modeling of Physical Systems*, Cambridge University Press, New York, 1998.

Christensen, U. R., Mantle rheology, constitution, and convection, in *Mantle Convection*, edited by W. R. Peltier, pp. 595–656, Gordon and Breach Science Publishers, New York, 1989.

Cryer, C. W., A comparison of the three-dimensional consolidation theories of biot and terzaghi, *Quart. Journal of Mech. and Applied Math.*, *16*, 401–412, 1963.

Daehlen, M., and E. A. Tveito, *Mathematical Models and Software Tools in Industrial Mathematics*, Birkhauser, 1997.

Dennis, J., and R. B. Schnabel, *Numerical Methods for Unconstrained Optimization and Nonlinear Equations*, Prentice Hall, New York, 1983.

Detournay, E., and A. H. Cheng, Fundamentals of poroelasticity, in *Comprehensive Rock Engineering*, edited by J. A. Hudson, pp. 113–171, Pergamon Press, New York, 1993.

Dubois, P., The future of scientific programming, *Computers in Physics*, *11*, 168, 1997.

Dullien, F. A. L., *Porous Media: Fluid Transport and Pore Structure*, 2nd ed., Academic Press, New York, 1992.

Fremming, N. P., O. Hjelle, and C. Tarrou, Surface modelling from scattered geological data, in *Numerical Methods and Software Tools in Industrial Mathematics*, edited by M. Daehlen and A. Tveito, pp. 301–315, Birkhauser, Boston, 1997.

Helmig, R., *Multiphase Flow and Transport Processes in the Subsurface*, Springer-Verlag, New York, 1997.

Hickman, S., R. Sibson, and R. Bruhn, eds., *The Mechanical Involvement of Fluids in Faulting*, Menlo Park, CA, U. S. Geological Survey, 1994.

Holmes, L. M., The future of scientific computing, *Computers in Physics*, *11*, 542–563, 1997.

Kaufmann, W. J., and L. L. Smarr, *Supercomputing and the Transformation of Science*, Scientific American Library, New York, 1993.

Koyama, J., *The Complex Faulting Process of Earthquakes*, Kluwer Academic Publishers, Boston, 1997.

Langtangen, H. P., *Computational Partial Differential Equations*, Springer-Verlag, New York, 1999.

McTigue, D. F., Thermoelastic response of fluid-saturated porous rock, *Journal of Geophysical Research*, *91*, 9533–9543, 1986.

Nedoma, J., *Numerical Modeling in Applied Geodynamics*, John Wiley and Sons, New York, 1998.

Pande, G. N., G. Beer, and J. R. Williams, *Numerical Methods in Rock Mechanics*, John Wiley and Sons, Inc., New York, 1990.

Pechenik, A., R. K. Kalia, and P. Vashishta, *Computer-Aided Design of High-Temperature Materials*, Osford University Press, Cambridge, UK, 1999.

Peltier, W. R., *Mantle Convection: Plate Tectonics and Global Dynamics*, Gordon and Breach Science Publishers, New York, 1989.

Place, D., and P. Mora, The lattice solid approach to simulate the nonlinear dynamics and heat of earthquakes, *Eos Transactions, American Geophysical Union*, *77*(46), 486, 1996.

Rice, J. R., Fault stress states, pore pressure distributions, and the weakness of the san andreas fault, in *Fault Mechanics and Transport Properties of Rocks*, edited by B. Evans and T.-F. Wong, pp. 475–503, Academic Press, New York, 1992.

Rice, J. R., and M. P. Cleary, Some basic stress diffusion solutions for fluid-saturated elastic porous media with compressible constituents, *Rev. Geophys. Space Phys.*, *14*(2), 227–241, 1976.

Rundle, J. B., Nucleation, growth and arrest of earthquakes; models and implications for general earthquake models, *Seismological Research Letters*, *69*, 143, 1998.

Toffoli, T., and N. Margolus, *Cellular Automata Machines: A New Environment for Modeling*, MIT Press, Boston, 1987.

Tompson, A. F. B., R. D. Falgout, S. G. Smith, W. J. Bosl, and S. F. Ashby, Analysis of subsurface contaminant migration and remediation using high performance computing, *Advances in Water Resources*, *22*, 203–221, 1998.

Walder, J., and A. Nur, Porosity reduction and crustal pore pressure development, *J. Geo. Res.*, *89*(B13), 11539–11548, 1984.

Wang, H. F., Poroelasticity, unpublished manuscript.

Wolfram, S., Computer software in science and mathematics, *Scientific American*, *251*, 188–203, 1984.

Wolfram, S., *Cellular Automata And Complexity : Collected Papers*, Academic Press, New York, 1994.

Woolley, B., *Virtual Worlds: A Journey in Hype and Hyperreality*, Viking Penguin, New York, 1994.

Yilmaz, O., R. Nolen-Hoeksema, and A. Nur, Pore pressure profiles in fractured and compliant rocks, *Geophysical Prospecting*, pp. 693–714, 1994.

Zeldovich, Y. B., and Y. P. Raizer, *Physics of Shock Waves and High-Temperature Hydrodynamic Phenomena*, Academic Press, New york, 1966, Nonlinear thermal waves discussed in Volume 2, pp652-676.

Zoback, M. D., and J. H. Healy, In situ stress measurements to 3.5 km depth in the cajon pass scientific research borehole: Implications for the mechanics of crustal faulting, *Journal of Geophysical Research*, *97*, 5039–5057, 1992.

W. Bosl, Geophysics Department, Stanford University, Mitchell Building, Stanford, CA 94305-2215. (e-mail: bosl@pangea.stanford.edu)

Crustal Fluids and Earthquakes

William J. Bosl and Amos Nur

Geophysics Department, Stanford University

The mechanical behavior of the upper crust over time periods of hours to thousands of years depends strongly on whether it is saturated with water or is primarily dry. There is much evidence to suggest that the crust, at least in some places, is saturated with water, possibly as deep as the brittle-ductile transition zone. If this is so, then the poroelastic behavior of the upper crust must be considered in studies of earthquake physics. The presence of fluid in an elastic, porous medium introduces a time-dependent element which cannot be dismissed as a secondary or insignificant factor. After briefly reviewing some of the arguments for the ubiquitous presence of crustal pore fluids, we will look at the physics of poroelastic materials, how it differs from elastic materials and the significance for earthquake physics. Computer simulations of poroelastic response following the large 1992 Landers, California earthquake will be presented and compared with postseismic data (aftershock occurence and deformation) to demonstrate the potential significance of pore fluids on the evolution of the physical state of the crust.

1. INTRODUCTION

One of the most important challenges in earthquake physics is to determine all of the significant physical processes that control stress evolution and seismic faulting in the upper crust. A fundamental question for earthquake physics is whether or not the upper, brittle crust is saturated with water and to what depth. The answers profoundly affect the physics of fault interaction and stress transfer in the crust. If water saturates the upper crust, that is, the seismogenic zone for shallow earthquakes, then the crust must be treated as a poroelastic material on timescales ranging from hours to millenia. The evidence for the pervasive presence of water in crustal rocks to at least 15 or 20 kilometers is substantial. *Walder and Nur* [1984] review some of this evidence and give a number of references. We will not review this evidence, but will instead examine the consequences of some possible scenarios for pore pressure development in the crust.

If water saturates crustal rocks, then it must be in a permanent state of disequilibrium. The source of this restlessness is the coupling between stress, pore pressure, and geochemical processes. If pore pressure is less than the local least principle stress, it will either dissipate entirely due to gradual inelastic pore closure or filling, leaving a dry crust, or rise until it reaches the least principle stress [*Sprunt and Nur*, 1977]. Geological evidence (see *Nur and Walder* [1992] for references) points to relatively low values of crustal permeability and widespread regions of high pore pressure in the middle and upper crust, suggesting that pore pressure in general will rise to the least principle stress. But when $P_p \rightarrow \sigma_{\min}$, the crust is on the verge of failure. Furthermore, *Gal, et al.* [1998] show that dissolution (and possibly precipitation) at a solid-fluid interface is governed by stress parallel to the surface. Thus, chemical equilibrium cannot be reached as long as the macro state of stress is nonhydrostatic . So, unless the crust is

GeoComplexity and the Physics of Earthquakes
Geophysical Monograph 120

dry, or at most contains only small separated fluid inclusions, it will remain 'restless', characterized by pore pressure bursts, earthquakes, and cracking and healing.

Metamorphic and deformation studies [*Bolton, et al.*, 1999; *Rye and Bradbury*, 1988] indicate that fluid flow is common throughout the brittle crust and that episodic rapid flow occurs in fractures. Mineral exchange rates can be used to estimate fluid fluxes and the time spent by fluids in veins. This is in agreement with the hypothesis of *Nur and Walder* [1992] that cycles of pore pressure buildup, followed by hydrofracturing and pressure release characterize crustal fluid dynamics. *Byerlee* [1996] and *Rice* [1992] posit similar episodic pore pressure increase and sudden release in fault zones as part of the earthquake cycle.

Pore fluids are therefore are likely to be involved in many dynamical processes related to seismic activity. These include the faulting process itself as well as postseismic phenomena caused by stress changes which result from seismic fault slip. Pore pressure buildup and release may play a significant role in the seismic cycle and particularly in the initiation of faulting. *Hickman, et al.* [1994] and *Evans and Wong* [1992] contain many papers which discuss the role of pore fluids in faulting. In this paper we consider only one particular aspect, namely, the fundamental behavior of the crust as a poroelastic material.

A better understanding of the crust as a fluid saturated poroelastic material is necessary in order to understand the physics of the entire earthquake cycle. Pore fluid pressure is often treated as a secondary effect when pre- and post-seismic stress changes are investigated. However, pore fluids have a significant effect on the elastic properties of the crust and must be included in stress evolution models. To do this, pore pressure effects must be included in simulations of crustal stress evolution. In addition, the mechanical and chemical effects of fluids within fault zones must be studied through computer simulation. The importance of pore fluids in earthquake physics can be estimated from comparison with available data.

In general, field measurements of physical quantities that are relevant to the physics of crustal dynamics are more readily available *after* an earthquake. Although an understanding of the role of crustal fluids in the whole process of stress buildup, faulting, and postseismic relaxation is desired, we simulate postseismic fluid flow and its consequences because data is available with which to compare the simulation results.

Observed seismically induced hydrologic effects include increased streamflow, water table changes, and well level fluctuations [*Roeloffs*, 1996]. Other observable coseismic and transient postseismic phenomena, such as vertical deformation [*Bosl and Nur*, 1998; *Peltzer et al.*, 1998; *Peltzer et al.*, 1996; *Zebker*, 1994] horizontal displacement [*Hudnut et al.*, 1994; *Wyatt et al.*, 1994] changes in aftershock location and frequency [*Nur and Booker*, 1972; *Booker*, 1974] and changes in seismic wave velocities in the fault region [*Nur*, 1972; *Li et al.*, 1997] are also likely to be or possibly related to pore fluid flow.

2. POROELASTICITY

The isothermal poroelastic equations are presented here for reference. A more thorough discussion of poroelasticity is given elsewhere in this volume, which includes thermal effects and the effects of geochemical evolution of crustal rocks. Isothermal conditions with static, spatially-heterogeneous coefficients are assumed here and equations are formulated in terms of coupled pore pressure and mean stress. Pore pressure and mean stress evolution in a poroelastic material are governed by two equations, a diffusion equation that derives from mass conservation and Darcy's law:

$$\beta\left(\frac{\partial p}{\partial t}+B\frac{\partial\sigma}{\partial t}\right)=\frac{\partial}{\partial x_i}\left(k_{ij}\left(x\right)\frac{\partial p}{\partial x_j}\right) \tag{1}$$

and an equilibrium equation that derives from strain compatibility conditions:

$$\frac{\partial^2}{\partial x_j^2}\left[Ap+\sigma\right]=0 \tag{2}$$

where

$$A=\frac{2\left(\nu_u-\nu\right)}{B\left(1-\nu\right)\left(1+\nu_u\right)} \tag{3}$$

and

$$\beta=\mu\phi\left(C_f+C_r\right) \tag{4}$$

$p = p_{total} - p_{ref}$ is the pore pressure deviation from a reference pressure (hydrostatic or pre-seismic, for example), $\sigma = \sigma_{kk}/3$ is the mean stress deviation from a reference mean stress state, B is Skempton's coefficient, ν_u and ν are the undrained and drained Poisson's ratios, μ is the fluid viscosity, ϕ is the porosity, C_f and C_r are the fluid and rock compressibilities, respectively, and $k_{ij}\left(x\right)$ is the spatially variable permeability tensor. The common assumption of a constant scalar permeability is avoided here, since heterogeneous permeability is likely to be significant for studies of crustal deformation. Initial and boundary conditions for both pore pressure and mean stress are required to complete this set of equations.

If deformation is of interest, the equations of linear elasticity may be solved as a separate system with pore pressure as an applied force.

$$\frac{\partial}{\partial x_i}\left((\lambda + G)\, u_{k,k}\right) + \frac{\partial}{\partial x_k}\left(G u_{i,k}\right) = \frac{\partial}{\partial x_i}(\alpha p) \qquad (5)$$

where u_i are the three components of the displacement vector, λ is the Lame coefficient and G is the shear modulus. Note that both of these may be spatially variable. The Biot-Willis parameter, α, is a function of Skempton's coefficient and the drained and undrained Poisson's ratios:

$$\alpha = \frac{3\,(\nu_u - \nu)}{(B\,(1-2\nu)\,(1+\nu_u))}$$

The right side of equation 5 is expressed in terms of the pore pressure deviation from a reference value, p_{ref}; that is, $p = p_{total} - p_{ref}$, as above. Gravitational forces have not been included here, but may be included as another body force.

3. FLUIDS IN FAULTS

Fluid pressure may be considered a seventh component of the stress tensor for a poroelastic material. In this sense, fluid pressure is part of the regional evolution of stress in the upper crust. Elevated pore fluid pressure within faults plays a particularly important role in earthquake physics. Elevated pore pressure acts to reduce friction between the opposing faces of a fault by opposing the local normal stress. Conversely, reduced pore pressure within a fault zone will strengthen a fault and resist slip.

When many processes occur simultaneously in a coupled fashion, surprisingly complex behavior can sometimes result. In fault zones, mechanical deformation, fluid flow, heat transport, and geochemical changes are all active in various ways [*Hickman, et al.* 1994; *Evans and Wong*,1992]. A striking example of surprising behavior is the coupling of fluid pressure and mean stress in an elastic porous medium. A comparison between simple fluid diffusion and coupled poroelastic diffusion was done by *Cryer* [1963] to illustrate the surprising behavior that can arise from simple linear processes that are dynamically coupled. The dynamical principles exhibited by this simple example demonstrate that in some circumstances the coupled effects of pore fluid pressure, elastic deformation, and heterogeneous properties must be considered if the dynamical behavior of the crust is to be properly understood. Thermal effects, chemical evolution of rock properties, and various nonlinearities due to the presence of fractures will not be considered here, though the addition of these processes will only serve to underscore the fundamental point of this paper: the known chemistry and physics of crustal processes, when considered as coupled processes with nonconstant coefficients, must be explored as a total system by extensive computer simulation and compared with the observed behavior of the real earth in order be fully understood.

3.1. The Mandel-Cryer Effect

Cryer [1963] compared the consolidation models of Terzaghi and Biot by considering a poroelastic sphere saturated with fluid at an initial pressure p_0. Terzaghi consolidation evolves pore pressure as a simple diffussion process, equivalent to solving equation 1 with $B = 0$. (Equivalently, if $A = 0$ in equation 2, then the time derivative of σ is zero in equation 1.) Biot's theory couples pore fluid flow with elastic deformation. The equations as originally written by Biot are equivalent to equations 1 and 2. The effect discussed by Cryer was first described in the soil mechanics literature by *Mandel* [1953]. The boundary of the sphere at time t_{o+} was set to zero pressure, with zero strain in the interior of the sphere and normal stresses on the boundary equal to the interior pore pressure. This is illustrated in figure 1. Two analytic solutions were presented to describe the pore pressure at the center of the sphere at $t > 0$. Discussions of the Mandel-Cryer effect in the literature are restricted to spherical and cylindrical geometries with homogeneous coefficients because analytical solutions for more complicated situations are not known.

Pore pressure at the center of the porous elastic sphere in the Terzaghi or uncoupled diffusion model remained at p_0 for a short time after the initial instant, then slowly decreased as the influence of the boundaries diffused inward. In the Biot model, the pore pressure at the center instantly began to *increase* above the initial pressure, then eventually declined as the boundary effects diffused inward. This surprising behavior is due to the initial reduction of the pore volume near the outer boundary of the sphere as pore fluid exits. In Biot's poroelastic theory, stress is transferred inward as the fluid near the boundaries escapes. Fluid in the center of the sphere is trapped and cannot escape immediately, so the pore pressure increases due to the compressive stress on the interior of the medium. Cryer noted that this phenomenon is not present in a one-dimensional case.

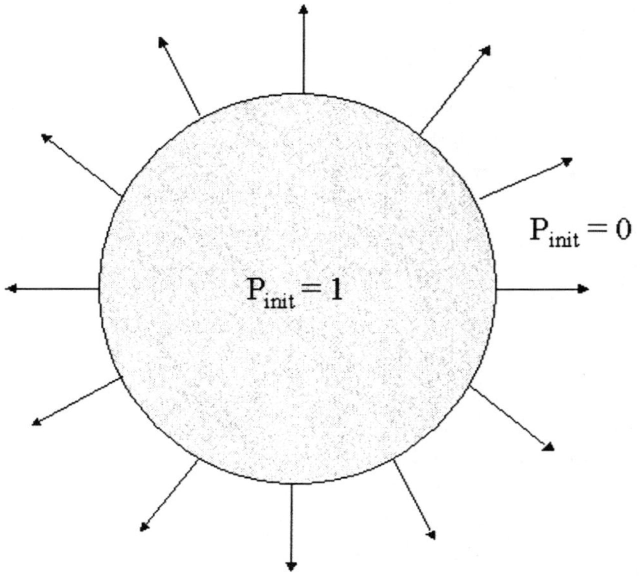

Figure 1. Cryer's sphere problem, used to compare the consolidation theories of Terzaghi and Biot. At the initial instant, pore pressure on the boundary of the sphere is dropped to zero, while mean stress is held constant within and on the boundary of the sphere. Fluid flows out of the boundary as time increases.

In order to study pore pressure in a coupled poroelastic medium, we simulated Terzaghi and Biot diffusion on a rectangular grid by solving equations 1 and 2 numerically. Both two and three dimensional simulations were carried out with similar results. The Cryer effect is larger in the three dimensional case, but qualitatively the same as the two dimensional case. We show the results of two dimensional simulations here. The pressure on the interior of the cube was initially 1 Pa and the side boundaries were held at 0 Pa. The undrained Poisson ratio ν_u was varied. When $\nu_u = \nu$, A in equation 2 is zero and the pore pressure follows a simple (uncoupled) diffusion curve and the pore pressure follows the simple diffusive flow model as in equation 1, with σ = constant (hence, $\partial\sigma/\partial t = 0$) at all times. As expected, the pore pressure in the center of the cube remains at the initial value p_0 for a period of time, then decays roughly exponentially. The rate of decay depends on the diffusivity. When $\nu_u > \nu$, A in equation 2 is nonzero and the pore pressure initially *increases*. The magnitude of this effect depends on the difference between the drained and undrained Poisson ratios. Figure 2 shows the effect on pore pressure of varying the parameter ν_u.

The abnormally high pore pressure caused by poroelastic effects can be attributed to the trapping of fluid in a region while mean stress is transferred from surrounding regions. Fluid flux out of a volume of material is controlled by the product of permeability, k, and the surface area of the volume. In a two- or three-dimensional volume, fluid inside concentric regions can only escape through a surface area that is smaller than the area of the external boundary. *Cryer* [1963] showed that anomalously high pressure cannot develop, even in Biot's coupled poroelastic equations, in the one-dimensional case. However, even in the one-dimensional problem, abnormally high pore pressure can be induced in regions of low permeability, which traps pore fluid. This may have important implications for fluid-saturated faults.

3.2. Anomolous P_p in Fault Zones

The San Andreas fault is weak [*Rice*, 1992; *Wang et al.*, 1995; *Zoback*, 1987]. By this it is implied that the fault is in a state of near failure. One likely explanation for the weakness of the San Andreas is the existence of high pore pressure within the fault zone [*Ingebritsen*, 1998]. Under some circumstances, pore pressure within a fault can rise even when no apparent source of fluid or applied stress exists if fluid pressure outside of the fault is dissipating. Low permeability, caused perhaps by the clogging of flow channels normal to the fault plane by

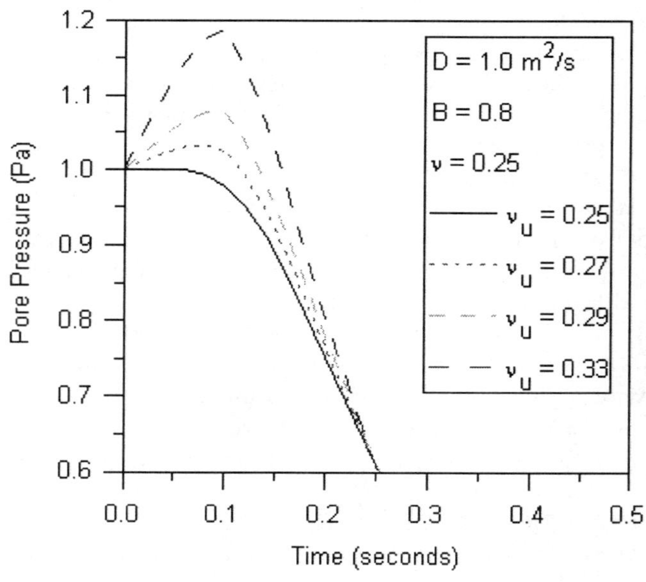

Figure 2. Pore pressure at the center of a square showing the Mandel-Cryer effect for several values of undrained Poisson ratio. The curve $\nu_u = \nu = 0.25$ is the Terzaghi consolidation result.

mineral precipitation, may lead to the development of abnormally high pore pressure and failure due to the Mandel-Cryer effect.

One model for a typical fault zone suggests that the permeability in the fault gouge is low, while the permeability in the surrounding damage zone is relatively high due to the presence of breccia and extensive microfractures [*Bruhn et al.*, 1994]. Byerlee's model [*Byerlee*, 1994] for the involvement of fluids in the earthquake cycle hypothesizes that permeability varies spatially in the region of a fault and evolves in time. Fluids saturated with minerals flow out of the fault zone and precipitate, clogging the pores and greatly reducing the permeability. Rice's model [*Rice*, 1992] for the San Andreas fault also assumes very low permeability perpendicular to the fault core.

A simple experiment with a two-permeability system as shown in the first frame of plate 1 was performed. A fault zone was delineated by assigning low permeability values on a narrow strip within a higher permeability region. Flow was allowed only in the x-direction, with zero flux boundary conditions on the other sides. Material parameters used were $\nu_u = 0.29$, $\nu = 0.25$, and $B = 0.8$, which are reasonable values for crustal rocks. Sedimentary rocks, such as might be found within a few kilometers of the surface, may have drained Poisson's ratios considerably lower (ν in the range of 0.12 to 0.15 and ν_u near 0.30 were determined by *Detournay* [1993]). The Mandel-Cryer effect would be greater near the surface in this case. Flow from the ends of a real fault would be restricted by the low surface area of the fault ends relative to the surface area of the sides of the fault. As pore fluids drain away from the sides of the fault, stress transfers to the fault core and pore pressure builds. This result is shown graphically in plate 1. Pore pressure in the low permeability fault zone rose to a maximum value of 1.11 Pa in the coupled poroelastic case, but never rose above the initial value in the uncoupled or simple diffusion case.

The abnormally high pore pressure that can develop in a coupled poroelastic medium can be large enough to cause the effective shear stress on the fault to exceed the frictional strength of the fault. For example, in our simulation with $\nu_u = 0.29$, $\nu = 0.25$, and $B = 0.8$, the pressure in the fault zone due to the Mandel-Cryer effect rose 11% higher than the initial pore pressure. On critically stressed faults, only a relatively small rise in pore pressure is needed to exceed the failure threshold. The essential point of these simulations is that pore pressure can attain values higher than expected if fluids are trapped by low permeability in the fault core.

Anomolous pore pressure that results from coupled poroelastic behavior can be simulated numerically and the occurrence of this phenomenon is not restricted to any particular geometry. Previously published discussions of the effect are restricted to spherical and cylindrical geometries because of the need to find analytical solutions to the governing system of equations. Computer simulation removes the domain restriction, and allows experimentation with heterogeneous property distributions. The fascinating Mandel-Cryer effect may have important implications for critically-stressed faults that are saturated with fluids, including faults in the vicinity of a large earthquake.

4. AFTERSHOCKS

Though aftershocks are clearly correlated to a large earthquake event, they have long been a puzzle. Essentially, they are earthquakes which have a rather obvious precursory signal in the form of a large, nearby earthquake. Clearly, the mainshock triggers an aftershock sequence in some sense. Yet, the aftershocks do not occur immediately. There is a time delay mechanism between the cause (the mainshock) and the effect (the aftershocks). The frequency of aftershocks decays like a diffusive process [*Nur and Booker*, 1972; *Scholz*, 1990], indicating that a time-dependent physical process is at work.

Nur and Booker [1972] proposed that pore pressure readjustment after an earthquake could explain the time delay in the occurence of aftershocks. Their idea was as follows: when an earthquake occurs, there is an almost-instantaneous modification to the regional stress field. The change in strength of a rock, determined experimentally by *Hubbert* [1959], was given by

$$\Delta S = \mu_f (\overline{\sigma} - p) \tag{6}$$

where μ_f is the coefficient of internal friction or simply the frictional strength, $\overline{\sigma}$ is the mean stress, and p is the pore pressure in the fault. They used the convention that compressive stresses are positive, a convention that will *not* be used hereafter in this paper. Aftershocks will occur on faults where the shear stress exceeds the strength of the fault.

Immediately following an earthquake, the pore pressure is changed by an amount proportional to the mean stress induced by the earthquake:

$$\Delta p = -B \Delta \overline{\sigma} \tag{7}$$

where B is Skempton's coefficient, an empirically determined constant that quantifies the fraction of a compressive stress on a porous rock that is transferred to the pore fluid. *Wang* [1993] gives values for crustal rocks in the approximate range 0.55 to 0.9. For many studies, a reasonable first-order assumption is $B = 1$. The condition that $\Delta p = -B\Delta\overline{\sigma}$ is referred to as the *undrained* condition. A poroelastic medium is said to be in the undrained state when the stress modification $\Delta\overline{\sigma}$ that results from, for example, a dislocation occurs so quickly that fluids in the pores are essentially stationary. After the initial dislocation, pore fluids will flow from regions of high pressure (compressional regions) to regions of low pressure (dilatational regions). The applied mean stress field will remain approximately constant, so the strength of the fault will change over time. The number of aftershocks, according to this theory, should be proportional to the time rate of change of pore pressure integrated over the region involved. Moreover, the theory predicts that aftershocks will occur where the pore pressure is *increasing*; that is, in regions of coseismic dilatation.

4.1. The Coulomb Fracture Criterion

The Hubbert criterion for rock strength, equation 6 does not take into account the anisotropic nature of a fault. The compressional stress normal to the fault plane controls the frictional strength of a fault rather than the mean stress. The Coulomb failure criterion, which was adopted from soil mechanics, where it was called the Drucker-Prager criterion [*Drucker and Prager*, 1952], is a more appropriate measure of fault strength. The Coulomb failure criterion may be written quantitatively as

$$\tau_c = \tau_s + \mu_f\,(\sigma_n + p) \tag{8}$$

where τ_s is the shear stress on the fault, μ_f is the coefficient of internal friction or simply the frictional strength, σ_n is the normal stress on the fault, and p is the pore pressure in the fault. Increasing pore pressure effectively reduces the normal stress and thus reduces the shear stress needed to overcome friction. The quantity $(\sigma_n + p)$ is often referred to as the *effective stress.* More accurately, the effective stress is $(\sigma_n + \alpha p)$, where α is the Biot parameter, but $\alpha \approx 1$ is frequently assumed in faults [*Nur and Byerlee*, 1971]. A fault tends to fail when $\tau_c > 0$. τ_c depends on the local stress field and the orientation of the fault of interest. Though the orientation of individual faults is usually unknown in a region, the average orientation of many faults in a region can often be inferred.

Coulomb stress is often treated as a static quantity following an earthquake. This assumes that pore fluids are stationary and that the undrained condition, equation 7 applies. This assumption allows pore pressure to be eliminated from the Coulomb stress criterion and an effective friction coefficient to be defined, $\mu' = \mu_f\,(1 - B)$, which incorporates pore pressure into an effective Coulomb stress:

$$\tau_e \equiv \tau + \mu'\sigma_n \tag{9}$$

The effective Coulomb fracture criterion is a static quantity that can be computed once, given a slip model or coseismic stress field for an earthquake.

Unfortunately, once pore fluids begin to flow appreciably, the undrained assumption is no longer valid; the undrained assumption is false on time scales over which aftershocks occur. *Harris* [1998] describes the use of an effective coefficient of friction:

> "This strategy is mostly an attempt to cover up our lack of knowledge about the role of pore fluids....Although it is convenient to lump our ignorance of pore fluid behavior into a redefined 'apparent coefficient of frction,' μ', we run the risk of missing some important clues in interpreting our data."

Typical diffusivities in the crust are on the order of 0.01 to 1.0 m^2/s [*Charlez*, 1997; *Wang*, 1993]; *Li et al.* [1987] used diffusivities in the range 0.1 to 10 m^2/s in their study of aftershocks. The time scale for diffusion over distances of 1 km can thus be as little as 1 days with the larger of these values to several years for the smaller values. Flow in fractured media can be much faster. The appropriate time scale for aftershock studies is not the rupture time of an event, but the time period over which aftershocks occur following an earthquake. For typical crustal diffusivities, the undrained assumption does not hold beyond the first few hours following an earthquake. Because dislocation-induced coseismic pore pressure changes occur all along a fault, pore pressure diffusion commences immediately following an earthquake throughout the region. Thus, the dismissal of pore fluid diffusion as an explanation for aftershocks by *Scholz* [1990], p. 209, is invalid. The computed Landers coseismic mean stress distribution, shown in plate 2, illustrates this point.

4.2. The 1992 Landers earthquake

In order to investigate the triggering of aftershocks by pore fluid diffusion, a poroelastic simulation of pore fluid flow following the 1992 Landers earthquake was performed. The 1992 Landers earthquake was a large and well-documented earthquake. The faults that ruptured in the Landers earthquake are located in the Mojave Block of southeastern California. The region is dominated by northwest-trending, right-lateral strike-slip faults. Rupture during the Landers earthquake occurred along several major faults that were previously thought to be disconnected. It thus provided an example of how stress changes caused by one fault can cause rupture along adjacent faults.

The Landers earthquake was preceded by two months of intense seismic activity in the region south of the mainshock fault trace [*Yeats et al.*, 1997]. The largest of these foreshocks was the Joshua Tree earthquake which occurred on April 23, 1998 with a magnitude of M_w=6.1. Aftershocks of the Joshua Tree earthquake migrated northward in the two months between the Joshua Tree and Landers earthquakes. This activity continued to migrate to the vicinity of the Landers hypocenter until only hours before the main shock. It is suggested that this time-dependent behavior requires time-dependent dynamics for a causal explanation and that pore-fluid induced stress evolution may account for this.

A major aftershock (or perhaps a separate earthquake) occured on the Big Bear fault approximately three hours later and 40 km to the west of the Landers rupture. The Big Bear earthquake had a different orientation and slip than the Landers earthquake; it was roughly conjugate to the Landers event. Thirteen additional aftershocks of magnitude 5.0 or greater occured in the year after the mainshock. The Landers surface rupture produced right-lateral displacement over a total rupture length of approximately 85 km. Average slip of two to four meters was measured along the surface and the overall magnitude was determined to be M_w=7.3 [*Hudnut et al.*, 1994].

4.3. Poroelastic changes at aftershock locations

Theoretical studies of poroelastic response following edge dislocations by *Booker* [1974] and *Li et al.* [1987] have shown that the Coulomb stress field in the region around a fault will change following rupture due to pore fluid diffusion. The complexity of real faults requires detailed numerical modeling to compare poroelastic simulation results with actual data.

Several studies of Landers aftershocks have looked at the static coseismic Coulomb stress field where aftershocks occured [*King et al.*, 1994; *Hardebeck et al.*, 1998]. *King et al.* [1994] computed the Coulomb stress at aftershock locations by assuming the faults were optimally aligned for failure and found that roughly two-thirds of the events were located where the coseismic Coulomb stress was positive. *Hardebeck et al.* [1998] determined the actual focal mechanisms for aftershock events and used these to compute the Coulomb stress. They found that approximately 85% of the events occured in regions of positive coseismic Coulomb stress. Both of these studies exclude aftershocks within 5 *km* of the Landers surface fault trace. This is reasonable, since the stress field might be expected to be highly variable near the fault and quite sensitive to details of the fault location. Both of these studies also used the fault model of *Wald and Heaton* [1994] which has complex vertical slip resolution, but relatively simple horizontal resolution.

To directly test the hypothesis that pore fluid flow is responsible for aftershocks, the stress and pore pressure field in the region of the Landers earthquake were simulated numerically. First, the stress field produced by the main Landers rupture was computed using an analytic dislocation model, similar to the previously cited studies. However, we use the fault slip model of *Hudnut et al.* [1994]. Plate 2 shows the computed coseismic mean stress, from which the initial pore pressures are derived. The coseismic Coulomb stress field was computed at aftershock locations, assuming faults were optimally aligned for failure, as did *King et al.* [1994], since we did not have focal mechanism data for the aftershocks. Unlike the previously mentioned studies, we included all aftershocks in our study, both near and far from the fault. Overall, 77% of the aftershocks occured where the coseismic stress was positive, which compares favorably with previous results.

Significantly, aftershocks that occured where coseismic stress was negative were not randomly scattered in the vicinity of the Landers fault trace. Rather, they were clustered in locations where there is reason to expect that the fault model is inadequate, such as near the Big Bear fault, in fault jogs, and near the Eureka Peak fault off the southern end of the Landers fault. This suggests that it is possible to model the stresses following a large earthquake even near a complicated fault system. More detailed discussion of models of the Landers fault and implications for aftershocks will be given in a forthcoming paper [*Bosl and Nur*, "Aftershocks and pore fluid diffusion following the 1992 Landers earthquake", submitted to JGR].

Though these results using static Coulomb stress fields indicate a fairly high correlation between coseismic Coulomb stress and the occurence of aftershocks, they do not explain the time delay nor the diffusive decay of aftershock frequency. If coseismic Coulomb stress changes cause aftershocks, why don't they happen coseismically? Though pore fluid flow is the dynamic force that causes poroelastic stress changes, the evolving Coulomb stress field driven by poroelastic diffusion that is the appropriate variable to compare with aftershock locations. Coseismic stress changes may bring regional faults closer to failure, but aftershocks will occur where the Coulomb stress *continues to increase* after the earthquake. As the rate of Coulomb stress increase slows, which is directly related to the decaying pore pressure gradients, the rate of aftershock occurence will also slow.

To test this hypothesis, the poroelastic equations were solved as a coupled system to simulate postseismic Coulomb stress evolution following the initial Landers fault dislocation. The change in Coulomb stress was computed at the time and location of each aftershock in the year following the mainshock.

For this calculation, an undrained Poisson ratio of 0.30 was used to represent the effective initial response of the crust. Coseismic pore pressure changes caused by the fault displacement were computed from equation 7. The pore pressure over time was computed by solving equations 1 and 2 as a coupled system and the stress deviation induced by pore pressure was computed from equations 5. Thus, the full initial and evolving postseismic stress field and pore pressure distribution were simulated.

The Coulomb stress was computed at the time and location of each aftershock and compared with the initial Coulomb stress at that location. The results are shown in plate 3. The events shown are only those for which the coseismic Coulomb stress is positive. Of these aftershocks, the Coulomb stress has increased since the mainshock at nearly three-fourths (70%) of the aftershock locations. The average computed *increase* in Coulomb stress, from the initial coseismic value, at aftershock locations is 0.155 MPa (1.55 bars). We found further that a significant number of aftershocks (10% of the total) occured where the initial (coseismic) Coulomb stress was negative, but that it had become positive by the time of the aftershock. We speculate that some aftershocks in the complex fault jog regions, where the initial Coulomb stress was negative, might become positive if pore pressure was to increase due to the Mandel-Cryer effect. This might occur if permeability in this region was low relative to the surrounding material. Such detailed information is not available for our simulations, but it is important to recognize that these physical effects are quite possible.

Aftershocks where the Coulomb stress has not increased from the coseismic value are not random scattered, but are clustered in two locations: near the Big Bear fault and in the vicinity of the Eureka Peak fault, off the southern end of the Landers fault. More information on the actual aftershock focal mechanisms, better resolution of the Landers fault segments, particularly near the Eureka Peak segments, and inclusion of stress changes due to large aftershocks, such as the Big Bear event, may give even better results.

Since permeability controls the direction and rate of fluid diffusion in a porous medium, more accurate knowledge of the permeability structure of faults and the surrounding regions will be required to make better calculations of the rate and magnitude of poroelastic stress changes due to pore fluid movement. The pressure dependence of permeability, spatial heterogeneities and anisotropy in and near the fault zone may significantly affect fluid flow patterns and the evolution of the postseismic pressure field. The magnitude of the poroelastic effect depends also on the elastic parameters of the material, particularly the difference between undrained and drained Poisson ratios, ν_u and ν.

The hypothesis that poroelastic processes following an earthquake can account for the time element in aftershock activity does not contradict previous theories that account for aftershock locations on the basis of correlations with coseismic Coulomb stress increase. Rather, it provides a first order augmentation of the static theory by providing a reasonable physical mechanism for evolution of the postseismic Coulomb stress field in the days and years following an earthquake.

5. AFTERSLIP

Pore fluids act like a shock absorber when a sudden dislocation imposes a load on an elastic material. The fluid resists deformation of the porous rock initially, but relaxes over time as it flows from regions of high pressure to regions of low pressure. When a fault dislocation occurs, the coseismic stresses are those of an undrained poroelastic material. An undrained poroelastic response is identical to a purely elastic response with appropriate elastic parameters. *Booker* [1974] suggested that pore fluid diffusion following a fault dislocation causes partial reloading of the stress on the fault. Another way to interpret this is that the presence of pore fluids in an elastic medium partially resists the stress imposed by the initial fault dislocation. This initial resistance then relaxes as pore fluids flow from compressional to dilatational regions.

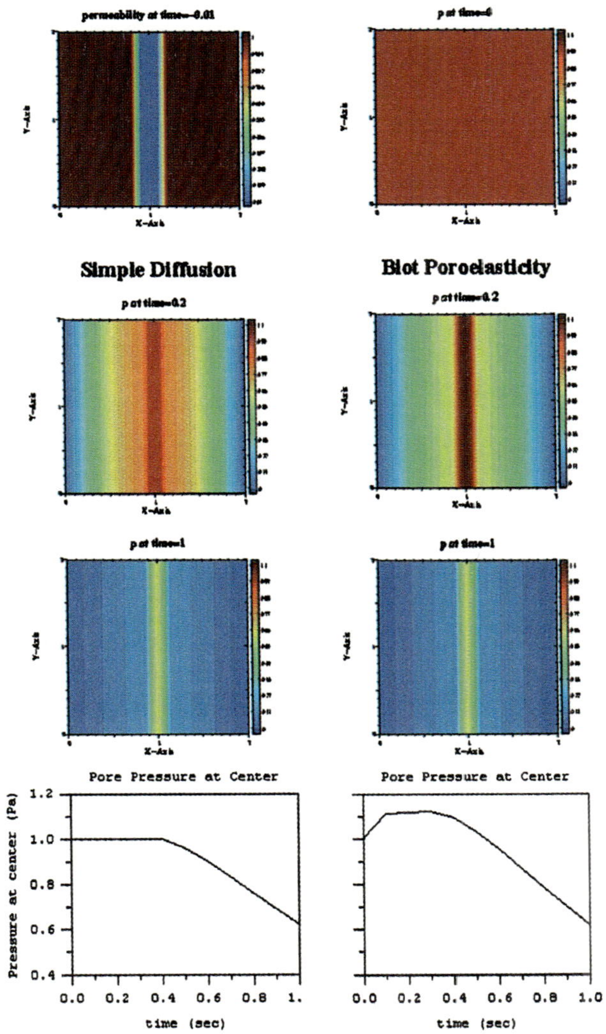

Plate 1. A low permeability fault zone (k=0.01) is embedded in a higher permeability (k=1.0) region. Boundary pressures are held at 0 on the left and right sides and no flow is permitted out of the top and bottom sides. As fluid diffuses out of the region, trap fluid in the fault zone rises in the coupled (Biot) case due to the Mandel-Cryer effect.

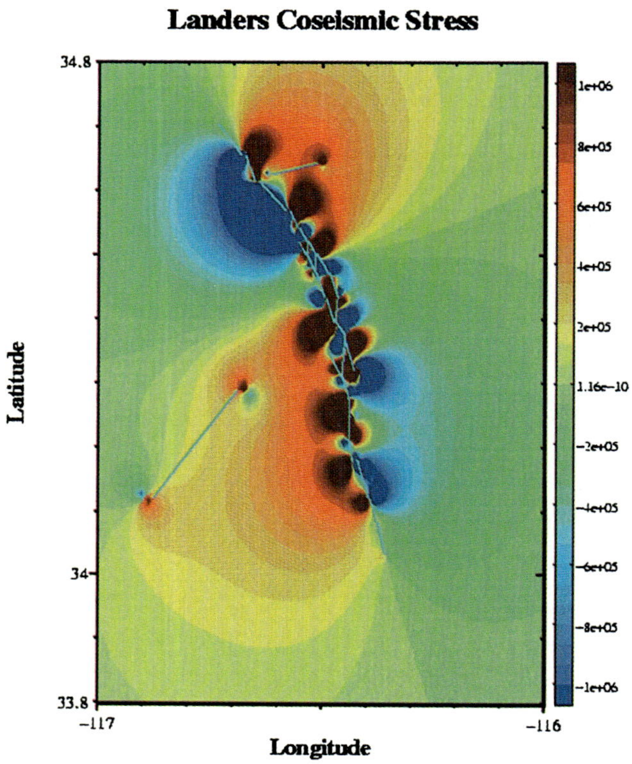

Plate 2. Coseismic mean stress field computed from Hudnut, et al. (1994) slip model. Fault trace is superimposed on the mean stress image. Negative mean stress is compressive.

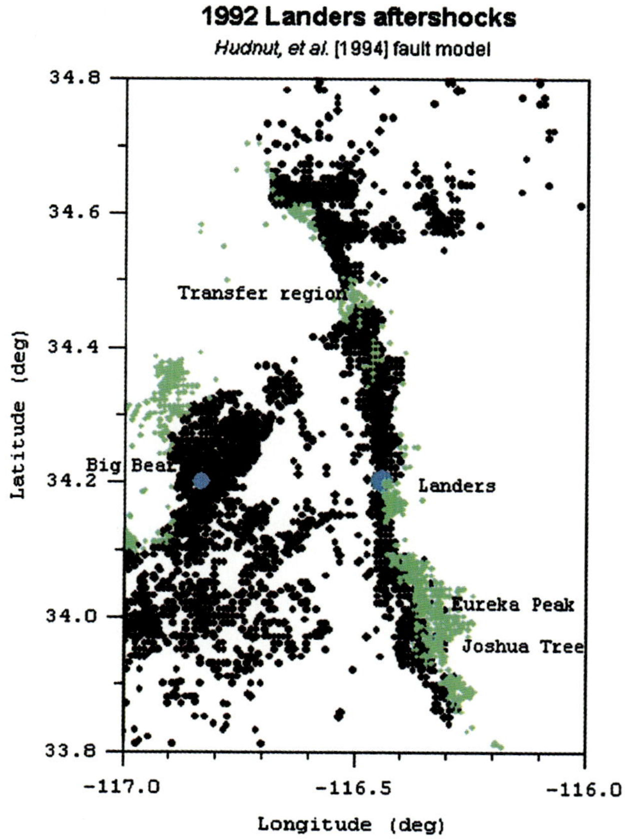

Plate 3. Aftershocks of the Landers earthquake compared to Coulomb stress tendency due to pore fluid diffusion. Black dots indicate aftershocks which occurred where the computed Coulomb stress at the time of the aftershock, $\tau_c(p)$, was greater than the coseismic Coulomb stress. Green dots are aftershocks where $\tau_c(p)$ was less than the coseismic value. The mean increase in Coulomb stress for black dots is 0.8 MPa (8 bars); the mean change for green dot locations is -0.2 MPa (2 bars). 73% of the 16661 events shown occur where the Coulomb stress had increased since the mainshock.

Measurements have shown that after an earthquake surface slip continues to increase for a year or so, with a time rate of decay that is similar to aftershock decay [*Scholz*, 1990]. For typical crustal diffusivities, relaxation time for a poroelastic material is on the order of one year [*Peltzer et al.*, 1996]. In order to estimate the magnitude of apparent postseismic stress changes, we assume pore pressure is induced by the initial dislocation stress and can be computed from equation 7. Stress changes that result from pore fluid diffusion must satisfy equation 2. Thus,

$$\Delta\sigma = (-A\Delta p) = (-A(-B\sigma_{init})) \Rightarrow \quad (10)$$

$$\frac{\Delta\sigma}{\sigma_{init}} = AB = \frac{\nu_u - \nu}{(1-\nu)(1+\nu_u)} \quad (11)$$

The postseismic mean stress change induced by pore pressure relaxation will have the same sign as the initial mean stress change. That is, mean stress will continue to increase or decrease where the initial mean stress caused by the dislocation increased or decreased, respectively.

For $(\nu_u, \nu) = (0.28, 0.25)$, which is reasonable for low porosity rocks, $\Delta\sigma$ will be approximately 6% of the initial coseismic mean stress. Taking more extreme values that might be typical for the upper crust when fractures are present, $(\nu_u, \nu) = (0.30, 0.15)$, we find that $\Delta\sigma \simeq 40\%$ of σ_{init}.

Since postseismic poroelastic mean stress evolution is in the same direction as the coseismic stress change, it is entirely possible that measured postseismic displacements caused by poroelastic relaxation might be interpreted as afterslip. Since crustal material near the surface typically has greater porosity, hence greater difference between undrained and drained Poisson ratios [*Detournay and Cheng*, 1993], greater postseismic displacements or apparent afterslip would be expected at shallower depths.

Plate 4 illustrates how postseismic fluid flow might be interpreted as afterslip. The images are simple calculations of displacements in a poroelastic material due to a single fault dislocation. The initial displacements, $u_i(t=0)$, were computed from the analytical formulas of *Mansinha and Smylie* [1971] , using an undrained Poisson ratio of 0.31 for the (effective) elastic medium. The domain is assumed to be an elastic halfspace in these formulas. The fault was a 10 km long vertical fault extending from the surface to a depth of 12 km, with right lateral slip of 1 $meter$. The domain is 100 $km \times 100\ km \times 12\ km$ and the grid spacing is 4 km in all directions. From the initial fault dislocation, the mean stress was computed. An initial pore pressure was induced, $p(0) = -B\sigma_{init}$, where σ_{init} is the mean dislocation stress and $B = 0.8$. The pore pressure was allowed to evolve according to equations 1 and 2, and the displacements due to pore pressure were calculated from equations 5. Material property values were reasonable values for the brittle crust: $\nu = 0.25$, $\nu_u = 0.31$, $\mu = 22\ GPa$, and diffusivity, $k/\beta = 0.1\ m^2/s$. The change in displacement over one year was computed by taking differences.

Though there are some minor variations in the shape of the displacement fields, most likely caused by the coarse grid and the difference between an analytic halfspace model and a finite element model with finite boundaries, the essential point to be seen in these images is that the postseismic displacements caused by poroelastic relaxation are in the same direction as the initial displacements and the magnitude is approximately 10% of the initial displacements. The magnitude of these simulated postseismic displacements is large enough to be interpreted as an apparent afterslip, particularly if measurements of displacement are made near the fault zone, where pore pressure changes will be greatest.

Fluid flow in a porous medium is controlled by the permeability. Permeability in real crustal rocks is both heterogeneous and anisotropic [*Tompson et al.*, 1998; *Hickman et al.*, 1994]. Permeability in fractured rocks is generally controlled by the fracture porosity rather than the matrix porosity and thus tends to be enhanced in the direction of the fractures. *Seeburger* [1981] and several papers in the proceedings edited by *Hickman et al.* [1994] indicate that permeability in fault zones may be low normal to the fault plane and enhanced along the fault plane. It is not our intention to discuss the complexities of the permeability structure of the crust here, but simply to emphasize that fluid flow in the crust is not generally uniform, but may follow quite complicated pathways. Correspondingly, the stress fields produced by pore pressure changes in the crust need not develop in the same patterns as initially introduced by a fault dislocation. Because stresses in equation 5 are due to *derivatives* of the pore pressure, shear stresses may develop as pore fluids flow in anisotropic pathways. Relaxation of pore pressure and the apparent induced aftership will not be uniform in this case. This mechanism provides a simple, consistent explanation for afterslip that can also be tested by numerical simulations compared with field measurements.

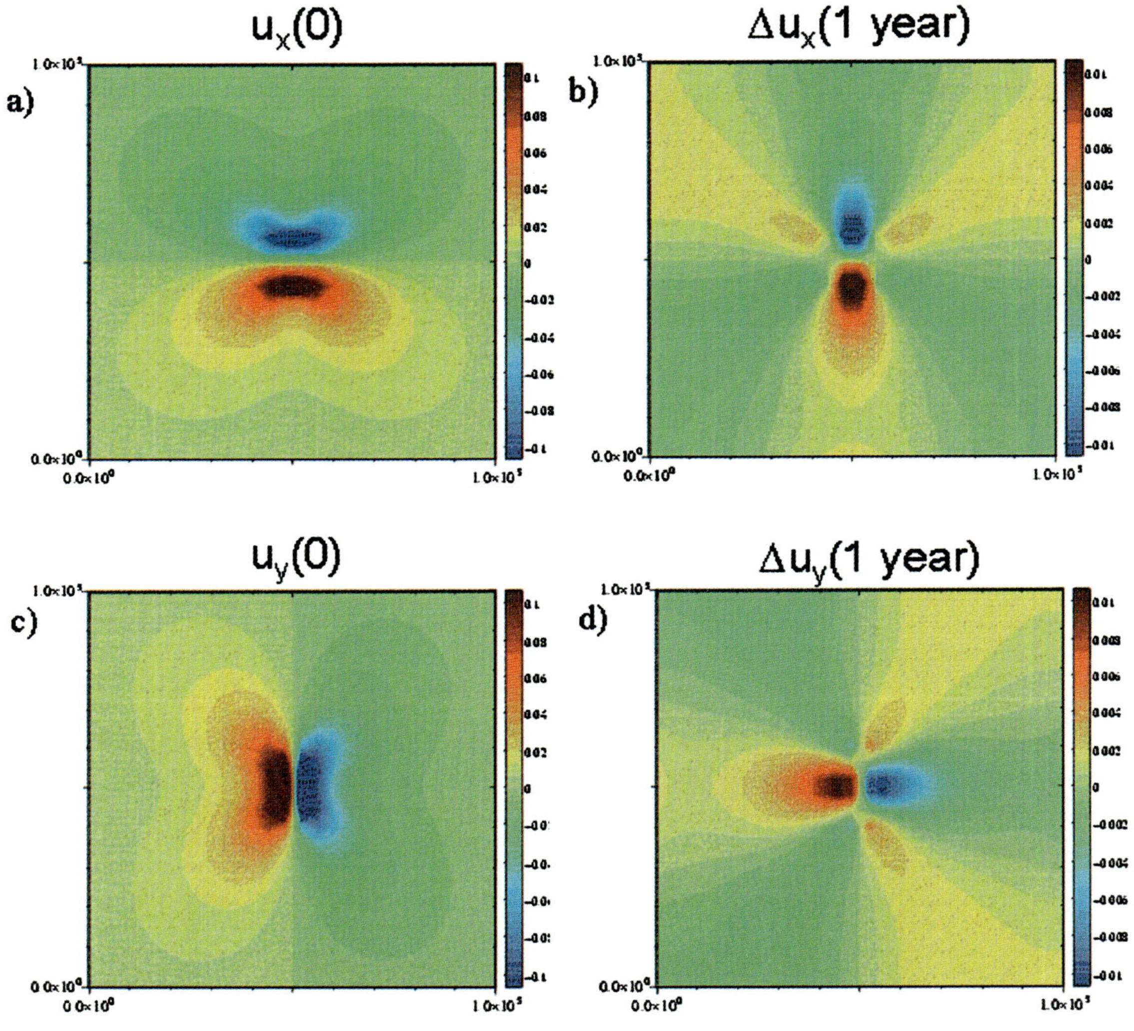

Plate 4. Computations of horizontal displacements due to an initial edge dislocation $u_i(0)$ and the change in horizontal displacement due to poroelastic relaxation after one year, $u_i(1\ \text{year})$. The dislocation displacements were computed using an analytic half-space solution. The poroelastic relaxation was computed using a finite element program.

More detailed computational studies of poroelastic materials with various combinations of material parameters and fault configurations are needed to understand more fully the role that poroelastic deformation might play in explaining observed afterslip. This simple simulation underscores the importance of treating the earth's crust as a poroelastic material rather than an elastic material in studies of time dependent crustal phenomena.

6. POSTSEISMIC DEFORMATION

Vertical and horizontal deformation of the crust following an earthquake are related to afterslip and may also have a component that is caused by postseismic pore fluid movement. Vertical deformation measurements following the 1992 Landers earthquake were made by *Peltzer et al.* [1994], *Peltzer et al.* [1996], *Peltzer et al.* [1998], and *Zebker et al.* [1994] , , and using synthetic aperature radar (SAR). *Peltzer et al.* [1996, 1998] suggest that pore fluid diffusion was responsible for the observed postseismic changes near the fault zone. Detailed numerical simulations of postseismic vertical deformation were carried out by *Bosl and Nur* [1998] and were found to have similar magnitude and spatial distribution to SAR measured values.

Horizontal deformation following the 1992 Landers earthquake has also been determined by a number of researchers, including *Shen et al.* [1994], *Savage and Svarc* [1997], and *Wyatt et al.* [1994] using GPS geodetic arrays, trilateration, and long-base strainmeters. Explanations for these observations include afterslip [*Shen et al.*, 1994; *Savage and Svarc*, 1997], viscoelastic flow in the asthenosphere [*Deng et al.*, 1998] and poroelastic relaxation of the crust. Since asthenospheric flow and poroelastic relaxation are physical processes that *must* happen to some degree following an earthquake, it remains only to estimate the magnitudes of these processes to determine their contribution to postseismic horizontal deformation. Afterslip is only a hypothesized displacement on a fault that is invoked to explain observations. Whether afterslip actually occurs, or is an apparent phenomenon that results from other physical processes such as pore fluid flow as discussed above, remains to be determined. Viscoelastic processes in the lower crust may cause postseismic displacements far from the fault (one to three fault lengths away) that are of magnitudes similar to observations [*Deng et al.*, 1998; *Nur and Mavko*, 1974]. Poroelastic effects would be expected to be greatest near the fault, where induced pore pressures are greatest. We present here the results of a simulation of poroelastic horizontal deformation as another demonstration that poroelastic effects are significant and must be considered as part of the physics of crustal stress evolution.

Shen et al. [1994] used GPS receivers at a number of sites in the region of the Landers earthquake to determine relative displacements between stations over a period of approximately three years. Figure 3 shows the locations of stations used. Red triangles represent the subset of stations that were selected for computer calculation.

Relative positions between selected pairs of survey markers at times up to approximately one year after the Landers earthquake event are plotted in figure 4. We computed horizontal displacements using the same parameters and fault model as for the aftershock study discussed previously. The analogous plots that resulted from computer simulation of poroelastic displacement are shown in figure 5.

The first three pairs of curves are for stations that are far from the fault and relatively far from each other (JPL1, PIN1, and DS10). Since pore pressure deviations are small far from the fault, relatively little change in position would be expected in the computed displacements. This is clearly seen in figure 5. The measured relative displacements for these three pairs is more significant in the northward direction, but relatively flat in the eastward direction. This indicates that some physical process in addition to poroelastic diffusion is operative far from the fault in the north-south direction. Viscoelastic relaxation or continuing movement of tectonic plates on either side of the fault, for example, might contribute to these north-south displacements.

The last four pairs of plots all involve the station PAXU, which is located very close to the fault near the epicenter at the southern end of the Landers fault. Pore pressure changes would be expected to be large at this location. Computed poroelastic displacements of four widely-dispersed stations (7000, PIN1, JPL1, and DS10) relative to PAXU are quite similar to the measured displacements in both magnitude and direction, suggesting that pore fluids might contribute significantly to elastic deformation near the fault.

Computed displacements for sets four through seven (7000-7001, 6052-DS10, Bear-JPL1, 7000-PIN1) deviate from measured trends in at least one component. Three of these four pairs involve stations that are very near fault segments (7000, 7001, Bear). The grid resolution for the computational model (4 km horizontally) is greater than the distance to the fault at these stations. Since there is a great deal of variation in pore pressures

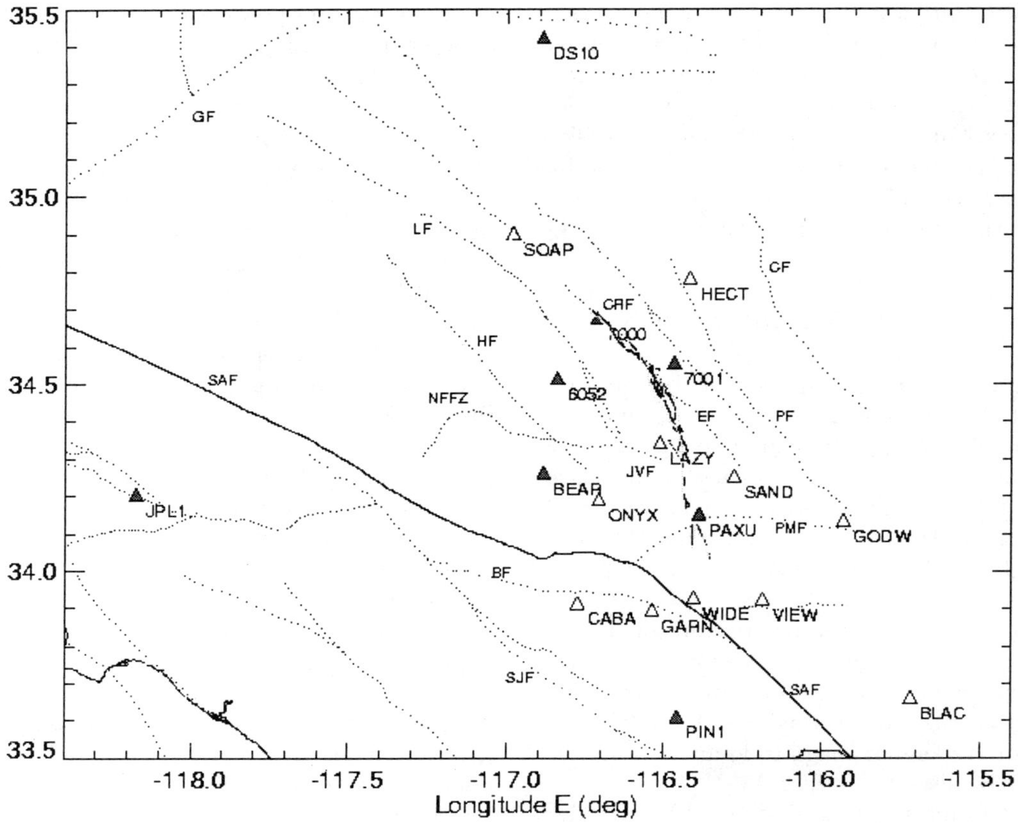

Figure 3. Map showing the surface trace of the Landers earthquake and GPS postseismic monitoring stations used to compute horizontal deformation. Figure is from Shen, et al. (1994). Used with permission of author.

along the faults at an apparently fine scale, calculations with finer resolutions might yield different results at these stations. It is significant, however, that the magnitude of simulated displacements at these stations is similar to measured values. We emphasize that if the crust is fluid-saturated, poroelastic relaxation must occur. Only the magnitude of material properties and details, such as anisotropies and heterogeneities caused by material property variability, are in question.

7. SUMMARY AND CONCLUSIONS

If the crust is saturated with water, and there appears to be considerable evidence that it is, then by the laws of physics and chemistry the crust must be in a continually evolving, restless state. The presence of water in the crust requires that models of the earthquake cycle consider the significance of hydrologic processes in seismic faulting. Moreover, the coupled effects of fluid flow, elastic deformation, heat transport, and geochemical evolution of crustal rocks must all be studied together as a complex system in order to test various theories about the earthquake process. The Mandel-Cryer effect is an elegant illustration of the kind of surprisingly complex behavior that can result when processes are coupled.

Sudden stresses applied to the crust, such as fault dislocations that cause earthquakes, are somewhat resisted by the presence of fluids in rocks. The fluid acts much as a shock absorber. Over time, induced fluid pressures tend to return to preshock levels. The resulting elastic deformation generally continues in the same direction as the initial deformation. One way of exploring the physical behavior of the poroelastic crust is to compare physically realistic computer simulations with observations of postseismic phenomena. Afterslip, aftershocks, and postseismic deformation all exhibit time dependent behavior with significant components caused by pore fluid flow and coupled stress relaxation. Simulations

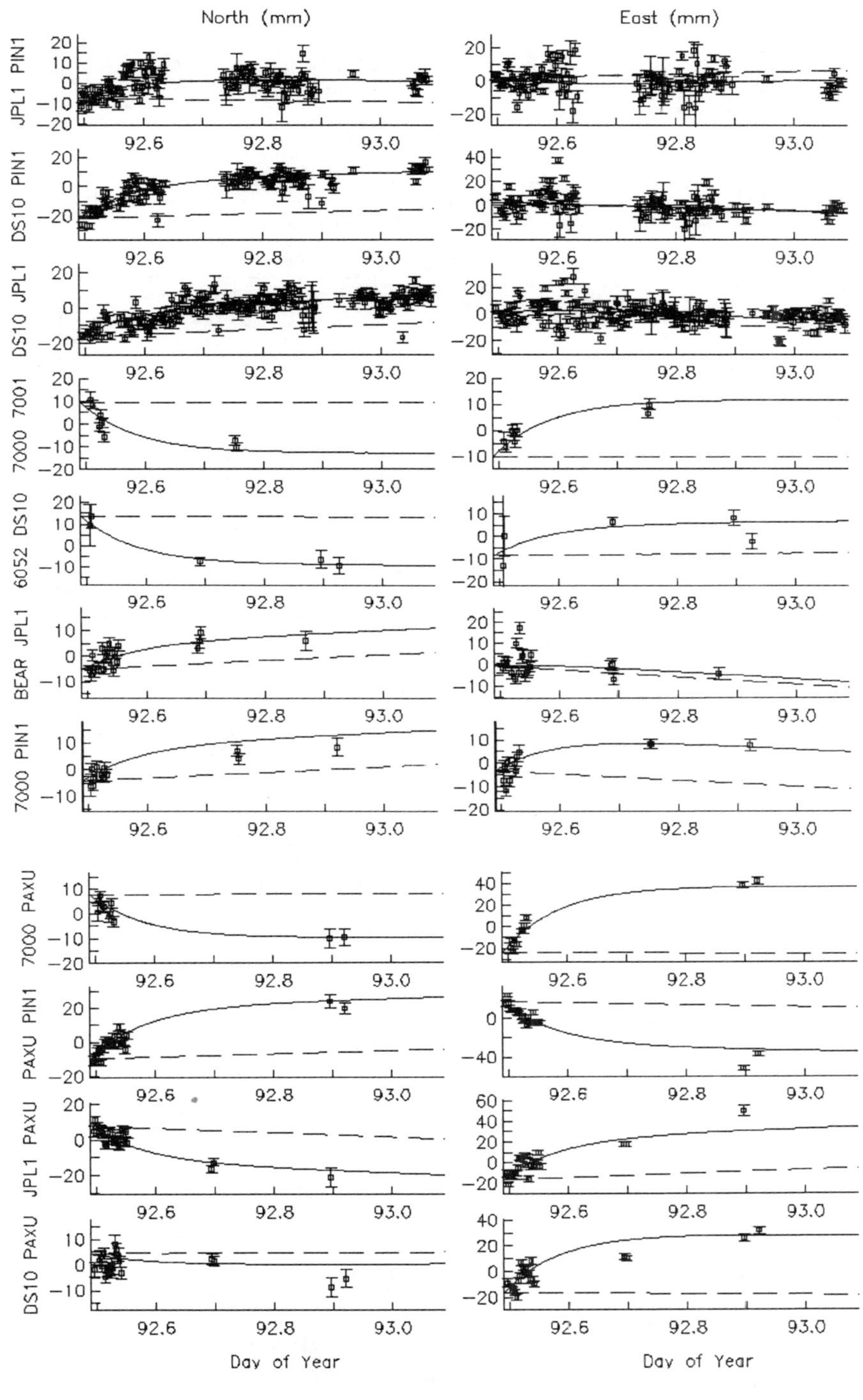

Figure 4. GPS baseline solutions from Shen, et al. (1994). Used with permission of the author.

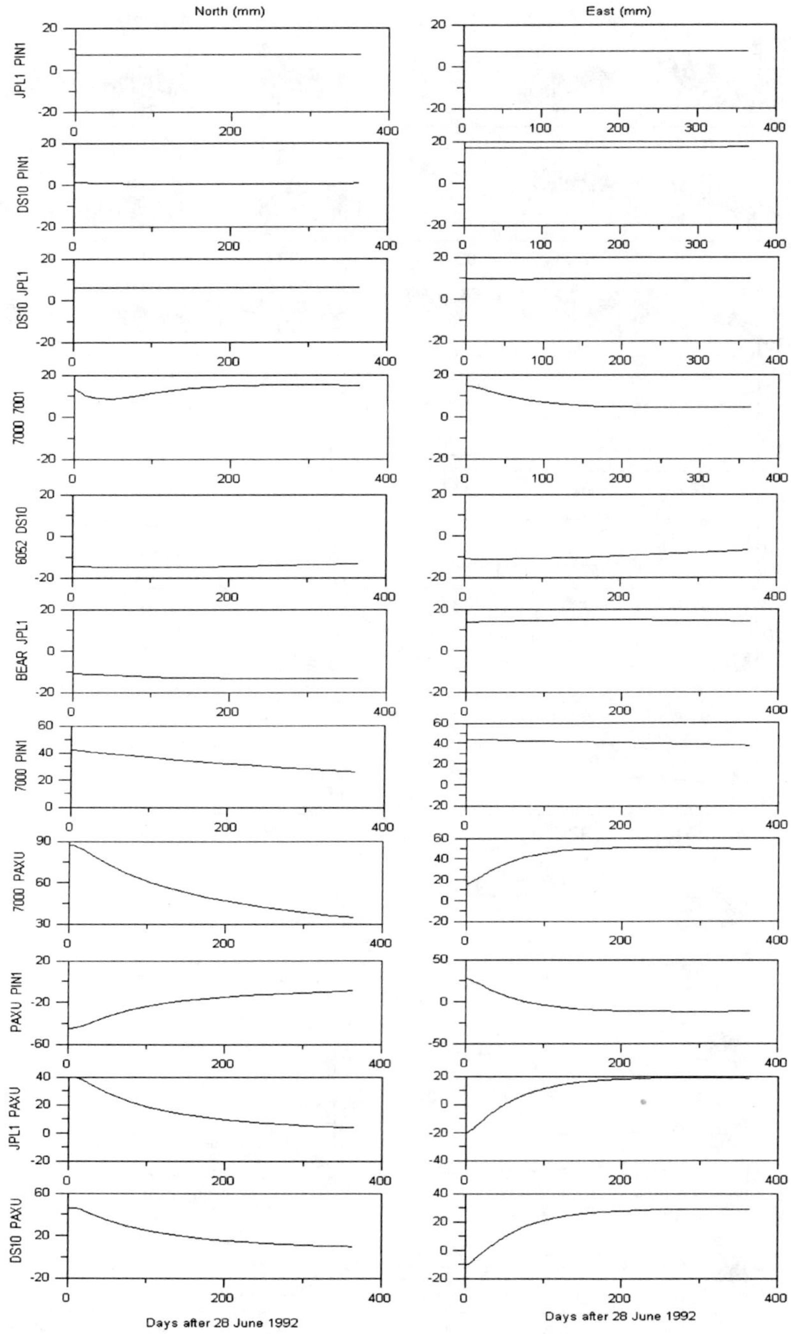

Figure 5. Computed horizontal (east and north) displacement differences. Stations are indicated on the vertical axis of the left (north) plot. These curves are analogous to the curves determined by Shen, et al. (1994) from GPS measurements.

of the 1992 Landers earthquake support this hypothesis. A number of current theories of seismic faulting invoke pore fluid processes. Because measurements of pore fluid pressure in the region of an earthquake, before and after an earthquake, are difficult to make, computer simulations to study the consequences of these theories are needed.

Because the coupled processes in the earth can exhibit unexpected dynamical behavior, significant physical processes cannot be ignored. This is particularly true of crustal fluids, which require that the crust be treated as an evolving poroelastic material, not simply an elastic material.

> "The crust of the Earth, working through hundreds of millions of years, has organized itself into a critical state through plate tectonics, earthquake dynamics and volcanic activity. The crust has been set up in a highly organized pattern of rocks, faults, rivers, lakes, etc., in which the next earthquake can be anything from a simple rumble to a cataclysmic catastrophe.... In the final analysis, the crust of the Earth can probably be thought of as one single critical system, in which the criticality manifests iteself in many different ways." [*Bak*, 1996]

General earthquake models must include coupled pore fluid processes if they are to accurately account for all the important physical processes that are involved in the complex phenomenon that we call earthquakes.

Acknowledgment. This work was supported by USGS / NEHERP Grant #1434-HQ-97-GR-03 and DOE Basic Science Program Grant #FG03-86ER13601.

REFERENCES

Bak, P., *How Nature Works*, Springer-Verlag, New York, 1996.

Bolton, E. W., A. C. Lasaga, and D. M. Rye, Long-term Flow/Chemistry feedback in a porous medium with heterogenous permeability: Kinetic control of dissolution and precipitation, *American Journal of Science*, *299*, 1–68, 1999.

Booker, J. T., Time dependent strain following faulting of a porous medium, *Journal of Geophysical Research*, *79*, 2037–2044, 1974.

Bosl, W. J., and A. Nur, Numerical simulation of postseismic deformation due to pore fluid diffusion, in *Poromechanics: A Tribute to Maurice A. Biot*, edited by J.-F. Thimus, pp. 23–28, A. A. Balkema, Rotterdam, 1998.

Bruhn, R. L., W. T. Parry, W. A. Yonkee, and T. Tompson, Fracturing and hydrothermal alteration in normal fault zones, *PAGEOPH*, *142*, 609–644, 1994.

Byerlee, J., Earthquakes, *Journal of Geophysical Research*, 1994.

Byerlee, J., Model for episodic flow of high-pressure water in fault zones before earthquakes, *Geology*, *21*, 303–306, 1996.

Charlez, P. A., *Rock Mechanics Volume 2: Petroleum Applications*, Editions Technip, Paris, 1997.

Cryer, C. W., A comparison of the three-dimensional consolidation theories of biot and terzaghi, *Quart. Journal of Mech. and Applied Math.*, *16*, 401–412, 1963.

Deng, J., M. Gurnis, H. Kanamori, and E. Hauksson, Viscoelastic flow in the lower crust after the 1992 landers, california, earthquake, *Science*, *282*, 1689–1692, 1998.

Detournay, E., and A. H. Cheng, Fundamentals of poroelasticity, in *Comprehensive Rock Engineering*, edited by J. A. Hudson, pp. 113–171, Pergamon Press, New York, 1993.

Drucker, D. C., and W. Prager, Soil mechanics and mechanical analysis or limit design, *Applied Mathematics Quarterly*, *10*, 157–165, 1952.

Evans, B., and T.-F. Wong, eds., *Fault Mechanics and Transport Properties of Rocks*, Academic Press, Inc., San Diego, 1992.

Gal, D., A. Nur, and E. Aharonov, Stability analysis of a pressure-solution surface, *Geophysical Research Letters*, *25*, 1237–1240, 1998.

Hardebeck, J. L., J. J. Nazareth, and E. Hauksson, The static stress change triggering model: Constraints from two southern california aftershock sequences, *Journal of Geophysical Research*, *103*, 24427–24437, 1998.

Harris, R. A., Introduction to special section: Stress triggers, stress shadows, and implications for seismic hazard, *Journal of Geophysical Research*, *103*, 24347–24358, 1998.

Hickman, S., R. Sibson, and R. Bruhn, eds., *The Mechanical Involvement of Fluids in Faulting*, Menlo Park, CA, U. S. Geological Survey, 1994.

Hubbert, M. K., and W. W. Rubey, *Geol. Soc. Amer. Bull.*, *70*(115), 1959.

Hudnut, K. W., et al., Co-seismic displacements of the 1992 landers earthquake sequence, *Bull. Seis. Soc. Am.*, *84*, 625–645, 1994.

Ingebritsen, S. E., and W. E. Sanford, *Groundwater in Geologic Processes*, Cambridge University Press, New York, 1998.

King, G. C. P., R. S. Stein, and J. Lin, Static stress changes and the triggering of earthquakes, *Bulletin of the Seismological Society of America*, *84*, 935–953, 1994.

Li, V. C., S. H. Seale, and T. Cao, Postseismic stress and pore pressure readjustment and aftershock distributions, *Tectonophysics*, *144*, 37–54, 1987.

Li, Y., J. Vidale, K. Aki, F. Xu, and T. Burdette, Evidence of shallow fault zone healing after the 1992 m7.5 landers, california, earthquake, *Science*, 1997.

Mandel, J., Consolidation des sols (etude mathematique), *Geotechnique*, *3*, 287–299, 1953.

Mansinha, L., and Smylie, The displacement fields of inclined faults, *Bulletin of the Seismological Society of America*, *61*(5), 1433–1440, 1971.

Nur, A., Dilatancy, pore fluids, and premonitory variations

of ts/tp travel times, *Bulletin of the Seismological Society of America*, *62*, 1217–1222, 1972.

Nur, A., and J. R. Booker, Aftershocks caused by pore fluid flow?, *Science*, *175*, 885–887, 1972.

Nur, A., and J. D. Byerlee, An exact effective stress law for elastic deformation of rock with fluids, *Journal of Geophysical Research*, *76*(26), 6414–6419, 1971.

Nur, A., and G. Mavko, Postseismic viscoelastic rebound, *Science*, *183*, 204–206, 1974.

Nur, A., and J. Walder, *Hydraulic Pulses in the Earth's Crust*, chap. 19, pp. 461–473, Academic Press, 1992.

Peltzer, G., K. Hudnut, and K. Feigl, Analysis of coseismic surface displacement gradients using radar interferometry: New insights into the landers earthquake, *Journal of Geophysical Research*, pp. 21971–21981, 1994.

Peltzer, G., P. Rosen, F. Rogez, and K. Hudnut, Postseismic rebound in fault step-overs caused by pore fluid flow, *Science*, *273*, 1202–1204, 1996.

Peltzer, G., P. Rosen, and F. Rogez, Poroelastic rebound along the landers 1992 earthquake surface rupture, *Journal of Geophysical Research*, *103*, 30131–30145, 1998.

Rice, J. R., Fault stress states, pore pressure distributions, and the weakness of the san andreas fault, in *Fault Mechanics and Transport Properties of Rocks*, edited by B. Evans and T.-F. Wong, pp. 475–503, Academic Press, New York, 1992.

Roeloffs, E., Poroelastic techniques in the study of earthquake-related hydrologic phenomena, *Advances in Geophysics*, *37*, 135–195, 1996.

Rye, D. M., and H. J. Bradbury, Fluid flow in the crust: An example from a pyrenean thrust ramp, *American Journal of Science*, *288*, 197–235, 1988.

Savage, J. C., and J. L. Svarc, Postseismic deformation associated with the 1992 mw=7.3 landers earthquake, southern california, *Journal of Geophysical Research*, *102*, 7565–7577, 1997.

Scholz, C. H., *The Mechanics of Earthquakes and Faulting*, Cambridge University Press, New York, 1990.

Seeburger, D. A., Studies of natural fractures, fault zone permeability, and a pore space-permeability model, Ph.D. thesis, Stanford University, 1981.

Shen, Z., D. Jackson, Y. Feng, M. Cline, M. Kim, P. Fang, and Y. Bock, Postseismic deformation following the landers earthquake, california, 28 june 1992, *Bulletin of the Seismological Society of America*, *84*, 780–791, 1994.

Sprunt, E. S., and A. Nur, Experimental study of the effects of stress on solution rate, *Journal of Geophysical Research*, *82*, 3013–3022, 1977.

Tompson, A. F. B., R. D. Falgout, S. G. Smith, W. J. Bosl, and S. F. Ashby, Analysis of subsurface contaminant migration and remediation using high performance computing, *Advances in Water Resources*, *22*, 203–221, 1998.

Wald, D. J., and T. H. Heaton, Spatial and temporal distribution of slip for the 1992 landers, california, earthquake, *Bulletin of the Seismological Society of America*, *84*(3), 668–691, 1994.

Walder, J., and A. Nur, Porosity reduction and crustal pore pressure development, *J. Geo. Res.*, *89*(B13), 11539–11548, 1984.

Wang, C.-Y., Y. Cai, and D. L. Jones, Predicting the areas of crustal faulting in the san francisco bay region, *Geology*, *23*(9), 771–774, 1995.

Wang, H. F., Quasi-static poroelastic parameters in rock and their geophysical applications, *PAGEOPH*, *141*(2/3/4), 269–286, 1993.

Wyatt, F. K., D. C. Agnew, and M. Gladwin, Continuous measurements of crustal deformation for the 1992 landers earthquake sequence, *Bulletin of the Seismological Society of America*, *84*, 768–779, 1994.

Yeats, R. S., K. Sieh, and C. R. Allen, *The Geology of Earthquakes*, Oxford University Press, New York, 1997.

Zebker, H. A., P. A. Rosen, R. M. Goldstein, A. Gabriel, and C. L. Werner, On the derivation of coseismic displacement fields using differential radar interferometry: The landers earthquake, *Journal of Geophysical Research*, *99*, 19617–19634, 1994.

Zoback, M. D., New evidence on the state of stress of the san andreas fault system, *Science*, *238*, 1105–1111, 1987.

W. Bosl and A. Nur, Geophysics Department, Stanford University, Mitchell Building, Stanford, CA 94305-2215. (e-mail: bosl@pangea.stanford.edu; nur@pangea.stanford.edu)